Thomas Kuhn

Thomas Kuhn

A Philosophical History for Our Times

Steve Fuller

The University of Chicago Press
Chicago and London

STEVE FULLER is professor in the Department of Sociology at the University of Warwick. He is author of a number of books, most recently *The Governance of Science: Ideology and the Future of the Open Society* (1999).

The University of Chicago Press, Chicago 60637
The University of Chicago Press, Ltd., London
© 2000 by The University of Chicago
All rights reserved. Published 2000
Printed in the United States of America
09 08 07 06 05 04 03 02 01 00   1 2 3 4 5

ISBN: 0-226-26894-2 (cloth)

Library of Congress Cataloging-in-Publication Data

Fuller, Steve, 1959–
    Thomas Kuhn : a philosophical history for our times / Steve Fuller.
        p. cm.
    Includes bibliographical references and index.
    ISBN 0-226-26894-2 (cloth : alk. paper)
    1. Science—Philosophy.   2. Science—History.
    3. Kuhn, Thomas S. Structure of scientific revolutions.
    I. Title.
Q175.F927   2000
501—dc21   99-056974

♾ The paper used in this publication meets the minimum requirements of the American National Standard for Information Sciences—Permanence of Paper for Printed Library Materials, ANSI Z39.48-1992.

To Sujatha Raman,
my fellow traveler who has often shown the way.

People surmount tragedy when they use themselves up fully, when they use what they have and what they are, whatever they are and wherever they find themselves, even if this requires them to ignore cultural prescription or to behave in innovating ways undefined by their roles. The tragic sense does not derive from feeling that people must always be less than history and culture demand; it derives, rather, from the sense that they have been less than they could have been, that they have needlessly betrayed themselves, needlessly forgone fulfillments that would have injured no one. The scientific enterprise, like others, becomes edged with a tragic sense when scientists suspect that they have wasted their lives. In confining work to the requirements of a demanding and unfulfillable paradigm, scientists are not using themselves up in their work and are, indeed, sacrificing, leaving unexpressed, certain parts of themselves — their playful impulses, their unverified hunches, their speculative imagination. When scientists commit themselves compulsively to a life-wasting high science model, they are making a metaphysical wager. They are wagering that the sacrifice is 'best for science.' Whether this is really so, they cannot confirm; but they often need no further confirmation than the pain this self-confinement inflicts upon them.

— Adapted from Alvin Gouldner, *The Coming Crisis in Western Sociology*

CONTENTS

	Preface: Being There with Thomas Kuhn	xi
	INTRODUCTION	1
I	THE PILGRIMAGE FROM PLATO TO NATO *Episodes in Embushelment*	38
II	THE LAST TIME SCIENTISTS STRUGGLED FOR THE SOUL OF SCIENCE	96
III	THE POLITICS OF THE SCIENTIFIC IMAGE IN THE AGE OF CONANT	150
IV	FROM CONANT'S EDUCATION STRATEGY TO KUHN'S RESEARCH STRATEGY	179
V	HOW KUHN UNWITTINGLY SAVED SOCIAL SCIENCE FROM A RADICAL FUTURE	227
VI	THE WORLD NOT WELL LOST *Philosophy after Kuhn*	260
VII	KUHNIFICATION AS RITUALIZED POLITICAL IMPOTENCE *The Hidden History of Science Studies*	318
VIII	CONCLUSIONS	379
	References	425
	Index	463

PREFACE

Being There with Thomas Kuhn

It is often said, even by those who should know better, that history leads from the front, with visionary geniuses showing the way for future generations to follow. Thomas Kuhn famously updated this elitist myth of humanity's collective quest by associating the great paradigmatic thresholds in the history of science with the names of revolutionary geniuses who set the pace for lesser worthies. "Newton," "Lavoisier," and "Einstein" appear in Kuhn's own text as the originators of paradigms. "Darwin" and other luminaries of the biological and social sciences have since been added by Kuhn's admirers. Indeed, much of the learned public today regards Kuhn as himself having revolutionized our understanding of the nature of science.

However, to stick to the letter of Kuhn on this point is to miss a somewhat less elevated reading of history that better captures the career of his own masterwork, *The Structure of Scientific Revolutions*. Call it *leadership from the middle*. It consists of a critical and a positive thesis. The critical thesis denies any facile correlations between intelligence and achievement. Instead, it holds that genius is an occult mental property superstitiously projected backward to explain the cause of deeds that have already had a remarkable effect on us. Our normally naive belief in the timely recognition of genius serves two complementary functions. It vindicates our sense of justice in the world, as reward is shown to follow merit. At the same time, it reaffirms the rationality of the value discriminations we manage to make of the surprisingly diverse things our fellows do. Thus, both Newton and his followers are flattered by honoring a scientific revolution with the name "Newtonian." In so doing, they constitute that peculiar mutual admiration society we call a community of inquirers.

The positive thesis associated with leadership from the middle is that to think ahead of the pack is just as bad as to think behind it. It is much more important to be *seen* as the first and the brightest than actually to be such things. Ironically, because attributions of genius are ultimately determined by collective response, it makes little sense to try to be the first and the brightest. What is true of innovation in technology is no less true in science: real pioneers take the bulk of the risks related to the introduction of new knowledge and bear most of the costs. In effect, they are living experiments from which their immediate successors vulturously benefit. Those entrepreneurs who follow up an original invention are able to stake the most comprehensive patent claims and hence corner the lion's share of the new market. In science, the corresponding individuals are assigned totemic status in a Kuhnian paradigm.

This begins to explain the success of Kuhn's *Structure*. Whatever originality adheres to the book lies not in the unique brilliance of its ideas but in the fortuitous timing and placing of its appearance. To be sure, as Newton would have it, Kuhn saw as far as he did because he stood on the shoulders of giants. Indeed, he inhabited a kingdom of giants. But it only takes a dwarf standing on the shoulders of giants to see beyond them, and from a distance, the dwarf may look like the head of the tallest giant.

Before continuing with what may strike readers as a disagreeable conclusion, let me explain my starting point. In 1992, through the good offices of Joseph Rouse and Brian Fay, I was asked to write an essay on the history of the reception of *Structure* for *History and Theory*, to mark the thirtieth anniversary of the book's first edition. The title of that essay, "Being There with Thomas Kuhn: A Parable for Postmodern Times," presented the core thesis of the book before you.[1] Lest I mislead philosophically informed readers, I should immediately add that the title does *not* refer to the interesting attempts to accommodate Kuhn's account of scientific practice to the existential phenomenology of Martin Heidegger.[2] Rather, it refers to the critically acclaimed motion picture, adapted from the novel by Jerzy Kosinski, starring Peter Sellers. The movie *Being There* may be seen as a "paradigm" for understanding the reception of Kuhn's *Structure*, an only slightly tongue-in-cheek gloss of Alan Musgrave's remark: "Perhaps the revolutionary never existed — but then it was necessary to invent him."[3]

At every opportunity, Kuhn disavowed all of the more exciting and radical theses imputed to him by friends and foes alike.[4] Indeed, he has even

---

1. Fuller 1992a.
2. See esp. Rouse 1987, 26–40.
3. Cited in Gutting 1979, 51.
4. See, e.g., Horgan 1991, 40. Disavowal is perhaps the most rhetorically consistent feature of Kuhn's self-referential remarks. It can be found as early as Kuhn 1970a; Kuhn 1977a, xxi–xxiii, 293. Here it is worth mentioning that I interviewed Kuhn at his home in Cambridge, Massachusetts, on 23 July 1993. While we both agreed not to reveal the exact contents of the interview, it is fair to say that he was much more interested in talking about his ideas than the

endeavored to demonstrate by example, in his subsequent work, that he could not have been capable of anything so exciting and radical.[5] It is doubtful that there has ever been another academic who has met the greatness thrust upon him with such ingratitude. Indeed, Karl Marx runs a close second to Kuhn in his desire to disown his admirers: "I've often said I'm much fonder of my critics than my fans," Kuhn declared to *Scientific American* on the thirtieth anniversary of *Structure*'s publication.[6] What is more remarkable, of course, is how, in full cognizance of Kuhn's disclaimers, so much of the academic world (I included) has nevertheless felt compelled to position itself with regard to his work. What might we learn from this very strange episode in the annals of scholarship? First, a quick trip to the movies . . .

The protagonist of *Being There* is Chance, whom the other characters repeatedly call "Chauncey." Chance is the ward of a wealthy Washingtonian whose death marks the beginning of the film. Chance is a kindly man of childlike simplicity who, in the course of settling the estate of his deceased employer, is subject to several misunderstandings that, by the end of the film, land him as a candidate for the presidency of the United States. The comic element of the plot is that Chance never quite realizes what it is that he is saying that makes all the bigwigs hold his opinion in increasingly high regard. Chance is largely uncomfortable with this newfound attention and makes periodic protests that he is not the man they think he is. Needless to say, these protests are either ignored or misinterpreted. From the audience's standpoint, Chance's interlocutors appear to take his quite literal references to gardening and television — the entire scope of his existence — as metaphors for various aspects of political life. Moreover, Chance unwittingly confers the utmost profundity on his utterances by speaking not very often,

---

contexts from which they came and in which they exerted influence. To be sure, Kuhn dropped hints about why he did and did not act as he did, and he was forthcoming on various issues that served to integrate his work over the last four decades. Indeed, his comments to me were entirely consistent with the record of his last extended interview. See Kuhn et al. 1997. I left convinced that Kuhn saw himself as engaged in a coherent intellectual project, the likes of which we rarely see today. As someone who continually struggles for an overarching sense of coherence and direction in his own endeavors, I found Kuhn's single-mindedness worthy of some admiration, though as will become clear my ultimate attitude toward the man is ambivalence.

5. See, in particular, the reception of his last sustained historical work, Kuhn 1978. A sense of the disappointment of the reviewers is captured by the title of the major symposium devoted to the book ("Paradigm Lost?"): Klein, Shimony, and Pinch 1979. This book seems to have pleased only positivists, internal historians of science, and practicing physicists — as well as Hacking 1984a, 114. Thanks to Skuli Sigurdsson for reminding me of the Hacking paper.

6. Horgan 1991, 49.

and then only in short and simple sentences. *Being There* is clearly a spoof on the superficiality of American politics, but it is more than that. Some of Chance's interlocutors have lingering doubts about his untraceable past and peculiar manner, but typically they give him the benefit of the doubt and try to make the most of the situation. In fact, the collective participation needed to propel Chance to ever greater acclaim is increasingly evident in the film, until he becomes a genuine rallying point for the disparate individuals involved.

I believe that a similar comedy of errors has marked the history of the reception of *The Structure of Scientific Revolutions* and its author, Thomas Kuhn, who, like Chance, tended to speak briefly and not often. However, the underlying causal structure of the two stories is markedly different. Kuhn's intellectual gestation at Harvard (1940–56) enabled him to acquire, with little effort of his own, what the economic sociologist Mark Granovetter has called "the strength of weak ties."[7] This enabled *Structure* to be accorded a charitable reading in diverse quarters once it appeared. Granovetter's original argument turned on the simple observation that the friendlier two people are, the more their networks of contacts are likely to overlap. Consequently, one's remote acquaintances often turn out to be more effective than close friends in landing oneself a job because the former are likely to travel in circles other than one's own and hence are better positioned to provide information about job opportunities one is not otherwise likely to run across.

In Kuhn's case, the power of his diffuse network of acquaintances came from a common culture that centered on the vision and actions of Harvard president James Bryant Conant. In Granovetter's terms, then, Kuhn had a singularly strong tie to Conant, who in turn had many weak ties to opinion leaders in American society. Indeed, Kuhn's dependence on the Conant network did not elude the Harvard General Education committee that denied him tenure.[8] Thus, if you are inclined to regard Chance's ascendancy

---

7. Granovetter 1973. Would it be unfair to mention that Granovetter developed this concept as part of his own Ph.D. at Harvard? In any case, Granovetter has drawn major social-theoretic lessons from his thesis that have yet to be fully exploited: "the personal experience of individuals is closely bound up with larger-scale aspects of social structure, well beyond the purview or control of particular individuals" (1377). The significance attached to Kuhn's work certainly fits this mold, a point that was perhaps first realized by his self-styled rival, Robert Merton. See Merton 1977, 71–108.

8. Since Harvard lacked an established history of science department, Kuhn's tenure prospects were bleak. Unlike his contemporary Gerald Holton, after 1950 Kuhn no longer published or taught the physics for which he received a Ph.D. The manuscript of Kuhn's first book, *The Copernican Revolution*, was available for inspection, but the committee deemed it a popularization, not a piece of original research. (For a recent reassessment of the significance of this book, see Westman 1994. Westman regards the book as the most influential

as allegorical testimony to America's democratic instincts that enable whatever he says to override his dubious past, then Kuhn's corresponding rise to greatness, on the contrary, should remind you of that country's elitist academic infrastructure that provides cultural ballast for utterances that, were they made somewhere else by someone else, would probably have had rather limited significance.

Aside from Kuhn's own sense of "being there," there is of course my own. I was originally attracted to the core disciplines of science studies— history, philosophy, and sociology of science—by the responses generated to *Structure* by the famous conference at Bedford College, London, organized by Imre Lakatos in 1965 and collected as *Criticism and the Growth of Knowledge*.[9] It remains the most heavily annotated book in my personal library. I purchased it as an undergraduate in 1977 and read it alongside Adorno's *The Positivist Dispute in German Sociology* and the first wave of French poststructuralism that had been recently translated into English. At that time, the words "critique," "criticism," and "critical theory" were held in high esteem, as thinkers attempted to outdo each other in just how "critical" they could be of any number of taken-for-granted beliefs, thereby opening the door to alternative ways of being and thinking. The odd man out was Kuhn, who not only failed to privilege criticism, but actually went so far as to argue that it should be avoided at all costs until a line of inquiry is saddled with so many unsolved empirical problems that it is forced to ask critical questions about the epistemological foundations of the entire enterprise.

Twenty years later, Kuhn's "acritical" perspective has colonized the academy. The progeny of Kuhn's incisive philosophical critics now presuppose the basic truth of his account of science. The radical skepticism of deconstruction has yielded to a postmodern pluralism that offends only by opening its doors to too *many* perspectives, courtesy of Kuhn. But perhaps most telling of all, the critical turn of mind has become so alien to my own field of science studies that Bruno Latour finds it a fit subject for anthropological—and, one suspects, taxidermic—inquiry.[10] All of this is a source of such profound disappointment that had I known when I began my aca-

---

general history of science written since the end of World War II.) It is clear that many members of the committee regarded Kuhn's protracted stay at Harvard to be the result of promise backed less by achievement than patronage—specifically that of James Bryant Conant, who (to Kuhn's misfortune) had two years earlier left the presidency of Harvard to become the U.S. High Commissioner for Germany. Indeed, Kuhn's vexed case is reflected in the minutes of the General Education Committee on his tenure proceedings being the longest of its kind. See *Minutes*, 8 November 1955.

9. Lakatos and Musgrave 1970.
10. Latour 1997.

demic career that things would turn out this way, I probably would have listened to my mother and gone into law. Nevertheless, I am not without hope. As a devout social constructivist, I believe that even disappointment can be used strategically to point out better paths that were originally not taken, but that (with some adjustment) may be taken up in the future. Certainly, the alert reader will be sensitive to these prospects in the pages that follow.

Kuhn himself held that history can be done only of periods and episodes whose defining debates have now reached closure, because the terms of the debates have been either resolved or superseded. In that case, the past acquires the sort of clear boundary that has traditionally made the metaphor of separation in time as if in space so compelling to relativists. However, closure is not something that happens only to other people, which historians then observe from a respectful distance. On the contrary, each historical inquirer participates in the closing—or opening—of the times she writes about. To deny the literal truth of this point is to betray an elementary constructivist tenet. In terms of the book before you, I hope to help put an end to certain trends traceable to *Structure* that remain very strong in contemporary understandings of science, while at the same time reviving the fortunes of other trends that many currently regard as being of "merely historical" interest. A fundamental premise of my book, then, is that the impact of *The Structure of Scientific Revolutions* has been largely, though not entirely, for the *worse*.

Taking this project forward has often meant treating the living as if they were dead. For the last five years I found myself participating in conferences where my natural mode of response has been to diagnose limitations in speakers' arguments that derived from the paradigms that grip their thought. My strongest response has been to recent paradigm converts who displayed, in most pronounced form, the historical amnesia and political inertia that Kuhn held to be conducive to "normal science." Moreover, I have come to distrust the Kuhnian nostrum that a research program with a robust empirical track record can ignore its serious conceptual problems with impunity. This principle makes sense only either as a make-work scheme for underemployed academics or, more sinisterly, a formula for rewarding intellectual conformity. Finally, in doing the research and thinking that have gone into this book, I have come to realize that a pincer attack from a dimly remembered past and a vividly imagined future is among the best means available to academics for displacing a sense of the present that refuses to plan its own obsolescence.

As I write this preface, I am about to leave Durham University, where I spent the last five years as Professor of Sociology in the most congenial and tolerant academic environment imaginable—at least to an American. However, much of my orientation to the topic evolved while at Virginia Tech, home to the largest graduate program in science studies in the United States and where I taught at the start of this project. My colleague Skip Fuhrman prompted me to give Alvin Gouldner a second—and third—look, and the result should be apparent in these pages. Stephen Turner provided the most useful clue for fathoming the source of Kuhn's influence, James Bryant Conant. Former students who encouraged me along the way include Steve Downes, Jim Collier, Kirk Junker, Joan Leach, Bill Lynch, Govindan Parayil, and Tim Rogers. Special thanks goes to Skuli Sigurdsson, who amazingly provided a detailed critique of the first draft of the manuscript during the 1996 meeting of the Society for Social Studies of Science. Other drafts were used as the basis for graduate courses at Gothenburg and Tel-Aviv Universities, from which I received many helpful comments that improved the writing of the final version. I also wish to thank the Secretary of Harvard College, the curators of the Harvard University Archives, and Theodore Conant for allowing me to examine the archival material cited in this book. None of the opinions expressed here represent those of the aforementioned individuals and institutions.

Susan Abrams has been exemplary as the main editor for this project. The book is much better because of her input. I am especially grateful for her providing me with Kuhn's last extended published interview, references to which are to be found in these pages.[11] The reader will find other expressions of gratitude in the footnotes. Here I would draw attention to the role of electronic listservs, especially HOPOS-L and SCI-TECH-STUDIES, in enabling me to present and hone my arguments, often to the irritation and consternation of various subscribers. On a personal note, Stephanie Lawler has been my main source of emotional support in the months leading to the book's completion. Finally, Sujatha Raman, to whom this book is dedicated, opened the letter of invitation that started me down this path. Much of her soul is to be found in these pages.

11. Kuhn et al. 1997.

Introduction

1. The Difference Kuhn Made

The Structure of Scientific Revolutions by Thomas Kuhn (1922–96) is probably the best-known academic book of the second half of the twentieth century.[1] Thirty-five years after its first publication, *Structure* has sold nearly a million copies and been translated into twenty languages. It remains one of the most highly cited works in the humanities and the social sciences, and certainly one of the few major works in these fields that have been received sympathetically by natural scientists. Moreover, the book has had this status for quite a while. When I entered Columbia University as an undergraduate in 1976, *Structure* was one of only two books by a living author that was required reading in the mandatory first-year course, "Classics of Contemporary Civilization."[2] But Kuhn's impact goes well beyond the academy:

> What is Al Gore's favorite book? According to a recent magazine profile, it's *The Structure of Scientific Revolutions*, by Thomas S. Kuhn. In a sense, it's an inevitable choice. Everyone of Gore's age, education and reformist outlook, certainly including Bill Clinton, knows about this book, which presents the classic model of the way in which prevailing ideas change. Its basic point is that people typically go for years and years believing one thing . . . despite mounting evidence to the contrary. Then all of a sudden they notice the conflicting evidence, change their minds and wonder why they ever believed otherwise.[3]

Thus begins a recent article by then *Atlantic* editor James Fallows, who goes on to show that President Clinton's economic plan can be regarded as a Kuhnian "paradigm shift," an expression so common inside the Capital

---

1. Kuhn 1970b.
2. The other book was Foucault 1967.
3. Fallows 1993.

Beltway that it has merited an entry in William Safire's *Dictionary of American Politics*. What exactly did Kuhn say that has gained him all this notoriety and influence?

Kuhn started with the idea that science is nothing more (and nothing less) than what scientists do. What Kuhn found most striking about natural scientists—especially in contrast to social scientists—was that, no matter how much they disagreed about other matters, natural scientists could usually agree on how to evaluate a piece of research, typically by reference to an earlier exemplar that the research was said to resemble. From this insight grew the protean term "paradigm," which Kuhn deemed necessary for the conduct of "normal science." Paradigm acquisition, the stuff of which scientific training is made, requires a commitment so deep that scientists cannot afterward readily change their research orientation in the face of mounting "anomalies" that resist paradigmatic treatment. Kuhn likens the inevitable shift between paradigms to a religious conversion or a change in worldview, the overall effect of which produces a "revolution" in the science. To underscore the difficulty of communicating across paradigms, or the "incommensurability thesis," Kuhn argued that the revolutionaries often turn out to be younger or marginal scientists who have little invested in the old paradigm. However, revolutions occur rarely enough to sustain a general faith in scientific progress. And despite the resemblance of his account to the cyclical histories commonly told of art and other cultural phenomena, Kuhn nevertheless held that scientists are distinctive in the amount of control they exert over the course of their inquiries and, more importantly, the public representation of those inquiries.

However, this bare summary of *Structure*'s thesis does not begin to explain the book's enormous impact, especially far afield from the natural sciences that figure most prominently in its pages. As the story of the reception of Kuhn's book is normally told, it liberated the academy from a "positivist" or "objectivist" conception of science that privileged the "hard" sciences at the expense of the other departments in the university. Kuhn supposedly showed that even the most rigorous natural sciences were constituted as communities and traditions that periodically were subject to ideological strife. Although Kuhn was never so bold as to say this, it seemed that the "logic of scientific inference" so lovingly cultivated by the logical positivists and their analytic philosophical offspring was little more than abstract rhetoric that gave structure to scientific papers but failed to predict or explain what scientists actually did in their research sites. All of these revelations induced a collective sigh of relief from practitioners of the humanities and the social sciences, who had a hard enough time making sense of each other, let alone agreeing on a common method. They quickly latched on to Kuhn's ideas and declared that they too were respectable knowledge producers laboring under paradigms.[4] Even inquirers on the fringes of science, such as parapsychologists and creationists, were inspired by Kuhn's suggestion that new paradigms can succeed by capitalizing on the "anomalies" that their more established competitors cannot explain.[5] Amazingly, Kuhn seemed to offer these radical insights in a way that natural scientists themselves found uncontroversial, if not exactly commonplace.

The overall effect of Kuhn's book—so the story goes—has been salutary. *Structure* helped to level disciplinary hierarchies and overturn inappropriate methodological standards, thereby contributing to the climate of pluralism that (at least for the time being) continues to flourish in most systems of higher education in the West. In addition, a band of historians and sociologists of science have turned *Structure* itself into a paradigm for further research. Thus were born the research program known as the "sociology of scientific knowledge" (SSK) and the interdisciplinary field of "science and technology studies" (STS), or "science studies" for short, departments of which have now taken root in Europe, North America, and Australia and will no doubt be institutionalized across the globe in the coming years. The example that Kuhn supposedly set for these scholars was to look beyond the positivist jargon that scientists use to justify their activities and to focus instead on what scientists actually do in their workplaces. Thus, the characteristic methodologies for this post-Kuhnian enterprise have involved histories and ethnographies of the research environment and deconstructions of disciplinary discourses.[6]

But as I have already suggested, the record also clearly shows that Kuhn disavowed every one of these appropriations of his work. Indeed, Kuhn's methodological predilections would seem to vary from the practice of his most prominent disciples. In particular, according to Kuhn, a history of the contemporary world—such as the one you are now reading—is impossible.[7] History can be written only about periods in which the historian does not have a stake in what happens because the issues that were of greatest concern to the historical agents no longer have the same relevance. In that sense, historians enter only once the case is closed. Thus, Kuhn accepts the familiar paradox that the historian cannot represent the agents on their own terms unless she is detached from their life concerns. Although

---

4. The best selection of these responses is still Gutting 1979.
5. Hess 1993, esp. 70–81.
6. The classic monographs are Barnes 1975, Bloor 1976, Latour and Woolgar 1979, Knorr-Cetina 1981, Collins 1985, and Shapin and Schaffer 1985. A volume of recent position papers in science studies is Pickering 1992. As it turns out, the field suffers from a dearth of reliable textbooks. Among the best is Hess 1997.
7. Kuhn 1977b, 3–20, esp. 16.

the prospect of incommensurable paradigms wreaks havoc on universal theories of scientific rationality, it enables the historian to do her job. As for contemporary paradigms, the historian's role is almost exclusively that of an archivist who ensures that a record of today's scientists is preserved for tomorrow's historians to analyze from the requisite distance. Kuhn has taken his own historiography to heart, writing about the history of physics no later than 1912 (roughly the end of classical quantum mechanics), while conducting interviews and collecting papers related to Bohr, Heisenberg, and the other great quantum physicists who flourished later in the century.[8]

A contrary position commonly taken in STS is that the inquirer should interact as much as possible with the scientists she studies in their research sites. Indeed, some have declared that *only* contemporary science can be reliably investigated, mainly because the scientists concerned have not yet had an opportunity to silence dissidents and forge accounts of their exploits that can then function as myths obscuring any hope of getting at what really happened.[9] Nevertheless, it is worth emphasizing that these STSers share with Kuhn the same end—an "objective" understanding of scientific knowledge—that they then pursue by divergent means, both of which have somewhat unfortunately traveled under the name of "relativism," specifically the *methodological relativism* to be discussed later in this chapter.

It is the rare author who does not complain about how her work is read. However, the monumental fecundity of misreadings attached to *Structure* begs for explanation. Not surprisingly, the only book that has attempted to present Kuhn's work as a systematic theory has proceeded largely by rational reconstruction, in which pieces of the reconstructed theory are supported by scattered quotes in Kuhn, to which lengthy footnotes are appended listing Kuhn's interpreters who have engaged in "remarkable misunderstandings."[10] What is the source of these messages that readers have picked up

---

8. Kuhn 1978. For Kuhn's extensive archival work on the great twentieth-century quantum physicists, see Kuhn et al. 1967, which was cosponsored by the American Physical Society, and received funding from the National Science Foundation.

9. This point has been most vigorously driven home by Harry Collins, mainly as his "Proposition Six" in Collins 1985, 76, 129, 170. The social relationship that typifies the interviews Collins conducts with scientists is the formal exchange. He wants to maintain a clear separation between the interests of the scientist and the historian, thereby ensuring the autonomy of both. In that case, the historian must provide an incentive for the scientist to speak with her at the length and depth that she requires. At an international workshop on the history of contemporary science in Gothenburg, Sweden, September 1994, Collins suggested that the historian can gain the confidence of her scientist-interlocutor by trading on what she has learned about the other scientists she has interviewed. Workshop papers are collected in Soederqvist 1997.

10. Hoyningen-Huene 1993. This careful and sympathetic text originally appeared in German in 1989, after Hoyningen-Huene had spent a year working with Kuhn at MIT. The book received Kuhn's official endorsement for obvious reasons, yet one wonders whether its

but that Kuhn did not intend? More generally: How did a book such as *The Structure of Scientific Revolutions*, written when it was, have the kind of impact that it has had? The fact that many Kuhn-like ideas were "in the air" both before and during the time *Structure* was written only adds to the mystery, especially since, in the early 1960s, Kuhn was probably the *least* prominent of the group of thinkers who are usually said to have advanced similar views of science. Listed in descending order of prominence at the time, these include Michael Polanyi (1891–1976), Stephen Toulmin (b. 1922), Norwood Russell Hanson (1924–67), and Paul Feyerabend (1924–94). To appreciate the uniqueness of Kuhn's work, as well as the vicissitudes surrounding its reception, *Structure* must be read as less *cause* than *symptom*.

As the first stage of this symptomatology, I urge that *Structure* be read as an exemplary document of the Cold War era.[11] In that context, Kuhn appears as a "normal scientist" in the Cold War political paradigm constructed by James Bryant Conant (1893–1978), president of Harvard University (1933–53), director of the National Defense Research Committee during World War II (which supervised the construction of the first atomic bomb), and chairman of the anti-Communist Committee on the Present Danger in the 1950s — as well as the person who introduced Kuhn to the historical study of science, and through whom Kuhn acquired his first teaching post. Shortly after the end of World War II, Conant designed the General Education in Science curriculum, in which *Structure* was conceived and elaborated. Kuhn returned the compliment by having Conant write the foreword of his first book *(The Copernican Revolution)* and dedi-

---

rational reconstructionist approach actually measures up to Kuhn's own historiographical scruples, which aim for a rendition of the historical agents "in their own terms." For, if Hoyningen-Huene is correct, virtually every substantive commentator on Kuhn's work has been in error, a conclusion that, under normal circumstances, would suggest that Hoyningen-Huene had violated the "principle of charity" in interpretation, according to which if everyone is interpreted as being in error, then it is the interpreter who has more likely erred, perhaps because she has interpreted the agents as doing something other than they intend. In other words, it may be that Kuhn seems to be so frequently "misunderstood" because his readers are really not trying to understand him to any great extent but to use his text as a token in some ongoing disputes. This would certainly explain why Hoyningen-Huene has no rivals in the Kuhn systematics sweepstakes. That Kuhn functions so admirably in this strategic capacity demands explanation, which will be attempted in the pages that follow. In any case, it may be impossible to render a single author's thought "sympathetically" and, at the same time, do justice to the collective thought of the community in which the author is situated. One will always appear as a means to the other's ends, and hence judged as such. This "complementarity" of the individual and the collective, to borrow a metaphor from quantum mechanics, would seem to vindicate a fairly strong sense of incommensurability in historical interpretation.

11. A good survey of Cold War science policy that stresses the continued reluctance of historians to see science as integral to the balance of power is Doel 1997.

cating his second book, *Structure*, to Conant. Kuhn first met Conant as a junior fellow at the Harvard Society of Fellows, whose not-so-hidden agenda included the inoculation of promising young scholars against the siren song of Communism, a point to which we shall return in chapter 3, section 5. All told, Kuhn seemed to consider Conant "the brightest person" he had ever known.[12]

It is tempting to conclude that *Structure* somehow transcended its origins and managed to grasp something essential about the nature of science. Intellectual affinities notwithstanding, *Structure* is clearly a more elegant and systematic account of science than anything Conant ever wrote. But beyond this aesthetic point is the book's proven persuasiveness in many academic precincts. Certainly, many of today's practitioners of the human sciences have a vested interest in promoting the canonical status of Kuhn's book. Nevertheless, I urge a less flattering interpretation, yet one more in keeping with the centrality of "normal science" in both Conant's and Kuhn's thought. It involves treating *Structure*'s elegance and systematicity as signs that Kuhn simply took Conant's politics of science as uncontroversial—indeed, as a taken-for-granted worldview. *Structure* does not so much transcend the Cold War mentality as express it in a more abstract, and hence more portable, form. (This point is further elaborated in chapters 3 and 4.) Whereas Conant's ragged presentation testified to his encounters with a wide range of often hostile audiences, Kuhn's sheltered institutional setting enabled him to articulate the Conant paradigm without having to register opponents or even the events of the day. Arguments that would be understood as contestable in a political setting, because agents see them as potentially affecting the course of events, can easily acquire the status of fact when transferred to the depoliticized environment of the academy, where agents are removed from the levers of change. In other words, we may have a situation in which abstraction implies less, not more, critical reflection on the material conditions of thought.

2. "A Philosophical History for Our Time"

No doubt, readers will want to know my motives for casting in such a diminished light a book as highly revered and quoted as *The Structure of Scientific Revolutions*. It is not that I have not been influenced by Kuhn. Indeed, I have drawn upon aspects of Kuhn's thought that both he and his admirers have wanted to downplay over the years. In my first book, *Social Epistemology*, I acknowledged that Kuhn's work led me to think seriously about the relationship between the "tacit dimension" of scientific practice

---

12. Kuhn et al. 1997, 146.

within a paradigm and the "incommensurability" that seemed to obtain between paradigms. Precisely because practitioners of normal science do not query each other's activities on a regular basis, it is possible for different research communities ostensibly operating under the same paradigm to develop divergent understandings of their supposedly common inquiry. Only once some outstanding problems force these communities to engage in "philosophical" debates of their field's principles and methods—a Kuhnian "crisis"—does it become clear just how great, and often unbridgeable, the disagreements are. Thus, radical conceptual difference in science can be explained in terms of the institutionalized communication breakdown that euphemistically passes for "autonomous research communities."[13] This conclusion has continued to inform my work, including the desire to develop a normative orientation to inquiry that does not presuppose that inquirers necessarily function with a common mind-set. Notwithstanding the efforts of Jürgen Habermas, Paul Grice, and others to demonstrate by a priori reasoning that there are incontrovertible foundations to communication, I take Kuhn to have shown that the goal of this quest for foundations is simply chimerical; hence, I have increasingly turned my attention to *rhetoric* as providing insights into how language can motivate collective action without requiring prior agreement on the point of that action.[14]

Without denying my own intellectual debt to Kuhn, I must nevertheless conclude that the overall impact of his book has been to dull the critical sensibility of the academy. In sociological shorthand and with the benefit of hindsight, *The Structure of Scientific Revolutions* unwittingly achieved much of what Daniel Bell's *The End of Ideology* tried to do around the same time, namely to alleviate the anxieties of alienated academics and defensive policy makers by teaching them that they could all profit from each tending to their homegrown puzzles. Good paradigms make for good neighbors. What dropped out of this picture was a public academic space where the general ends and means of "science" (or "knowledge production" or "inquiry") could be debated just as vigorously and meaningfully as the specific ends and means of particular disciplines or research programs. Social epistemology is devoted to recovering that lost space in the academic sphere

---

13. Fuller 1988, 85–89, 97–98, 111–19, 142–69, 219–24. To his credit, one historian of science has begun to explore this side of Kuhn that most philosophers still regard as tabooed. See Biagioli 1990, 1996. It is worth pointing out that while the idea of incommensurability is usually introduced via Kuhn's account of how he resolved the apparent incoherences in Aristotle, Galileo, and Newton (see chapter 4, section 4), his earliest appeal to the doctrine involved roughly contemporaneous communication breakdown rather than a failure to retrieve the meaning of the past. See Kuhn, "Energy Conservation as an Example of Simultaneous Discovery" (1959), in Kuhn 1977a, esp. 72.

14. This perspective is most fully developed in Fuller 1993b.

and contributing to a democratization of science in the public sphere.[15] Much of the present book is concerned with fleshing out the larger sociohistorical context that has enabled the closure of that space, a task that will take us beyond the Cold War back to the primordial myths of Western civilization. However, before plumbing the depths of the West's collective unconscious, a good point of entry into the mentality that informs *Structure* is the distinctly *pedagogical* context of its composition.

Although some brief autobiographical remarks in the preface to *Structure* hint at the importance of Kuhn's teaching experience to his own understanding of the history of science, the rest of the book fails to do justice to the role of education in the transmission, advancement, legitimation, and, yes, even negotiation of scientific knowledge claims in precincts outside the vicinity of specialists.[16] A naive reader of *Structure* may be excused

---

15. My own efforts at the democratization of science are perhaps best epitomized in Fuller 1997b. The book that includes this article was written and edited by a former student of mine. It aims to insert a critical perspective on the ends of science into the undergraduate courses in the humanities that are normally required of science and engineering majors in the United States.

16. Although Kuhn's pedagogical orientation is due primarily to his experience in the Harvard General Education curriculum, he was also influenced by Ludwig Wittgenstein (1889–1951) and Karl Popper (1902–94), two Viennese thinkers who emigrated to Britain to become the most influential figures in anglophone philosophy in the twentieth century.

Kuhn generally cites Stanley Cavell (b. 1926), America's most creative aesthetician, as having introduced him to Wittgenstein's later work, which turns out to be the source of his modeling of paradigm shifts on the gestalt switch: i.e., incommensurable interpretations of the same data. See Kuhn et al. 1997, 177. Cavell, in turn, studied under Morton White (b. 1917), whose catholic philosophical interests in an era of competitive parochialisms have relegated him to the margins of his field. For an early attempt to incorporate the later Wittgenstein into a jealously guarded, Harvard-based, American pragmatist tradition, see White 1956, esp. ix, where White detects signs of Wittgensteinian influence in the tendency of positivistically inclined logicians—Nelson Goodman and W. V. Quine appear the most likely candidates—to speak of the "role," "function," and "job" of words. (Note White's explicit effort to distance this convergence from their divergent sociological roots, which in both cases have unsavory normative implications: the positivist legitimation of Cold War calculations and the "ordinary language" reinforcement of gentility. I regard White's refusal to allow the sociology of knowledge to influence the terms of rapprochement between British and American thought as a serious intellectual weakness in an otherwise unique attempt to bridge philosophical differences.)

For Popper's influence on Kuhn, see the very perceptive Jarvie 1988, esp. 314 ff. Kuhn attended Popper's William James lectures at Harvard in 1950, but probably learned only Popper's negative lesson, namely, the entrenchment of science's institutionalized biases, which tend to undermine science's critical mission. Whereas Popper diagnosed this tendency as a personal weakness on the part of scientists, Kuhn tried to give a more panglossian explanation—namely, that no general theory with a modicum of support would flourish, were it not sheltered from relentless criticism in its early stages. (Unfortunately, Kuhn failed to say when "early" ended, aside from noting the accumulation of unavoidable anomalies.) I am happy to let the reader decide whether Popper or Kuhn held the more cynical view about humanity's ability to fathom reality. Thanks to Francis Remedios for bringing Jarvie's article to my attention. As it turns out, both Wittgenstein and Popper also qualified as secondary school teachers (with Popper actually taking his Ph.D. in educational psychology under Karl Buehler), an

for believing that a "paradigm shift" is fully accomplished once the major researchers in the relevant science have decided to pursue the new paradigm, and that adoption of the new paradigm by professional teachers, scientists in other fields, popularizers, and the public at large after that point is a more or less instantaneous trickle-down effect. In this respect, Kuhn simply repeats the popular historiography of science as the succession of trailblazers at the research frontiers, except that the heroic genius is replaced by the self-perpetuating cult.[17] In Kuhn's hands, education appears as the vehicle of acculturation, which is indeed the only social process that is ever adequately explored in his account of science. As a result, education is treated as merely the means by which a paradigm is reproduced in successive generations of dutiful researchers.[18]

In contrast, Kuhn's mentor, Conant, was fully cognizant of the special character of science's pedagogical mission to *nonspecialists*. He believed that American-style democracy itself was at stake. Specifically, Conant wanted the future managers of America to be taught how to tell the difference between good and bad science in what he envisaged to be an increasing number of science-based policy proposals. These decisions would be made in an age for which the image of the Bohr atom stood as an ever-present Janus-faced symbol for humanity's ultimate destruction or ultimate salvation. Conant's original teaching staff included freshly minted Ph.D.s like Kuhn who are among the founders of the history of science profession in the United States.[19] Most held their terminal degrees in a science, not

---

aspect of their lives that carried over to their philosophical practice. For a penetrating analysis, see Bartley 1974.

17. Thus, Kuhn 1957, his first book, ends with Newton's vindication of Copernicus 150 years after the latter's death. As Robert Westman has astutely observed, the providential character of this conclusion could not have been stated better by Newton's own eighteenth-century enthusiasts. See Westman 1994, 113. Nevertheless it would be another fifty to a hundred years after Newton's *Principia* before Copernicus would be widely taught in the schools. The historiographical problem highlighted here—that is, taking events at the cutting edge as definitive of the entire knowledge enterprise—has been most vividly raised in the context of science and technology's role in industrial development and wealth production. Here historians have shown that the fixation on innovation has led to misleading accounts of economic growth and decline, as they draw attention from the mechanisms of diffusion. See Edgerton 1996.

18. This point comes out especially clearly in response to critics. See Kuhn 1977b.

19. Here it is natural to ask about George Sarton (1884–1956), the Belgian humanist who, shortly after the end of the First World War, started editing from Harvard what has become the premier history of science journal, *Isis*. Sarton managed to wangle some office space in Harvard's Widener Library for his archival and editorial work without holding a tenured post, in part courtesy of Carnegie Institution trustee Andrew Dickson White, the historian most responsible in the United States for portraying science as being in a perennial struggle for enlightenment against religion. See Thackray and Merton 1975. Upon becoming Harvard president, Conant continued to indulge Sarton and even imagined that his work might be important in preserving the collective memory of civilization in the event of a nuclear war. Nevertheless, despite Sarton's petitions, Conant never saw the history of science as a field

history, yet their facility in science was primarily in the kind of "tabletop" experiments that prevailed before military and industrial factors converted science into the scaled-up enterprise with which we associate it today. Not surprisingly, the last experiments covered in Conant's courses were Pasteur's. Professionally ill equipped and ill disposed to pursue research in an era of "Big Science," Kuhn thus took refuge in the history of science.[20]

From this standpoint, *The Structure of Scientific Revolutions* is an exercise in wish fulfillment, as it proposes a general model of scientific change based on examples drawn almost entirely from the three hundred years of European physical science prior to World War I.[21] Moreover, despite its reputation for having propelled the sociology of science into the academic forefront, the book says nothing about the transformation that the massive infusion of labor, capital, and technology had wrought on the conduct of science starting with the First World War. In Kuhn's telling, the core cognitive and social processes have remained unchanged, at least to the extent that he seemed to believe that studying the science of the last three hundred years can illuminate contemporary practices. Certainly, this was a point that Conant was keen to drive home in his General Education courses. For the more that students were able to see a Faraday or a Maxwell in every Bohr and Einstein, the less likely they were to be swayed by the hyperbolic claims made both for and against the dawn of the "atomic age" in the popular media.[22] These claims typically focused on such "applications" of science as magic-bullet medicine, extraterrestrial space travel, and unlimited nuclear power reserves — not the science done in the laboratory

---

worth pursuing for its own sake, say, with its own departmental autonomy and degree structure. Rather, as Conant's pedagogical interests should make clear, a certain sense of the historical trajectory of science was to be introduced throughout the curriculum, which made it imperative that the field not become another specialist subject. Given science's intimate involvement in the social upheaval produced by the two world wars, Sarton felt that the time had come to preserve the glories of an endeavor that may never be repeated again. See Sarton 1948, 168–86. On Sarton's correspondence with Conant on these matters, see Hershberg 1993, 407–9. I happen to sympathize with Conant's institutional vision of the history of science, though for ideologically opposed reasons. See Fuller 1993b, 220–24.

20. Moreover, Kuhn and his Harvard cohort were not the only trained physicists who gradually moved in a more critical-historical direction with the advent of the Cold War: Paul Feyerabend and Stephen Toulmin would also have to count in their number. Though less overtly critical in their outlook on science, Derek de Solla Price (who pioneered the use of quantitative methods in the study of scientific change) and John Ziman (who stressed the consensual character of scientific knowledge) also fall into this category.

21. The restriction of Kuhn's examples to physics and chemistry reflects the experimental emphasis that Conant placed in designing Natural Sciences 4. Most of these examples are given more extended treatment in Conant 1947, which was later expanded (with Kuhn's acknowledged help) in Conant 1961.

22. Conant 1952a, xiii–xv.

or on the blackboard. An autonomous science, so Conant and his cohort thought, is a science safe both *from* and *for* democracy.

These features of Kuhn's background begin to explain a striking feature of his public statements, namely their marked insensitivity to his own *historicity*, that is, the elementary hermeneutical point that one's understanding of the past is predicated on living when and where one does.[23] That Kuhn published *Structure* in the United States in 1962—and not Germany in 1862, or France in 1762, or Britain in 1662—did not seem to tell him anything special about where science might be heading.[24] The message is the one originally promulgated in the General Education curriculum: The scientific process remains essentially the same whenever and wherever it occurs. While Kuhn is hardly the first historian to overlook his own historicity, the contagiousness of his perspective has been most remarkable. More importantly for what follows, Kuhn's historical sensibility, while providing inspiration for many historians and historically minded thinkers, remained just as locked in the present as the "Whigs" and "rational reconstructionists" who are usually regarded as the Kuhnian's sworn enemies. In the process, Kuhn's work decisively contributed to reversing a 150-year project of *philosophical history*, which promised to bring the politics of knowledge production to the center of the public arena by making "the lessons of history" relevant for defining the contemporary scene and its prospective transformation. In Hegel's memorable phrase, this project presented history as philosophy teaching by examples.

The period in question is bounded by Hegel and Comte, on the far side, and Popper and Feyerabend, on the near. Among the illustrious and eclectic set of thinkers in this tradition would have to be counted William Whewell, Herbert Spencer, Charles Sanders Peirce, Ernst Mach, Max

---

23. Here I must take exception to the excess hermeneutical insight that Richard Bernstein ascribes to Kuhn. See Bernstein 1983, esp. 20–34. Although Kuhn periodically invoked something he called the "hermeneutical method," it is clear that his conception of it owes more to nineteenth-century textual analysis than the reflexive considerations of twentieth-century Continental philosophy. Thus, for Kuhn, to understand the historical agent sympathetically, the historian must bracket her presentist prejudices, but without making the exposure and critique of those prejudices themselves part of the inquiry. In practice, this meant that Kuhn strove to render a scientist's work "internally coherent," even at the expense of the work's removal from any relationship to its cultural context, which could indirectly raise reflexive considerations (i.e., differences between the historian's and scientist's contexts). When applied to living scientists, as in Kuhn's oral history of quantum mechanics (see note 8 above), this dogged pursuit of internal coherence could lead in some perverse directions, such as restricting the scientists' interview responses to a preordained template. See chapter 4, section 2, for more details. On Kuhn's pseudohermeneutics more generally, see Hacking 1979b, esp. 226.

24. The four dates and places allude to the faux-Hegelian "moments" in the institutional history of science charted in Ben-David 1984.

Weber, Ernst Cassirer, Pierre Duhem, Emile Meyerson, Gaston Bachelard, Alfred North Whitehead, John Dewey, Bertrand Russell, and John Desmond Bernal.[25] Each thinker tied the fate of humanity with the direction that he divined in current (techno)scientific tendencies, toward which he would then display a characteristic attitude of approval or disapproval, sometimes with specific policy recommendations. By weaving their abstract arguments around a historical narrative punctuated by recognizable landmark episodes in "Western civilization," these thinkers enabled a relatively large intellectual (or at least transdisciplinary) public to find points of contact and contestation with what they had to say.

The spirit of which I speak survives in such recent philosophical histories of *politics* as Paul Kennedy's *The Rise and Fall of the Great Powers* and, better still, Francis Fukuyama's *The End of History and the Last Man*. It is a perspective that confers epistemic privilege on "being later," though it need not presume that what comes later is better than what came before. (But when such a presumption *is* made, the history becomes "Whiggish.") Rather, it prods the audience to convert a usable past into a viable future. The proof of the historiography contained in such a book is less in the ingredients it mixes from the past than in the pudding it serves up as policy for the future: Does it inspire action? Is it history with a *point*?[26]

---

25. I would be remiss to leave out such encyclopedic, albeit now unfashionably pessimistic, purveyors of philosophical history of science as the righteous Pitirim Sorokin (1889–1968) and especially the apocalyptic Oswald Spengler (1880–1936), the author of *The Decline of the West* (1918–22), a best-seller whose avid readers have included not only Adolf Hitler but also the German quantum physicists of the 1920s and even some recent sociologists of scientific knowledge. See Restivo 1985.

26. Recently, a much-needed introduction to philosophical history has been published: Graham 1996. Among the isolated cases of philosophical history in our own time, one of the very best has been Toulmin 1990. Toulmin's thesis turns out to be complementary to several pursued in this book. In the first chapter, I stress how "Platonization" has downgraded the epistemic value of easily changed beliefs. Complementarily, Toulmin has emphasized the way Plato's legacy has demeaned the epistemic value of tolerance for beliefs that contradict one's own. Toulmin points to 1570–1640 as pivotal in the revival of this particular Platonic legacy. During this period—which separates Montaigne and Descartes as the respective exemplars of tolerance and intolerance—the prosperity of the Renaissance and the Age of Exploration yielded to economic uncertainty and depression, which were scapegoated through a series of religious wars. A good benchmark here is the fact that Copernicus was more tolerated in his day than Galileo was in his, despite Galileo's better-developed arguments for the Copernican perspective. The only way to stabilize society (so thought Descartes) was to provide indubitable foundations, ones on which all "rational" beings would have to agree lest they forfeit their rational status. This relatively subtle threat to the potential dissenter soon became bare-knuckled in the terms set out in Thomas Hobbes's social contract theory. Among recent philosophical histories of science that have been specifically written as wake-up calls for practitioners of a specific discipline, one stands out: Mirowski 1989. Mirowski not only shows the steady decline of neoclassical economic theory as it develops a metaphor drawn from defunct physics, but he also points the way toward redeploying resources from the institutionalist tradition to constitute a new social theory of economic value.

3. Philosophical History: A Coroner's Report

But how did *Structure* manage to kill the historicist impulse?[27] In the first place, the book's plot portrays science as passing through a developmental sequence that does not lead in any particular direction, save a more intensified version of the same cycle: a progress *from* that is not a progress *to*. The surest historical measure of progress in Kuhn's account of science is the increased specialization of disciplinary research agendas. In my own less charitable gaze, this appears as a cancerous growth of mutually impenetrable jargons that obstructs the search for a more holistic—if not "unified"—understanding of reality, a project that would carry on the spirit of *episteme, scientia, Wissenschaft,* and *science,* as these concepts have been understood throughout most of the Western tradition.[28] In any case, while some philosophers of science still claim to have been left untouched by Kuhn, nevertheless over the past quarter-century the field has come to bear the marks of Kuhn's impact. In short, philosophy of science has become "Kuhniferous" in both structure and content.

In structure, the philosophy of science has itself exfoliated into "philosophies of X," where "X" is the name of a special science, under which philosophers labor (a point to which I shall return in chapter 6).[29] On the one

---

27. Readers who doubt that philosophical history is dead should consult the justly celebrated White 1973, which offers a logically closed taxonomy of philosophical histories that a cynic could easily regard as an exercise in *taxidermy,* given White's judicious silence on the relevance of his analysis for our own times.

28. The fragmentation of inquiry that I describe below, which is normally thought to be the distinct product of our own "postmodern condition," has in fact been the norm throughout most of the history of humanity. The challenge for the West has been to pursue a unified vision of knowledge without lapsing into religious authoritarianism. The invention of the university as a legally autonomous unit in the Middle Ages was crucial in that development. For the implications of this view of the history of science, in light of cross-cultural comparisons, see Fuller 1997d; Huff 1993.

29. It is commonly thought that this tendency predates Kuhn, going back to the logical positivists, who virtually equated philosophy with the logic of the sciences. While the positivists did periodically recommend that philosophers attend to foundational issues in the special sciences (such was the nature of the International Encyclopedia of Unified Science, of which Kuhn's *Structure* was the final volume), they rarely followed their own advice, being more concerned with laying down normative scientific principles against which claimants to the title of science could be evaluated. The positivists who had significant formal training in physics, especially Rudolf Carnap and Hans Reichenbach, closely studied the structure of inference in classical and relativistic quantum mechanics because they held these to be the most mature sciences and hence harbingers of what other disciplines could become under a positivist regime. The interchangeability of physics and science occurs even in Carnap's last published writings. A good example is Carnap 1966, his University of Chicago lectures in the philosophy of physics. (Significantly, Carnap's editor, Martin Gardner, attended Carnap's 1946 course and subsequently became the *Scientific American* journalist best known for his wide-ranging campaigns against "pseudoscience.") Perhaps the most influential example of this physics imperialism of the positivists is the "deductive-nomological" (or "covering law") model of explanation that was originally championed by Carl Hempel in the context of

hand, this has served to curb the excesses of philosophers inclined to dismiss entire fields of inquiry (including branches of biology) for their failure to live up to the standards of another such field (usually physics); on the other, it has, as noted above, removed any public space for discussing the overall ends of science.[30] In content, post-Kuhnian philosophical defenses of scientific progress have been disconnected from any substantive ends that science might be presently pursuing. It has thus been possible to argue about the character of science's progressiveness without passing judgment on the activities of contemporary scientists or, indeed, having anything very illuminating to say about the trajectory that science should follow in order to realize a preferred theory of scientific progress.[31]

The philosophical naif may be forgiven for questioning the value of an account of science "as it ought to be" that refuses to pass judgment on what scientists actually do and, indeed, implies nothing in particular about what scientists ought to do. Even if we grant David Hume's point that "ought does not imply is," we should at least expect that Kant's "ought implies can" holds. Otherwise, what is the point of doing epistemology or the philosophy of science, when instead we could be doing the history or the sociology of science, which typically eschews normative issues and provides a more empirically adequate account of what scientists do?[32]

---

explaining *historical* events, even though it was clearly drawn from Newton's mechanical explanation of planetary motion. See Hempel 1942. However, in recent years, historians of twentieth-century physics have begun to challenge the grounds on which the positivists (and Popperians, it might be added) relied on physics as the fundamental science. Philosophical judgment seems to have been based less on the actual track record of physics than on the distinctive accomplishment of Albert Einstein's theory of relativity, a clear case in which theoretical vision outpaced empirical testing by many years — in other words, a philosopher's sense of science that was by no means characteristic of the reasoning patterns of physicists in general. See Sigurdsson 1992.

30. Perhaps the most notorious philosophical dismissal of a science in recent years is Karl Popper's charge that evolutionary theory is unscientific because it contains no falsifiable predictions. See Popper 1963, 340. However, it is often forgotten that even the relatively mild-mannered positivist Carl Hempel argued that evolutionary theory was misnamed, since the "theory" is more a systematic description of natural history than an adequate explanatory account of why the history is as it is. See Hempel 1965, 370–71. In this respect, Hempel is at one with the creationist critiques of Darwinism. On Kuhn's role in encouraging an underlaboring role for the philosophy of science, see Callebaut 1993, esp. 41–45. On the cultural implications of Callebaut's book, a set of interviews with prominent philosophers of science, see Fuller 1994h.

31. The situation is even worse in epistemology, where we witness elaborate debates over definitions of knowledge that are purportedly of a normative character, but which have failed to demonstrably improve the conduct of inquiry. This should come as no surprise, since the hypothetical cases and thought experiments around which these debates turn bear little resemblance to questions of evidence and inference that concern scientists and other first-order knowledge producers. For an explanation of this peculiar turn of events, see Fuller 1992b.

32. Often the pointedness of this question is finessed by claiming that philosophers of science "describe the normative structure of science," which is increasingly called science's

To appreciate our distance from the world of the philosophical history of science, consider the transformed division of labor among scientists and historians and philosophers of science that followed in the wake of Kuhn's book:[33]

> *Before Structure:* Philosophers examine past science and declare who was right and wrong as a means of exemplifying a historically emergent conception of scientific rationality, sometimes with the aim of making explicit that rationality in terms of formal logic. Historians examine contemporary science and declare whether or not the dominant research programs deserve their status by considering how they dealt with their precursors and competitors. Philosophers and historians are thus engaged in overlapping and mutually enriching activities. Scientists themselves might engage in either or both activities and consider them integral to the conduct of their first-order inquiries.
>
> *After Structure:* Philosophers ignore the past and focus only on the contemporary sciences, in order to discern the implicit logic of their methods and theories. Historians ignore the present and focus only on past science, in order to understand it on its own terms. Philosophers and historians are

---

"cognitive structure." This is to suggest, in Platonic fashion, that the norms of science are sufficiently independent of what scientists do that the philosopher is never forced into the politically awkward position of pronouncing on the normative status of science as it actually exists, that is to ask: *Is science as it ought to be?* In chapter 1, section 5, we shall see that Feyerabend raised precisely this complaint after reading the first draft of *Structure*. If anything deserves to be called the "Curse of Kuhn," it is the tolerance of this finessing of descriptive and prescriptive matters. To his credit, the French philosopher of science whose structuralist account of conceptual revolutions most resembles that of Kuhn's, Georges Canguilhem (a student of Gaston Bachelard and teacher of Louis Althusser and Michel Foucault), had no trouble identifying this problem in *Structure*. See Bowker and Latour 1987, 725.

33. Because *Structure* straddles the disciplinary writing conventions of the history and the philosophy of science, and Kuhn rose to fame while affiliated with Princeton's history and philosophy of science department, it is often assumed that Kuhn actually welcomed the integration of the two fields into one "HPS." However, unlike the other "historicists" in the philosophy of science with whom he is grouped—including Toulmin, Hanson, Shapere, and Laudan, as well as Popper, Feyerabend, and Lakatos—Kuhn was very clear in his *opposition* to any interdisciplinary merger: see esp. Kuhn 1977a, 4. To appreciate the contrast even in 1961, the year before *Structure*'s publication, Hanson 1962 argued as follows in his vice-presidential address to section L of the American Association for the Advancement of Science: "complete cleavage of [the history and philosophy of science and scientific practice] would ultimately result in the death of each." He summarized his thesis: "Using one's head to maximum effect requires *learning* how most effectively to use one's head." This "learning," which is largely of the structure of past scientific debates, is the subject matter of the history and philosophy of science. A very colorful and energetic figure who in his short life earned the conspicuous admiration of his distinguished contemporaries Stephen Toulmin and Hilary Putnam, Hanson helped establish history and philosophy of science programs at Cambridge, Indiana, and Yale Universities. In addition, he could be found on the pages of American kiosk periodicals (especially *The Nation*) during the late 1950s and early 1960s. Even pages from his most important piece of scholarship, Hanson 1958, were excerpted in *The Saturday Review*. Interestingly, however, Hanson stopped short of arguing a case *contrary* to Kuhn's, namely, that scientists should be *required* to study the history and philosophy of science. This view, which I happen to hold, is most consistently attributed to Ernst Mach in chapter 2, where the issue reappears.

thus engaged in mutually exclusive and autonomous activities. Scientists reject—or more precisely, ignore—most professional history and philosophy of science, while making up their own versions in popular works that end up enjoying a wider circulation than the professionally crafted ones.

Progress can be made, in Kuhn's view, only once a group of inquirers have taken full control of their research agenda and are no longer beholden to the experiences or interests of those who are not part of the inquiry. In practice, *it is a world in which the natural sciences and their historians exist, side-by-side, equal but separate: scientists without historians, and historians without scientists.* It is a peaceful coexistence born in Plato's Heaven.[34] This point is often lost today because we rarely consider that the autonomy of the natural sciences may have been *enhanced* by the historians of these fields seceding to form a field of their own, called "history of science."[35]

Contrast this situation with the one that obtains in the humanities and social sciences, in which historical courses are still often required for degree majors and the people who teach them consider themselves primarily practitioners of those disciplines. Students in these fields are thus routinely exposed to professors who are living reminders of a past that most contemporary practitioners would rather suppress. Consider, say, all the behaviorists in psychology and the institutionalists in economics who have taken refuge in teaching the required "history and systems" courses, where they reveal to students the infighting, subterfuge, and downright ignorance that enables cognitivism and neoclassicism to maintain their hold over their respective disciplines. Under such circumstances, the past is never forgotten but in continual competition with the present, no more vividly than in

---

34. It also presupposes, rather unrealistically, that "scientists" and "historians" are self-selecting groups, as opposed to "historians" being merely the name given retrospectively to those scientists whose interests are excluded from their discipline's constitutional convention. A good way to chart the "polite" transition of a disagreeable past out of a science's present is to say (delicately) that "philosophically inspired" criticism is beside the point once a research tradition has solved enough of the empirical problems that it has set for itself. At that point, criticism that originally appeared foundational is now merely "external" to the proper business of the science. In recent times, this view has been most clearly associated with Larry Laudan but, as we shall see in chapter 2, it is traceable to Max Planck.

35. For a trenchant critique of this tendency, see Forman 1991. Moreover, the research agenda of historians is not only separate from but also parallel to that of the scientists under study. Thus, historians routinely take their marching orders from the disciplinary character of the sciences, thereby normalizing their role in the university, even though in most cases the academic lineage goes back no further than a century. For example, Holmes (1997) writes hopefully, as the current academic arrangements of the sciences may turn out to have been an extended perturbation, if the current calls to privatize, or at least de-academicize, scientific research are heeded. Not surprisingly, today's historians find it difficult to make sense of the norms surrounding the production and distribution of knowledge in nonuniversity settings. After all, what is the exact role that science plays in government agencies and industrial firms, places where scientific research is heavily supported, yet where the scientist is not primarily focussed on contributing to a discipline's theory and practice?

the footnotes and bibliographies of humanities articles, where Plato and Derrida jostle for space in bearing witness to the latest development.

Two styles of doing the history of science are traceable to Kuhn's work, one roughly associated with how historians do it and the other with how philosophers do it. Both derive from Kuhn's experience in the Conant curriculum. One style harks back to Kuhn's fledgling efforts, the now frequently recounted story of how Kuhn's physics training originally made the texts he had to teach by Aristotle and even Newton appear unintelligible.[36] This characterizes the "relativistic" style that conceptualizes the past as a foreign country separated by time as if by space, requiring acculturation into native customs quite unlike our own. We shall have more to say about this orientation shortly (and in chapter 4, section 4). However, once Kuhn became a proficient instructor, a second style predominated, one more closely associated with Piaget and other developmental psychologists who focus the student's attention only on the salient aspects of scientific episodes that were necessary for recognizing and resolving some paradigmatic tension.[37] This perspective enabled the students in his course to acquire the kind of "understanding" of science that Conant sought. Although few professional historians have adopted this style, it is well represented by philosophers and psychologists who profess a "cognitivist" approach to the history of science (not to mention popularizers of Kuhn's ideas).[38] According

36. See, e.g., Kuhn 1977a, xi–xiii.

37. Kuhn's debt to Piaget appears most directly in a stylistic tic of *Structure*, namely, the frequent use of "assimilation," which in Piagetspeak refers to the process by which children incorporate new experiences into their conceptual scheme. Interestingly, Kuhn makes little use of "accommodation," Piaget's term for the reverse process by which the child's scheme is adjusted to new experiences, since for Kuhn that is always a matter of last resort, the precipitant of a conceptual revolution. See Kuhn 1970b, vi. His first encounter with Piaget was in a footnote in Merton's doctoral dissertation, *Science, Technology, and Society in Seventeenth Century England* Merton 1970. See Kuhn 1976, viii. The link that Kuhn implicitly draws between a Piagetian child advancing to the next stage of cognitive development and a student reenacting a gestalt switch from the history of science ("ontogeny recapitulates phylogeny," in Ernst Haeckel's slogan) comes out most clearly in his Festschrift piece for his intellectual mentor, Alexandre Koyré. See Kuhn 1977a, 240–65. Kuhn seems to have regarded Piaget and Koyré as holding complementary views on cognitive development, which together underscored his commitment to the idea that there is a "natural" trajectory "internal" to science. Kuhn "said to [Koyré] that it was Piaget's children from whom I had learned to understand Aristotle's physics. His response—that Aristotle's physics had taught him to understand Piaget's children—only confirmed my impression of the importance of what I had learned" (Kuhn 1977a, 21–22). As it turns out, Bruner's and Kuhn's mutual admiration did not inhibit Bruner from proposing the termination of Conant's General Education in Science program. See chapter 4, section 7, below.

38. The most comprehensive attempt at a "cognitivist" understanding of science in the Kuhnian spirit is still De Mey 1982. The best single historical monograph in this vein is Margolis 1987. Today, "cognitive history of science" covers a wide variety of approaches, most of which are rather un-Kuhnian in that they draw more on artificial intelligence than on developmental psychology. A good sense of the range is provided in Giere 1992.

to it, as the scientific mind-set moves from one paradigm to the next, it repeats a fixed sequence of stages: puzzle solving, anomalies, crisis, revolution, new paradigm, etc.

4. Relativism's Uncanny Legacy to Historiography

Interestingly, when both historians and philosophers want to highlight Kuhn's pernicious influence, they fixate on a normatively charged sense of *relativism*.[39] Relativism becomes problematic once it is raised from a methodological principle for understanding alien cultures in their own terms to an unconditional principle that values cultural difference for its own sake, regardless of its consequences. Because Kuhn himself endorsed methodological relativism without convincingly denying stronger forms of the doctrine, he has been widely understood as having accepted relativism as an unconditional evaluative principle, or "judgmental relativism," the Pirandellist philosophy of "It is so, if you think so."[40] While superficially a permissive doctrine, judgmental relativism ultimately condescends to the peoples under study by refusing to hold them accountable to standards by which we ourselves supposedly place such great store.

---

39. The first and still typical systematic pursuit of this charge by a philosopher is Scheffler 1967. For latter-day versions of this attack on Kuhn, consult the many works of Scheffler's student Harvey Siegel, esp. Siegel 1991, 91–115. A good example of a historian prosecuting the case against Kuhn's relativism is Cao 1993. My thanks to Skuli Sigurdsson for drawing this provocative article to my attention. A popular version of the critique has been recently put forth in Appleby, Hunt, and Jacob 1995, esp. 160–97. It is worth noting that these antirelativist opponents of Kuhn regard themselves as liberal-to-leftist intellectuals who identify with the general goals of the Enlightenment.

40. Here I allude to the title of a work by one of the great untold sources for the peculiar style that STS relativism has often taken: Italian playwright Luigi Pirandello (1867–1936), winner of the 1934 Nobel Prize in Literature. Pirandello's name is historically associated with the rise of expressionist and absurdist drama in Europe between the two world wars. Nowadays he is conveniently seen as a "proto-postmodernist." The title of Pirandello's most famous play, "Six Characters in Search of an Author" (1921), encapsulates the nature of his influence on STS, albeit usually through his Argentine admirer Jorge Luis Borges. Pirandello was concerned with the constraints that conventional dramaturgical techniques placed on character development and hence on the multiple perspectives from which dramatic action may be presented. His signature way of dealing with these matters was to superimpose multiple takes on the same scene and to have characters break frame by interjecting reflexive comments in defiance of the playwright. STS applications of Pirandellist techniques may be found in self-styled "new literary forms," which aim to demonstrate by example the consequences of taking seriously the socially constructed nature of scientific facts. Thus, the distinction between fact and fiction is blurred, just as Pirandello blurred the boundary separating action on- and offstage. A convenient source for this Pirandellist sociology is Woolgar 1988. A direct link between this literature and Pirandello is provided by the ethnomethodologist Pollner 1987, 58. A still authoritative source for Pirandello's own dramaturgy is Bentley 1946. An important point of convergence between Pirandello and his sociological followers is an explicitly skeptical attitude toward not only the feasibility but also the desirability of resolving the multiple perspectives and restoring the broken frames.

A rarely articulated background assumption here is that differences of opinion among people should be presumed deep if they have persisted over a sufficiently long period. Not surprisingly, relativism finds its earliest expression in eighteenth- and nineteenth-century "racialist" accounts of human history, especially when the biological component of race was sufficiently plastic to allow for the genetic transmission of environmentally acquired traits, what by the early nineteenth century had come to known as "Lamarckianism."[41] Yet the presumption that another culture is unlikely,

41. Once Lamarckianism was empirically discredited as a theory of genetic transmission with the rediscovery of Mendel's experiments in the early twentieth century, relativism divided into two doctrines: on the one hand, an environmentally incorrigible, "hard" racism; on the other, an environmental, or more specifically geographical, determinism. The former was associated with claims made (not only by Nazis) for "Aryan science" and "Jewish science"; the latter with the ethnoscience orientation championed by the German neo-Kantian émigré Franz Boas (1858–1942) and his American students Edward Sapir, Alfred Kroeber, Ruth Benedict, and Margaret Mead. Two excellent accounts of the traffic between biological and social thought during this period, especially the peculiarly supportive role played by "progressive" ideologies, are Smith 1991 and Stocking 1968.

This dual genealogy of relativism helps explain a peculiar feature of the sociology of knowledge as formulated by its most famous practitioner, Karl Mannheim (1893–1947). Prior to the Second World War, Mannheim was reluctant to extend the purview of his sociology of knowledge—underwritten as it was by the relativization of thought to social conditions—to mathematics and the natural sciences because he did not want it to be associated with doctrines of racially grounded science. This was a fair concern, as the expression "sociology of knowledge" had been coined in German by Wilhelm Jerusalem, a champion of the French philosophical anthropologist Lucien Lévy-Bruhl (1856–1939), who is best known today for having depicted the "primitive mentality" as given to "prelogical" reasoning patterns associated with children's fantasies. Mannheim objected to this formulation of the sociology of knowledge for its lack of historicity, which strongly suggested that mentalities were innate or at least unchanging. (It is worth noting that Mannheim's critique applies equally to *both* strands of relativism identified above. Thus, cultural preservation, a goal of post-Boasian anthropology, is tantamount, in biological terms, to the sort genetic inbreeding that contributes to the culture's racelike characteristics.) See, e.g., Mannheim 1936, 310.

More sympathetic to Lévy-Bruhl was Kuhn's avowed sociological precursor, the physician Ludwik Fleck, whose *Genesis and Development of a Scientific Fact* mentions all the major sociologists of knowledge of the day *except* Mannheim, even though the book appeared five years after *Ideology and Utopia*. Fleck, like today's sociologists of scientific knowledge, was less disturbed by the possible racialist overtones of mentalities than with its implicit judgment that "primitives" would improve their lot by adopting "civilized" patterns of reasoning, regardless of whether the latter bore any relation to the needs of their life world. See Fleck 1979, 174. Why was Fleck better disposed to the mentalities approach to the sociology of knowledge than Mannheim? One important reason was that, like Lévy-Bruhl and his followers, Fleck interpreted the "social conditions of thought" primarily in terms of the face-to-face interaction of a community's members in their local environment, be it a tribal village or a laboratory research team.

In contrast, Mannheim generally had in mind a geographically dispersed group, such as a political party, that related to the world in a common fashion because of a shared ideology, which in turn rested on an understanding of history that was shared by members of the same generation. (Moreover, Mannheim was not alone in grounding his sociology of knowledge in cotemporality rather than cospatiality. His neoliberal contemporary Friedrich von Hayek (1899–1992) based his normative theory of democratic representation on just such a conception of "natural constituency." See Hayek 1978, 95–96, 105–10.) Mannheim's ur-model was

or incapable, of change from encounters with outsiders — except by force — implies a stubbornness on their part that we would refuse, in principle, to attribute to ourselves as a liberal, open-minded people.[42]

But, of course, as my appeals to symmetry here suggest, the deeper reason for today's easy acceptance of judgmental relativism is that, in practice, it absolves *us* from ever having to prove our open-mindedness, as relativism demands nothing more from *our* affections than bare tolerance — just as it involves nothing more from another culture.[43] If we are prohibited from converting the natives, they must likewise respect the integrity of our own practices, however vile ours may appear to them (perhaps because they are third-party recipients of the effects of our culture, such as industrial pollution or commercial television). As we shall see in chapter 3, section 6, when

---

the Weimar Republic, in which many parties read the same events (most notably, Germany's defeat in World War I) against radically different ideological assumptions, which made negotiation impossible, even though the members of these parties were to be found throughout Germany. See Frisby 1992, 19–21.

Thus, Mannheim's sense of "relativism" did not have the strong "local knowledge" character common to Fleck and more recent sociologists of scientific knowledge. (It would be interesting to examine the role that "mass media" like national newspapers and broadcast radio may have had on Mannheim's thought, as opposed to Fleck's.) Although Kuhn's stress on the communitarian, craftlike character of science places him for the most part closer to Fleck than Mannheim, a key Mannheimian element in *Structure* is the "Planck effect," roughly the idea that the difference between scientists who are open- and closed-minded to a new paradigm depends on their age, and hence the amount of effort they have already invested in developing the old paradigm. For an explanation of the expression, see chapter 6, note 33.

42. It is not by accident that the early formulations of relativism in late nineteenth-century anthropology and linguistics took either a geographically isolated tribe or a dead language as their paradigm case of a "culture." Neither was likely to be subject to external interference. See Fuller 1995a, 1996d. Interestingly, Paul Feyerabend, though often stereotyped by philosophers as the exemplar of relativism, seems to have realized this point toward the end of his life. See Feyerabend 1991, 151–52. An argument of this sort had also been used by Ernest Gellner against Peter Winch's influential Wittgensteinian defense of relativism in the social sciences. See Gellner 1968. John Preston deserves credit for pointing out these last two references to me.

43. In ethnomethodology and phenomenological sociology, the idea that the inquirer should be able to return from a native culture to her own without undergoing any substantial epistemic transformation is known as the principle of "alternation." This principle, which corresponds to Kuhn's invocation of the hermeneutical method (see note 23 above), has been upheld by Harry Collins and more philosophically conservative practitioners of the sociology of scientific knowledge. It has recently been challenged — though more on philosophical than explicitly political grounds. See Fuller 1996f. A more politically interesting challenge to relativism is posed by Third World scholars who regard Western deference to "multiculturalism" and "postcolonialism" as just so much pharisaic piety that does little either to understand or to improve the lot of the "developing" world (though it may cause major cultural reverberations in the "developed" world). For the clearest and most learned statement of this position, see Ahmad 1992, which argues, among other things, that Marx's notorious lapses from "politically correct" understandings of India and the "Orient" may not have been so far off the mark after all. Thanks to the late Michael Sprinker for having the courage to recommend the publication of this work and alerting me to its significance.

discussing Kuhn's Cold War context, such a relativism goes hand in glove with a "realist" political sensibility that presumes all nations to be equally tainted, and hence recommends a defense posture that prepares for the worst possible scenario. Relativism, then, is the ideal position to ensure the containment of cross-cultural conflict among strong and weak nations, as it appeals to a sense of "culture" in which all nations are said to partake equally in the abstract, regardless of the differences in power they concretely exert over one another.[44]

From what I have said so far, it may seem that only the politically powerful have benefited from judgmental relativism. But of course, this position has also benefited contemporary scholars dedicated to understanding past scientists in their original contexts. Sometimes this achievement is portrayed as a victory for the scientists under investigation, since a relativist reading would seem to treat their words and deeds more sympathetically than today's scientists would. What remains unsaid, however, is that today's scholars treat these denizens of the past with a charity that was absent *even at the time that the original scientists lived*. In other words, the "contextual sensitivity" underwritten by relativist readings of the past reflects more the guild mentality of contemporary historical scholarship than a cross-culturally valid principle of social knowledge.

Of course, there has been a tradition of scholars since the European Renaissance devoted to recovering original sources. But for all their meticulousness, these scholars were motivated more by an interest in legitimating their own knowledge claims—which typically supported the legal entitlements of their political patrons—than those whom the claims were supposedly about.[45] After all, "history as it actually happened" is an attractive methodological principle only for those who have access to the sort of evidence that normally establishes "actuality," such as official documents that are careful to register exact time, place, and witnesses. For those without access to such evidence (or who were not in the room when such evidence was under construction), "necessity" and "possibility" turn out to be friendlier modalities, in that they undercut the superficial advantage of an empiricist methodology focused exclusively on what actually happened. Thus, a "necessitarian" historian can argue that the available evidence obscures the deeper processes, while the "possibilist" can claim that the evidence suppresses other events that escaped official registration.

---

44. Sujatha Raman first raised this objection against a defense of the sort of relativism described here ("contextualism") that appeared in the oral presentation of Smocovitis 1995.

45. My sensitivity to the politics of archival history has been informed, over the years, by the works of Donald Kelley, former editor of *Journal of the History of Ideas*. See esp. Kelley 1970, 1984.

To be sure, the grounds for these suspicions are themselves amply documented. German historical scholarship in the nineteenth century reveals very clearly the compatibility of archivally informed accounts of the past and contemporary political designs. For example, Heinrich von Sybel's *History of the Revolutionary Period*, published in 1853, was for many years the most extensively documented work on the impact of the 1789 French Revolution. Nevertheless, its purpose was transparent, namely to demonstrate the evils of political radicalism. Likewise, the valuable grammatical reconstructions of dead Indo-European languages performed by German philologists from the late eighteenth century onward are now regarded as the academic phalanx of Aryan racial ideology.[46]

To put this scholarly activity in contemporary focus, we might say that in the nineteenth century the amassment of archival sources in large learned tomes served to create a presumption of legitimacy, a "forward momentum," that in the twentieth century has come to be identified with the investment of large amounts of public funds in Big Science projects. Whereas today we are inclined to think that the amount of material resources bound up in a particular research trajectory makes its progress irreversible, a similar hope animated the feverishly scribbling scholars who held professorships in the national university systems of Europe in the nineteenth century. The only significant difference is that in the nineteenth century the material resources mobilized were rather elusive by any ordinary accounting scheme (i.e., the time and energy expended by scholars) though the intended outcome was politically concrete (typically the maintenance and extension of national boundaries). In contrast, today the material resources can overrun large national budgets even though the intended outcome has become rather abstract, if not metaphysical (e.g., "the ultimate unit of matter and energy" as the stated goal of building the bigger particle accelerator). This particular reversal of means and ends in the modern world remains one of the great mysteries for the sociology of knowledge to fathom.

5. Whig History and Its Discontents: Tory and Prig

To sum up the argument of the preceding section, what is now derided as *Whig history*, the recounting of the past to vindicate the present, was in fact behind the first flowering of archivally based historical scholarship, a practice that was only much later colonized by the species of relativism associated with the contextualist orientation of professional historians. The

46. See Bernal 1987.

most engaging account of how this scholarly sensibility spread from Germany to the rest of Europe, especially Great Britain, was written by the man who coined the dreaded expression "Whig history" and subsequently (along with Alexandre Koyré) raised the "Scientific Revolution" of the seventeenth century to a significance rivaling that of the emergence of Christianity: Herbert Butterfield.[47] Butterfield, in turn, credits Lord John Acton (1834–1902), the founding professor of modern history at Cambridge (whose chair Butterfield subsequently occupied), with having first recognized the great transition to documented Whiggery.

Acton is today remembered for little more than his quip "Power tends to corrupt, and absolute power corrupts absolutely." But his astute observations of nation building reached back to his student days. The specific role of the archives became evident to him during the Italian revolution of 1860, in which dispossessed dynasties located documents that established their right to rule. Acton believed that once historians were allowed some access to national political records, it quickly became necessary to allow virtually complete access, "for the actions of a government wear a better appearance in the documents of that government than as they are exposed in the papers of other countries. To bar historians access to the archives would be tantamount to leaving history to one's enemies."[48] As a result, problems of national identity were easily converted into ones of historiography. Such is the root of the Whig interpretation of history, which Butterfield had first discussed in 1929. Although Butterfield originally, and most famously, portrayed Whiggism as a methodological fallacy, after World War II he typically judged varieties of Whiggism for their effects on the current political scene. Thus, he credited the Whiggish portrayal of British history as "the history of liberty" for bolstering the British spirit in the face of an external authoritarian threat, whereas he criticized German Whiggism for having fostered a sense of inevitable expansionism that eventuated in the two world wars.[49]

Of course, this politically checkered past does not make archive-based

---

47. Butterfield 1931. On Butterfield's coinage of "Scientific Revolution," see I. B. Cohen 1985, 398–99. It is worth observing that the religious revolution to which Butterfield compared the Scientific Revolution was more at the level of *ideas* than of *institutions*, given the lengthy and painful incubation Christianity underwent before becoming "Christendom." This seems to be forgotten whenever people are surprised to learn that the natural sciences became proper university subjects only in the second half of the nineteenth century. Christianity eventually succeeded because it took control of the standards by which it was judged. The same is true of the natural sciences in the period of Kuhnian normal science.

48. Butterfield 1955, 79 (quoting Acton).

49. Butterfield 1955, 27. One contemporary thinker who regards the value of Whiggism in similar terms is Jürgen Habermas, especially in his critique of revisionist tendencies in recent German historiography of the Nazi period. See Habermas 1989.

scholarship any less useful to today's "contextualist" scholars. However, the result is an unrequited transhistorical affection. While today's historians credit, say, Pierre Gassendi with reviving Greco-Roman atomism in the seventeenth century and Pierre Duhem with publicizing medieval writings on mechanical principles in the early twentieth century, Gassendi and Duhem themselves would be disappointed to learn that this admiration comes from historians who pursue their calling for its own sake, removed from debates about the direction that contemporary (i.e., late twentieth-century) science should take. Gassendi and Duhem might reasonably conclude that, since the times of their deaths, science had managed to institutionalize historical amnesia, which has resulted in the alienation of historical inquiry from matters of science policy. As we shall shortly see, they would need only turn to *Structure* to vindicate this hypothesis.

Therefore, it will be paramount for what follows to distinguish the two alternatives to Whig history adumbrated here. The historiography that Gassendi and Duhem would prefer, one that vindicates them over their rivals, deserves the name of *Tory history*. It is the mirror opposite, or *contrary*, of that espoused by the Whig historian. The Tory historian literally believes that the historical figures under study got it right (or at least more so than their historical opponents), and that we ignore their lessons at our peril today. In chapter 2, I shall attribute this view to Ernst Mach. It is a view for which I have considerable sympathy, despite its politically conservative provenance. However, most of today's contextualist historians are best seen as making claims that merely *contradict* those of the Whigs. In other words, they prefer to keep their own politics hidden but simply wish to refute, at an empirical level, what the Whig historian takes to be fact. Such historians defend their guild values as ends in their own right, above and beyond what other causes their insights might serve. A good name for this attitude, the one that I see *Structure* as having popularized, is *Prig history*.[50]

In essence, Prig history is the result of Tory history forgetting its own historical origins. It would thus be a mistake to claim that the relativism behind much contemporary historiography of science simply recounts the past from the standpoint of the losers. On the contrary, most relativists seem to be devotees of Prig history who use past historical figures merely to shore up the historian's own authority. Indeed, strictly speaking, the Prig historian

---

50. My source for the expressions "Tory history" and "Prig history," in contrast with "Whig history" (though without the distinction drawn here), Brush 1995, 219. In formulating the tripartite distinction of Whig, Tory, and Prig history, my thanks go to Daniel Garber, Brian Baigrie, George Gale, Roger Ariew, and others, who in February 1994 helped to clarify my understanding of their respective historical sensibilities during a debate we had on HOPOS-L, the History of the Philosophy of Science electronic mail network.

denies the historical agents a second chance to win, which is probably what they would have wanted. The Prig historian forbids them this pleasure because of an accompanying moral psychology that perversely assumes the ultimate futility of reasoned disagreement. This attitude assumes a Zen-like stance toward intellectual disputes, whereby the historian's point becomes one of showing that each party's arguments make sense in their own terms and that dispute arises only because each side refuses to accept the other's terms.

Of course, such quintessential losers in the history of Western science as Descartes and Leibniz (in the sense of their having lost to Newton as the source of the modern physical paradigm) can be rehabilitated quite neatly in this fashion. But given their own less than generous attitudes toward history as a form of knowledge, Descartes and Leibniz themselves would have been the last to appreciate the restoration. From the historian's standpoint, the rehabilitation of these philosophical titans of yore would simply be part of redressing a historical imbalance, the historian's version of Wittgenstein's "letting the fly out of the bottle," that is, to show the ultimate pointlessness of precisely the sort of "metaphysical" disputes that Descartes and Leibniz thought were worth having. Even if Prig historians cannot always quite achieve Zen mastery, they can at least demonstrate the impossibility of a "clean win" in intellectual disputes. This sentiment can be dramatized by writing in an ironic mode that serves to attenuate the winners' success. Thus, the winners may simply have benefited from the errors of their opponents (whom the historian officially supports), or their win was really pyrrhic as they were forced to sustain some long-term costs, or the winners manifested some patently undesirable qualities that tarnish the luster of their success.[51]

The distinction between Prig and Tory historiography is not easy to draw in practice. Both lines of historians claim the spiritual paternity of Thucydides in that their urge to write history is born of loss and disappointment.[52] Specifically, the anti-Whig mentality characterizes the moral psy-

---

51. A more charitable reading of the Zen historiographical perspective has been associated with Herbert Butterfield's Christian sensibility, whereby we all lose in the long term of secular history, by virtue of the human fallibility engendered by Original Sin. For more than simply an ephemeral sense of success in our endeavors, we must therefore turn to God. See Berlin 1969, 82–85. In the concluding sections of chapter 7, I attribute a more cynically Zen attitude to contemporary STS researchers, who actually appear to benefit by the uncertainties surrounding science-based political and intellectual disputes.

52. Thucydides (460–408 B.C.), often dubbed the father of scientific history, wrote his celebrated *History of the Peloponnesian Wars* after having lost his commission as a general in the Athenian army, following a humiliating defeat early in the war. He was then forced into exile from his native city, which allowed him the opportunity to cultivate contacts among the enemy Spartans, who turned out to be victorious. A good general history of the sentiment that

chology of people who turn to write the history of a discipline in disillusionment after having pursued degrees, and sometimes even careers, in that discipline. However, while Prigs may believe that the history of their discipline had taken a wrong turn, they nevertheless hold that not much can be done about it now. In the case of the physical sciences, the turn is typically identified with science's involvement in the military-industrial complex, which is variously traced to the Franco-Prussian War, World War I, or World War II. (Over the course of this book, the reader will see why any of these wars might be considered relevant thresholds.) The Tory mentality, in contrast, is ever hopeful that justice will be delivered and science regain its original mission. In chapter 1, I shall associate this mentality with the Platonic tradition in philosophy. In psychoanalytic terms, it amounts to a belief in a return of the repressed. However, what marks Kuhn and his emulators as Prigs rather than genuine Tories is their "repression" (ignorance? amnesia?) of the circumstances under which *they themselves* have come to do history. I shall elaborate on this claim in chapter 8, but for now let me dwell on the common ground shared by Tory and Prig.

6. The Tory Worldview and Kuhn's Place Therein

An especially extreme version of Tory history is the "deconstructionist" approach that upends the "rational reconstructionist" approach traditionally favored by philosophers of science and roughly aligned to a Whig historiography. The deconstructionist historian wishes to show that most major debates in philosophy and science could have been avoided had the original parties attended more closely to their interlocutors' contexts of action and utterance—in other words, *had they behaved more like the latter-day historian.* Thus, had Descartes understood his Scholastic forebears with the scruples that characterize today's seventeenth-century historians, he would not have exaggerated the differences between his views and theirs, and consequently would not have resorted to the polemics that inform the systematically distorted views of the seventeenth century that underwrite the "modernist" perspective in contemporary philosophy and science, especially the oversharp demarcation of the human mind from its ambient material reality.[53] Of course, one can take this thesis back to the very beginning

---

unites Prig and Tory historiography is Herman 1997. A provocative feature of this work is its inclusion of ecologism and multiculturalism as the latest offspring of what he calls the "declinist" historiographical sensibility.

53. I shall return to a version of this argument in chapter 4, section 4, when considering Koyré's and Kuhn's interpretation of Galileo.

of the Western philosophical tradition and claim that the pre-Socratics have been misunderstood to such an extent that we have lost touch with the original philosophical impulse.[54]

A bit more hermeneutically merciful—but no less Tory—is the case prosecuted by Leo Strauss (1899–1973) and his fellow carriers of the Platonic torch.[55] They believe that the past is systematically misunderstood not simply because philosophers are poor communicators, but more importantly because they had to camouflage their insights so as not to upset the pieties of secular authorities whose mass-marketed myths kept the rabble at bay. This perspective will prove crucial to fathoming the motivation for Kuhn's *historiographical segregationism*, namely the idea that history of science for scientists and for historians are equally valid but mutually exclusive activities. In effect, it creates a *double truth* doctrine, one for the elites and one for the rabble.[56] Kuhn's twist on the history of this doctrine is that the "rabble" turn out to be the scientific community (and most of their philosophical well-wishers), and the "elite" the historians. Put it this way: before Kuhn, the keepers of esoteric truths huddled out of public view for fear of persecution; after Kuhn, they tell inside jokes to each other when faced with a public display of scientific authority. Look at the back row of any public lecture given by a prominent scientist. There you will see historians and sociologists (and a few enlightened philosophers) having a good laugh, deconstructing any authoritative historical reference that the scientist makes. Never has the light of truth been so gleefully hidden under a bushel basket![57]

The following quote from Kuhn, where he comes closest to admitting his segregationist impulses, is the prism from which my interpretation of his significance is projected:

> When it repudiates a past paradigm, a scientific community simultaneously renounces, as a fit subject for professional scrutiny, most of the books and articles in which that paradigm had been embodied. Scientific education makes no equivalent for the art museum or the library of classics and the result is a sometimes drastic distortion in the scientist's perception of his discipline's past. More than practitioners of other creative fields, he comes to

54. Indeed, this is not an unfair characterization of Martin Heidegger's life project, and if we shift the original fall from grace a couple of centuries to classical Athens, we come close to Alasdair MacIntyre's sensibilities about the history of ethics. See Heidegger 1996 and MacIntyre 1984.
55. Strauss 1952. An excellent introduction is Drury 1988.
56. On the historical significance of the double-truth doctrine, see Fuller 1997d, 20, 114–21. An excellent survey of the period of the doctrine's strongest influence, thirteenth- to seventeenth-century Western philosophy, see Pine 1973.
57. For the significance of this metaphor, see chapter 1, note 1.

see it as leading in a straight line to the discipline's present vantage. In short, he comes to see it as progress. No alternative is available to him while he remains in the field.

Inevitably those remarks will suggest that the member of a mature scientific community is, like the typical character of Orwell's *1984*, the victim of a history rewritten by the powers that be. Furthermore, that suggestion is not altogether inappropriate. There are losses as well as gains in scientific revolutions, and scientists tend to be peculiarly blind to the former.[58]

That Kuhn would appeal to 1984 to characterize the captivity of scientists to the history they learn in their textbooks gives away the Platonic sensibility. While *The Structure of Scientific Revolutions* begins by claiming that the history of science has the potential for radically transforming "our" understanding of science, by the end it becomes clear that "we" are the historians who must learn to cope with the ambivalent legacy of a social practice — science — that progresses by systematically forgetting its past. According to Kuhn, scientists need to tell stories of collective progress in order to motivate labors that may look trivial when taken on their own terms or, if not quite trivial, still may never issue in the results being sought. The professional historian armed with detailed knowledge of episodes from the scientists' stories is wont to recount so many accidents and failures that it would dispirit fledgling natural scientists.[59] Thus, every scientific revolution must be followed by a Whiggish rewriting of the discipline's history to make the victorious party appear the discipline's natural heirs, thereby motivating the specialized work on which they and their students are about to embark. To his credit, Kuhn acknowledges the power politics involved here. In the sentences following the ones quoted above, Kuhn admits that might does indeed make right in science, except that those with the might are recognized members of the scientific community. *In short, the progressiveness of science lies not in the character of the work scientists do, but in the control they exercise over how they recount their collective history.*

In evaluating the Tory perspective, the first point to grant is the strength, if not incontrovertibility, of its empirical basis. Everything we know about the semantic drift of natural languages, the cognitive biases and infirmities of individuals, and the ideological streamlining of institutional memory casts substantial doubt that anything like an intention, idea, proposition, or message could have ever been transmitted intact over any great expanse of

---

58. Kuhn 1970b, 167.
59. Thus was Kuhn's thesis popularized in Brush 1975, which provocatively asked, "Should the history of science be rated X?" Given Brush's intended audience (scientists), the answer was clearly yes.

time and space.[60] The only thing that prevents this point from having more impact on the conduct of philosophy and science is that virtually all theories of progress and rationality presuppose the contrary to be the case.

A venerable rhetorical strategy for inhibiting Tory inclinations, first popularized by Immanuel Kant over two hundred years ago, is the *transcendental argument*, which attempts to convert an impoverished imagination—specifically, our inability to envisage what the world would be like if progress and rationality turned out to be complete myths—into a guarantee that our faith in these myths is well-placed. In short: "X must be true, for were it not, then nothing would make sense." (Not making sense is apparently not an option!) Admittedly, philosophers who have squarely faced the mythical status of our most cherished epistemic principles — Friedrich Nietzsche and Jacques Derrida come most readily to mind — have met with a mixed fate, including glib rejoinders to the effect that our ability to understand *their* claims of global misunderstanding undermines the truth of those very claims. (But if one *really* understood the content of those claims, would it make sense to try refuting them by appealing to the pragmatic paradox involved in their assertion? Hardly. Rather, the popularity of this "refutation" shows that the thesis of radical misunderstanding is not well understood, thereby proving the Tory case!)

However, once we grant the empirical basis of the Tory perspective, we are still left with evaluating its significance. Three conclusions are possible. The first is that the misunderstandings are not inevitable but can be corrected by greater regimentation of language or some other form of psychosocial hygiene. This was the view taken by the logical positivists, or at least as they were popularized in the general semantics movement in such works as Alfred Korzybski's 1933 magnum opus, *Science and Sanity*, whose introduction thanks the Vienna Circle's American observer, Willard van Orman Quine.[61] Nevertheless, it is clear that the positivists regarded the emergence of cumulative knowledge growth as a remarkable occurrence in the history of humanity that requires special institutions whose long-term mainte-

---

60. For an analytic-philosophical treatment of this matter, see Fuller 1993a, esp. xv–xvi, where I endorse the "metaphysics of entropy" over the "metaphysics of inertia." It is unfortunate that an ideological version of the "two cultures" split prevents Tory historians and social-scientific students of human fallibility from recognizing their mutually reinforcing arguments. One person who has recognized the affinity of the two camps is the political theorist Jon Elster. See esp. Elster 1979, 1983.

61. The best-selling book in popular front positivism has probably been Hayakawa 1939. It is worth recalling that among the intermittent contributors to this genre of paraphilosophical literature have been New Dealer Stuart Chase and rhetorical theorist I. A. Richards. Quine discusses his mixed feelings toward the general semantics movement in his autobiography: Quine 1985, 139 ff.

nance cannot be taken for granted but must be subject to eternal vigilance.[62] The second possible conclusion is simply to resign oneself to the inevitability of misunderstanding, strategic or not. In that case, the positivists would appear to be engaged in a noble but futile (if not naive) gesture. Here the difference between the Tory and Whig interpretations of history comes into sharpest relief. It enables the Tory to invoke a *prisca sapientia*, an original moment of hermeneutical grace in the distant past, when we understood the secrets of the gods, but from which we have since fallen. Thus, to a Straussian, the logical positivists are little more than an unwitting parody of the master, Plato.

The third conclusion, the one I favor, is that "understanding"—in the strong cognitive sense presupposed equally by the Tory, Whig, and positivist—is *not* necessary for explaining the interest and concern that humans naturally express toward each other. In other words, the "understanding" I show of someone is not predicated on my having grasped some set of propositions they happen to hold. Grasping propositions—understanding exactly what people mean for purposes of evaluating what they say—is a relatively restricted activity, even in science, where the need to affirm one set of propositions over another occurs only when a scientist stakes a knowledge claim in a public forum, typically a professional journal, which is the moment that science becomes most gamelike and hence artificially severed from the cognitive flux that normally characterizes the human condition. Only philosophers engage in this kind of activity on a regular basis. And as so often happens, a guild virtue of the people writing the history—in this case, philosophers—has been mistakenly raised to the level of a categorical imperative, which leads us to assent all too easily to the idea that making sense of people is tantamount to making sense of their utterances.[63]

7. *Structure:* The Ultimate Servant Narrative?

If we assume that *The Structure of Scientific Revolutions* is meant to provide—at least in general outline—the Tory's "hidden history" of science, in contrast with the Whiggish history portrayed in the science textbooks, then is it not strange that Kuhn's account should become such an open

---

62. On this feature of positivism, see Fuller 1996e.

63. In a revisionist history of philosophy (which clearly cannot be done in these pages), the perennial quest for the "propositional content" of language would be portrayed as an elaborate rationalization for the fact that we never know exactly what people are talking about. But because the content of people's utterances may not matter very much, we relegate it to an ideal realm, where its contact with mundane reality is episodic and inconsequential to a definition of its nature.

secret, nowadays perhaps even *more* popular than the textbook histories themselves? Here an argument for Kuhn's utter lack of historical self-consciousness becomes most plausible. A key blind spot occurs in the earlier "Orwellian" quote, as Kuhn fails to identify Conant's educational strategy as the scientific analogue for "the powers that be" who in 1984 are entrusted with continually rewriting history. Had Kuhn reflected on the larger ideological mission that informed the courses in which he had been teaching the history of science for the decade prior to the writing of *Structure*, he might have anticipated the extent to which his own account could contribute to a new kind of Orwellian history, one well suited to the students in the humanities and social sciences who were the original clientele in Conant's General Education in Science program. In that case, Kuhn would not have been so surprised by his popular acclaim in these fields, especially the tendency for humanists and social scientists to adopt his account of science wholesale without closely analyzing or developing its constituent ideas.

For those inclined to extract a silver lining from any cloud, my last remark tags *Structure* as the consummate postmodernist work, one whose cross-disciplinary appeal is founded on its ability to compel readers without demanding too much engagement in return. It is not a master narrative but a *servant narrative* that is indefinitely adaptable to user's wishes. Although the death of the grand narrative (a vivid name for "philosophical history") and the rise of relativism are usually associated with Jean-François Lyotard (1924–98) and allied French thinkers, Kuhn's stylistic achievement was to ease not only humanists and social scientists, but more importantly (often in spite of themselves) practicing scientists, analytic philosophers, and other congenital Francophobes into a postmodern mind-set.[64] Certainly, the book itself does not encourage a deep reading. That Kuhn's preparation for writing *Structure* was more the classroom than the archives is revealed by its nonthreatening prose style, which contains relatively little technical language—and, for that matter, relatively few footnotes to other authors, no matter how much they may have influenced him.[65] By refusing to cloak

64. Lyotard 1983, 26, 43, 61 (references to Kuhn). Someone who appreciates Kuhn's ability to make "deep" antitheoretical points by simple means is Rorty (1979, 320 ff.; 1995). In the latter work, Rorty argues specifically against Francis Fukuyama that it may not be so bad that with the fall of the Soviet Union (a.k.a. "the end of history") there is nothing for theorists to do. Yet Rorty fails to see that Kuhn is so user-friendly that he serves Fukuyama's purposes as well. See chapter 5, section 5, below. I argue against the Rortyesque denigration of theory in the postmodern world (as epitomized by Stanley Fish) in Fuller 1993b, 347–76.

65. Indeed, the Harvard General Education curriculum, in which Kuhn honed the book's thesis, was quite explicitly regarded by its main participating faculty as just that—devoted to education, *not* original research. See the remarks of Clyde Kluckhohn upon the establishment of the General Education curriculum. See *Minutes*, 30 October 1947.

imperfections in jargon, *Structure* invites the reader to participate in correcting its flaws and completing its argument. But this invitation is less to interpret than to apply the text. Insofar as the invitation is to interpret at all, the object of interpretation turns out to be more inkblot than palimpsest.

Thus, a common thread that runs through the formal and informal comments that people make about *Structure* is that it is quite thin on matters in their own field of expertise, but truly enlightening in some other field, one in which they have long had an interest but could not locate a suitable point of scholarly intersection.[66] It might be said that *Structure* has a philosopher's sense of sociology, a historian's sense of philosophy, and a sociologist's sense of history. A text with such holographic qualities is assured a good reception just as long as the practitioners of these different disciplines continue talking only to their own colleagues and not those of the field that Kuhn supposedly represents so well for them. In that respect, Kuhn's success is symptomatic of the much larger problem of interdisciplinary communication breakdown that afflicts contemporary academic life. This breakdown helps explain why Kuhn has garnered many grateful users but few devoted followers. In the pages that follow, I trace this state of affairs to the successful promulgation of Kuhn's double-truth doctrine of historiography. It ends up insulating not only the elites from the masses but also one set of elites from another—assuming that it is still fair to include practitioners of the various academic disciplines among the elites in today's world.

My narrative consists of a series of overlapping accounts that divide roughly into two sections. The first four chapters are concerned with the background to the writing of *Structure*, which extends back to Plato's invention of the double-truth strategy, through its materialization in nineteenth- and twentieth-century humanistic scholarly practices, to Conant's General Education in Science program. The last four chapters cover *Structure*'s impact on the social sciences, philosophy of science, and sociology of science, along with a strategy for overcoming what I regard to have been the overall conservative nature of the book's influence. Readers wanting a more detailed account of the circuitous route on which they are about to embark should turn to chapter 8, section 1. What follows is the saga of how the West developed so as to make *The Structure of Scientific Revolutions* the most "natural" way of understanding the premier knowledge production practices of the late twentieth century.

---

66. For one historian's detailed complaint about Kuhn's influence on how nonhistorians understand his field, see Reingold 1991, 389–409.

8. Summary of the Argument

This account of the curious origins and even more curious reception of *Structure* has begun by considering the historiographical traditions to which it was affiliated and opposed. I have argued that while much has been made of Kuhn's opposition to Whig historiography, little has been made of his affiliation to Tory and Prig historiographies, which operate from the assumption that science has either declined or failed to display the degree of progress popularly accorded it. The Tory pedigree is associated with disappointment that a preferred outcome was not historically realized, the Prig with the professionalism of the historian who refuses to take sides in any past dispute. Given the choice of Tory and Prig, I stand firmly behind the Tory (its unfortunate political overtones notwithstanding). However, the dominance of the Prig interpretation of Kuhn has contributed to the death of what I call "philosophical history," the point of which was to make an explicit normative pronouncement about the course of history.

Next, I trace Kuhn's normative reticence to the double-truth doctrine, whose roots reach back to Plato's reservations about the public display of critical reason following the fall of Athens. In chapter 1, I coin the term "embushelment" to capture this fear of publicly contradicting received opinion, because of its potentially destabilizing social consequences. The signature response to this fear has been to imagine that all significant cultural artifacts are doubly encoded, with one message intended to appease the masses by reinforcing their prejudices and the other meant only for elite inquirers who are mentally prepared to assimilate a strongly counterintuitive truth. In the modern period, the idea of doubly coded artifacts was turned into an academic discipline with the emergence of iconography as the premier school of art history. The iconographic perspective infiltrated the history of science mainly through Kuhn's historiographical mentor, Alexandre Koyré.

However, Kuhn's own personal history with Conant gave this legacy a distinctive twist, one that may have led Kuhn to conclude that regardless of how far Big Science has fallen from its normative ideal, more good is done in our volatile world by continuing to support science, even in its corrupt form, as a principle of legitimation than by rejecting it for some unknown alternative. The turn of mind I impute to Kuhn, characteristic of the "dirty hands" thinking of Cold War political realism, is explored in more detail in chapter 3, which explicitly traces Conant's (and more generally, Harvard's) efforts to establish and protect the autonomy of science as a cultural defense strategy against various threats, especially New Dealers and Marxists. Emblematic of this concern was Harvard's Pareto Circle, which is probably the source of Kuhn's restorationist conception of revolutions in science.

The rest of chapter 1 explores the historical construction of "scientist" as a unique social role requiring protection and "autonomization" from the rest of society. Here I present the conditions that led William Whewell to coin the term "scientist" in the 1830s, thereby constituting the object of inquiry for future historians, philosophers, and sociologists of science. Accompanying this coinage was a distinction between what are now called the contexts of "discovery" and "justification," the transition between which marks the rational reconstruction of the history of science into an edifying and progressive myth. Although Kuhn is normally seen as having subverted this distinction, a closer look reveals that his real interest was in showing that all the key processes of science—including its messy discovery phase—could be explained in terms of science's self-organizing principles. Those who called for a distinction in contexts typically admitted that much of the inspiration or initial ideas for doing science came from outside science itself. This is what Kuhn wanted to deny.

I pick up on this point again in chapter 4, when discussing the transformation of Kuhn's pedagogical charge in Conant's curriculum into a general research strategy for doing the history of science. Here I show how the so-called internal/external history of science distinction was manufactured in the presentation of course materials for students, the conduct of interviews with great scientists, and the cultivation of appropriate historiographical attitudes in historians of science. In this chapter, I also sketch a "Hegelian" account of a sequence of moments in the history of science's self-understanding, at the end of which Kuhn's theory of paradigms and revolutions would appear quite natural to believe. The key to this account is that various competitors and contaminants of science—specifically religion, technology, and history—were successfully excluded from the scope of scientific inquiry.

Chapter 2 is designed to remind readers what it was like when philosophical histories of science mattered to science policy debates, as scientists themselves argued over the future direction of their fields. I focus on the debate between Max Planck and Ernst Mach in the decade preceding World War I. At that time, the experimental physical sciences had proved themselves to be an academic, military, and industrial force, but questions remained as to their long-term institutionalization. Planck argued for an autonomous disciplinary orientation, one that presaged Kuhn's discussion of paradigms in many respects, whereas Mach supported a more scaled-down and user-friendly vision of science that would be adapted to human ends. Their debate spanned many important epistemological and political issues, the most lasting of which has probably been realism versus instrumentalism in the philosophy of science.

Mach's appeal to a "critical" approach to the history of science as a source of dissenting voices to the orthodoxies of his day is much like the one taken in this book and very much against the historiographical spirit that has followed in the wake of Kuhn's influence. Nevertheless, Planck's generally recognized victory over Mach paved the way for the acceptance of a Platonic vision of the experimental scientist, one guided by the sort of overarching commitment to the truth that is associated with a Kuhnian paradigm. The main witness in this transition was the chemist Michael Polanyi, who saw firsthand the horrific consequences of a purely instrumentalist approach to science in the role that his own discipline played in World War I. Polanyi is also interesting for his frequent appeals to anthropological studies of indigenous peoples in order to model the self-contained integrity of the scientific enterprise, a strategy that would be eventually used to ironic effect by practitioners of science and technology studies.

Chapter 5 begins, like chapter 2, by recounting what the world was like before the influence of *Structure* was in place. Here I consider social-scientific attitudes toward the natural sciences, noting a very strong critical orientation in the work of Alvin Gouldner and C. Wright Mills, and even Karl Mannheim's later writings. However, *Structure* unwittingly defused these critical sentiments that had emerged from science's increasing involvement in the military-industrial complex. I say "unwittingly" because the very features of Kuhn's account that had enabled him (and Conant especially) to distance the nature of science from its most destructive contemporary manifestations — namely, the omission of science's technological and economic dimensions — were precisely those that emboldened social scientists to think that they could reinvent themselves as "real scientists."

The objective reality of Kuhn's theory was bolstered by the development of science indicators tied to particular phases in his scheme, and by 1970 sociologists and political scientists had come up with competing accounts of how to render themselves as paradigms. However, a survey of the legitimatory uses of *Structure* reveals almost entirely conservative effects. For example, Daniel Bell used Kuhn's theory to reinforce the role of disciplines over interdisciplinary research in the besieged universities of the late 1960s, while more recently Francis Fukuyama has invoked Kuhn for the view that science's autonomous development has enabled it to be the motor of global wealth production. Even supposedly radical appropriations of Kuhn, such as those of the German finalizationist movement and the American countercultural theorist Theodore Roszak, turned out on closer inspection to either mute or divert any "revolutionary" impulse.

The post-Kuhnian fates of philosophy and sociology of science are the topics of chapters 6 and 7, respectively. As for philosophers, they have come to openly embrace the role of science "underlaborers," relinquishing the prescriptive, legislative, and critical attitudes that had marked philosophy's traditional relationship to the natural sciences, even well into the twentieth century. Philosophers nowadays are content to attend to the norms implicit in the particular sciences they study, which are presumed to proceed in a normatively desirable fashion. As to be expected in a Kuhnified world, much of this change in philosophical orientation has been accompanied by a rewriting of the field's own history. Indeed, the title of chapter 6 alludes to Richard Rorty's now-famous attempt to draw Kuhn's philosophy of science much closer to mainstream analytic philosophy than either Kuhn or analytic philosophers would have imagined possible a decade earlier.[67]

This underlaboring mentality is sometimes dignified as a philosophical position, "naturalism." However, a search for a "missing link" between classical naturalism and its Kuhnified counterpart reveals that the Harvard pragmatist C. I. Lewis anticipated many of Kuhn's most radical statements concerning the incommensurability of worldviews, yet without relinquishing the strong normative position that philosophers traditionally adopted toward the special sciences. Lewis notwithstanding, terms relating to "reason" and "rationality" have undergone substantial revision, with the marginalization of Karl Popper, Imre Lakatos, Paul Feyerabend, and especially Stephen Toulmin paying witness to *Structure*'s redefinition of the philosophical agenda, such that issues relating to the argumentative and rhetorical sides of scientific inquiry are now occluded. Most striking in this context is that radical philosophical criticism of science has come to be associated with irrationalism.

The post-Kuhnian sociology of science, the core of the interdisciplinary field of science and technology studies, is probably the most self-consciously Kuhnified of all fields. Chapter 7 begins by recounting the conditions under which *Structure* was read by the founders of the "Strong Programme" in the sociology of scientific knowledge at Edinburgh University in the late 1960s. Whereas Conant's curriculum aimed to make nonscientists receptive to science, the Edinburgh curriculum operated in reverse to make scientists more receptive to nonscientific concerns by sensitizing them to the various social and cultural environments in which their

---

67. Rorty 1972.

work was embedded. This initiative was part of the general attempt to address what C. P. Snow had called the "two cultures" problem that first embroiled British intellectuals in the 1950s, when "technocrats" started to replace Oxbridge humanists in the civil service.

However, as the Edinburgh Science Studies Unit shifted its charge from the undergraduate teaching of scientists to the graduate teaching of STS specialists, it increasingly acquired the trappings of a Kuhnian paradigm. This has had several long-term effects, which may be encapsulated as intellectual tunnel vision and depoliticization, be it in STS accounts of its own history, the style of writing its practitioners adopt, or the difficulty that the field's practitioners have had in confronting their political differences in professional and public forums. It is here that readers will see most clearly the Zen-like qualities of Prig historiography transferred to a contemporary research setting. Chapter 7 ends with a diagnostic explanation of the leading school of STS today, whose much vaunted policy relevance has been based on successful "parasitic" relations with competing interest groups and institutions in French society.

In chapter 8, the conclusion, I begin with the recapitulation of the book's main argument, focusing on whether Kuhn was cognizant of the sociohistorical factors that shaped the vision of science presented in *Structure*, and coped adequately with how his book was taken up in academic and policy circles. Kuhn withdrew from the challenges posed by his work in a manner familiar from the lives of saints who passively renounce their past. However, beyond simply passing judgment on Kuhn, I want to overcome *Structure*'s effects on its readers. This implies reversing the value orientation that Kuhn's book has promoted in the organization of knowledge production. Specifically, I argue that paradigms should be seen, not as the ideal form of scientific inquiry, but rather an arrested social movement in which the natural spread of knowledge is captured by a community that gains relative advantage by forcing other communities to rely on its expertise to get what they want. My proposed transvaluation should be seen as the latest moment in the West's historic project of secularization. The university has traditionally played an important role in this project, especially as a critical clearing house for new ideas, facilitating their uptake by as many different groups as possible. Toward this end, I would reinterpret what philosophers call "the context of justification" as just this practice of public dissemination. It amounts to a reintroduction of historical contingency into the history of science that scientists learn.

I
The Pilgrimage from Plato to NATO
Episodes in Embushelment

1. Embushelment: The Closing of the Western Mind

One of the most vivid metaphors that Jesus used to address his Apostles was of the lamp hidden beneath a bushel basket, a situation that of course only served to subvert the lamp's illumination. By this metaphor, Jesus meant to decry the reluctance of Christian converts to spread the Gospel, for fear of persecution as they inevitably upset the social order. I have coined the word *embushelment* to recall this iconic episode in one of the founding myths of Western culture.[1] The episode's exemplariness comes from revealing the dual-tracked character of the Western conception of Reason. The first track extends from Socratic questioning in the Athenian forum through the Enlightenment to Ernst Mach and Karl Popper. It is critical, libertarian, and risk seeking—and it also seems to be the track that Jesus himself espoused. The second track extends from the cloistered setting of Plato's Academy through positivism (probably in all of its incarnations but certainly in Auguste Comte's)[2] to Max Planck and Thomas Kuhn. It is foundational, authoritarian, and risk averse—and it also characterizes the track with which institutional Christianity, especially the

---

1. The original version of the metaphor is probably Mark 4:21–25. Interestingly, Jesus dovetails this metaphor with one concerning return on investment: the more light one gives off, the more one gets in reflection. Elsewhere in the New Testament, an opposing spin is put on the metaphor, such that the value of effort becomes entirely dependent on divine approval: the more goodness one receives from God, the more one will continue to receive. This is the biblical root of Robert Merton's "principle of cumulative advantage," which he sometimes calls the "Matthew effect," after Matthew 13:12, where the metaphor appears. In that sense, the difference between my own normative sociology of knowledge, or "social epistemology," and Merton's may be reduced to the passage in the Gospels from which we choose to proceed: I from Mark and Merton from Matthew. In chapter 8, section 1, I argue that an "aristocratic" interpretation of the Matthew effect goes a long way toward explaining the success of Kuhn's *Structure*.

2. On the various thinkers who have either called themselves "positivist" or have had the label thrust upon them, one need look no further than Kolakowski 1972.

Roman Catholic Church, has often identified. This telegraphic summary of the contrast in positions will be useful in understanding the sensibility that informs these pages.³

It would be easy to say that the two tracks are "complementary." In fact, it would be too easy, since each regards the other as its worst nightmare. In this respect, the Western conception of Reason is schizoid. From the standpoint of the first track, the second appears to be throwing up obstacles in the way of inquiry, the removal of which then becomes the task of Reason. However, seen from the second track, the first appears to be preventing inquiry from ever taking a clear course, the securing of which then becomes the task of Reason. This book is effectively an account of the second track written from the standpoint of the first. Consequently, its plot will center on tasks relating to the closure of the field of inquiry, the presumption being that inquiry remains open to many alternative extensions, until they have been expressly closed. These tasks of closure have been described — in ascending order of abstractness — as *professionalization, disciplinization, internalization, normalization*, and most simply, *containment*. They are terms that are valued positively by denizens of the second track but at best ambivalently by those who, like myself, dwell in the first track.

The reader should now be put in the right frame of mind to approach this chapter, which treats the fall of Athens in the Peloponnesian Wars as the original traumatic moment in the Western psyche. The resulting collective defense mechanisms have restricted the realm of the rational so that the possibilities for change are defined in advance of any actual change. In philosophical shorthand, this aspect of Reason has been variously called *innate, a priori,* and *paradigmatic*, depending on whether the vehicle of Reason is a mind, a logic, or a community of inquirers. Accordingly, novelty is presumed to threaten this preestablished order, and hence appear as irrational, unless the vehicle of Reason has already become a house divided against itself, be that state called, Platonically, an *aporia* or, Kuhnianly, a *crisis*.⁴ The first task for us, then, is to sample the genealogy of embushelment that connects Plato's and Kuhn's coinages.

It would be nice to think that were the icons of Western civilization miraculously to grace us with their presence today, we would continue to hold them in high esteem. Unfortunately, the respect accorded to these icons often varies inversely with our knowledge of the real-life situations in

---

3. At this point, a reader with literary leanings may wish to turn to "The Grand Inquisitor" in Dostoevsky's *The Brothers Karamazov* (1880). Sociologists may wish to revisit Max Weber's distinction between prophetic and priestly modes of religious legitimation.
4. An exceptional attempt to cast the major systematic philosophies in the Western tradition as expressions of aporia by self-fashioned "heretics" is Ross 1989.

which they put forth their seminal ideas. A distinctive feature of the Platonic tradition—especially of those Platonists who take the *Republic* as the cornerstone work—is its active attempt to carry over into the present day not merely the content, but the original *context* of Plato's thought. Thus, preserving the attitude surrounding the fall of democratic Athens in Plato's youth becomes at least as important as preserving the all-too-familiar realm of pure ideas that epitomized his mature thought.[5] If Plato's generalization from his own experience in Athens is to be believed, democracies result when the people take revenge on rulers who betray their noble upbringing. But in their revenge, the people are prone to throw the baby out with the bathwater—the very idea of nobility with those who feign it. Who, then, will save the baby? Plato's question has been answered over the last two millennia by a series of self-appointed guardians of Western civilization— philosopher-kings-in-waiting, so to speak. Thus, it is easy to imagine Leo Strauss's most colorful student, the late Allan Bloom, the American translator of Plato's *Republic* and author of *The Closing of the American Mind*, as sounding much as Plato would, were he to return for a visit.[6]

Bloom held a politicized (i.e., left-wing) professoriat responsible for the moral relativism of contemporary youth, which Bloom observed over the course of his teaching career, from revolts in the 1960s to complacency in the 1980s. Bloom blamed academics for introducing the radical views of Marx, Freud, and Nietzsche without having first nourished them on the works of Plato, Aristotle, and other stalwarts of Western civilization. It was not that these chaired politicos uttered falsehoods when they proclaimed, "God is dead! Morality is a sham!" Rather, they acted irresponsibly in that they knew how these normally esoteric truths would be likely received by the philosophically unsophisticated. Guardians like Bloom typically populate a virtual, even hypertextual, reality designed to rehearse and extend the thoughts of those noble people who have taken it upon themselves to fathom the mysteries of the human condition. As Plato would have it, these mysteries rarely soothe or flatter the populace, and in fact have the potential for greatly disturbing them. Consequently, the guardians have had to live in obscurity, so as to preserve the mysteries for the mentally prepared to receive them. This, then, is the Enlightenment stood on its head: an illuminati who jealously guard the light of their wisdom under a bushel

---

5. Fuller 1998b. Perhaps the most vivid attempt to build a collective memory based on the conservation of context is the living history of Jewish persecution.

6. Bloom 1987. According to the *New York Times*, this book was 1987's best-selling nonfiction book. A good sense of Bloom's intellectual pedigree, including his indebtedness to the French Hegelians (Kojeve and Koyré, about whom more in section 4 below) can be gleaned from his collected essays, memoirs, and book reviews, Bloom 1990.

basket, a republic of letters whose preferred mode of discourse is the private conversation and the double entendre.

Strauss credits the Arab philosopher Alfarabi (878–950) with introducing this interpretation of the Platonic legacy.[7] It is worth recalling its formal characteristics at this point, since Alfarabi's take on Plato epitomizes the mind-set that finds it necessary to communicate a *double truth* in its narratives. It is a mind-set that would receive its elaborate articulation two centuries later by Averroes (1126–98) and Moses Maimonides (1135–1204) in defense of a protected space for critical philosophical inquiry in a religiously grounded social order, Islam and Judaism, respectively.[8]

According to the double-truth doctors, Plato concealed his radically elitist politics and rigid sense of social stratification in a metaphysics consisting of multiple levels of reality dominated by a realm of pure ideas. This metaphysics—what is normally called "Platonic idealism"—is a rhetoric that operates on two levels. On the surface, Plato can be read as literally providing the structure of reality, as revealed to the philosopher (Socrates) through divine inspiration. But even here Plato has a trick up his sleeve, since he never provides a sufficiently full and consistent theology to explain God's relationship to his (or her) Creation. For Alfarabi and his followers, this omission, the defining difference between metaphysics and theology, was intended so that common readers would be led to accept the legitimacy of Plato's world order—that it is grounded in the nature of things—without losing their sense of personal responsibility for their actions, a political nightmare that could easily result if people believed that whatever they felt like doing was divinely preordained. Moreover, for Plato's sophisticated readers (those who read him as a tactful atheist), any explicit appeal to theology would appear insulting, thereby potentially threatening his credibility with this crucial audience. Because Plato presented the work as a metaphysics instead of a theology, his deeper readers could more easily read it as an allegory for the ideal constitution, as well as a model for how to persuade the populace that such a constitution is a just one.

Alfarabi first observed that the two-tiered reasoning exhibited in the Platonic corpus was also characteristic of how the foundational Western religions—Judaism, Christianity, and Islam—managed to become the basis for legislation covering culturally disparate populations. He further argued that Plato's use of rhetoric could be seen as appealing to a human sensibility that was located somewhere between brute sensory perception and pure

---

7. The most lucid recent elaboration of the Straussian understanding of Plato through the original Arab commentators is Parens 1995.
8. On the similar social contexts (Cordoba, Spain) in which Averroes and Maimonides developed their double-truth doctrines, see Collins 1998, 437–46.

intellectual understanding. To the thirteenth-century Christian Scholastics who were the long-term beneficiaries of the Arab intellectual legacy, this sensibility became *faith*, which implied that the human roots of metaphysics had been forgotten. Thus, one simply had a feeling that begat commitment, even though it was based on an incomplete understanding of verbal objects (e.g., passages from the Bible) rather than the normally sanctioned modes of knowledge, namely incomplete perception of natural objects (i.e., an inductively supported belief) and complete understanding of verbal objects (i.e., a deductively valid proof). By the eighteenth-century Enlightenment, philosophical reaction against theology led to a rejection of any via media between sense and reason.[9] Symptomatic is David Hume's claim that books that were neither empirical nor mathematical should be cast to the flames, an attitude that would be reincarnated during the critical phase of logical positivism in the twentieth century.[10]

Yet we need to recover precisely this third genre so quickly discarded by Hume in order to understand the ultimate source of Kuhn's appeal. Kuhn's recruitment to the Platonic cult came, institutionally, through James Bryant Conant, and intellectually, through Alexandre Koyré. Like Plato, both mentors were obsessed with the specter of total war consuming the achievements of the West, and hence with preserving them in the face of an ever-impending Dark Age.[11] The crucial difference between Conant and Koyré is that Conant stressed Plato's strategy in the *Republic* of promulgating "noble lies" on behalf of a still nobler truth by establishing the General Education in Science program, in which Kuhn was first employed. For his part, Koyré focused on imparting the "still nobler truth" to an elite men-

---

9. The partial exception to this stricture was the suggestion put forward by "critical-historical" theologians from Spinoza to David Friedrich Strauss (about whom more in chapter 2, note 59), that one could have "symbolic knowledge" of the Bible that enabled the devout to project major religious significance onto fairly ordinary events. This symbolic capacity provides the cognitive basis for faith. See Harrisville and Sundberg 1995.

10. On the origins of this claim and the rhetoric of book burning in Western thought more generally, see Fuller and Gorman 1987. It is worth noting that Hume's bold gesture found little favor in Britain until the 1870s, when his work was simultaneously appropriated by T. H. Huxley and T. H. Green to support complementary forms of skepticism: the one of religion (from an empiricist standpoint) and the other of empiricism (from a religious standpoint). Until that time, it had been generally agreed that Hume's rejection of a third form of knowing, whatever its strict logical merits, was psychologically eclipsed by our possession of a "commonsense" faculty. As might be expected, the philosophical champions of this perspective, led by Thomas Reid (1710–96), were trained in theology. See Passmore 1966, 40, 57–58; Collins 1998, 617, 667. The critical phase of logical positivism may be seen as marked by Ayer 1936. Ayer, then a recent Oxford graduate, was attracted to the Vienna Circle's declaration of Hume as a philosophical precursor.

11. An excellent work, both historically informed and up-to-date in its applications, that contrasts the vicissitudes of Plato's cultural-preservationist approach to education with the more civic-minded orientation of the rhetorical tradition is Oakley 1992.

tally prepared to receive it. To a large extent, they constituted the first cohort of professional historians of science in the United States.

To understand the motivation uniting Conant and Koyré, we must first examine the role of warfare in prejudicing the Western mind against conflict and volatility as properties of beliefs. This prejudice has intensified since Plato's day, as material (including human) resources have come to be increasingly bound up with the pursuit of inquiry, which, in turn, have heightened the level of risk involved in any fundamental change of mind. This cultural preservationism became self-conscious with the so-called *iconographic* approach to art history that crystallized during the First World War, just when Conant's science, chemistry, reached its peak as a science of destruction. The iconographic focus then shifted from self-sustaining traditions of craft to ones of discourse. In this context Koyré becomes the central figure, especially after he migrated to the United States at the end of World War II at the behest of the leading émigré iconographer, Erwin Panofsky. Koyré's brand of intellectual history, which Kuhn so much admired, stressed the continuity of physical theory with philosophical cosmology, while divesting it of its modern associations with instrumental control of the world—especially those that would come to be associated with the atomic bomb. According to this account, Kuhn and Conant represent, respectively, an unconscious and self-conscious realization of the Platonic mission. Finally, I examine the history of the cardinal representation of the Platonic mission in the philosophy of science, the distinction between the contexts of discovery and justification.

2. How Reason Came under House Arrest

Few people today realize that many of the words used by the ancient Greeks to refer to warfare were also used to refer to the verbal exchanges that originally characterized the polis as the "cradle of Western democracy" and later Plato's Academy as the first collectively organized form of inquiry. It is easy to dismiss the agonistic overtones of "dialectic" as little more than a dead metaphor, much like our expression "verbal jousting." Nevertheless, it would not be too far off the mark to say that, until the late eighteenth century, warfare was regarded as having many of the formal properties of dialectical engagement—and the Greeks themselves may have well thought about warfare as dialectics continued by other means.[12] The

---

12. There are other ways of observing the intimacy of the relationship between inquiry and warfare. One, of course, is to note that the rhetoric of science tends toward military metaphors as it becomes more popular. Thus, we have not only the "war on cancer" that U.S. president Richard Nixon declared in 1971, but also the fights to eradicate ignorance and

following features common to normative theories of warfare before 1800 are worth bearing in mind.[13]

First, when done right, warfare had very much the character of games: activities taken up for their own sake, subject to their own rules, and potentially corrupted by the introduction of considerations "external" to the actual fighting. The primary aim of warfare was to get the opponent to recognize one's own superiority. Thus, "honor" was often said to be at stake in wars. However, to secure this recognition, one could not simply annihilate — or even humiliate — the opponent. One reason was the lack of elegance involved in such a strategy. If nothing else, warfare was an art, and art requires an efficient use of means to achieve an end — not overkill. Moreover, a completely vanquished opponent will not respect his oppressor and may even seek to escalate the level of violence should he ever get a chance at revenge. This would invariably ruin the gamesmanlike quality of warfare by jeopardizing the lives of the people and resources under contention. The classic treatises on warfare, from Thucydides to Machiavelli, are basically extended morality tales of what happens when these normative constraints have been exceeded.

The case of Athens brings the relationship between the dialectic and the ancient conception of warfare into especially sharp focus because the very Athenians who would argue in the forum one moment may be off to war the next moment. The Athenians were obsessed with showing off their dialectical skills, to each other and especially to passing foreigners. They routinely placed their honor at risk by having others dare them to prove the most outlandish points in the fewest and cleverest number of steps. This taste for risky arguments translated into increasingly volatile foreign policy decisions and military engagements, which eventually led to the downfall of Athens recounted in Thucydides' *Peloponnesian Wars*.[14] Observing the

---

eliminate disease (especially in its microbial form), both of which date from the previous century. On the latter, see Montgomery 1995, 134–95. An especially revealing connection is that whenever a non-Western country has revamped its military forces in response to a threat from the West, it has always had to introduce technical scientific training into the educational system. Specifically, in the cases of Russia, the Ottoman Empire, Egypt, China, and Japan, military threat occasioned at least the limited institutionalization of science as a Westerner would recognize it. See Ralston 1990.

13. A popular source for what follows is Keegan 1993, which emphasizes the traditionally limited and competitive character of warfare — again in the spirit of games. Similarly, one might argue, experimental natural science began (in the seventeenth century) as a "gentleman's game," only to mutate to the point of engulfing the whole world in the maintenance of its activities. In this respect, the "information infrastructure" associated with the Internet marks the culmination of this symbiosis of military and scientific trajectories, as it was the result of a joint initiative of the U.S. Defense Department and National Science Foundation to ensure communications between scientists in the event of a nuclear war.

14. See Gouldner 1965, which is among the most sensitive treatments of the lessons to be learned from the fall of Athens.

unfortunate train of events, Plato concluded that the main culprits were the Sophists, themselves mostly foreigners, who earned a living by teaching Athenians how to refine their dialectical skills, especially "how to make the lesser argument appear the greater." Not so well known is that they also taught some martial arts, in which they showed how, say, speed could make up for lack of size under the right conditions. But be it in the verbal or the martial arts, the student learned how to defend and attack "positions." Plato believed that the Sophists indulged the Athenians in their worst tendencies by encouraging the dialectics to spill over into a frenzied call to arms. This led to irreversible damage, most poignantly (for Plato) in the trial and death of his teacher, Socrates. Socrates had unsuccessfully tried to turn the arguments of the Sophists against them in an effort to curb the relativism that had reduced his fellow citizens' sense of commitment to whatever could be defended on a dare.[15]

Plato's solution, as detailed in the *Republic*, was to sequester students of the dialectic in the tranquil setting of the Academy until they had sufficiently matured to entertain alternative lines of thought without feeling a need to call for action on each of them. At that point, around age fifty, they would be fit to be "philosopher-kings." Yet, to a large extent, Plato's prescription was merely one of the first flights that the Owl of Minerva has taken at dusk. By the time Plato came on the scene, there was not much more an aspiring philosopher-king could do than to sublimate his desire for power. The defeat of Athens at the hands of Sparta, and the subsequent conquest of both city-states first by Alexander and then by Rome, meant the end of Greek democracy. The dialectical skills that were in such great demand in the open forum no longer applied to the backroom politics of the imperial court. Consequently, the quality of the rhetorician most often stressed by the Sophists—*kairos* ("timeliness")—gradually shifted in connotation from "opportunistic" to "decorous," from one who constructs occasions for action to one who adapts to actions already taken. For the first time, rhetoricians selected students on the basis of their "character," which is to say, their amenability to this newfound sense of decorum.[16]

---

15. An interesting attempt to understand the long-term cultural significance of the Greek love-hate relationship with dialectics is Elkana 1980. Elkana distinguishes between *metic* and *epistemic* rationalities. The former is the Sophist's cunning reason that aims for maximum impact from minimum input, whereas the latter is Plato's demonstrative reason that aims for an impact proportional to input. This distinction has been formalized in the history of the scientific method in terms of induction (or, more precisely, abduction) versus deduction, or the logic of discovery versus the logic of justification. In each case, a fallible form of inference that promises to increase the store of knowledge is contrasted with an infallible form that promises only to preserve the current store of knowledge.

16. Kinneavy 1986. Aristotle and, especially in his day, Isocrates were most influential in effecting this transition. See Conley 1990, esp. 13–26.

Plato's counsel of patience for dialecticians eager to rule was certainly not without practical merit under the circumstances. The main way of instilling the requisite patience was to have students write down their thoughts before opening their mouths. This form of discipline was introduced in quite explicit opposition to the practice of the Sophists and even of Socrates himself. In Plato's day, writing was associated with the royal edicts and legal tablets of the Near East. They used language noncommunicatively to assert authority and display power. In a political environment that had been symbolized by the open forum, those who wrote before speaking lacked spontaneity, wit, and perhaps even honesty. Nevertheless, Plato held that in a world where fondness for free speech can have such disastrous consequences, it might not be such a bad idea to create a little distance between one's thoughts and one's words.

By the time the Stoics became the leading philosophical movement of the Alexandrian and Roman Empires, writing had become the defining skill of reflective people. Many Stoics found themselves employed in imperial administration, a realm where tact was always the order of the day. To retain their sanity, the Stoics took their duties in stride while filling pages upon pages with complex observations. But unlike the battle manuals written by Julius Caesar and other great generals, these observations were not made in the spirit of guiding future action. On the contrary, they were made in a spirit of catharsis, so as to dampen the desire for action in a world where decisive action was unlikely to have beneficial consequences. Thus, the seemingly perennial distinction between "theory" and "practice" became psychologically ingrained in the West only once people developed a practice that forced them to think long enough about something to render it too complex to provide a clear basis for action.[17] The word "forced" is meant as a reminder that writing began its career as a manual art akin to those practiced by slaves and craftsmen. Sometimes, of course, the coercion was privately administered, as when Marcus Aurelius, who, as Roman emperor, was otherwise accountable to no one, held himself accountable to the soul whose traces he recorded in his commonplace book. But most of the time, clients dictated the terms in which the manual art was to be

---

17. This, in a nutshell, is the role that hermeneutics has played in stabilizing and ramifying cultural commitment, especially as institutionalized in the difference between legislator and judge as legal roles and between prophet and priest as religious roles. In each case, the former "creates" and the latter "interprets" authoritative texts. A brief but synoptic history of the increasing distance between the creation and interpretation of texts is Tompkins 1980. To bring the history up to date, stressing the moral ambiguity and ultimately apolitical character of contemporary "cultural criticism," see Siebers 1993. For the lessons that these accounts have to teach the "interpretivist" turn in the sociology of scientific knowledge, see Fuller 1994c.

applied, as scholars performed routine accounting functions at the behest of rulers whose genuinely free status limited the use of their hands to the arts of war.

Nevertheless, the subservient status of the literary arts turned out to have its own political benefits, once economic decline made it difficult for rulers to maintain their domains. By the twelfth century A.D., it became common to issue charters to self-governing groups that gave them an effective monopoly for the application and transmission of a recognizable manual art. The typical *universitas*, as such a group was called, was the town guild devoted to a specific craft that required novices to undergo several years of apprenticeship before being granted a license to practice the craft. After some controversy, the medieval Scholastics managed to make a persuasive case that their writing skills similarly deserved institutional autonomy, and with that began the "university" in its current usage.[18]

But what exactly was protected by a university charter? The art of writing? These legal questions became increasingly difficult to answer because, unlike most other manual arts, writing gradually lost its guild mystique, as literacy became widespread. Eventually, large segments of the population were able to write for themselves without having passed through Scholastic tutelage. Academic staff no longer had to be seconded in order to sort out some wealthy illiterate's finances. Indeed, the eighteenth-century Enlightenment ideal of the *public sphere* reflected the emerging contradiction of a professional class—the latter-day residue of the Scholastic guild—that supposedly spoke for a public that had begun to acquire the skills needed to speak for themselves.[19] It was only with the rebirth of the university system in early nineteenth-century Germany that a distinctive way of writing emerged that merited special state protection, *scholarship*, which became the source of the guild rights that have come to be collectively known as *academic freedom*.[20]

But before leaping ahead to the modern era, let us review the profound effect that the fall of Athenian democracy has had on Western thought. As the social psychologists say, the trauma caused by this event led to an "adaptive preference formation" whereby an apparent liability is converted into a hidden virtue.[21] To those unschooled in the ways of philosophers, it might seem strange that so much ink is spilled by epistemologists and

18. For an elaboration of this thesis, see Fuller 1994d.
19. A historically sensitive account of this tension is presented in Broman 1998.
20. For a history that shows the continuity between the rise of the modern "Humboldtian" university in Germany and American principles of academic freedom, see Hofstadter and Metzger 1955.
21. Festinger 1957. For some brilliant applications of this idea to political theory, see Elster 1983.

ethicists over whether people "really believe" what they ought to believe. Why is it not enough for people to act in a way that accords with what is true or good? Why must philosophers be preoccupied with things called "intentionality" and "consciousness"? Historically, of course, these concerns reveal the continuity between Judeo-Christian theology and modern secular psychology. But, as we have seen, the legacy reaches back to the Stoics and maybe even to Plato, as writing manufactured a sphere of delayed responses and deferred gratifications that has since become "the life of the mind," a place where the conscious and the intentional are allowed free rein.

Platonism has always appealed to people living on two extremes of the political continuum: either those who think they can bend the world to their will or those who think they must bend their own will to the world. The former captures the totalitarian rulers who have fancied themselves descendents of Plato's philosopher-kings-in-waiting. The latter captures the Stoics and their offspring in both scholarship and administration. For them, writing has been the preferred means of addressing what Leo Strauss regarded as "the political question *par excellence*," namely, "how to reconcile order which is not oppression with freedom which is not license."[22] Writing treads the Straussian middle ground by forcing all parties—both articulators and interpreters—to be more circumspect in their words than they would be in speech. One of the great values of Strauss as a political theorist is that he always recognized that the search for a middle position was grounded not only in a sense of justice for prospective citizens of the ideal society but also in a sense of prudence for the utopian theorist struggling to flourish in her own society.

To understand why the life of the mind should be so closely identified with the pursuit of truth and goodness, we must recall that what bothered Plato most about the democratic attitude toward belief was its *volatility*. Moreover, this volatility may well have been an Athenian social invention. At least, the Athenians had managed to develop ways of keeping people's beliefs permanently up for grabs. Leaders were selected by lot, and policy debates were resolved by voice vote.[23] The Athenians prided themselves on their ability to change course quickly, which they took to be an advantage against larger and slower foes like the Spartans and the Persians. Today's democratic theorists often suppress this embarrassing legacy, as it suggests that democracies can work only if people are sufficiently impressionable to be swayed by public debate. Had the Athenians followed the advice of one

22. Strauss 1952, esp. 37.
23. See Ober 1989.

of today's leading democratic theorists, Jürgen Habermas (b. 1929), who recommends that people determine their "real interests" before entering the forum, they would have probably been so preoccupied with defining and protecting those interests that they would have never been receptive to negotiating their differences in the name of collective action. This is an important indication of the conceptual distance that has been traversed from the original Athenian situation to its modern rehabilitation.[24]

Athens continued to suffer from Plato's verdict that it was decadent, volatile, and unduly commercial until the rise of Whig ideology in eighteenth-century Britain. The dual root of "speculation" in intellectual fantasy and venture capitalism sums up the matter here. The Whigs reconnected commercial culture and civic republicanism, in part with the help of a newfangled moral psychology whereby the calculation of risks and utilities in the pursuit of commerce was said to discipline the passions in ways that benefited the individual and society simultaneously. However, the downside of this conversion of passions to "interests" was that the idea of "the public good" that had traditionally focused the civic republican polity was gradually abandoned as unintelligible.

The Athenian forum could afford to be an endlessly volatile place because (so believed the Whigs) citizens were equally secure in their property rights and hence could not be economically humiliated by saying wrong or unpopular things. Moreover, they were all commonly beset by the same external foes—Sparta, Persia, etc.—and hence concerns about national security never lurked far from the surface of debates about the public good. However, as the eighteenth century wore on, Britain became a country of increasingly mobile capital where political misjudgment could easily incur substantial financial disadvantage. One could not be sure that, in Popper's paraphrase of Goethe, our ideas would die in our stead. The dispossession of the aristocracy had not led to any greater economic security for the remainder of the population, including those who bought the land to build factories. In this emerging world, one had to keep an eye on one's personal interests in any public forum and hope that a remnant of the public good emerges as a positive by-product, via some "invisible hand" process, as we would now say. In this way, republicanism yielded to more familiar forms of liberalism by the early nineteenth century.[25]

24. A good account of the sociopolitical contexts from the 1960s to 1980s that have shaped Habermas's conception of the public sphere in democratic politics is Holub 1991.
25. Perceptive accounts of the brief revival and ultimate decline of republicanism in the eighteenth century are provided by Pocock 1985, esp. chap. 6; Polanyi 1944; Hirschmann 1977. A recent attempt to reinvent civic republicanism as a political theory for our own times is

That the Whig revival of Athens was well received by the French and American Enlightenment thinkers is no doubt due to the modern conception of probability, which started to refer to a relationship between belief and reality only in the seventeenth century. Before that time "probability" meant "credibility" in the sense of having authoritative approval, in which authorities functioned less as repositories of truth than as belief-fixing institutions, once again reflecting the reluctance to have individuals bear their own epistemic risks in the face of uncertainty.[26] Nevertheless, Jean-Jacques Rousseau remained a partisan of Spartan virtue who accused the Athenians of hypocrisy for basing their "democracy" on a slave-driven economy. While Rousseau's scruples would not have particularly moved Plato, they did defer any wholesale embrace of Athens as the model society. Thus, the permanently revolutionary character of the Greek city-state played no role in the democratic imagery of the French Revolution, which was supposed to have been the revolution to end all revolutions. Only once the Prussians required an allegorical foil for Napoleon's imperial image did Athens finally match Rome as an icon of "classical civilization."[27] But even then, the renovated university system, Prussia's institutional response to the *polytechniques*, Napoleon's training ground for "civil engineers," harked back to the secluded Groves of Academe, not the public square frequented by Socrates and his Sophistic interlocutors. Plato, so it would seem, had the last laugh in Prussia, especially given the fates of such academically led attempts to radicalize reason as the idealist world systems of Fichte, Schelling, and, of course, Hegel.[28]

Because the lust for change had led to a series of catastrophes for his city, Plato concluded that beliefs that changed too often are unlikely to be true or good. Instead, he claimed that the true and the good are likely to be found in beliefs around which *commitments* have been built. Writing is the instrument for creating such commitments, as a trail of text constrains and focuses thought—this in marked contrast to the convenient forgetfulness of memory that enables speech to make such an immediate impact. And so, given a challenge to their beliefs, Platonic inquirers do not simply produce a "fight or flight" response; rather, they deepen and refine their commitment, render themselves accountable, "put it in writing"—even if that

---

Pettit 1997, whose ideas are given some historical legitimation in Skinner's (1998) fascinating set of lectures. Here I must acknowledge Brian Baigrie's role in first identifying my civic republican streak (in a critique of my social epistemology): see Baigrie 1995. Since then, I have further developed a republican theory of science in Fuller 1999b, esp. chap. 1.

26. See Hacking 1975, esp. chap. 3; Daston 1987.
27. On the Prussian self-fashioning, see Montgomery 1994, 284.
28. On the intellectual and institutional politics of this period, see Beiser 1987, 1992.

means courting a stronger challenge in the future. No one has ever empirically demonstrated that Plato's hypothesis about the need for commitment is correct, but without it, the pursuit of academic life—be it as religious orders or Kuhnian paradigms—would make little sense.[29]

3. Platonism in Practice = Science as Art

Once Prussia and France assumed the respective mantles of Athens and Rome to fight the Napoleonic Wars at the dawn of the nineteenth century, the European powers spent nearly the next century and a half routinely invoking the "fate" of Western civilization as a pretext for going to war.[30] Beyond the loss of human life, these wars have also frequently damaged, or threatened to damage, the very achievements of Western civilization in whose name the wars were supposedly fought. In the aftermath of World War I, a wide range of humanists took it upon themselves to construct an institutional memory of those achievements, just in case all of Europe

---

29. For a spirited attack on the (often implicit) centrality of commitment to religious and scientific epistemology, see Bartley 1984. Those skeptical of my line of argument here should consider the reluctance of most professional academics to take the metaphor of "the marketplace of ideas" literally. Since the days of Adam Smith, many economists have pointed to markets as efficient knowledge systems that do not require elaborate academic mediation for their success. Traders have a vested interest in adjusting their prices in response to signs of the going rates of exchange. But the success of this system presupposes an ability to respond quickly to changes in evidence and a willingness to negotiate virtually any term in the exchange. Even if the traders are themselves pleased with the outcomes that the market delivers, the system does not lend itself to the maintenance of constant standards of exchange—"criteria" as the Stoics originally called them. Thus, what passes as a "fair price" now may not pass in the next exchange. Academics from the time of Plato's student Aristotle have complained about the market on these grounds. In the modern period, these criticisms have grown into charges of "exploitation" and calls for "social justice" to redress the damages done in past exchanges.

The charges are undoubtedly well grounded, but an unusual virtue of our "postmodern condition" is its refusal to succumb to the Platonic impulse to redress these injustices by slowing down the pace of exchanges, subjecting them to greater scrutiny, or, in the limit, "universal standards." An interesting witness here is the financier George Soros, who, as a student of Karl Popper at the London School of Economics, learned that a sustainable environment for bold conjectures in both the intellectual and economic sphere requires substantial public regulation at the international level. Soros 1998. Consider the case of the natural sciences, in which "commitments" are built by investing so much time, effort, and money in certain lines of research that their reversal becomes nearly impossible (in the Machiavellian sense that politics is the art of the possible). Indeed, such irreversibility may be the cynic's operational definition of "progress." See Fuller 1997d, 50–52. However, these deepened commitments need not lessen the level of danger in the world, as evidenced by the threats that such well-endowed fields as particle physics and molecular biology pose to the general public in the form of nuclear weapons and genetic engineering, respectively. In that case, the postmodernist argues, injustice results not from the market being too free and open, but from its not being free and open *enough*. Certain terms in the scientific exchange would seem to have become nonnegotiable.

30. Lambropoulos 1993, 79–80.

should be engulfed in total war. If the French Revolution marked the triumph of Humanity over human beings, these humanists were keen on saving Humanity from the human beings who originally spoke on its behalf.

At this point, the reader may be forgiven any suspicions that these humanists displayed little interest in the preservation of rank-and-file human beings whose names were not attached to the production of Humanity's signature artifacts. Our story begins with Aby Warburg (1866–1929), who, despite being the senior sibling in his generation, was the only one not to help administer the family's international banking empire. Warburg was an art collector and amateur scholar who introduced an understanding of the history of Western civilization—*iconography*—that enabled humanists to participate in the Platonic cult.[31]

When Aby announced his turn away from banking to art, his brothers vowed to fund what would become a massive personal library, which was transferred from Hamburg to London after the Nazi occupation of Germany. It then became the core holdings of the Warburg Institute of Art. Reflecting on the history of iconography, the Warburg's most prominent director, Ernst Gombrich (b. 1909), argued that the uniquely "progressive" element of Western art is its increasing ability to transcend the limitations of artistic media by simulating features that cannot be directly represented.[32] This ability consists mainly in anticipating what the observer is likely to contribute toward the construction of an image from the work put before her. Thus, corresponding to the role of *rhetoric* in language is the role of *perspective* in painting and the plastic arts, which, once taken to its high-tech conclusion, brings us to the realm of holograms and virtual reality machines.[33]

---

31. "Cult" is not a word I use loosely to describe Aby Warburg's bibliophile sensibilities. In his library, the philosophy books were catalogued next to the astrology and magic books. This proved to be an eye-opening experience for Ernst Cassirer (1874–1945), who was professor of philosophy at Hamburg before the Nazis came to power. It would influence Cassirer 1963. See Ferretti 1989. Kuhn, in turn, openly admitted his methodological debt to Cassirer as a historian of philosophy and science. See Kuhn 1977a, 108.

32. Gombrich reveals these insights in Miller 1984, 212–31, an interview.

33. The American rhetorical theorist Kenneth Burke (1897–1993) has probably been most attuned to the relationship between rhetoric and perspective, especially his concept of "perspective by incongruity," whereby something is understood by observing what its opposite omits. See Burke 1969.

It is rather surprising that no one seems to have picked up on the debt that Kuhn's sense of paradigm owes to art-historical conceptions of "perspective." The most forthright philosophical attempt to overcome Kuhn's incommensurability of worldviews understood as perspectives has been Donald Davidson's 1973 presidential address to the American Philosophical Association: see Davidson 1982. Davidson's efforts notwithstanding, art historians have long recognized that perspective—that is, linear perspective—originated in Greece and was transferred from the plastic arts to painting in sixteenth-century Europe. The sort of relativism

In a Hegelian mood, we can interpret this overall movement as the gradual release of the spirit from the material world—what led Gotthold Lessing and other Enlightenment critics to rate the expressive capacities of the literary arts over those of the fine arts.[34] The other side of the coin, of course, is that precisely because writing is such a pliable medium of expression, it can convey meaning to a general audience while encoding messages that only the cognoscenti—the iconographers—can read. Not surprisingly, in true Platonic fashion, Gombrich once defined aesthetic experience as inhabiting the semantic space between boredom and confusion: the former occurs when the elites confront the popular message, the latter when the populace confront the elite one.[35] Mutatis mutandis, this also defines the space of scientific experience captured by Kuhn's double-truth conception of the history of science, if we take the "elites" to be historians of science and the "populace" to be practicing scientists. But before turning to the iconographic residue in Kuhn's work, let us consider in more detail its source in Warburg and his direct descendants.

Warburg instructed scholars to rise above dwelling on the political

---

associated with Kuhn, whereby one's perspective determines one's view of the world, makes sense only if the world is projected from a single, stationary eye point and the observer's line of vision eventuates in converging parallels on the horizon. Under those circumstances, it is possible to specify a set of objects that appear clearly to the observer, thereby satisfying the minimal conditions of relativism.

However, classical Egyptian, Chinese, and Indian art lacked perspective in this sense—as did the pictorial arts in pre-Renaissance Europe. Instead, they presupposed a viewer whose traveling eye perceived parallel lines to diverge, rather than converge, in the distance. This meant that the viewer did not perceive an outer boundary to the world projected by the artist. Absent such a boundary, it is impossible to assign objects to unique worlds, thereby violating a necessary condition for relativism. Consequently, a contemporary viewer of these art forms would tend to experience not only optical distortion but also a sense of what Gilbert Ryle would call "category mistakes," as objects pertinent to one domain appear to be interacting on the same plane with objects from another. (The most obvious case in point of this would be the interaction between humans and such divine entities as angels or even God.)

If art history were taken as a validity check on epistemology—which is to say, if the visual metaphors associated with perspectival knowing were treated literally—we might suspect that our instinctive offense at witnessing such categorical promiscuity were nothing more than a sign of our own provincialism. However, given the renewed life that Kuhn has given to perspectivalism, this subtle form of Eurocentrism reigns supreme. I must confess to having flinched at the sight of category mistakes; hence, my endorsement of a "sociological eliminativism," which calls for the reduction of all putatively nonsocial categories to social ones, so as to provide philosophically adequate grounds for treating objects as if they were interacting on the same visual plane. I defend this position on pragmatic, not metaphysical, grounds in Fuller 1993a: placing objects on a common line of vision renders them more tractable to the observer's ends. Perhaps the only work in the philosophy of science that takes the cultural sensitivity of perspective with the requisite seriousness is still Heelan 1983. I discuss this book's significance in Fuller 1988, 122–27.

34. The locus classicus of this viewpoint is Gotthold Ephraim Lessing, *Laokoon* (1766), esp. chap. 16.

35. Gombrich 1979, 9.

exigencies that have perverted interpretations of the West's enduring works. They were to focus instead on the ways in which these artifacts have embodied, conserved, and renovated signature Western themes, or "coded passion" in Warburg's term. This cryptographic exercise would finally permit the great artists to address each other freely across the centuries. The only obstacles of note would be between artists and their craft tradition, as artists struggled with the precedents set by the works of their illustrious predecessors.[36] In Warburg's hands, then, the Platonic faculty of nous metamorphosed into a form of historical understanding that enables the artist to distill the legacy of previous generations into a culminating work, thereby rendering it a "classic" in the dual sense of producing a work of lasting value that at the same time renovated perennial themes in the history of Western culture. For example, paintings of nature owe more to other paintings than to nature itself.[37]

Perhaps the most interesting feature of Warburg's brand of connoisseurship was the way it converted a naturalistically based "distancing" method into a new aristocracy of taste.[38] From the second half of the nineteenth

36. The similarities between these functions performed by a classic in art and a paradigm in Kuhn's conception of science was not lost on the aesthetician Stanley Cavell, another Harvard product of the period who was Kuhn's colleague at Berkeley from 1957 to 1961. He has been among those most consistently sensitive to the similarity between Kuhn's vision of scientific and artistic change—indeed, in spite of Kuhn's own reluctance to admit the connection, as we shall see below. Cavell highlights the fact that a "master of the science" will recognize a truly revolutionary theory not as a threat but a transformation of already existing traditions and hence a worthy successor to the reigning paradigm. See esp. Cavell 1979, 121. Kuhn acknowledges Cavell's role as a "sounding board" in Kuhn 1970b, xi.

37. Gombrich attributed this insight to Alexander Pope's *An Essay on Criticism* (1711), line 130 ff., as transmitted by his teacher, Heinrich Woelfflin, about whom more below.

38. Warburg's historicized sense of Platonic vision was influenced by Darwin's later work, *Expression of the Emotions in Men and Animals* (1872), which proposed to interpret human and animal behavior, not via superficial analogies between the creature observed and the observer's own emotional states, but by casting the creature's behavior as responsive to the environment defined by the creature's perceptual horizons. Applied to art interpretation, this "ethological" strategy shifted the locus of aesthetic competence away from the ability to derive feeling or function from an artwork, which locates the work in the observer's life world, and toward an ability to situate the work within the notional world defined by the artist's original aesthetic horizon. Genius would thus be recognized, not by the indelibility of the impression that the work leaves on its audience, but rather by the subtlety with which the artist managed to shape long-standing traditions to her own ends. For Warburg's debt to Darwin, see Ginzburg 1989, 20–30.

While Warburg's Darwinian inspiration lent a "naturalistic" feel to his proposal, in fact it stood in sharp contrast to the more popular forms of naturalism emanating from French literary and art criticism in the second half of the nineteenth century. Two are worth noting here because of their eerie analogical relevance to subsequent developments in the sociology of scientific knowledge (to be discussed in more detail in chapter 7). The first version, associated with Charles-Augustin Sainte-Beuve (1804–69), judged the value of an artwork by the politics and personality of the artist, which Sainte-Beuve would have discerned from his own personal (ethnographic?) encounters with artists in their haunts.

The second was associated with Hyppolite Taine (1828–93), who kept his distance from

century to the end of World War I, the rhetoric surrounding the distinction between "scientific" and "folk" knowledges typically identified "scientific" with an intellectual vanguard drawn from the middle class and traditionally less privileged sectors of society, whereas religious authority and aristocratic prejudices were pejoratively lumped under the rubric of "folk." However, the pivotal role of chemistry in World War I revealed the disastrous consequences of an explicitly artificial science untempered by the natural virtue traditionally associated with aristocrats and peasants. This served to scramble the class allegiances of inquiry in the 1920s. Thus, self-professed "scientific" modes of understanding increasingly sought protective coloration under a priestly worldview that before the war would have been regarded as aristocratic. Instead of creating new artifacts, the renovated scientific attitude was oriented toward conserving old ones, specifically by separating artworks from their appreciators, only to reacquaint the two groups by the mediation of iconographic inquiry.[39]

A good example of the discourse surrounding what turned out to be a temporary return to the nineteenth-century humanist model of science is the correspondence between Erwin Panofsky (1892–1968) and Karl Mannheim in the 1920s, when the former was laying the groundwork for "iconology" and the latter "sociology of knowledge."[40] Influential in both cases were the art historians who before the war had brought iconography within

---

artistic culture but argued that the relevant context in which to place artworks was national-cultural (which still traveled under the name of "racial") rather than "traditional" in the sense that would come to mark an iconographical approach. A convenient source for representative writings by Sainte-Beuve and Taine is Adams 1971, 555–62, 601–14. Warburg's main objection to these contemporary politicizations of art was *not* that they wrenched artworks from their original ideological contexts, but rather that they treated the artist as if she operated *only* at the level of politics—as might the ordinary observer or hack critic—and not "art proper." Such an error is comparable to a biologist thinking that she can grasp a frog's view of the world simply by squatting on a lily pad and looking out for flies.

39. On the efforts by natural and social scientists to cultivate a "folk" image for purposes of legitimation, see Herf 1984.

40. The correspondence between Panofsky and Mannheim is analyzed in Hart 1993. The crucial term common to their efforts at carving out a specialist niche between art and *both* artists and art appreciators was *culture*, which Panofsky and Mannheim understood in a depsychologized sense to mean the collective inheritance of humanity. The trick for both thinkers was to postulate a sufficiently clear boundary around this heritage to constitute a proper object of scientific inquiry, but without attributing to that object a fixed essence that could be interpreted in racialist terms. For more on the struggle to retain what was basically a Lamarckian conception of culture in a world that had become thoroughly Darwinized, see introduction, note 41. Indeed, the idea behind the verbal innovation of Panofsky's "iconology" was that radical changes in the history of art were emergent from previous developments and not predetermined in the teleological fashion that seemed to be implied in the *Kunstwollen* idea pursued by iconography's founders (about which more below). See Ginzburg 1989, 30. Panofsky's concerns anticipate Kuhn's open-ended account of scientific change, a progress *from* that is not a progress *to*.

the ambit of *Wissenschaft*, "science" in the sense of a body of knowledge closed under a set of technical methods and theories. Alois Riegl (1858– 1905) and Heinrich Woelfflin (1864–1945) held that art possessed a *Kunstwollen* (a "will-to-art") comparable to the spiritual tendencies of the individual soul that, in a past age, would have been the object of priestly ministrations. These tendencies defined the limits of possible artistic expression in a given age, from which actual artworks would then be constructed. The key rhetorical move here was the Kantian one of ceding "actuality" to the naïve art observer (including artists' consciousness of their own work), while reclaiming "possibility" for the historically minded critic (who articulated the artistic unconscious). Woelfflin crystallized this rhetoric in the distinction between "internal" and "external" histories of art, which still provides the inspiration for cognate distinctions in the history and philosophy of science.[41]

The link that was needed to forge a renovated scientific attitude to art and culture more generally—the precedent for redefining the domain of iconography from actual to possible experience—was the neogrammarian approach to philology, which since the end of the eighteenth century had provided the most sustained example of historical inquiry as *Wissenschaft*.[42] Significantly, the neogrammarians studied *dead* languages, each of which was treated as an extinct organism, whose activity was captured by its phonetics, anatomy by its syntax, and physiology by its semantics. The texts used to reconstruct these different linguistic layers were thus comparable to fossils.[43] The completed grammar of each language projected a unique closed world that could be, to some extent, reexperienced by the philologist. This "world" was most naturally characterized in Kantian terms as everything that *could* be said in the language, regardless of whether it was actually said. Moreover, the dead languages that most concerned the

41. Woelfflin 1932, 226. Kuhn 1989 is most explicit in identifying a paradigm (or disciplinary matrix) with the limits of possible experience. Among philosophers of science, Ian Hacking has been virtually alone is seeing the connection between the Kuhnian concern for possible experience and "style of thought." However, it is clear that Hacking's immediate inspiration was not art history but Alastair Crombie's reduction of the history of Western science to six successive but temporally overlapping styles of reasoning. See Hacking 1982, 1992.

42. On the place of the neogrammarians in the history of linguistics, see Dinneen 1967, 176–91. For their more general significance in the history of representational practices, see Foucault 1970, 280–94.

43. And lest we think that the neogrammarian conception of languages as discrete organic species was a mere metaphor, it is worth recalling that when the historian Johann von Droysen canonized the distinction between "interpretive" and "explanatory" sciences (*Geistes*- versus *Naturwissenschaften*) in the mid–nineteenth century, he placed biology on the interpretive side of the divide, on the strength of philology's demonstration that anything conceived as an organism could be understood only in terms of a largely preordained developmental scheme. See Apel 1984, esp. 1–4; Cassirer 1960.

neogrammarians—Sanskrit, Hebrew, Greek, Latin—were taken to be the original sources of Western culture, with modern national tongues seen as offshoots, seeds that have taken root in different soils.[44]

While the neogrammarian project implied an evolutionary view of linguistic history, the conception of evolution owed more to Lamarck than to Darwin. Specifically, linguistic forms directly bear the marks of usage. For example, commonly used words tend to acquire an irregular and truncated syntax, while accumulating meanings that could be distinguished only by familiarity with the exact context of use. The external world does not directly determine appropriate usage; rather, it simply provides an occasion for the language to reorient or reorganize its inherent tendencies. Thus, when the speaker regards her language as improving its representation of reality over time, the philologist sees instead the speaker's language becoming self-conscious of its expressive capacity.

An important but subtle legacy of these philological roots appears in Kuhn's insistence that the limits of one's paradigm are indeed the limits of one's world.[45] Here Kuhn breaks most clearly with the Harvard logician Willard van Orman Quine, about whom more later in this chapter. Despite their mutually acknowledged affinities on many matters, Kuhn adamantly denied the Quinean presumption that anything can be expressed in any language (albeit sometimes with great effort and ingenuity). According to Quine, as long as we suppose that everyone inhabits the same world, we can resolve any lingering difficulties in translation by observing the context of usage.[46] In contrast, Kuhn held that we understand each other—insofar as we really do—only because we have at least partially mastered each other's languages.[47]

The difference in the normative implications of these two attitudes toward translatability could not be more striking. Whereas in principle Quine could grade languages according to the ease with which they represent a common reality (here the superiority of a language whose logic is transparent would become apparent), Kuhn would counter that each language has its own expressive capacity, which can be judged only in its own terms, such that to judge in any other terms would be to import an alien standard that misrepresents what the speakers are trying to convey. Thus, what Quine might regard as an obscurantist formulation in ancient science would be, for Kuhn, quite appropriate given the communicative aims and

---

44. The same technique was applied to a living language, Arabic, which was regarded by "Orientalists" as having become arrested in its development because of its self-imposed isolation from the West. For a deconstruction of this sensibility, see Said 1978.
45. The allusion here is, of course, to Wittgenstein 1922, proposition 5.6.
46. This may be seen as the main point of Quine 1960.
47. Kuhn 1989, 11.

means of the ancients concerned. I raise this difference in attitudes toward language because it was presaged by the break that professional (i.e., aristocratic) art historians saw themselves making from the naturalism that characterized popular (i.e., bourgeois) art criticism, which presupposed that all artists are trying to capture the same reality.[48]

As the decades leading up to the Second World War increased the perceived level of irrationalism in European society, iconographers pushed the idea of *Kunstwollen* further so that the artistic impulse metamorphosed into a kind of cultural unconscious that resisted the attempts of authoritarian forces—be it the medieval Church or the recent twin threats of Nazism and Bolshevism—to impose their own sense of order. For example, the English translation of the epic history of the Latin Middle Ages by Ernst Robert Curtius (1886–1956) appeared in a book series devoted to promoting Carl Jung's analytic psychology of collective memory. In Curtius's case, the key mnemonic elements were the *topoi* of classical mythology and biblical symbolism that repeatedly surfaced in medieval literature to keep the candle of learning burning during a "dark age." Befitting someone who regarded himself as a denizen of a dark age, Curtius invoked the Platonic method of participation, whereby the scholar becomes a living reminder of the past in defiance of the historical myopia of aspiring totalitarians.[49]

However, Curtius's dark age did not come to an end with the fall of the Nazis, the period of captivity in which his book was written. Reviewers of the English edition of *European Literature and the Latin Middle Ages* still regarded it as espousing a scholarly stance fit for what had become the Cold War era.[50] Curtius's extreme form of Platonic participation supported the view that any artist of lasting aesthetic value is radically misread as an innovator; rather, the radicalness lies in the culturally amnesiac character of

---

48. I discuss the naturalistic critics in note 38 above. The explanatory strategy common to the naturalists—be they followers of Taine or Quine—is that one's (in)ability to capture an intended piece of reality is attributed to features of one's background over one has have little or no direct control. In contrast, in proto-Kuhnian fashion, Woellflin argued that each style has its own form of "imitation" and "decoration." The former, "the idea of reality," is object oriented; the latter, "the idea of beauty," is subject oriented. What naturalists take to be the decadent period of a particular style, a "silver age," was reinterpreted by Woellflin and his followers as a reflexive application of the style's inherent tendencies, such that the style's implicit aesthetic ideal is finally raised to self-consciousness. The corresponding phenomenon in Kuhn's account of scientific change is the high level of technical sophistication that characterizes the mature stages of a paradigm, as it tackles problems that fail to yield easy solutions. The Hegelian provenance of the vision of history shared by Woellflin and Kuhn should be clear. For an articulate and authoritative defense of this perspective, see Schapiro 1953.

49. Actually, Curtius's record is not quite so exemplary. He was sufficiently willing to oblige the Nazis by declaring sociology a "Jewish" science and Karl Mannheim, in particular, a "mischievous" theorist. See Woldring 1986, 36, 148.

50. On the metamorphosis of Curtius's reception from Nazi resister to Cold War stalwart, see Cantor 1991, 189–97.

the artist's admirers, who fail to appreciate the full repertoire of *topoi* at the artist's disposal. Dante was Curtius's case in point. Shakespeare received a similar treatment at the hands of Frances Yates, a scholar affiliated with the Warburg Institute under Gombrich's directorship. Her reputation largely rests on having introduced the idea of *Kunstwollen* as repressed cultural unconscious to the history of science.[51]

In the 1960s Yates produced a series of remarkable books revealing the indebtedness of the seventeenth-century Scientific Revolution in England to religious roots that extended beyond Puritanism back to messianic, heretical, and even ancient pagan sources, much of which was grounded in the practice of magic.[52] Yates was catapulted into controversy with historians and philosophers of science because she suggested that such "hermeticism," her name for these alternative religious sources, constituted a self-contained tradition of scientific thinking in its own right, distinct from the so-called internal history of science that students learn early in their careers and professional historians and philosophers of science devoted most of their energies to articulating.[53]

Unlike historians and philosophers of art, whose Hegelian provenance enabled them to entertain the existence of several dialectically counterpoised traditions, historians and philosophers of science still presupposed a simple model in which everything that stood outside the one internal tradition of science was regarded as "external," and hence merely accidental to the general march of progress presented by the internalist account. Kuhn himself, though appreciative of Yates, ultimately rejected the idea of multiple internal histories of science in favor of the more conventional view that the scientifically relevant parts of the history were eventually channeled into a single mainstream history; hence, in Kuhn's hands, hermeticism becomes an impure source of the experimental tradition in modern science.[54] Starting in section 7 below, we shall critique the image that implicitly informs this model, namely that the history of science is a river into which various traditions flow as tributaries.

Kuhn's clearest link to iconography is his source for the idea of a paradigm as a self-sustaining scientific mind-set. Ludwik Fleck's concept of "thought-style" *(Denkstil)* came from the early twentieth-century art-historical debates over the unit of craft transmission across generations of

---

51. Yates 1975.
52. Yates 1964, 1966.
53. For a positive and negative assessment of Yates's claims, respectively, see Hesse 1970b and Rosen 1970. It is worth noting that Hesse's respondent, Arnold Thackray, did not seem to grasp the idea of an alternative tradition internal to science that was not simply the usual object of "externalist" historiography.
54. Kuhn 1977a, 54–55.

artists.[55] Clearly, this perspective was not originally intended for the natural sciences at the center of Kuhn's account, namely, physics and chemistry. Indeed, Fleck himself found *Denkstil* an attractive notion because he was talking about *medicine*, a discipline that was only at that time exchanging its artisan roots for a more experimentally based future.[56] The appositeness of *Denkstil* to the natural sciences was felt only gradually, as it became clear that the positivist view of science as a theory-neutral instrument had enabled scientists to subserve their intellectual calling to destructive ends. As we shall see in the next chapter, on the eve of World War I, Max Planck worried about the potentially negative consequences—for both science and society—of science being seen as a mere means without ends of its own. Three other figures turn out to be significant in forging the link between iconography's applied Platonism and Kuhn's distinct internalization of the history of science: Alexandre Koyré, James Bryant Conant, and William Whewell.

4. Koyré's Iconographic Turn in the History of Science

Alexandre Koyré (1892–1964) was plagued by the specter of total war throughout his life. His move to France had been precipitated by the Rus-

---

55. Fleck 1979. A considerable mythology has developed around this book's allegedly formative role in Kuhn's thought, despite Kuhn's own characteristic attempts to minimize the extent of influence by claiming that he discovered the book by accident (in a footnote in Hans Reichenbach's *Experience and Prediction*) after his basic ideas had been already formulated. See Thomas Kuhn, foreword to Fleck 1979, vii–xii. Relevant here is Kuhn's late admission that he could not follow Fleck's Polish-inflected German. See Kuhn et al. 1997, 165. Nevertheless, Fleck's book is now often presented as the long-lost philosophers' stone, if not the veritable genie in a lamp, when it comes to unraveling the Secrets of Science. Fleck's life and his book's reception are discussed in Fleck 1979, 149 ff. No doubt the mystique surrounding Fleck's book has been enhanced by his internment in a Nazi concentration camp during World War II, after which he migrated to the newly created state of Israel. Nevertheless, contrary to learned opinion, Fleck 1979 was not ignored when first published in 1935. In fact, it generally received favorable reviews as a piece of science popularization. That the book was neglected by historians, philosophers, and sociologists of science until Kuhn's rediscovery can be explained simply in terms of its intended audience, which was not science studies professionals. And while the book certainly resonates with many themes subsequently developed by Kuhn, especially the ones drawn from collective psychology and art history, those themes—albeit contextualized somewhat differently—were standard fare in the Weimar culture of Germany that preceded the rise of Nazism. Even the concept of *Denkstil* had been used by Karl Mannheim, ten years before Fleck, in his seminal essay, "The Problem of the Sociology of Knowledge" (1925). During this period Mannheim was in correspondence with Erwin Panofsky, as discussed in note 40 above. However, Fleck failed to cite Mannheim's prior usage. See Fleck 1979, xv.

56. Fleck's early sociological reviewers raised this point, arguing that such an art history–based analysis was unlikely to be generalizable to the harder sciences. See Fleck 1979, 164. On the "science vs. craft" battles in the last two centuries of medical history, see Matthews 1995.

sian Revolution of 1917; his move to America by World War II. But once in America, Koyré was most closely associated with Princeton's Institute for Advanced Studies, where he came at the invitation of the great émigré iconographer Erwin Panofsky. Koyré's affinities with Panofsky are most immediately apparent in the historical period over which their work ranged. While both were very much men of their times, with a few interesting exceptions, neither wrote substantially about science or art after the seventeenth century.[57] This can be explained, at least in part, by the basic premise of the iconographic method, namely that exemplary works are more or less knowing transformations of classical themes in new settings. Once factors external to those themes become essential to understanding a work's significance—say, the instrumental or ideological value served by the work—the work loses its transcendent character, the ultimate object of iconographic inquiry, and simply becomes another product of its time. Of course, developments in both science and art over the last 150 years have provided serious challenges to this strong sense of Platonism. However, the challenge has been more often registered in art, with the rise of avant-garde movements that have had the express intent of breaking with traditional forms of composition and perception. In response, museums have become increasingly important as institutions that provide the space needed to nurture the public acceptance of revolutionary art forms.[58]

As we shall see in this and the next section, a notable consequence of Kuhn's work was to inhibit a similar awareness from arising with regard to science, despite the tensions arising from science's increasing involvement in political and economic goals since the seventeenth century, at the same time it has loudly claimed the virtues associated with self-determination as its own.[59] Indeed, in a famous encounter with art historians who saw in *Structure* the basis for arguing that the histories of art and science took a similar form, Kuhn insisted that the big disanalogy was that science lacked an avant-garde impulse, that is, a sense that revolutions must come from

---

57. As discussed below, Koyré wrote several essays on Hegel's conception of time, as well as the Slavophile movement in nineteenth-century Russia. Panofsky virtually founded American film studies in the 1940s with a work on the iconography of silent movies. See Cantor 1991, 176.

58. For an interesting discussion of this phenomenon, coupled with a strong denial that museums function similarly for the sciences, see Kuhn, "Comments on the Relations of Science and Art," in Kuhn 1977a, 340–51, esp. 345–46. Kuhn's point is that science museums do not facilitate change but rather help the public catch up with change that has already occurred. What, then, should we make of the Smithsonian Institution's 1995 exhibition, "Science in American Life," which drew a strongly polarized response comparable to that elicited by galleries that displayed the impressionists in the 1860s, the surrealists in the 1910s, and pop art in the 1960s?

59. The best history of this development is Proctor 1991.

outside the establishment.⁶⁰ In that respect, Kuhn seemed to imply that iconography was *better* suited to science than to art in today's world. Koyré would have been pleased.

In the 1930s, the decade before he joined Charles de Gaulle in the Nazi resistance, Koyré and his fellow émigré Alexandre Kojeve (1902–68) launched what would become the "French" interpretation of Hegel that took the German philosopher at his word about humanity verging upon "the end of history," the triumph of the human spirit that Hegel associated with the French Revolution (or more specifically, Napoleon's transmission of the revolutionary spirit to Germany by defeating the Prussian army at Jena). The rest of recorded history would simply be a matter of the world catching up with this "universal and homogeneous state." A century after Hegel's prophetic pronouncements, his theory acquired a new vogue for the members of Kojeve's and Koyré's seminars given at the Ecole Pratique des Hautes Etudes, which included such ideologically disparate leaders of the postwar French intelligentsia as Maurice Merleau-Ponty, Georges Bataille, Jacques Lacan, and Raymond Aron. (Leo Strauss had signed up for the seminar but did not attend.) Although Hegel's prediction was literally false, there were some (like Kojeve) who believed that a version of it would come true this time around, and others (like Koyré) who believed that any steps taken toward the end of history could be reversed at any moment.

However they interpreted Hegel, French intellectuals were generally agreed that any attempt to impose a rational regime was subject to a "return of the repressed" that had the potential to turn reason into an instrument of violence, and hence irrationality. This distinctly "dialectical" way of thinking testified to Hegel's continued utility in explaining the formidable challenge posed by totalitarian states such as Nazi Germany and the Soviet Union, which achieved a unity of political will and technoscientific advancement heretofore unseen in the history of the West. Yet the result was not civilization but barbarism. Nothing of this massive paradox could be captured by the neat division between matters of epistemology and ethics (or metaphysics and politics) that had come to characterize French academic philosophy in its neo-Kantian captivity during the first three decades of this century. It almost goes without saying that anglophone analytic philosophy remains captive to the neo-Kantian divisions from which the French intellectuals had begun to wrest themselves in the 1930s.⁶¹

---

60. Kuhn 1977a, 350.
61. A good account of the transition from the neo-Kantian to Hegelian sensibility in French philosophy (along with the succession of a Heideggerian sensibility after 1968, which accounts for the rise of Jacques Derrida, Gilles Deleuze, and other currently fashionable thinkers in the English-speaking world), see Descombes 1980, esp. 9–54.

A brief contrast of Kojeve's and Koyré's views on the end of history would be instructive at this point. Kojeve was notorious for shifting ground.[62] Initially, he regarded Stalin as a latter-day Napoleon and the Soviet Union as the long-awaited terminus, but by the 1950s the apparent effectiveness of U.S. Cold War policy, combined with its booming domestic economy, led him to believe that the Soviets had met their dialectical foil in the Americans. At that point, he set his sights on the emerging European Union (which he helped forge through his work in the French Ministry for Economic Affairs) as a plausible synthesis. By the 1960s, Kojeve was looking toward Japan. Thus, the political shape of the elusive Platonic paradise metamorphosed in Kojeve's mind from Communism to corporatism to ceremonialism (i.e., Kojeve's understanding of the vestigial elements of samurai culture in contemporary Japan). Kojeve believed that Hegel's *Phenomenology of the Spirit* (1807) held the keys of human salvation — the only problem left was to identify the savior. In contrast, Koyré drew on a wider range of the Hegelian corpus and came to the conclusion that human freedom is fundamentally incompatible with historical finality.[63] The end of history could not be foreseen, but backsliding into barbarism was freedom's prerogative, one already adumbrated in Plato. As Koyré said of Plato's account of the decline of Athenian democracy, *"de nobis fabula narratur"* ("our own story is being told"). The year was 1945.[64]

Koyré follows Plato in identifying science with *theoria*, which, for the Greeks, connoted not merely an abstract, systematic understanding of the world, but one in secret communion with the gods. Koyré's innovation was to induct natural scientists into the Platonic cult. Indeed, he even suggested that the truly great scientists have seen themselves as members of this virtual community. In his influential studies on Galileo, Koyré argued that

---

62. An excellent analysis of Kojeve's thought that sharply focuses on the "end of history" theme is Drury 1994.

63. For a comparison of Kojeve and Koyré as readers of Hegel, see Anderson 1992, 314–16. The most recent proponent of the "end of history" thesis has been, of course, Francis Fukuyama, whose appropriation of Kuhn will be discussed in chapter 5, section 5. Kojeve had taken seriously Nietzsche's objection to the very idea of an end to history, namely, that it would strip humans of their humanity by removing the need for struggle that has been the hallmark of our species. Far from fulfilling human destiny, then, being one of the "last men" would be a fate to dread. Kojeve's own way around this problem was to argue that Nietzsche's sense of struggle would be sublimated as "gratuitous negativity": random acts of violence that assert the spontaneity of the human will but do not upset the overall order of the universal and homogeneous state. (For a glimpse at the "ideal" Kojevian polity, see the Anthony Burgess novel, *A Clockwork Orange* (1962) and the major motion picture based on it.) Writing at the Cold War's conclusion, Fukuyama has proffered a more popular resolution, yet one still rooted in Hegel's original vision: Humans in the ultimate state would still desire the respect and recognition of their fellows, the conditions for which will be subject to ever-changing market forces.

64. Koyré 1945, 110.

the Platonic urge created the need for science to remain autonomous from the rest of society. Consequently, despite its centuries of intellectual domination, Aristotelianism had failed to make progress as a system of thought because it insisted on saving the phenomena of ordinary experience. Galileo's wisdom lay precisely in his overstepping the sensations of the rabble in an effort to commune with the ancients, specifically by situating his seminal insights in the context of a Platonic metaphysics. What of Galileo's famous experiments? According to Koyré, they never really occurred but were fabricated for public consumption, that is, for minds not strong enough to reason through the fundamental mysteries of space, time, matter, and motion from first principles. Although the diagnosis of Galileo may be novel, the line of reasoning would have been familiar to the seventeenth-century natural philosophers (Descartes, Hobbes, Leibniz) who today we are more inclined to classify as "philosophers" than as "scientists."[65]

Though less famous than his seminal studies of Galileo and other luminaries of the "Scientific Revolution" (a Koyréan coinage), Koyré's remarks on Plato are very revealing of the sensibility that has enabled the Greek philosopher to recruit so many disciples down through the ages. Like the Straussians above, Koyré reserved a special animus for the Sophists, who flaunted their dialectical skills in the Athenian forum. They permitted the critical use of reason, a necessary means for removing the prejudices of the rabble, to become an end in itself. This conversion of means to ends only served to level all standards of judgment and remove whatever inhibitions these had provided to rule by the vulgar. For Koyré, original sin came from binding the deployment of reason to the material conditions of social life. The Sophists, of course, charged for their services, which they made available to whoever could pay. And later, Aristotelianism would be deeply implicated in the ideological maintenance of Christendom. More generally, Koyré acquired his reputation as the arch "internalist" historian of science by repeatedly stressing that science does not require specific social conditions for its existence—aside from the wealth needed to sustain the leisure of the individuals who decide to pursue a life of reason.[66]

65. Koyré 1978. Koyré's view of Galileo turns out to have been influential on both phenomenology and psychoanalysis. Koyré is credited in Edmund Husserl's account of scientific idealization (Husserl 1970, xix and section 9). Jacques Lacan used Koyré's account of Galileo's practice as a model for the subversive character of psychoanalysis as a science. See Brennan 1993, 68–69. Koyré's cynicism about the run of humanity may be usefully contrasted with Paul Feyerabend's justification for what he too judges to be Galileo's fabrications. See Feyerabend 1975. I return to this topic in chapter 6, section 7. Ironically, at least for someone with the reputation for negativity, Feyerabend stressed Galileo's ability to see a truth that transgressed the best methods of his day by an effective yet unexalted sense of cunning, not a godlike penetration of appearances.
66. Especially revealing is Koyré 1963. See also Elkana 1987.

That the life of reason should first arise in the culture that it did, a small democracy sustained by a slave-based economy, could not have been predicted. Such could have been the material conditions for any number of leisure activities that could have occupied the best Athenian minds. In asserting that science is a sociologically undermotivated activity, Koyré was partly exploiting a version of what philosophers of religion call the "ontological argument"—only this time for the existence of science rather than God. According to the argument, first proposed in the eleventh century by Saint Anselm of Canterbury (the subject of one of Koyré's early works), God's existence was warranted by the inconceivability that an all-powerful being would not also exist. There was thus no need to justify God's existence by such "external" rationales as the universe's need for a creator. In Koyré's case, science (as the pursuit of pure reason) is a concept that is so unlikely to emerge in *any* culture that if it does, then it must be the genuine article, and not the portmanteau pursuit of wealth, fame, or power. This helps explain Koyré's tendency to blur the distinction between what we might call "first-order" and "second-order" reflections on science—that is, doing science and thinking about doing science—even after experimentalists in the seventeenth century like Robert Boyle had begun to draw that distinction openly. If one even has the idea of doing science, then one is most probably already doing science.

However, Koyré should not be seen as remarking only on the contingency of science's origins. More subtly, yet importantly, he was commenting on the precariousness of science's *continuation*. A combination of greater insight into the ancient civilizations of Asia, the reversal of European imperialist expansion, along with the steady rise of the United States and Japan, had led most historians by the early twentieth century to conclude that science was not a uniquely European achievement, or at least not one in which Europe's continued leadership was assured.[67] However, Koyré would stand this observation on its head, claiming that science could disappear just as mysteriously as it appeared by becoming bound up with

---

67. Pyenson 1993. Once the history of science could not be subsumed unproblematically under the cultural history of Europe, two somewhat opposing tendencies followed. On the one hand, it opened the door to explorations of the non-European origins of science; on the other, it enhanced the image of science's history as an autonomous trajectory, which invariably strengthened the scientific community's authority over its past. Moreover, this latter tendency has not necessarily been to the detriment to the sort of philosophical history promoted in these pages—nor, for that matter, to more traditional archival history. Pivotal was that these scientists' "internal" histories were written with an eye to the future, which gave them the freedom (and courage) to be selective about which aspects of the past they highlighted, namely, those that they wished to see continued. While this practice led to relatively little agreement over how the history of science should be told, it nevertheless made a strong case for history's relevance to contemporary science policy. This point is elaborated in chapter 2.

the material conditions of life, be they the means of mass production or mass destruction.[68] Thus, Koyré's intellectualist disdain for experimental practice was, at the same time, a denial of the technological imperative that many advocates of science in the present century have claimed to be characteristic of progressive inquiry.[69] Typically these advocates—notably John Dewey and the American pragmatists—have come from places that did not suffer the ravages of the First World War. But true to Platonic form, Koyré never publicly engaged these ideologues. Instead, he simply refused to comment on the fate of science after the ascendancy of Newtonian mechanics, aside from a few remarks about how true scientists have kept the theoretical and practical sides of their activities separate, despite the increasing demands of bureaucrats for there to be a "payoff" for scientific work.[70]

Despite the temptation to read Koyré's silence as signs of incompetence or lack of interest in the last quarter millennium of scientific activity, a more reasonable hypothesis is that Koyré judged the "Scientific Revolution" a failure that ultimately yielded to forms of inquiry better suited to military and industrial concerns than truly epistemic ones—in which case, historians of science are entrusted to carry the torch for an epistemically

68. I continue this theme of the possible disappearance of science, but due more to economic than military factors, in Fuller 1997d, chap. 7.

69. Evidence that Kuhn shares Koyré's disdain for the technological imperative is his claim that sciences that cease to yield new research problems devolve into "tools of engineering": see Kuhn 1970b, 79. Thanks to Set Lonnert for pointing this passage out to me. Throughout *Structure*, technology is the "other" of science: either the application of science once it is no longer progressive or the reason a form of inquiry diverts from the puzzle-solving trajectory that Kuhn prescribes for "true" sciences. Interestingly, *ideology* never functions in the same capacity as technology in these contexts, though it easily could. As we shall see in chapter 2, this point signals Kuhn's siding with Planck over Mach in their debate over the future of science. Nevertheless, in keeping with the promiscuous application of Kuhn's model, even historians of technology have tried to show that innovation adheres to a Kuhnian dynamic. See esp. Constant 1973.

70. Koyré 1963, esp. 852–56. As evidence of Koyré's brave face, consider that after criticizing the current tendency to justify scientific projects by their practical payoffs, he says: "Yet, even today, at least so it seems to me, it is not the practical breakthroughs. Applications result from discoveries—they do not inspire them. Wireless telegraphy was not the goal pursued by Maxwell, though it resulted from his work. No more than the construction of the atomic bomb was that of Einstein and Bohr. It was understanding, not practical application that they sought" (856). Yes—that is, until Einstein wrote President Franklin Roosevelt advising him to initiate a U.S. atomic bomb project because the basic physical principles for building such a bomb were already known, and the Nazis (under Heisenberg's leadership) might be in the process of realizing that prospect themselves. That Koyré could conveniently ignore the direct involvement of the founders of atomic physics in the construction of the first nuclear weapons, by virtue of their not having been originally inspired to build them, may be attributed to his speaking casually about a field that lay beyond his expertise. Nevertheless, these occasions—when one has strong opinions about areas one knows only superficially—are precisely the ones that reveal most nakedly the predilections of the scholar that would normally be sublimated in a network of footnotes and judicious phrasing.

pure science.[71] We shall return to this interpretive problem in the next section, when confronting Kuhn's failure to include examples of science conducted after the 1920s in his theory of scientific change.

Koyré himself seemed to have been influenced by Edward Burtt's curious work from the 1920s, *The Metaphysical Foundations of Modern Physical Science*.[72] Burtt articulated a theme that in our own day is most closely related to the phenomenon of "Kuhn Loss," namely Kuhn's claim that a new paradigm need not save all the phenomena of its predecessor because fields of inquiry may not develop at the same rate, and hence to make substantive breakthroughs in one field, one may need to leave behind other fields traditionally associated with it.[73] Thus, Newtonian mechanics became the basis of a paradigm, even though it addressed only that fraction of Aristotle's conception of motion that concerned changes in location. Attempts were subsequently made to assimilate the neglected aspects of motion—including biological and psychological phenomena—to Newton's mechanistic model, most notably by eliminating the moral character of mental life. However, the partial character of Newton's original success failed to be recognized, except by those whose own work turned out to be marginalized by the Newtonian paradigm.[74] Instead, either it was prema-

71. For a contentious account of Koyré that stresses the extent to which he *deliberately* steered his followers away from examining contemporary science, see Crowther 1968, 286–87. According to Crowther, Koyré found it natural to think of the history of science as a cult activity because his own earliest historical work was on German and Russian mystics, especially the followers of Jakob Böhme (1575–1624), who maintained the heretical view that even God contained both good and evil. Might Koyré's orientation toward science represent a secularization of this viewpoint—one that, after Jean-Paul Sartre, has been called "the problem of dirty hands"? In chapter 3, note 63, I return to this problem in the context of Conant's political realism.

72. Koyré 1965, 24–25; Burtt 1954. One of Burtt's most vivid examples of the asynchronous character of scientific change is that the basic framework for the theory of mind has remained the same since the seventeenth century, despite the revolutions in twentieth-century physics. In this regard, Burtt had considerable sympathy for such latter-day metaphysicians as Alfred North Whitehead, who tried to come up with categories for understanding reality that would resynchronize the development of the sciences. On Koyré's debt to Burtt, see Cohen 1994, 86–92.

73. On "Kuhn Loss," see Fuller 1988, 223 ff. The origin of the expression is often credited to Post 1971. Kuhn describes the phenomenon in Kuhn 1970b, 104. One philosophically respectable strategy for legitimating Kuhn Loss is to say that an earlier paradigm's sense of ontological unity has now been revealed as a product of semantic sloppiness. Thus, Newton's purely physicalist account of motion does not merely refine a fraction of the domain covered under Aristotle's theory of motion; rather, it reveals that domain to have been ill conceptualized. See Fuller 1988, 163–74.

74. Denis Diderot (1713–84), the mastermind of the Enlightenment's single greatest contribution to Western culture, *The Encyclopedia*, is one figure whose criticism of Newton consigned him to the margins of the history of science. For an attempted revival, see Prigogine and Stengers 1984, 79–85. Johann Wolfgang von Goethe (1749–1832) is perhaps the paradigm case of a marginalized anti-Newtonian, as Ernst Mach made evident (see chapter 2, sections 5–6, below).

turely generalized to domains where it did not obviously apply or it engendered a reaction against mechanical explanations altogether. The former led to a pan-instrumentalism in which Newton's mathematical sophistication devolved into a fetish for measurement as a vehicle for prediction and control (a common reading of "positivism"), the latter to a mystical holism in which mathematics was itself taken to be the source of human alienation from nature, the hallmark of what the French called "Bergsonism."[75]

At this point, before turning to Kuhn himself, it is worth examining the extent to which the Platonic historical sensibility has been misunderstood by contemporary analytic philosophers, even ones of broadly historicist sympathies. This exercise will allow us to consolidate the train of thought pursued up to this point. In a survey of broadly "Platonist" accounts of the scientific revolution, Gary Hatfield considers several figures discussed in this chapter, including Burtt, Cassirer, and especially Koyré.[76] Hatfield targets these philosophers for their focus on the *presuppositions* rather than the *arguments* of the historical figures they studied. In Hatfield's words:

> [T]he treatment of past texts as repositories of presuppositions can divert attention from the integrity of their argumentative structure. It invites one to consider the individual psychologies of the various authors: to ask whether a particular metaphysical position such as Platonism, as expressed in various books known or conjectured to have been in their libraries, might have been in their thoughts but only indirectly or elliptically expressed in their works. It encourages the search for hidden influences behind the text, and often takes a biographical approach. By contrast, to examine metaphysics as it was historically conceived is to examine it as argument: as something put forward with conviction, in order to evoke conviction. Such an approach cannot be satisfied with the charting of influence; it requires seeing how a text hangs together and develops its force. Extended beyond metaphysical texts, it seeks to be sensitive to the distinctive styles of argument that philosophers and scientists have employed, whether metaphysical or not.[77]

According to Hatfield, Koyré could have strengthened his case for Galileo's Platonism by attending more to what Galileo actually said than to what he supposedly meant by what he said. Perhaps not surprisingly, Hatfield's Galileo looks more like an analytic philosopher of science than the undercover metaphysical *revanchiste* portrayed by Koyré. In particular, Hatfield's Galileo seems to be responding to his immediate interlocutors in a piece-

---

75. Kuhn apparently discussed these larger cultural implications of the mechanistic worldview with Stanley Cavell, especially in terms of the positivist impulse to regard the "artificial languages" of science as exerting normative force over natural languages. See Cavell 1976, 42 n. 38. On Bergsonism, see Descombes 1980, 9–12. It is worth noting that Henri Bergson (1859–1940) was himself originally trained as a mathematician.

76. Hatfield 1990a.

77. Hatfield 1990a, 147–48.

meal fashion within the scientific conventions of his day, appealing to Platonism in the broad, decontextualized sense of a belief in a realm of entities, access to which requires mental powers that transcend sense perception.[78] The point that Hatfield neglects at his peril is one that was very clearly stressed in the methodological writings of another of his targets, the Oxford archaeologist and idealist philosopher of history Robin Collingwood (1889–1943).[79]

Collingwood observed that the presupposition of an author's text may consist of logically distinct premises that are nevertheless historically conjoined. The text itself then may be seen as extending or contradicting the presupposition. Nowadays, we would simply drop the misleading logical baggage associated with the term "presupposition" and say that Collingwood was talking about the *context* that gives meaning to the text. By invoking context in this sense, one can explain why an argument seems either belabored or elliptical at various points—namely, in terms of the beliefs that the author takes the reader already to hold.[80] At the very least, this

---

78. Even after we ignore the anachronistic character of Hatfield's own sharp distinction between presupposition and argument, some specifically historiographical points are worth making. The first and most general point is that focusing on "the force of the text" as carrying the weight of its author's argument is no more valid than fixating on the payload of the weapons used by a combatant to explain the combatant's fate in warfare. At the risk of belaboring the obvious, it is worth recalling that an argument is a form of social interaction that transpires in a context that may not be immediately transparent to the historian because some of the relevant parties go unmentioned. Within this only partly disclosed context, argumentative texts function as tokens that are generally understood by the arguers not as ends in themselves, but as means for getting across certain points and promoting certain agendas.

Admittedly, once a scholar like Hatfield enters the field, his own interest and expertise may favor a primarily or purely textual study, but that should not be confused with the attitudes of the original arguers themselves. (Could Galileo accept Hatfield's understanding as his own?) More specifically, Koyré was committed to Platonism, in true Hegelian fashion, as a historically self-conscious phenomenon. In other words, Koyré realized that Platonism's philosophical sensibility was not itself born in Plato's Heaven but in response to the decline and fall of Athens, which Plato interpreted as resulting from the democratic subversion of reason (mainly through the rhetoric of the Sophists). Given the significance of these origins, and the desire not to repeat them in the future, it was important that a reminder of them always be included in the transmission of the Platonic doctrine. A comparable mentality can be found among Judaic scholars reflecting on the traumatic history of the Jewish people. (Not surprisingly, many Jews—Koyré himself included—have been distinguished Platonists.) This defined the sense in which Koyré would allow Galileo to stand for Platonism. As an analytic philosopher of science, Hatfield is used to thinking of problems of knowledge and existence as distinct from those of values and ends. However, Koyré (arguably like Plato himself) always linked the two and for this reason never openly endorsed the rather perverted sense of Platonism that has dominated Western scientific thought after Newton, which stresses the instrumental side of mathematical reasoning above all else—including the sanctity of human life itself.

79. Collingwood 1972, pt. 1.

80. I have called this phenomenon "the inscrutability of silence": see Fuller 1988, chap. 6. Collingwood's appeal to context should be understood as a generalization of his training as an archaeologist of Roman Britain: the inquirer faced with an incomplete set of artifacts at a site

shows that the texts are not self-sufficient to supply their own meanings but require an understanding of the author's rhetorical situation.[81] We shall return to this point in chapter 6, section 7, when discussing the work of Kuhn's major rival, Stephen Toulmin, a critical admirer of Collingwood.[82]

5. Nationalizing the Iconographic Vision: Kuhn and Conant

Kuhn acknowledged Koyré as his benchmark of historical scholarship, even though Conant's stress on the role of experimentation in science ultimately enabled Kuhn to mitigate Koyré's improbable intellectualism.[83] Moreover, Kuhn postulated several scientific revolutions, not simply the one that Koyré said began with Nicholas of Cusa, culminated in Galileo, and ended with Newton.[84] Nevertheless, Kuhn sustains the Platonic impulse, the need to tell a noble lie in order to protect a still nobler, but less palatable, truth. As the first step to unraveling the lie, consider the curious role that the social dimension of science plays in *Structure*. It would seem that Kuhn managed to "discover" that science flourishes in self-governing communities just at the time that the democratic instincts of American politicians and political commentators insisted on greater public accountability from scientists.[85]

As in so many other cases, a sense of "community" emerges among disparate individuals as soon as they face a common foe.[86] To put the point in boldest relief, the construction of *science* as *social* served much the same role as earlier constructions of the *scientist* as *individual* — only now reflecting the potentially more widespread societal opposition that scientists faced. In both cases, the uniqueness of science is highlighted as demanding

---

must reconstruct the material culture in which they figured. The sphere of that culture's possible thought and action depends on the interpretation given to the absence of otherwise expected implements. I raise this point because much of the skepticism surrounding Collingwood's method of reenacting past thought has been due to false associations with his "idealist" philosophy. Collingwood's sense of idealism is perhaps better seen as a commitment to producing an interpretation that coheres across the widest body of evidence, attaching potentially equal significance to what is present and absent from the historical record.

81. In recent years, this turn in the history of thought is most closely associated with Quentin Skinner (b. 1940). See note 25 above. For a critical review of his work, see Tully 1988.

82. Toulmin's debt to Collingwood is explained in the new introduction to Toulmin 1986. His adaptation of absolute presuppositions appears in Toulmin 1972, 118 ff.

83. On Kuhn's debt to Koyré, see Kuhn 1970b, vi. On his recognition of Koyré's limitations as a historian, see Kuhn 1977a, 150.

84. This trajectory comes out most clearly in Koyré 1957.

85. Hollinger 1990.

86. I portray the so-called Cognitive Revolution in psychology in the late 1950s as another instance of combining to defeat a common foe (the behaviorists) in Fuller 1993b, 139–85. This revolution was occurring in New England (Harvard, MIT, Dartmouth) virtually under Kuhn's nose. A good account, written by one of the Cognitive Revolution's most visible legatees, see Gardner 1987.

special treatment.[87] Nevertheless, Kuhn failed to recognize that the scientific community itself is subject to conflicting interests, each of which potentially represents a different direction in which science may go. Ultimately, Kuhn tried so hard to minimize the presence of disagreement, or division of any sort, within a paradigm that he ran perilously close to producing an image of the scientist that, after the sociologist Dennis Wrong, is called "oversocialized."[88] Telling in this regard are the tasks that Kuhn sets for sociological inquiry in the postscript to *Structure*. He concludes the agenda:

> What does the group collectively see as its goals; what deviations, individual or collective, will it tolerate; and how does it control the impermissible aberration? A fuller understanding of science will depend on answers to other sorts of questions as well, but there is no area in which work is so badly needed. Scientific knowledge, like language, is intrinsically the common property of a group or else nothing at all. To understand it we shall need to know the special characteristics of the groups that create and use it.[89]

The urgency with which Kuhn recommends the study of science as a unique and autonomous enterprise prompted Paul Feyerabend to wonder whether Kuhn was not implicitly smuggling a prescription under his description: Should Kuhn be taken to mean that science *ought* to be maintained in its current state?[90] Kuhn responded, somewhat jesuitically, that

---

87. Perhaps the most explicit objector at the time to the Polanyi-Kuhn idea that scientists form a natural community was Stephen Toulmin, who argued that while the interests of individual scientists may converge on various policy issues, it was unlikely — given the variety of social roles they occupy — that scientists constituted a unified interest group. See Toulmin 1964, 350 ff.

88. An oversocialized conception of people sees individuals entirely in terms of their group characteristics, as in the Kuhnian tendency to see all scientists who work in a given paradigm as sharing the same mind-set. In effect, this conception collapses the "micro" and "macro" levels of social reality by making it seem that an entire community can be studied simply by regarding one of its members. See Wrong 1961. Oversocializationists tend to regard acculturation — the acquisition of membership rights in a community — as the most important social process. But upon admission to the group, sociology ends and individual discretion begins — at least, that is the story that will be told once Kuhn's ties to Polanyi and the scientific community are introduced. Postmodernism, with its predilection for fractured identities and contingent associations, has done much to mitigate this totalizing sense of Kuhn's paradigms. However, recently the pendulum has swung back, as some sociologists now suggest that there are common terms, or "schemes or meaning," in terms of which a community will relate their otherwise disparate activities. In this context, the sense of paradigm as "worldview" has come back into vogue. See Arditi 1994. For a wide-ranging discussion of what may be the source of commonality when it is claimed that scientists share a paradigm, see Fuller 1988, 207–32.

89. Kuhn 1970b, 209–10.

90. Feyerabend 1970. After Feyerabend died in 1994, Paul Hoyningen-Huene found among Feyerabend's personal effects two letters addressed to Kuhn dated 1960–61, just after Feyerabend had received the final draft of Kuhn's *Structure*. In them Feyerabend provides a detailed critique of the book, which he sums up as "ideology covered up as history." In particular, Feyerabend was concerned with how Kuhn's blurring of the is/ought distinction

by describing the implicit normative structure of science, we would acquire a better sense of how best to govern science. Consider Kuhn's actual words:

> The structure of my argument is simple and, I think, unexceptionable: scientists behave in the following ways; those modes of behavior have the following essential functions; in the absence of an alternate mode *that would serve similar functions*, scientists should behave essentially as they do if their concern is to improve scientific knowledge.[91]

It is natural to read Kuhn at this point as thumbing his nose at the is/ought distinction. However, this may be one of those deconstructive philosophical moves that looks bold on paper but effectively serves to "leave the world alone" in true Wittgensteinian fashion. In that case, Kuhn would have simply confirmed Feyerabend's suspicion that Kuhn wants to understand science in order to determine what it will take to maintain its status quo. While this is probably the best interpretation of the short-term implications of Kuhn's call for a sociology of science, I also think there is a deeper and more radical reading, one that suggests that Kuhn wants a sociology of science that is empirically adequate to science *only when* it is functioning in a normatively desirable fashion. This leaves open the question of whether science is currently functioning in such a fashion.

Despite the presumptive universality of a title that proclaims that all scientific revolutions have the same structure, *Structure* may have been conceived to apply *only* to the disciplines and periods from which its examples are drawn. After all, it is not necessary that science be portrayed as having an uninterrupted historical trajectory to be retained as a normative ideal. In other words, the history of science may be more like the history of *democracy*, the subject of which is an ever-present political possibility whose conditions of realization have varied across cultures and over the centuries.[92] In that case, Kuhn's historical sociology would be written—in the spirit of other Platonist works—with the tacit understanding that the history is being written in a period of captivity.

Whoever still doubts Kuhn's covert commitment to Platonism would need to explain the skew of his historical examples, especially if we envisage

---

discouraged scientists from *trying* to overturn the dominant paradigm. Because anomalies arise for Kuhn as the unintended consequence of normal science research, scientists are effectively enjoined to adopt an "if it ain't broke, don't fix it" mentality. Thus, Feyerabend viewed Kuhn in (right-wing) Hegelian terms as encouraging the idea that whatever is, is rational. See Hoyningen-Huene 1995.

91. Kuhn 1970a, 237. Italics in the original.

92. The problem, of course, is that many democratic theorists following in the lead of Jean-Jacques Rousseau have in fact argued that democracies can exist only in certain spaces—small homogeneous communities—that have all but vanished. I consider whether the same principle applies to science in Fuller 1994f.

them as contributing to a normative sociology of science. The data are taken almost exclusively from the history of the physical sciences in Europe from about 1620 to 1920 — and the cases from chemistry stop after the mid-nineteenth century. Given *Structure's* lack of examples from the history of other disciplines and other periods — especially from the rest of the twentieth century — one may be forgiven for concluding that Kuhn does not believe that much of what now passes for "science" matches up to the norms that he finds implicit in the earlier history, the empirical basis for his theory of scientific change.

Grounds for such a conclusion may be found by taking very seriously the idea that normal science is dictated by the "logic of puzzle solving" — and *not*, say, the need to utilize existing laboratory facilities and personnel as cost-effectively as possible. The fact that Kuhn occasionally refers to instruments and techniques as "internal" to the sociology of *past* science does not preclude that, in their capital-intensive scaled-up form, they have *now* become "external" and hence distortive of the normative character of science. This "deeper" reading would render Kuhn quite subversive of the status quo, a point that Jerome Ravetz (b. 1929) has been virtually alone in following up in his concept of "post-normal" or "industrial" science.[93] Indeed, had Ravetz's radical reading of Kuhn been as influential as the more familiar domesticated ones, it could easily have served as a rhetorical springboard for abandoning most of what is now recognized as science, especially so-called Big Science. Ravetz and his comrades in the British Society for Social Responsibility in Science attempted to deploy just such a rhetoric but failed because they could not square the call for the deindustrialization of science with the interests of technical workers who wanted to refashion for socially productive ends the very skills that had been honed in the industrial regime.[94]

---

93. See esp. Ravetz 1971. For a charitably radical view of Kuhn on this point, see Ravetz 1991. Ravetz, an expatriate American who has lived in the UK for over thirty years, originally studied the history and philosophy of science in the early 1960s with Stephen Toulmin when he was at Leeds University and had begun to take an interest in science policy, largely, as we saw above, in reaction to Michael Polanyi's views about the reality of the scientific community. Ravetz subsequently taught at Leeds for over twenty years, after which he became a science policy analyst, concerned especially with the role of uncertainty and ignorance matters of risk assessment. His more recent essays on the politics of science are collected in Ravetz 1990a.

94. For an honest appraisal, see Ravetz 1990b, 905. An excellent selection of essays on the transformative role of Big Science in both science and the larger society is *Osiris*, n.s., 7 (1992). A source of friction between radical science intellectuals and the labor movement in the late 1960s and early 1970s may be seen in what Ravetz and his colleagues were calling "critical science," which had a strong environmentalist bent under the leadership of the U.S. ecologist Barry Commoner, a devotee of the "small is beautiful" ideology. See Ravetz 1971, 424 ff. A famous worker-led revolt against the military-industrial complex that did not require turning

In Kuhn's case the risk of a wholesale denunciation of science may have seemed so great that he could not admit—even under Feyerabend's stiff cross-examination—*that science as he knew it had indeed come to an end*.[95] Moreover, the risk must have seemed all the greater once *Structure* had become a blockbuster success in the late 1960s. Thus, a Faustian bargain may have been struck in Kuhn's mind: What science lost in its normative desirability, it replaced by providing a stable military-industrial infrastructure and virtually the only source of legitimate authority for an increasingly fragmented and volatile populace. In light of those stakes, a policy of strategic vagueness on Kuhn's part would allow for a tame misreading of the book that would help ward off the drastic calls for the disestablishment of science that by the end of the 1970s would come to characterize Feyerabend's own much more open rhetoric.[96]

Thus, whenever his name was linked with that of Herbert Marcuse as gurus of the worldwide student revolutionary movement, Kuhn would point out, quite rightly, that his view of science was very conservative, basically an extended exercise in showing how innovation could result from a highly disciplined, perhaps even authoritarian, social system. (This point received its most sustained analytical treatment in Kuhn's strong sense of the distinction between science and art, which, as we saw in the previous section, includes the former's lack of an avant-garde.)[97] At the same time, however, Kuhn refused to acknowledge the point that Ravetz and the

---

one's back on high technology is described in Wainwright and Elliott 1982. Although Wainwright and Elliott made a valiant effort at trying to associate the intellectual left with the defense conversion strategy of the Lucas aerospace workers, in fact radical scientists and academics were slow in coming to their assistance when their support was first actively solicited. One of the workers' leaders has bitterly observed that only 4 (including Elliott) out of 180 responded to the initial call: Cooley 1980, 66. Thanks to Sujatha Raman for bringing this telling example to my attention.

95. This point constitutes the critical theme of Fuller 1999b.

96. Much of Feyerabend's writings along these lines are collected in Feyerabend 1979. It is worth noting that, despite his image as an establishmentarian philosopher, Feyerabend's teacher and Kuhn's nemesis Karl Popper could match the outspokenness of his student when it came to the pernicious effects of the emerging military-industrial complex on the integrity of science. In an address entitled "The Moral Responsibility of the Scientist," first given in 1968, Popper called for scientists to design their own version of the Hippocratic oath, in which the overriding loyalty that doctors are supposed to have toward patients is replaced by the loyalty that scientists should have toward humanity—not just their immediate colleagues and superiors. Here Popper makes a clear connection between the loss of this universalistic sense of moral responsibility in science with the rise of Kuhnian puzzle solving, which discourages science students from asking about the larger ends their research might serve. See Popper 1994, chap. 6. I shall return to this point in chapter 4, section 3.

97. Kuhn's scruples equally extended to discouraging those who wished to manipulate the course of science, post-Sputnik-style, in order to boost America's fortunes in the Cold War. This came in the form of a skeptical lecture on the prospects of improvements in science education making students more innovative. Kuhn 1977a, 225–39. See chapter 3, section 5, below for a fuller discussion.

student radicals had extracted from *Structure*, namely, that science as a self-determining — even if authoritarian — form of inquiry had come to an end, having been replaced by one driven by military-industrial imperatives. Characteristically, in the brief period he engaged in sustained dialogue with the radicals, Kuhn focused on correcting misreadings of the letter of his doctrine, while remaining silent on the understanding that his readers had of the spirit and import of what he said when applied to the contemporary scientific scene.[98] But if the Faustian bargain was evaded, suppressed in the body of Kuhn's writings and thought, it emerges very clearly in his Harvard mentor's speeches and actions.

Despite being a staunch supporter of scientific autonomy, James Bryant Conant had to admit a profound difference between the role of science in the First and Second World Wars. In the former, the industrial applications of science produced instruments of unprecedented destruction, but no new discoveries were made or theories tested; in the latter, science's military-industrial applications unleashed even greater destructive capabilities, while serving as laboratories that generated phenomena that fed the conduct of inquiry. Conant's own words about the transition, in which he played a major administrative role, are worth seeing again:

> Chemical warfare, radar, proximity fuses, underwater warfare, jet planes, new mines and missiles—all these developments depended on the special application of the publicly known facts and principles of physics and chemistry; what new scientific knowledge was gained in these military developments had no revolutionary consequences in relation to the forward march of science. How completely otherwise it was in the manufacture of the atomic bomb! No one else could be sure that the phenomenon known as the "critical mass" was an experimental reality until a large-scale manufacturing operation was put into motion. In 1940 the physicists were in possession of the results of certain experiments with microscopic amounts of materials; they likewise held in their hands powerful theoretical concepts in the new field of nuclear physics and chemistry. By an enormous extrapolation they predicted the operation of atomic piles and the explosion of an atomic bomb. Most of these predictions, however, could not be tested by any large-scale laboratory experiments or even by the erection of small-scale pilot plants. The flower of this whole new field of science was dependent on a large sum of tax-payers' money; this expenditure could be justified in the 1940s only in terms of the destructive power of a weapon required in a desperate global struggle.[99]

In the nineteenth century, it was common for academic scientists to derive theoretical conclusions from agricultural and industrial innovations, even

98. Kuhn et al. 1997, 185–86.
99. Conant 1952b, 12–13.

though the scientists were rarely involved in their original invention. The classic case in point is the development of thermodynamics from attempts to understand the workings of the steam engine for purposes of improving its efficiency in large-scale settings. However, a mark of a mature science was that the innovations—regardless of their significance for the larger culture—did not fundamentally alter the science's research trajectory, which was set entirely by the scientific community. Thus, in Kuhnian terms, one could tell that most branches of physics were paradigm based, while most branches of biology were not.[100]

The case of chemistry, Conant's home discipline, proved awkward for this scheme in ways that anticipate the postwar Faustian bargain of physics. In the course of the nineteenth century, chemistry seemed to backslide from a paradigmatic to a preparadigmatic state, as its research trajectory came to be oriented to government and business concerns to such an extent that the field's research activities required facilities that only a large corporate actor could provide. This phenomenon goes very much against the spirit of models of scientific change like Kuhn's, which assume that, other things being equal, the sciences become more autonomous over time as they gain greater control of their research agendas. Backsliding into a state of extrascientific dependency is thus unthinkable. Such an assumption *would* make some sense if all of science's potential distortions were merely ideological, that is, matters of unduly influencing beliefs. The remedy would be the sorts of activities that sociologists of science nowadays associate with "boundary construction" and "boundary maintenance": the proliferation of jargons, credentials, and all the other barriers that experts place between themselves and the public. Matters change, however, once the "normal" conduct of inquiry itself requires a massive provision of human and material resources. In that case, it becomes difficult to see how science can ever be autonomous from the rest of society.

Perhaps the most important sign that physics had forever lost its traditional sense of autonomy was that, even after the end of the war, and against

---

100. The exclusion of biology from paradigmatic status may seem unusual, considering that Kuhn excuses his failure to deal with that discipline on grounds no more principled than sheer ignorance: see Kuhn 1970b, xi. However, if Darwin is treated as comparable to Newton in providing the exemplar for the pursuit of normal science, then a strict Kuhnian would place the origins of paradigm-driven biology no earlier than the 1930s, the dawn of the so-called neo-Darwinian synthesis. Before then, genetics and paleontology, the two disciplines most directly implicated in Darwin's achievement, had been antagonistic in both method and theory. (Roughly speaking, genetics tried to move biology toward physics, whereas paleontology tried moving it toward history.) For an analysis of some key texts that forged the neo-Darwinian synthesis, see Ceccarelli 1995 and Journet 1995. It is worth mentioning here that Kuhn refused to grant medicine any paradigmatic status because he deemed it as having always been driven by external social needs.

his own earlier judgment, Conant was forced to embrace large-scale military support for basic research, a mutual dependency that only escalated over the Cold War.[101] Given the pivotal role of the atomic bomb in ending the war, the top brass were more positively impressed with academic scientific achievements than other leaders in government or academia, and hence were less inclined to meddle in the scientists' day-to-day activities — something that the U.S. Congress, armed with social scientists, had threatened with its "oversight" panels.[102]

As it turns out, the military applications developed in the Second World War fed back into the basic research agenda across *both* the natural and social sciences, suggesting new work patterns and ideas, not only in physics but also in experimental psychology, computer science, and the cluster of disciplines associated with "cybernetics."[103] This development brought to light two rather different senses of autonomy that had been lurking throughout the history of science: a *complete* sense, in which scientists pursue their own ends by their own means; and a *partial* sense, in which scientists pursue someone else's ends by their own means.[104] The distinction has been blurred in the minds of scientists and philosophers because, even when only partially autonomous, scientific inquiry has often unintentionally generated findings that are relevant to the scientists but not to their clients. We shall return to this point in chapter 3, when considering various attempts by scientists — on both the left and right of the political spectrum — to institutionalize some appropriate sense of autonomy in the interwar years of the twentieth century. But for now, let us backtrack to the moment when science was first constructed as autonomous to the extent of requiring one's professional accreditation as a *scientist*.

6. Once upon a Time: When Science First Required Scientists

In a world made for aristocrats, coming from the right background is the key to having a good image. In the case of the natural sciences, whose experimental lineage in the past had yielded uncouth mechanics and heretical magicians, this meant providing these disciplines with a suitably

---

101. On Conant's change of heart toward military support of science, see Hershberg 1993, 573-77.
102. See Reingold 1991, 284-333.
103. For a good account of the military's role in seeding American social science after World War II, see Heims 1991.
104. The difference between complete and partial autonomy corresponds to the classical Greek distinction between *praxis* and *techne*, respectively. I elaborate this point in chapter 2, note 70, in order to show that the concept of "practice" in contemporary sociology of science is just as blurred as that of "autonomy" in contemporary philosophy of science.

edifying history. This goes beyond simply showing that science has been done now and again over the course of history—a point nobody denies. In addition, all of these doers of science must be engaged in an activity that adds up to something of transcendent significance that can be isolated from ambient social processes. This, in turn, suggests that there are ways of evaluating historical episodes in terms of their contribution to an overall trajectory of progress. We have here the elementary insight behind the *internalist* orientation to the history of science. But more than merely licensing the view that scientists are autonomous inquirers, this orientation implies that scientists gain their autonomy through membership in a virtual community ("invisible college," as Francis Bacon put it) with a common set of problems and techniques. From here it is a short step to argue that this virtual community ought to be rendered real and its members' autonomy converted into a guild right requiring formal credentials. In a nutshell, a certain idealized image—"myth" is a better word—of science's standing in Western culture was institutionalized in the nineteenth century, as the natural sciences slowly won a place in the university curriculum. Befitting a simpler time, one person simultaneously played the role of Kuhn's two mentors, Conant and Koyré: William Whewell (1794–1866), master of Trinity College, Cambridge University.[105]

As late as the third quarter of the nineteenth century, the natural sciences did not constitute the single-minded pursuit of the ultimate truths of nature that we imagine it to be today. That pursuit was still very much in the hands of the theologians who presided over the universities. Rather, the sciences were a source of helpful hints for improving inventions and controlling limited parts of the environment. The mathematical elegance of Newton's account of the physical universe played well in Sunday sermons as evidence of God's design, but it was rather uncharacteristic of the piecemeal grubbiness of most actual science. Contrary to Kuhn's famous image, physicists and chemists during this period were not primarily devoted to solving well-defined problems according to the dictates of the Newtonian paradigm. In fact, strictly speaking, there were no "physicists" and "chemists," in the sense of professional practitioners of these fields. Rather, the sciences were taken up in the course of doing other things, be it for leisure or profit—with the odd academic puttering around in a makeshift workshop. While the word "science" had long been in use (mainly to refer to

---

105. Amid the recent renaissance of historical interest in Whewell, a good kickoff point is Fisch 1991. A smorgasbord of perspectives on Whewell's career and significance is provided in Fisch and Schaffer 1991. Clearly anticipating the dual role I assign to Whewell as a proto-Kuhnian is Hodge 1991, 255. Hodge specifically compares Whewell to a combination of Conant and Cassirer, the latter being reasonably close to Koyré for the purpose of my argument (see note 31 above).

the traditional liberal arts subjects), the word "scientist," meaning a full-time practitioner of the natural sciences, did not enter English usage until Whewell introduced it in 1833.[106]

The casualness with which science was pursued up to this point cannot be overstated. It continues to bedevil historians interested in tracing the flow of information during that original watershed of "R and D" called the Industrial Revolution. We can get some perspective on this matter by comparing the social character of science two hundred years ago to what we find nowadays in the emerging personal computer culture. Here we see a torrent of activity, but very little of it passing through the gatekeeping mechanisms of academia. To put it bluntly, the most interesting developments in computer culture are happening outside the computer science departments. However, there are so many "hackers" operating in such diverse contexts contributing to these developments that it is difficult to determine, say, where a new application originates, how it spreads or even whether it is very reliable. The popular computer magazines, with their overriding interest in publicizing noteworthy "phenomena," are themselves just as much a source of myth and error as of reliable computer knowledge. Still, the trial-and-error, decentralized, hands-on nature of personal computing hardly serves to deter hackers; on the contrary, it is an inspiration for newcomers to branch off into their own sectors of cyberspace for possible fun or profit. Yet it would be seriously misleading to say that these devotees of computer culture are collectively engaged in a disciplined quest to solve the fundamental problems of virtual reality.

If the natural sciences two hundred years ago looked a lot like computer culture does today, then one can easily imagine the uphill public relations battle that lay ahead to convert them into proper university subjects. Suspicions were rife on all sides. Many scientists were self-styled religious freethinkers who accepted the Enlightenment view that the universities were the last bastions of outdated dogma. The clerics returned the compliment by disdaining laboratory work for its unseemly resemblance to a factory shop floor. Into the breach stepped William Whewell, an Anglican priest who held a chair in mineralogy, an upwardly mobile Lancashire lad whose accent betrayed his humble origins in the elite circles where he traveled. As critical correspondent of Faraday, Mill, and Darwin, he was the uncrowned king of cross-disciplinary kibitzing. More importantly, Whewell invented the history of science in the image and likeness of the Christian salvation story. Yet, for all his influence on modern scientific sensibilities,

---

106. A panoramic sweep of what we would now regard as "scientific" activity in eighteenth- and nineteenth-century Britain is provided in Inkster 1991, 80–86.

Whewell was probably best known in his day as a defender of the compatibility of science and theology. Like many Creation scientists in our own time (who are more likely to be chemists and engineers—experts on "design"—than biologists), Whewell held that the natural sciences reverse-engineered the design of God's clockwork universe. However, his claims of compatibility did more for science than for religion: As theology lost its monopoly over the ultimate questions, scientific pursuits basked in the glow of religious high-mindedness.[107]

In a series of pamphlets and reviews published in 1830s and '40s, Whewell made it clear that if the theologically based "old universities" (Oxford and Cambridge) failed to incorporate the sciences into the liberal arts curriculum, their students would be ill prepared to assume the helm of an increasingly industrialized society. Indeed, several wealthy industrialists had already taken the initiative to endow what, by the end of the century, would become the large "red brick" metropolitan universities of England. Central to these new universities was an industrialist's sense of what science had to offer humanity. Their curricula were organized around manufactures-based skills, not theory-driven disciplines.[108] It was up to Whewell to reverse this trend by promoting an image of scientific development that would profoundly influence all subsequent historians and philosophers of science.[109]

Indeed, the image has become so much second nature that it is worth dwelling on how it served the immediate needs of Whewell's old university constituency. The crucial move was to make science theory driven.

---

107. The idea that "science" and "religion" have been in perennial institutional conflict is a product of the late nineteenth-century historical imagination. Only once the natural sciences had begun to assume religion's role as the seat of authoritative knowledge in Western society did the previous history start to be written in terms of science's deliberate attempt to wrench that role away from religion. This was also a case in which newly available archival material—the Vatican's release of Galileo's papers starting in 1850—helped magnify the institutional significance of Galileo's trial. See Laudan 1993, 15–16. Key texts included John Draper's *History of the Conflict of Science and Religion* (1875) and Andrew Dickson White's *A History of the Warfare of Science with Theology in Christendom* (1895). As points of contrast, consider that even heretics such as Galileo never renounced their faith, but asserted that the Church fathers, given the chance, would uphold their heretical views over the current orthodoxy. For its part, the Enlightenment typically supported science, not as an institutional alternative to religion, but as an exemplar of "Reason" that can demystify the more superstitious aspects of religion. An excellent historical guide to all these matters is Brooke 1991.

108. The best short account of the development of alternative institutes of higher learning in Britain devoted to the natural sciences is still Cardwell 1972. A good summary of these developments seen in terms of the emerging role of education as a principle of social stratification is Simon 1987. In the late twentieth century, as the universities once again are seen as failing to adapt to a changing political and economic climate, the challenge originally posed by the red brick universities is now provided by vocational schools, community colleges, and especially polytechnics.

109. The most comprehensive treatment of what Erving Goffman would have called Whewell's "impression management" of science is Yeo 1993.

"Theory" meant systematic knowledge abstracted from the concrete contexts in which science was then normally pursued. Thus, it was more closely associated with philosophers and theologians (note the Greek root of *theos* common to "theory" and "theology") than with the mechanics and other protoscientists. Nowadays, following Kuhn, we glibly speak of normal science being conducted under the aegis of a "paradigm," which is basically a comprehensive theory that functions as a blueprint for further research. However, in accepting this picture, we are implicitly granting Whewell's point that proper scientists must be trained in a relatively traditional academic setting before being given a license to practice science.[110]

Because science up to that point had been practiced in diverse settings, mostly outside the university, Whewell needed a story for how a paradigm could come into being. Here the image of tributaries flowing into a major river served as a rhetorically effective image.[111] Whewell retrieved Newtonian mechanics from the Sunday sermons to show how the principles discovered by Copernicus, Galileo, and Kepler fed into it, thereby providing a focused basis for subsequent developments in the history of physics. Whewell suggested that now that science had begun to fragment into many different specialties, it was more than ever imperative that universities take the lead in channeling this torrent of activity in an edifying direction. Not only would universities have to provide curricula for the training of scientists, but they would also have to be ready to pass judgment on the validity of alleged discoveries, according to their deducibility from the larger body of common scientific knowledge. Thus began the familiar philosophical separation of the *context of discovery* from the *context of justification*.

---

110. Cardwell 1972, 242, observes that the distinction between "pure" and "applied" science arose in this setting and suggests that the pride of place given to the pure side over the course of the nineteenth century may have resulted from the influence of Germany's two-tiered higher-educational system—universities and polytechnics—since it is not clear that the traditional British emphasis on empiricism and utilitarianism would have necessarily privileged the pure over the applied. In terms of the philosophical ideology used to legitimate distinction, Whewell was among the first generation of British scientists to take Kant seriously, whose works had been first popularized in English by the poet and critic Samuel Taylor Coleridge (1772–1834).

111. A clear reaffirmation of the Whewellian faith in an overall direction to science, despite what would now be called, after Gilles Deleuze, its "rhizomatic" tendencies, is the following by Alexandre Koyré:

> No, the *itinerarium mentis in veritatem* is not a straight path. It twists and turns, is occasionally blocked, and turns back upon itself. It is not even one path, but many. That of the mathematician is not that of the chemist, nor that of the biologist, nor even that of the physicist. We are therefore forced to pursue all these paths as we find them, and that is to say connected or separated, as they occur in history, and to resign ourselves to writing histories *of sciences* before writing the history of science into which they will finally join together like tributaries of a river. (Koyré 1963, 857; italics in original)

According to Whewell, a discovery does not become properly scientific until it has been justified by an appropriate academic authority, who can then chart it on the "map of knowledge," another of Whewell's images.[112] This mentality would soon permeate the civil service, leading to the establishment of the Patent Office as the department of the British Museum devoted to the registry of new inventions and the adjudication of intellectual property claims. Eventually, the discovery/justification distinction would become the cornerstone of the "received view" in the philosophy of science, the alleged common ground between the logical positivists and the Popperians that Kuhn is said to have overthrown.[113]

7. Reinventing the Need for Scientists in Twentieth-Century Philosophy of Science

Nearly a century separates Whewell from the first modern renditions of the discovery/justification distinction, and in that period the grounds for separating the two contexts had deepened.[114] Whewell's concerns about institutional status of knowledge claims and, ultimately, the class origins of their claimants were superseded, in the 1930s, by a sense that scientific discovery may be explainable entirely in "psychological" or "sociological" terms, which had become polite ways of suggesting that it might be the product of, respectively, unrepeatable genius or unacquirable racial characteristics. For its part, justification was no longer limited to elite settings like the university or public ones like the national patent office; rather, the scope of a justified knowledge claim was potentially available to anyone who could follow a pattern of deductive reasoning and relate its conclusions to empirical observations.

Thus, the politics of the distinction had radically altered—from a revanchiste aristocratic strategy for containing the advancement of the industrial bourgeoisie to a liberal Enlightenment response for containing the excesses of authoritarian romanticism, ultimately as represented by the Nazis.[115]

---

112. For a discussion of Whewell's interest in treating reality cartographically as domains of inquiry, see Schaffer 1991.

113. The most comprehensive account of the "received view" and its aftermath remains Suppe 1977, 1–243.

114. The modern starting point for most anglophone discussions of the distinction is Reichenbach 1938, esp. 5–8, 381–84. (Although Popper's *Logic of Scientific Discovery* drew a somewhat similar distinction three years earlier, the book was originally published in German and not translated into English until 1959.)

115. Ronald Giere has come closest to recognizing this connection. In his 1994 presidential address to the Philosophy of Science Association, Giere suggested that these considerations may have motivated Reichenbach's modern formulation of the distinction. See Giere 1995, 4–5. Perhaps then it is not surprising that Reichenbach envisaged the context of discovery as being as much, if not more, a matter for the sociology of knowledge as for psychology.

The missing link in this transition is the popularization of the discovery/justification distinction in the form of the *genetic fallacy*—the (deductively invalid) inference from a claim's origins to its truth-value. Although nowadays enshrined in introductory logic textbooks alongside the other forms of faulty reasoning originally identified in Aristotle's *Sophistical Refutations*, not only had Aristotle himself failed to recognize the fallaciousness of genetic arguments (indeed, he frequently indulged in them), but the expression "genetic fallacy" first appears in English only in 1934.[116]

Despite its grip on the minds of elementary logic teachers, the genetic fallacy has been inconsistently enforced by professional epistemologists and philosophers of science. A good case in point is the so-called historicist turn in the philosophy of science, allegedly spearheaded by Kuhn (allied with Hanson, Toulmin, and Feyerabend) in the early 1960s and eventually culminating in the work of Imre Lakatos and Larry Laudan in the mid-1970s.

A common target of the historicists was what Lakatos sarcastically called "instant rationality," the idea that two research programs can be comparatively evaluated simply by formalizing their theoretical logics and identifying (either prospectively or retrospectively) a "crucial experiment" where they predict contradictory observations. The logical positivists, to whom the caricature of "instant rationalists" most readily stuck, were taken to believe that anything of value from a theory's past would remain as propositions located somewhere in the latest logical reconstruction of the theory. In contrast, the historicists insisted that comparative judgments must be grounded on the actual track records of the two programs. Yet, strictly speaking, to forgo instant rationality in this manner is to court, if not outright commit, the genetic fallacy. Indeed, it would seem to suggest that a research tradition enjoys intellectual property rights over the knowledge claims it originates. Thus, if a scientist working in, say, the Newtonian research tradition happens to make an important finding, then the finding counts as a reason for continuing to promote the tradition. Soon, the impression is given—especially in textbooks—that the finding could have been made *only* by someone working in that tradition. In short, priority quickly becomes grounds for necessity.

---

116. At least, such is the testimony of the *Oxford English Dictionary* and Hamblin 1970, 45.

The genetic fallacy was originally presented in Cohen and Nagel 1934. In judging genetic arguments fallacious, Cohen and Nagel were challenging the dicta of their teacher John Dewey, who, in continuing deference to Hegel, presupposed that the historical career of an idea is relevant to its evaluation, in that an idea's past typically constrained the contexts in which it would be applied in the future. For a latter-day development of this strategy, which is very much against the grain of both contemporary deductivist and inductivist thought in scientific methodology, see Will 1988.

Moreover, the historicists tended to commit the genetic fallacy in an especially asymmetrical fashion. For, if we wish to credit, say, the Newtonian tradition with solving certain empirical problems, why not also continue to hold it responsible for ignoring its major conceptual problems — indeed, the ones that would eventually lead to its demise at Einstein's hands? As we shall see in chapter 2, this was a question that Ernst Mach repeatedly raised against Max Planck and other defenders of the Newtonian orthodoxy at the dawn of the twentieth century. Mach astutely realized that a research program can erase the memory of its deep conceptual problems by accumulating enough striking empirical successes. While Mach meant this point as a criticism of the collective memory of the physics community, Laudan, drawing on Kuhn and Lakatos, unwittingly followed in Planck's footsteps and converted this apparent liability into an influential criterion of a research tradition's progressiveness.[117]

The key to why few logicians have accused historicist philosophers of science of elaborate commissions of the genetic fallacy lies in the typical textbook example of the fallacy. It is a (positive or negative) judgment based specifically on the cultural or racial origins of scientists, such as "Aryan science" or "Jewish science." Interestingly, however, other sociohistorical features of scientists, such as their research traditions, have *not* been counted as falling under the fallacy's purview. In this respect, the political scars of Nazism have lingered in contemporary philosophy beyond what can be justified on strictly logical grounds.[118] After all, if the race or nationality of a scientist is irrelevant to the truth of her discoveries, then why should not other contingent features of her discovery be equally irrelevant — such as the fact that she trained and practiced within a certain research tradition (which may itself be characterized as having a "national style")?[119]

This inconsistent application of strictures against the fallacy could be avoided if, instead of treating each scientific discovery as a basis for credit-

117. See Laudan 1977, esp. chap. 3.
118. A striking example of an article that has received much more praise than criticism, but in which the origins of knowledge claims are presumed quite clearly to contribute to their justifiedness, is the cornerstone of modern sociology of science: Merton 1942. Here a link was explicitly asserted between democratic political regimes and good science, on the one hand, and authoritarian political regimes and bad science, on the other. We shall resume with this thread, especially in the Harvard context common to Kuhn, Conant, and, for that matter, Merton, in chapter 3.
119. The appeal to "national styles" to justify knowledge claims can be found in many, if not most, of the classic works of historicist philosophy of science from the twentieth century, esp. Duhem 1954, chap. 4. Although historicism has steadily declined in the philosophy of science since about 1980, commissions of the genetic fallacy continue apace in the annals of "naturalized epistemology," as when Philip Kitcher deems that the religious commitments

ing one tradition and stigmatizing others, the occasion were treated as an opportunity for cross-tradition fertilization, with the variety of traditions that can incorporate the discovery used to measure the discovery's justifiedness.[120] At this point, the much-vaunted thesis of the *underdetermination of theory by data* enters the argument. According to this thesis, most closely associated with Quine, any empirical claim is, in principle, deducible from a large number of mutually incompatible sets of theoretical claims. Put a bit more loosely, just because a discovery was made by someone working with one set of assumptions, it does not follow that it could not have been made — or, more importantly, understood, used, and extended — by someone operating with an entirely different set of assumptions. Therefore, from what Quine called "a logical point of view," the fact that one research tradition managed to arrive at an important empirical finding before its competitors may well be little more than a historical accident.[121]

To the philosophically innocent, this may seem an obvious point, but the fact that Quine's thesis has acquired a luminous status over the past half-century reflects, at least in part, the extent to which philosophers of science had hoped that some formalization of the scientific method would focus and streamline the course of inquiry, the very goal sought by Whewell (albeit often neither by his means nor for his reasons). For example, over the last 150 years, many philosophers have tried to use probability theory to make good on Whewell's intuition that there is something epistemically significant about a theory that can account for phenomena that were not

---

of certain groups, such as believers in the divine creation of species, render them psychologically incapable of doing science and hence their scientific pronouncements should be treated with deep suspicion. See Kitcher 1993, 195–96.

Ironically, this comes from the author of the most eloquent philosophical book in opposition to sociobiology, itself a speculative science of the genetic capacities and incapacities of various socially significant traits. See Kitcher 1985. Following G. E. Moore's (1873–1958) critique of ethical naturalism in *Principia Ethica* (1903), logicians have christened a "naturalistic fallacy," i.e., a deductively invalid inference from what people tend to want to what they ought to want. The naturalistic fallacy may be regarded as the genus of which the genetic fallacy is a species, given that someone who commits the genetic fallacy derives normative conclusions from specifically historical facts.

120. If this proposal sounds familiar, it is because it has been suggested as the main characteristic of most of the generally agreed upon "advances" in the *social* sciences. Thus, the value of "class" or "bureaucracy" as a conceptual advance has not depended on the fortunes of, respectively, the Marxian and Weberian research traditions. (I shall return to this important point in chapter 8, section 5, where I attempt to resurrect the discovery/justification distinction.) For more on the proposal, see Deutsch et al. 1986. The proposal builds upon an article that Karl Deutsch, a Harvard political scientist, published in *Science* (Deutsch, Platt, and Senghaas 1971) under the influence of Kuhn. Thus, his own emphasis was on the robustness of the advances — that is, their status as permanent contributions to knowledge — rather than their utility in a wide variety of research and practical settings.

121. Quine 1953. A good source for clarification of the underdetermination doctrine is Roth 1987.

part of the evidence base used in the theory's formulation (especially when no rival theory can do likewise).[122] The underdetermination thesis can be read as reducing this line of thought to mere superstition, since whether or not a theory includes a particular phenomenon in its evidence base will depend on the exact formulation chosen, which in turn will depend on who formulates the theory, when, and where. Bluntly put, a potentially good theory suffering from a lack of sufficiently clever defenders at the moment of decision is doomed by the historicist criteria developed by Whewell's progeny.

Of course, I do not deny the many occasions when scientists have had to make largely irreversible decisions from among two or more theories. Unsurprisingly, then, counterfactual speculations concerning how some other theory could have arrived at the same discoveries are often relegated to the realm of the "Monday morning quarterback." But to keep the ledger balanced, I would not accord these "irreversible" decisions too much metaphysical weight, since their exact moment is determined, not by the dictates of the scientific method, but by factors external to the scientific enterprise that forced scientists to place a greater value on solidarity than diversity, and hence to mobilize around a common research trajectory. Of course, after the fact and once an Orwellian historiography takes over, it is easy to make it appear as though the moment of decision were itself determined by the very methodology that was then used to justify the direction taken. In effect, two conceptions of time are conflated, which I call *mythos* and *kairos*: that is, how a decision is shown to follow from what preceded it versus why the decision was taken at the time it was, and not earlier or later. I shall reflect on the implications of this distinction for the historiography of science in the final section of this chapter.

With the Orwellian moment, we enter specifically Kuhnian terrain. Perhaps the most striking feature of the interpretive trail that has followed the underdetermination thesis is its pronounced conservative bias, to be sure, a bias Quine himself encouraged.[123] At first glance, this is strange, since

---

122. Laudan 1981, 163–80. The most sophisticated defense of this line of thinking in recent years has been Leplin 1997. Leplin has the courage of his convictions to grant that if a later theory can predict everything an earlier one could (and more), then the earlier theory loses its unique epistemic warrant. My major disagreement with Leplin is that he presumes that the superseding theory must be a proper successor of the earlier one rather than a contemporary that is recast so as that it can predict all that its competitor can, but in a context that gives its partisans certain practical advantages, which thereby erases any intellectual property rights acquired by the earlier theory without assuming them for oneself. The significance of this point will become apparent in chapter 8, section 5, where I recast the distinction between the contexts of discovery and justification.

123. It is worth mentioning, given their reputation for antiphilosophical radicalism, that even sociologists of science appeal to the underdetermination thesis in a conservative manner.

once the conventionality of theory choice is revealed, one might have supposed that the door was opened to pursuing empirically equivalent theories that promise to push inquiry in bold new directions. Nevertheless, according to Quine and his Harvard colleague Nelson Goodman, if no foolproof method is available to determine which of two empirically equivalent theories should be pursued, then one should stick to the theory that has worked so far, rather than risking it on the theory that has not yet had a chance to work. This is what Goodman called "entrenchment" and Quine "conservatism."[124] It involves two presuppositions.

The first presupposition is that only one theory can be pursued at any given time, which given the empirical equivalence of the theories, suggests that the need for choice arises in a period of material scarcity or collective threat to the scientific community, which then demands a univocal response by its members. Otherwise, relieved of these two external demands, would one not simply tolerate inquiry proceeding in many directions at once?[125] Yet this presupposition accords with one of Kuhn's grounding intuitions, namely that there is only one paradigm in a science at a given time and that the dynamics of scientific change is largely motivated by the need to maintain consensus under all but the most extreme circumstances. As will become apparent in the pages that follow, the intuition can be traced to the concerns that governments have expressed about the costs and benefits of public support for science in the twentieth century. In short, whenever it appeared that scientists could not put their own house in order, there were plenty of offers on hand to do it for them.

The second presupposition of the conservative reading of underdetermination is that there is much to lose if the better theory is not chosen at

---

A contributing factor is that the thesis implies that locally occurrent social factors enter scientific deliberations only after it has been shown that general methodological rules cannot determine the decision one should make. In other words, acceptance of the thesis presumes that sociology is relegated to what statistical methodologists call "explaining the variance" around a central tendency, which is not so far from the role philosophers have traditionally assigned to sociologists under the rubric of the "arationality assumption." For further discussion, see Fuller 1990. An example of a sociologist whose endorsement of the underdetermination thesis leads him to conclusions rather like those of the conservative Michael Polanyi is Harry Collins, who is discussed below in chapter 2, note 96.

124. Goodman 1954, esp. chap. 3, sec. 4; Quine 1960, 20. A thorough discussion of their joint conservatism is provided by one of their Harvard colleagues: see Scheffler 1963, pt. 3, esp. secs. 1–2, 7–10.

Some commentators have made much of the fact that Goodman wrote specifically of entrenched predicates (i.e., terms referring to properties of entities specified in scientific theories), whereas Quine was concerned with conservatism as a feature of the decisions rendered between entire theories. Those eager to follow these niceties should peruse Stalker 1994.

125. This turns out to be the stance of Paul Feyerabend, about whom more in chapter 6, sections 5–7.

the moment a decision is taken. Such a risk-averse sensibility makes sense only if irreversible damage is thought to follow from incorrectly informed scientific decision making. In contrast, where decisions are reversible, risk taking would be encouraged. However, because decisions concerning the future of science involve more than simply pursuing one of a number of theoretically inspired lines of inquiry—the decisions also incur substantial material consequences—then an exaggerated sense of caution is to be expected.[126]

In short, even though Quine, Goodman, and Kuhn never directly addressed either the economics or the politics of scientific research, it would be difficult to motivate their jointly held presuppositions were it not further assumed that science exists in a political economy with the following three features:

1. that the evaluation of science is largely centralized in one setting, such as a government funding agency;
2. that science is regarded as an activity that substantially impacts not only on itself but also on its environment (specifically, the public who must support and endure it);
3. that science plays a sufficiently important role in legitimating the social order that its trajectory should be altered only if an improvement is clearly foreseen.

To be sure, these three conditions capture the state of science in the most economically developed countries in the second half of the twentieth century. Moreover, they amount to a big shift from the political economy of science in Whewell's day. However, this shift is easily obscured because of the tendency—much more in Quine and Goodman than in Kuhn—to accept sheer familiarity as a prima facie mark of validity, as if in the long term, the truth is bound to make the most lasting impression on inquiry. Yet, as Karl Popper so clearly understood, if the quest for knowledge simply aimed to maximize the survival value of beliefs, then science would have

126. Here I should mention a rhetorical strategy used to bolster this risk-averse sensibility, namely a promiscuous appeal to the royal "we" as the subject of scientific beliefs, as in "We believe in relativity theory today in the way that nineteenth-century physicists believed in Newtonian mechanics." Richard Rorty and Hilary Putnam have been perhaps most susceptible to this rhetoric as a stylistic tic. Presupposed is that, at least in scientific matters, there is a universal community of inquirers (to which the reader is presumed to belong or, more to the point, *ought* to belong) whose common sense of "track record" would enable them to agree on which theories are entrenched at any given time. However, given that a knowledge claim typically acquires "universal" status in science only after having passed muster in a handful of Euro-American test sites, such universality is at best "notional." Indeed, someone with an especially suspicious turn of mind might argue that the Western preoccupation with enforcing ceteris paribus clauses in the validation of scientific knowledge claims is a ploy to prevent intercultural differences from influencing the context of testing. See Fuller 1993b, 108–10.

never developed in the first place, since the aim of survival would be best met by stressing applications that showcase a belief's strengths, while avoiding any test of the belief's limits.[127] In cognitive-psychological terms, Quine and Goodman fostered a "confirmation bias" in inquiry that would serve in the long term to inhibit the growth of knowledge.[128]

Although Kuhn denied the key Popperian premise that scientists proactively test the limits of their beliefs, at least he defined the puzzle-solving domain of normal science as pushing the frontiers of the paradigm as far as it will go and hence occasionally, albeit unintentionally, finding itself in Popper's desired cognitive state. In effect, Kuhn condensed Goodman and Quine's temporally extended sense of epistemic entrenchment into a spatially extended community united under one paradigm of inquiry at a given point in time: i.e., instead of believing X because one's predecessors did, X is believed because one's neighbors do.[129]

While the wedge that Whewell drove between the contexts of discovery and justification was designed to ensure that inventors could not explain their innovations without the mediation of trained scientists, Kuhn's attempt to blur the two contexts reflected his interest in showing that discoveries do not occur in a sociological vacuum, but rather against the background of a paradigm within which they could be, at least in principle, justified.[130] Nowadays Kuhn's perspective has been normalized in the sociology of science by claims that the contexts of discovery and justification are "co-produced," in that a scientific discovery is not recognized as such unless it can be justified as issuing from the received history of the relevant science. The standard case in point is Gregor Mendel (1822–84), the Moravian monk whose pea-breeding experiments were neglected for forty years, after which Mendel was retrospectively canonized as the "father of modern genetics."[131] While coproduction characterizes Kuhn's treatment of discovery and justification in his early papers on instances of so-called simultaneous scientific discoveries, it does not capture the main thrust of his discussion in *Structure*:

> [I]ndividual variability in the application of shared values may serve functions essential to science. The points at which values must be applied are invariably also those at which risks must be taken. Most anomalies are

---

127. See esp. Popper 1972.
128. On the significance of confirmation bias for the philosophy of science, see Fuller 1993a, 108 ff.
129. One of the more interesting treatments of Kuhn's views on scientific theory change in relation to Goodman's and Quine's is Hacking 1993, esp. 305.
130. See Kuhn 1970b, 8–9; Kuhn 1977a, 325–30. This perspective has been most thoroughly developed in Nickles 1980a.
131. See esp. Brannigan 1981.

resolved by normal means; most proposals for new theories do prove to be wrong. If all members of a community responded to each anomaly as a source of crisis or embraced each new theory advanced by a colleague, science would cease. If, on the other hand, no one reacted to anomalies or to brand-new theories in high-risk ways, there would be few or no revolutions. In matters like these the resort to values rather than to shared rules governing individual choice may be the community's way of distributing risk and assuring the long-term success of its enterprise.[132]

Here Kuhn seems to have taken the metaphor of natural selection from evolutionary biology with deadly seriousness. Individuals function as generators of novelty for the scientific community, as modeled on a population-based conception of species. Just as an individual organism that successfully reproduces itself increases the overall fitness of the species, if a novelty is accepted as a genuine discovery, then it strengthens the community and, in that sense, the discovery is justified. But if the community fails to incorporate the novelty, then only the individual who proposed the novelty suffers.[133] For Kuhn, then, discovery is a process that normally occurs within the scientific community for purposes of extending and reinforcing the community's boundaries, even at the expense of its members. The significance of this general line of thinking should not be underestimated. It would be difficult to motivate recent debates over scientific realism without presupposing Kuhn's paradigm-based redefinition of scientific discovery. Next chapter, we shall see Max Planck's role in establishing what Kuhn would regard as the sociological precondition for such realism.[134] Moreover, in chapter 3, sections 5–6, we shall revisit this mind-set as Conant's "political realism," only there the citizenry serve as guinea pigs and pawns in the strategic designs of state policy makers.

A view that differs significantly from Kuhn's is that discovery is a process designed to *challenge* the constitution of the scientific community, which

---

132. Kuhn 1970b, 186. Kuhn's coproductionist view of discovery and justification is perhaps best seen in Kuhn 1977a.

133. I have characterized the negative case—the failure of a novelty to be accepted as a genuine discovery—as *disutilitarianism*, the thesis that everyone may benefit from one person's misfortune. See Fuller 1993b, 333–34.

134. "Scientific realism" means here the treatment of current scientific theories as representations of ultimate reality. The two main attitudes toward scientific realism among contemporary philosophers of science can be seen as alternative responses to Kuhn's understanding of scientific discovery as a paradigm's boundary maintenance process. On the one hand, philosophical resistance to realism stems from a fear that an indefinite extension of current theories will not necessarily lead to truth, but simply a self-reinforcing dead end. On the other hand, philosophical enthusiasm for realism rests on the intuition that the degree of success enjoyed by current theories is a sign of their future success in new domains. Bas van Fraassen and Dudley Shapere reproduce these respective positions in Suppe 1977, 598–99.

implies that the boundary between science and nonscience is an ongoing construction. This view, associated with Popper's rather counterintuitive test-oriented conception of justification, will be discussed in more detail in chapter 6. For our purposes here, it is worth noting that Popper used Whewell's means for counter-Whewellian ends. As we saw in section 6 above, Whewell appealed to a sharp discovery/justification distinction in order to restrict entry to the scientific community to those who presented claims that were justifiable in terms of academically sanctioned forms of knowledge. Writing more than a century after Whewell and nearly a century after most of the natural sciences had become academic departments, Kuhn had no need to police the boundaries of science so overtly. Or, rather, the need for boundary maintenance occurred elsewhere, namely, in terms of who is eligible to steer the course of science—in which respect Conant updated Whewell's role for the twentieth century.

Consequently, for Kuhn, at this stage in the history of science, to divide the contexts of discovery and justification so sharply would be to draw undue attention to the artificial character of science's boundaries, which could serve as a diversion from the taken-for-granted attitude that scientists need to have toward their paradigm in order to carry on with the focused puzzle solving of normal science.[135] However, for Popper, such attention was merited precisely because of the institutionally entrenched character of science in our time. For him, a stable scientific community is valuable only insofar as it enables the expression of individual differences of opinion that then opens the community to a full range of possible futures, much like the ideal of civic republicanism discussed in section 2 above. Indeed, this is the core idea behind Popper's "open society."[136]

---

135. Another potential problem that arises from revealing the artificial character of the boundaries of science is that it could encourage groups to simulate the trappings of science in order to accrue some of science's social authority. This has been the unwitting legacy of positivist and Popperian attempts at prescribing "demarcation criteria," namely to reduce science to a rhetorical formula that virtually anyone can imitate. Kuhn avoids at least this problem because only other scientists can decide whether someone is doing science. It is not sufficient simply to follow a recipe. See Fuller 1998f. Of course, as we shall see esp. in chapter 5, Kuhn's model runs into its own problems of encouraging unwanted fellow travelers, namely by setting the institutional threshold for science sufficiently low that social scientists and humanists can meet it.

136. The priority given to the individual vis-à-vis society accorded in Popper's conception of the open society was indebted to Bergson 1935, esp. 230–31. See Popper 1945, 1:202–3. It is worth noting that in Bergson's original formulation, the open society is marked by a periodically realized impulse to return religion to humanity in reaction to its institutional captivity. Bergson explained this tendency not in terms of some preordained goal of openness, but a revolutionary individual who intuits the need for reformation of the religious community. This point will acquire special relevance in chapter 8, where I suggest that *movement* replace *paradigm* as the default social vehicle for organized inquiry.

TABLE 1. *Whewell's Bifurcated Legacy*

How to deal with the discovery-justification distinction?

	NORMATIVE REASON	EMPIRICAL REASON
Use Whewellian means to un-Whewellian ends: enforce the distinction (Popper)	The distinction shows that science is open to things outside itself.	It shows that science always needs to construct its boundaries, so that whether a discovery constitutes an error or a revolution is left open.
Use un-Whewellian means to Whewellian ends: eliminate the distinction (Kuhn)	The distinction denies the self-organizing character of scientific activity.	It ignores that potential discoverers already orient themselves to what it takes to justify knowledge claims scientifically.

Kuhn's and Popper's alternative appropriations of Whewell's distinction in the contexts of justification and discovery in science are represented in table 1.

8. The Platonic Legacy: From Double Truth to the Dual Rhetorical Structure of History

In the end, what is so potentially subversive about scientific traditions that explains why scientists and their philosophical well-wishers have acted as cultishly as they have? The general answer is a Platonic concern for the peculiar position of elites in a democratic culture. Elites must publicly legitimize their activities in terms that are, at best, orthogonal, and perhaps even contradictory, to their self-appointed missions. For example, democracies lavishly fund scientists because science is persuasively portrayed as the pinnacle of collective human achievement. But scientists themselves are called to their vocation by a commitment to fathoming mysteries that, by their nature, must remain remote from the concerns of ordinary mortals. The continuity of science is maintained, on the one hand, by the respect (if not awe) that laypeople have for the technical wizardry of science, and, on the other hand, by the care with which scientists monitor each other's actions so that the relevant traditions are transmitted without corruption. The threat of corruption is ever present because scientists are always tempted to believe their public image, succumb to hubris, and invariably suffer a tragic fate. Thus, in World War I, the German scientific community erred in thinking that intellectual superiority conferred some natural advantage in combat. The German public all too willingly indulged these delusions and then all too willingly blamed the scientific community once Germany lost the war. We have seen the humanist response to this tragedy in the form of iconography. In the next chapter, we turn to the German debate in the years leading up to World War I over whether science has intrinsic or merely instrumental ends that took place.

Platonism's commitment to a double-truth doctrine is inscribed in a dual rhetoric of time that we have already seen in the practice of iconography and shall subsequently see (especially in chapters 4 and 7) in the attempts to construct an "internal" history of science. One reason Plato is not normally seen as a theorist of the historical method is that he would have regarded most historians as taking their own mythical constructions too much at face value—that is, unreflexively—as if they had come to forget their origins in the pacification of potentially unruly audiences. Thus, historians have tended to stress narrative time to the near exclusion of real time—what a classical rhetorician might call a very high *mythos*-to-*kairos* ratio. In other words, a higher premium is placed on the internal logic of events recounted in the historical narrative *(mythos)* than on why the sequence transpired at the pace and over the length of real time that it did *(kairos)*.[137]

Closely related to the failure to distinguish clearly mythos from kairos is the fundamental ambiguity about how "the social" should be conceptualized in historical explanation. Much of the so-called social history of science has been really a history of scientific traditions, very close in both intent and execution to iconography. Thus, episodes and artifacts in the history of a given science are explained as contextualized reproductions and transformations of prior episodes and artifacts in the same history: that is, new twists or combinations to time-honored themes.[138] Opposed to this mythos-based approach are social histories of science that stress the exigent character of what scientists do, namely, their need to construct opportune responses to situations that are largely *not* of their own creation. The tradition, such as it is, no longer appears as a continuous medium of cultural

---

137. The locus classicus for the mythic logic of narratives is Frye 1957. Its application to historiography (via the concept of "emplotment") is due to White 1973. It is worth noting that *mythos* has crept into self-styled "contextualist" histories, ones laden with "thick descriptions," which tend to reduce temporality to overlapping cultural tendencies in which even the actual order of events no longer seems important. See Bender and Wellbery 1991. Perhaps the most fecund reification of the mythos/kairos distinction is Fernand Braudel's conception of history as a set of overlapping durations, ranging from the kairotic *evenement* (the clearly dated events that punctuate political history) to the mythic *longue durée* (long-term environmental factors that characterize ecological history, to which economic history then appears as a series of local adaptations). Generally speaking, a historian can be placed on the mythos-kairos continuum by the extent to which counterfactual historicizing (thought experiments in the "what if" mode) is taken as a legitimate method. Here the kairotic historian will prove most enthusiastic.

138. Thus, Peter Galison has presented the "experimental tradition" in high-energy physics as "autonomous" from the tradition of theorists and others. See Galison 1987, esp. 6–12.

transmission, but rather a stockpile of contestable resources. Here science is described as the art of using intellectual means to solve social problems.[139] The scientists always need to repair the integrity of their practices by reframing outside pressures as issues that can be legitimately treated within their own domain.

Among the few histories of science that have been acutely sensitive to kairos is Paul Forman's account of the adoption of the indeterminacy interpretation of quantum phenomena by physicists in Weimar Germany, which is perhaps the single most influential article on twentieth-century history of science.[140] Nevertheless, Forman's mode of inquiry remains alien to the way historians and philosophers of science have traditionally framed the problem of theory choice. Of course, social historians of science have demonstrated the artificiality of the very idea of a "problem of theory choice," insofar as it assumes two or more well-defined groups of scientists lined up behind fully articulated research programs. History is rarely that neat and clean. But setting that issue aside, one can certainly detect many long-standing theoretical disputes in the history of science. Yet it is often unclear what is at stake, aside from a point about the right way to talk about a certain class of phenomena. Thus, these disputes can simmer for long periods, causing little disruption to the business of doing experiments and collecting data. One might be tempted to claim that the relative autonomy of theoretical disputes demonstrates their relative *lack* of influence on the course of scientific research.[141]

Yet eventually a time comes when the dispute reaches a climax—the paradigm is in crisis, a revolution ensues, and closure is finally reached. Traditionally, historians and philosophers have presumed that the time for

---

139. Here I allude to the explicit thesis of Shapin and Schaffer 1985.

140. Forman 1971. Perhaps unsurprisingly, Forman's thesis tends to be misinterpreted. He is often taken to have argued simply that the doctrine of quantum indeterminacy was a "reflection" of the irrationalist tendencies in Weimar culture. In response, one could easily argue that ideas about the indeterminacy of the physical universe predate Weimar culture, thereby establishing the relative detachment of those ideas from any specific social setting. However, Forman actually acknowledged this point at the outset of his inquiry, since the indeterminateness of the thing-in-itself figures prominently in most nineteenth- and twentieth-century followers of Kant, not least Ernst Mach's positivism, which emerged over the half-century prior to the adoption of the indeterminacy interpretation of quantum mechanics. But he was less concerned with underlying causes than with occasioned action: Granted that the crucial ideas had been debated inconclusively for at least fifty years, why do physicists suddenly close ranks behind the indeterminacy interpretation in 1927? Forman purported to show that no new arguments or experimental findings justified the quick closure of debate. Instead, one needs to look at the growing cultural hostility to determinism and materialism (both associated with the German loss in World War I), which was about to threaten the funding of physics research. Even if the physicists were not quite as calculating as Forman makes out, he nevertheless deserves credit for asking a rhetorically astute historical question. See Fuller 1988, 233–50.

141. The idea was the basis of the influential Hacking 1983.

closure is always appropriate, and that the only interesting question to ask is what the decisive arguments and evidence that favored one side over the other(s) were. By contrast, a historian like Forman begins by asking why the "moment of decision" occurs at this point in time and not some other? After all, had the moment been a bit earlier or a bit later, the relative standing of the parties to the dispute — not to mention the actual composition of the parties — might have been quite different. Left in the realm of dialectic, for any two competing theories, every new argument or piece of evidence can be met, if only by slightly adjusting the background conditions of the theory challenged by the argument or evidence in question. (This is the essence of the so-called Duhem-Quine thesis that is related to the underdetermination of theory by data raised in the previous section.) Given these slight changes in context, the weaker argument may have been, indeed, the stronger. In short, those who control *when* a decision is made control what decision is made.[142]

142. Social-psychological evidence for this claim is provided in Levine 1989.

11
The Last Time Scientists Struggled for the Soul of Science

1. Imagine a Time when Philosophy Mattered to Scientists . . .

In the introduction, I claimed that one way in which Kuhn's general picture of scientific change is taken for granted by recent philosophy of science is that philosophical defenses of progress never seem to make reference to any specific direction science should be taking. This becomes evident whenever the philosophically innocent have difficulty seeing the difference between the two main positions in contemporary philosophy of science: *realism* and *instrumentalism*. It would seem natural to assume that science's increasing ability to predict and control phenomena ("instrumentalism") implies that it has a more comprehensive understanding of nature ("realism"). If so, then does not instrumentalism simply presuppose realism—and maybe even vice versa?

Prior to the exchanges analyzed in this chapter, this conflation was quite common even among sophisticated philosophers and scientists. However, an awareness that one need not imply the other arose only once a scientific research program had been allowed to develop in a sufficiently autonomous fashion for a sufficiently long time. At that point, just under a hundred years ago, people began to discern that the search for truth ("realism") and the search for utility ("instrumentalism") might involve trade-offs and hence could not be jointly maximized. A philosophical dispute about the nature of science thus became a genuine concern for science policy. Thus, the line was drawn between the realist Max Planck (1858–1947) and the instrumentalist Ernst Mach (1838–1916). The outcome of their dispute gradually came to be taken for granted by philosophers of science, culminating in the Orwellian sensibility that Kuhn highlights as so integral to science's sense of its own history.

Nowadays the realism-instrumentalism debate (RID) is often seen as the core debate in the philosophy of science, yet increasing numbers of philosophers have followed Arthur Fine's lead in questioning the point of this dispute.[1] Fine's largely unchallenged rendition of RID makes it easy to see why. RID appears doomed to stalemate. According to Fine, realists and instrumentalists are both trying to account for the string of progressive episodes in the history of science. The two sides are said to agree on what those episodes are and that they constitute progress. However, it seems that every realist story of one such episode can be matched by an instrumentalist one—and vice versa. As Fine sees it, any such story is merely an attempt to extract surplus philosophical value from the historical labors of scientists. It only adds a misleading air of inevitability to their original efforts. These labors can be more simply and adequately captured by examining the norms implicitly governing the scientists' actual practices, which, if one insists on a philosophical label, may be called the natural ontological attitude (NOA). Admittedly, the content of NOA varies over time and place, but then the philosophical supposition that the history of science would deliver an overarching logic of inquiry was false from the start.

Philosophers who embrace the recent turn to "historicism," "contextualism," and "practice" in science studies take their lead from Fine in critiquing this Whiggish exploitation of the historical record. But ironically, Fine's "surplus value" account of RID appears impressive just as long as we suppose that the context surrounding the debate is the same now as it was when versions of the realist and instrumentalist positions were first articulated a little over a century ago. One would expect, then, that the debate was originally conducted by professional philosophers in technical journals. But, of course, this was not the case. On the contrary, the prototypes of these positions—often better illustrated and more engagingly expressed—were to be found in the popular writings and lectures of the leading scientists of the day. Professional philosophers, insofar as they followed these matters at all, did so as kibitzers, not as major players.[2] Not surprisingly, the philosophers were able to detect nuances in position that seemed to have eluded the participating scientists, mainly because the philosophers had abstracted RID from the pragmatic contexts in which advancing a common agenda was more important than sheer intellectual differentiation. Once the logical positivists had succeeded in converting RID into a primarily "metaphysical" disagreement, the scientists started to be portrayed as careless reasoners swayed by the heat of polemics, the antidote for which was to be found in a philosophically inspired linguistic therapy.[3]

1. Fine 1984. Perhaps the philosopher who has most explicitly taken up Fine's lead in this regard is Joseph Rouse. See Rouse 1987, 1996.
2. A good case in point is Cassirer 1923.
3. See esp. "Pseudoproblems in Philosophy" in Carnap 1967, 305–46.

This transition was initiated by Moritz Schlick (1882–1936), a philosopher whose unique position owed much to his having done a Ph.D. in physics under Planck, while proceeding to occupy Mach's chair in the history of the inductive sciences at the University of Vienna. In that capacity he convened the Vienna Circle, from which sprang logical positivism.

Historians have been much quicker than philosophers to pick up on why RID has mattered to scientists.[4] When, say, Boltzmann and Ostwald, or Planck and Mach, argued about the existence of atoms, they were not merely trying to second-guess what empirical research would eventually show; rather, they were trying to influence the direction that such research should take and the way it should be evaluated. Like judges, they would appeal to history to show precedent for alternative science policies. "Realism" and "instrumentalism" were thus comparable to schools of legal interpretation. They captured alternative strategies for extending the methods, findings, and worldview of the physical sciences into new domains. These new domains included not only the research practices of the newer academic disciplines, such as the biological and social sciences, but also, as we shall see here, educational practices at the university and subuniversity level. In chapter 6, sections 5–7, we shall return to this treatment of philosophies of science as akin to alternative legal traditions, especially with regard to "the roads not taken" since the triumph of Kuhn's vision.

Given a shrinking public research budget combined with calls for research accountability, the rarity with which scientists nowadays polemicize against *each other* about the ends of science is truly striking.[5] The reason lies less in any putative convergence of ends among the sciences than a common appreciation for Ben Franklin's counsel, "Let us hang together, lest we shall hang separately." Such have been the diplomatic consequences of science relying on peer consensus as the united front against the government watchdogs and a barbaric public. The closest one recently got to the original public debate over realism and instrumentalism was when Nobel laureates such as Steven Weinberg defended continued support of the superconducting supercollider on realist grounds, for

---

4. See esp. Laudan 1993.
5. The discreet silence of scientists on this matter is all the more striking because scientists tend to be disciplinary imperialists in their popular writings. Physicists are undoubtedly the worst offenders, as they have confidently claimed that the earth's ecological problems will be solved by extraterrestrial settlements (Freeman Dyson in *Disturbing the Universe*), while our health is placed in the hands of chemists and robotics experts (J. D. Bernal, in his youthful work, *The Word, the Flesh, and the Devil*). Not much of a role for biology, as conventionally understood, so it would seem. For a trenchant critique of such hubris, see Midgley 1992, esp. 165–82. However, at both ends of this century, biologists have outstripped physicists in their claims that many of these problems can be addressed by selective breeding (for the first decade of this century) and genetic engineering (for the last decade).

contributing to explanatory unity in physics, which would in turn illuminate the foundational problems of the rest of the sciences.[6] Philosophers are prone to dismiss such an argument as unsophisticated and self-serving, and hence feel no need to contest it in the public forum. Yet this philosophical disdain unwittingly serves to lower the standards of public science policy debate, as science popularizations come to be seen in extreme terms, as either authoritative accounts of how science works or dubious cases of special pleading.

However, the tact that marks cross-disciplinary relations today is a distinct product of the twentieth century. In the first decade of the twentieth century, the engineer and amateur historian John Merz was able to present nineteenth-century science in terms of several contesting worldviews that at first laid explanatory claim to all phenomena, generating several decades of heated debate, but resulting in a gradual retreat behind relatively clear intellectual and institutional positions, typically academic disciplines that varied somewhat from nation to nation.[7] An apt metaphor for this development is the nebular hypothesis that Kant and Laplace offered for the origins of the universe, whereby an original swirl of burning gases eventually stabilizes into galaxies and solar systems. However, one of the last bits of the swirl to settle down was the foundational debate between physics and chemistry. More than merely a realist philosophy defeating an instrumentalist one, Planck's triumph over Mach marked the physics community's success in imposing its vision of science upon all the other disciplines, especially its most serious challenger (given its track record of service to the state and industry), chemistry. Despite Mach's own training in physics, the chemists were among his biggest supporters, for reasons that will become clear in what follows.

2. A Return to the Repressed: The Hidden Constructivist History of Chemistry

Chemistry is rarely accorded much philosophical respect these days because it is seen as "applied physics," physics being the ultimate science of matter. In their philosophical (including *naturphilosophische*) moods, chemists have claimed theirs to be a science of "secondary qualities," that

---

6. Weinberg 1992. On the ill-fated history of the supercollider, see "Preface 1995" to Kevles 1995. I recommend Kevles's masterwork for its sensitivity to the tension that informs this chapter, but applied to the American context: an increasingly esoteric and elitist scientific community that is, at the same time, increasingly beholden to the general public for its economic and political resources.

7. Merz 1965.

is, of things as perceived by the human senses. Chemistry's definitively subordinate status is often traced to Einstein's discovery of Brownian motion in 1905, generally regarded as the experimental resolution of the debate over the existence of atoms, the scientific pretext of the Mach-Planck debate. Nevertheless, taken in its entirety, the history of chemistry—the science most closely associated with the properties of ordinary materials—reveals rather different metaphysical preoccupations and political sensibilities from those of the history of physics.

Consider the idea of matter itself. For physicists, the atomic character of matter implied that nature provided an ultimate barrier to human construction with, so to speak, "a mind of its own," thereby engendering a fundamentally contemplative attitude toward nature, even when laboratory experiments had to be performed. In contrast, for chemists, matter was raw material, an Aristotelian potential *(dynamos)* that can be given any number of forms, according to human design. Whatever limits there are to matter's constructibility would become apparent only as construction proceeds, as the pursuit of one project precludes the subsequent use of that matter for other projects. Whereas the physicists treated laboratory experiments as occasions for manifesting nature's hidden structure (preferably, as anticipated by a theory), chemists regarded experiments as opportunities for realizing nature's potential in some humanly desirable fashion. For Pierre Duhem (1861–1916), Wilhelm Ostwald (1853–1932), and other opponents of the dominant physics worldview who sided with Mach over Planck, this constructivist conception of matter often went by the name of *energy*.[8]

---

8. In his inaugural lecture as professor of physical chemistry at the University of Berlin (1905), Ostwald's student Walther Nernst recast the history of both physics and chemistry as being of "constructive sciences," in terms that resonated with the pragmatist philosophy flourishing in the United States at the time. See Buchwald 1995. By the time of Duhem 1954 in 1914, we witness the extension of constructivism from the natural to the human realm, through the "underdetermination of theory by data," a thesis we attributed in chapter 1, section 7, to Quine, who in turn traced it back to Duhem.

Though Duhem was himself a practicing physicist (in thermo-, hydro-, and electrodynamics), his chair in Bordeaux ensured a marginality to the French physics establishment that enabled him to embrace a chemist's understanding of the history of physics. This understanding turns on attributing progress in physics to its increasing artificiality. Specifically, physicists have come to reproduce theoretically pure conditions in laboratory experiments to such an extent that the artificial settings have effectively replaced natural ones as the arbiters of physical reality—much as synthetic materials have increasingly replaced natural ones in the chemical research and industry. Therefore, only the ends of the physicist—especially her commitment to the theory under investigation—can determine the appropriate response to any data generated in such artificial settings. Nature as such offered no direct assistance, but merely reminded the physicist of the need for overall consistency in her resolution of theory and data. But clearly, some resolutions required more effort than others, and so the need to "conserve energy" provided an intuitive sense of direction to the physicists' inquiries.

As constructivism moved more completely into the human sciences, the artificiality of physical discourse and its experimental applications was generalized to all of social life. Thus,

A sign of the failure of this "constructivist" or "energeticist" perspective — at least in its effort to become the metaphysical foundation of the sciences — is that chemistry is today a philosophically underappreciated science, and the contours of its history are still not widely known, despite the enormous work that historians have done to articulate them. There are four major phases of that history, which reveals successive reconstructions of the concept of material resistance, but without presupposing that they constitute a search for the ultimate, "atomic," unit of matter.

The first phase, corresponding to the period of alchemy, consisted of (largely unsuccessful) efforts to transform one substance into another with rather different physical properties, typically by appealing to forces able to transcend the differences. The second phase, the chemical revolution in the late eighteenth and early nineteenth centuries, introduced techniques for obtaining "pure" samples of the substances defined by modern chemical theory as "elements." Both the purifying agents and the purified substances were soon seen to have commercial value. The third phase, marked by the industrial domination of chemistry from the mid-nineteenth to the early twentieth century, focused on synthesizing new materials by rearranging the molecules of naturally occurring substances. These "synthetics" were designed to heighten certain humanly relevant properties, such as resilience to weather changes (in the case of synthetic fabrics). The fourth phase, which covers the rest of this century, has involved the gradual replacement of naturally occurring substances by synthetic ones, but more

---

an ordinary language-in-use became the paradigm case of a socially constructed world, a raw material that enables and constrains its inhabitants in various ways. Thus, while it may be a strict grammatical fact that all languages can convey all thoughts, it is also true that for any given thought, some languages have traditions of speaking and writing that make the thought's conveyance relatively easy. Imagine, say, the difference between talking about automobiles in English and in Latin. No doubt, automobile talk can be translated between the two languages, but given the complexity and clumsiness of the Latin expressions, it is unlikely that the automobile would have ever been invented by anyone who thought exclusively in Latin. It could be done, of course, but it would be an uphill battle. In Quine's hands, the energeticist imperative to "conserve energy" was more specifically focused on logic, as one decided on the theory that best fit the data by considering which would cause the fewest changes in background beliefs. A introductory philosophy text that revolves around this notion is Quine and Ullian 1978.

A key moment in the transition of constructivist-energeticist thinking from the material to the linguistic realm came with the original German publication of Ludwig Wittgenstein's *Tractatus Logico-Philosophicus* in the last volume of Ostwald's journal, *Annals of Natural Philosophy* (1921). See Hakfoort 1992, 532. Given the tendency to see the influence of the natural sciences on the human sciences as almost entirely negative, it is worth elaborating on how some of the latter's most characteristically "antinaturalistic" doctrines were originally found in the natural sciences. One article that drives this point home especially well by examining the way in which debates between historicists and naturalists cut across both the natural and human sciences is Wise 1983. Any future philosophical-historical study into the connection between constructivism and energeticism will need to focus on the term common to the two movements: *work*. A good start is Rabinbach 1990.

with an eye to satisfying commercial imperatives (i.e., consumer demand and corporate profits, though not necessarily in that order) than to reproducing the physical properties of natural substances. Indeed, chemistry's role in the depletion of natural resources in our time has arguably put it in a position of eliminating nature in the course of replacing it.[9]

Not surprisingly, given chemistry's impressive trajectory as a science that successfully intervened and even replaced parts of what commonsensically passes for "natural reality," it was only in the twentieth century that physics came to be the indisputable foundation of the sciences, out of whose domain the objects of all the other sciences were said to be constructed. As we shall see in more detail below, Planck's arguments for the foundational status of physics rested—as such arguments do today—on the field's supposedly unique ability to define the limits of human comprehension and control. On the one hand, physics could encompass the intellect with its unified worldview; on the other, it could determine the final point at which matter resists the will.

However, scientists less ensconced in the physics establishment than Planck ridiculed these grandiose claims when they were first presented. Philosophers of science are familiar with Duhem's critique that physicists are concerned only with the "hows" but not the "whys" that characterize the great theological and metaphysical worldviews.[10] Less known, but more pertinent here, is the argument of Mach's fellow traveler, Ostwald, that the advent of synthetic materials has rendered the search for immutable atoms obsolete because any resistance in nature can be either overcome or circumvented through a chemist's intervention. Ostwald regarded the idea of a final "end" to inquiry as a vestige of earlier eras when people expected to run up against some insurmountable barrier in nature.[11] The Promethean image of science projected by chemists like Ostwald figured prominently in German rhetoric during World War I. Unfortunately, Germany's humiliating defeat led to a classic case of Platonic disenchantment in the nobility's betraying its birthright. In this case, the fallen nobles were the chemists who openly embraced German militarization in World War I. If ever there had been a modern case of tragic hubris, this is it. "The kaiser's chemists" clearly felt that Germany's scientific superiority necessitated a military victory. Once necessity evaporated into surrender, however, the German public reacted—per Plato's prediction—by going to the opposite

9. My potted account of the history of chemistry is distilled from Bernal 1971 and Knight 1994.
10. Duhem 1954, 320–35.
11. Ostwald 1910, 13–15.

extreme of demonizing the cultural achievement of the natural sciences during the Weimar Republic.[12]

3. Mach and Planck at the Crossroads

Mach and Planck offered bold and conflicting visions of the future of science at a time when the future seemed up for grabs. The natural sciences had been incorporated into the university system only in the previous generation, and academics were ambivalent at best about the industrial model of research and the credentialist model of education associated with those fields—though most acknowledged emerging tendencies in those directions. The lesson to be learned from the brief rise and precipitous fall of Mach's favorite discipline, chemistry, is that a science that does not postulate its own intrinsic, or "internalized," ends runs the risk of becoming captive to its context, much as an organism can become "overadapted" to its environment. Planck consistently maintained that the ends of science must be served before science can serve the ends of society, or neither ends will be achievable. In this way, scientific realism acquired sociological status.

However, it would be a half-truth, at best, to conclude simply that science maintains its integrity by successfully imposing its demands on society. After all, the other major sectors of society have not automatically complied with science's wishes. Thus, science has also had to prevent their demands from being superimposed on its own. Herein lies the Faustian bargain that Planck and his successors have struck that has enabled the insinuation of science in the regulation—and not merely the facilitation—of societal functions. As we shall see, perhaps the most vivid legacy of this bargain is the control that academics hold over the credentials that people need for holding jobs that have little or no specifically academic content. But before we catalogue the temptations of the "knowledge society" in which many think we live, let us return to the beginning . . .

Suppose, on the one hand, that we have a physicist who spends the bulk of his career as an elite functionary for the scientific establishment. He identifies "the ends of science" with the interests of professional scientists, yet he continually needs to demonstrate the relevance of arcane and expensive research to a public who still think of Galileo and Newton as their paradigm cases of scientists. Then, suppose, on the other hand, we have a physicist who spends a good part of his career as a parliamentary champion of democratic education. In his view, "the ends of science" are nothing

---

12. See Forman 1971; also Herf 1984.

more than a repertoire of techniques for ameliorating the human condition, one of whose burdens may turn out to be the scientific community itself. The two exemplary science policy spokespersons are Max Planck and Ernst Mach, respectively.

The Planck-Mach exchange arose at a specific point in history when the demands of research and education had to be met simultaneously. Between the end of the Franco-Prussian War (1871) and the start of World War I (1914), the scientific community embarked on a period of unprecedented research growth and specialization, just as the imperial powers of Europe were expanding and modernizing citizen education, so that the populace could be more easily mobilized for economic and military purposes, as the need arose.[13]

The original theorist of this mobilization was the political economist Friedrich List (1789–1846), who, already in the second quarter of the nineteenth century, had instructed the German principalities on the example set by the United States, a confederation of former colonies that were swiftly constituted into a single-minded republic. (List left America just before slavery began to tear at the union's fabric.) List argued that free trade benefited only the country that entered the market first (an unsubtle reference to Great Britain), as it would then be in a position to flood its less developed trade partners with relatively cheap goods, which would, in turn, inhibit the development of native industries. The appropriate protectionist policy for a united Germany was to extend Adam Smith's concept of the division of labor from the factory to the nation as a whole.[14] This "whole"

---

13. Albisetti 1983. In light of chapter 1's discussion of Aby Warburg's significance for extending Plato's double-truth doctrine into the humanities and ultimately Kuhn's historiography, it is worth noting that the Warburgs set up the subscriptions for France's payment of war reparations to Germany after the Franco-Prussian War. As the Renaissance historian Jakob Burckhardt observed at the time, the House of Warburg probably had devised the most "civilized" means of profiting from war—certainly much more civilized than the "warmongering" tendencies of bankers such as the American John Pierpont Morgan, who would later facilitate the sale of munitions to countries opposed to his own nation's interests. See Rosenbaum and Sherman 1979, esp. 70–72.

14. List 1904. On List's significance for past and present science policy, see Freeman 1992, 62–68. List's long-term impact has been felt in two arenas. One (though still often unacknowledged) is in the theory of Third World economic dependency and development. See Senghaas 1991. The other is in that staple of elementary sociology, Emile Durkheim's claim that modern societies benefit from an "organic" division of labor. Long before he held the first chair in sociology, Durkheim had been commissioned by the French government to discover how a newly unified Germany could so quickly and decisively defeat his own nation in the Franco-Prussian War. List's image of the nation-state as a self-sustaining organism seemed to play an important motivational role, one that led Durkheim to justify the introduction of sociology courses as a means of consolidating France's national identity. See Jones 1994. The fact that the Germans encouraged a wider segment of their population to seek scientific and technical training was not lost on Durkheim, who argued that the French should forsake the British model of schooling the nation's future leaders in the literary classics. See Paul 1985, 40.

would be defined by the myth of unified German *Kultur*, as coined by the Romantic philosopher of history Johann Gottfried von Herder (1744–1803). Herder invoked common "blood and soil" to foil economically inspired arguments—often traveling under the guise of "cosmopolitanism"—that would reduce differences in cultural values to negotiable degrees of utility. Over the course of the century, such cultural differences were actively reinforced by nationalist policies, mainly by homogenizing regional dialects into a "mother tongue" and instituting the study of "national history" as the cornerstone of compulsory citizen education.[15] Such education, in List's words, enabled a country to develop its "mental capital" (we would now say "human capital") alongside "capital" in its usual sense of factories and raw materials.

After the Franco-Prussian War, and especially after the imperialistic "scramble for Africa" in the 1880s, the incommensurability of national policies became the major premiss in any "realistic" assessment of international relations. The combination of expanding military-industrial might and irresoluble differences of interests made the indefinite postponement of an "inevitable" world war the primary item on the diplomatic agenda. It was in this context that the mastermind of Germany's victory in the Franco-Prussian War, Baron Helmut von Moltke (1800–91), concluded that a healthy nation was always ready for "total war," that is, not merely strategic engagement with a definite goal in sight (the classical aim of warfare), but rather the elimination of a direct threat to the nation's very survival. In other words, the nation needed to be in a *permanent state of emergency*, a concept that has turned out to have remarkable transnational appeal and was especially well ingrained in the Cold War mentality. Moltke was the first to realize that once state-of-the-art technology becomes decisive to winning wars, it will always be in the national interest to engage in peacetime research and development for the purpose of producing armaments that will deter prospective foes. For were a nation to wait for a high-tech attack, it would probably be too late to manufacture an effective counterattack.[16]

15. For a more detailed philosophical reflection on the European construction of "culture" as a projection of the nation-state, see Fuller 1999e. Given the constructivist impulse imputed here to the German identity builders, the reader might conclude that I have anachronistically read a postmodern sensibility back into the nineteenth century. If so, then Arnold Toynbee stands similarly accused. In one of the earliest uses of the term "postmodern" (1956), Toynbee claimed that the Franco-Prussian War marked the beginning of the postmodern era, as Germany became the first country to constitute itself as a nation-state by selectively extending Euro-American modernization policies. This practice of what Jean Baudrillard would eventually call "simulation" and "hyperrealization" was soon adopted by Japan, once Germany proved its military superiority over one of the imitated nations. See Anderson 1998, 5–6.

16. Drucker 1993, 127. Drucker claims that the deterrence mentality first affected the German navy, the branch of the armed forces that initiated the arms race.

This vision took hold with a vengeance. Eventually politicians and ideologues of all persuasions portrayed themselves as living in "mobilized" societies propelled by "preparedness" economies, even when it was difficult to identify the exact source of the threat.[17] For many nations, the Great Depression of the 1930s crystallized the idea of economic problems as matters of "mobilization." Not surprisingly, the leading architect of this mentality, John Maynard Keynes (1883–1946), was himself instrumental in managing the British treasury during World War I. Once America's industrial base had been consolidated through government intervention during the New Deal (which already mobilized large numbers of people in the construction of roads, utilities, and public buildings), the infrastructure was set in place for rendering the United States an effective military force in the European and Pacific theaters during World War II, which then carried over to the sense of preparedness that motivated research and educational policy during the Cold War, leading to today's "postindustrial" or "knowledge" society.[18]

However, the intimacy of the link between the promotion of basic research and the emerging image of the "security state" runs still deeper. With the advent of mass consumer society in the twentieth century, scientific leaders like Conant gradually came to believe that a business would invest in research only to the extent that it would be likely to increase its

---

17. Daniel Bell has probably been the most active in promoting this interpretation of contemporary society. See Noble 1991, esp. 18.

18. This is the main thesis of Hooks 1991. Thanks to Sujatha Raman for bringing this important text to my attention. At first glance, it may seem paradoxical that the New Deal should be the crucible in which the military-industrial complex was forged, since much of its success was due to the support of "isolationist" politicians outside of FDR's northeastern U.S. base who were attracted to the New Deal's overall vision of national self-sufficiency. However, Germany itself provides historic precedent for this link between the "welfare" and "warfare" states, a point that was presciently made and generalized by Herbert Spencer in the first volume of *The Principles of Sociology* (1872), written at the dawn of the imperial age. In this work, Spencer distinguishes between what he calls "militant" and "industrial" societies. He regards the latter as more highly evolved, with their autonomous individuals and laissez-faire capitalism. However, Bismarck's Germany represented the former type of society, which makes people feel dependent on the state for their welfare. Specifically, the state ensures the social security of its citizens, so as not only to create an obligation for them to defend the country in times of war but also to guarantee that these duty-bound troops are healthy enough to fight effectively.

The expression "welfare-warfare state" to describe this situation is coined in Gouldner 1970b, 500. Many of the great medical advances in minimizing the threat of disease and improving nutrition that were made in this period (especially the work of Robert Koch) originated amid concerns of soldier maintenance in the battlefield, and were then transferred to the general population (regarded as potential troops in future wars) during peacetime. In chapter 1, note 9, we have already had occasion to note the militaristic imagery associated with medical science over the past hundred years. For an account of the U.S. military's investment in professional science education so as to ensure a steady stream of defense-related researchers, see Leslie 1993. Also see Noble 1991, esp. 11–68.

market share. Products too ahead of their times are just as unprofitable as products that fall behind consumer demand. Thus, industry had little direct incentive to make major revolutionary breakthroughs in research. In contrast, the security state was dedicated to mastering and overcoming the military capabilities of the Communist bloc through the design of ever "smarter" weapons, whose potential for "mutually assured destruction" would supposedly be sufficient to prevent war indefinitely.

This search for what might be called "the military sublime" was the very antithesis of the run-with-the-pack market mentality of business, but suited to the sense of "ultimacy" that has animated the basic-research mind-set. Moreover, based on the performance of academic scientists during the Second World War, the defense establishments of most Western powers seemed content to continue allowing scientists to work with minimum oversight. In this way, the military became the main external supporter of basic research in the Cold War era, with 20 percent of all scientists worldwide working in defense-related research by the 1970s (30 percent in the United States and United Kingdom).[19] The current plethora of researchers is not a "natural" outgrowth of the search for knowledge but a vestige from the upscaling of universities that accompanied the Cold War and virtually every other period of military buildup since the Franco-Prussian War.

Moltke launched the West on this trajectory by taking to heart List's view that capitalist competition would never replace the need for warfare, but rather the two would use the same means. Toward that end, Moltke envisaged a citizenry technically equipped to switch the nation's infrastructure between industrial and military purposes on a moment's notice. The most obvious examples of this convertibility lay in the ease with which factories could produce munitions and trains could transport troops and supplies to the frontlines. Germany was seen as enjoying an advantage over France and Britain in the efficient integration of scientific developments into the national infrastructure. Although formal scientific knowledge was arguably still more relevant for explaining and legitimating new technologies than actually creating them, the German chemical industry had already established a precedent for science-based technological innovation.

Not surprisingly, and contrary to the historical myopia of some of today's science policy analysts, by the turn of the last century Germany had already designed hybrid research institutes that were jointly funded by the state, industry, and the universities: the Kaiser-Wilhelm Institutes, the forerun-

---

19. Proctor 1991, 254. It is worth pointing out that Vannevar Bush and his allies were more successful in diminishing science's public accountability by having much defense research reclassified as "basic" than by boosting actual research funding. The latter started only with the launching of Sputnik in 1957. See Reingold 1994, 367–68.

ners of today's Max Planck Institutes.[20] In contrast, on the one hand, France suffered from a surfeit of state-commissioned projects that stifled independent innovators; on the other, Britain's decentralized approach allowed for multiple sources of innovation but no efficient means for developing and utilizing those innovations. Thus, throughout the nineteenth century, it was common for aspiring German scientists to apprentice with solitary British geniuses, only to return to Germany to set up a proper laboratory and maybe even create an academic discipline.[21]

There is a real issue as to whether "technoscience," understood as an all-encompassing behemoth to which all of society needs to adapt, would have come into being had the Germans not found a permanent state of emergency such a compelling description of their situation. It is difficult to ignore the historical coincidence between the introduction of scientific and technical education at a mass level and the conscription of adult men into the armed forces. The practical question confronting the development

20. An excellent analysis of this phenomenon, as well as parallel developments in Britain, France, and the United States, is Harwood 1994. The best source in English for the importance of the German chemical industry in pioneering the hybridized forms of research that enabled science-based technological innovation is Johnson 1990. Johnson deserves credit for highlighting the international pressures that conditioned this development. On the pressures that forced scientists to "cooperate," if only to ensure that their nations were not falling behind the race to global supremacy, see Schroeder-Gudehaus 1990.

Ironically, much of this sense of history seems to have been lost on contemporary European science policy analysts who herald a transition from "mode 1" knowledge production, which is largely university based and paradigm driven, to knowledge production of the "mode 2" type, which involves extensive state-industry-university collaborations. See Gibbons et al. 1994. I have dubbed this a *stereoscopic* view of history, for it gives the illusion of diachronic depth to two roughly synchronous developments. See Fuller 1999b, chap. 5. On the view of these "modists," mode 1 dates back to the seventeenth-century Scientific Revolution, whereas mode 2 is traced to the end of either World War II or the Cold War. Yet, as we have seen, the two modes were in fact institutionalized only within a generation of each other — the third and fourth quarters of the nineteenth century, respectively — as the compatibility of the laboratory sciences with academic norms opened the door to more extended collaborations between the universities and industry, where laboratories in the past had usually been housed. Indeed, the Kaiser Wilhelm Institutes predated today's "triple helix" institutions by an entire century. A characteristic recent set of essays on the university-state-industry research triad is Etzkowitz, Webster, and Healey 1998.

By stereoscoping history as they do, the modists obscure the crucial point that the natural sciences have been historically the main vehicle by which industrial and commercial concerns have permeated academic culture, largely as a reaction to universities, defined as liberal arts institutions with no particular interest in scientific inquiries that strayed beyond the traditional bounds of natural philosophy. A subtler but equally important reason for the modists' stereoscopic historical vision is that they treat the history of science as if theoretical physics has always been the vanguard discipline. Consequently, they regard the large temporal gap between, say, Newton's professorship in mathematics at Cambridge (mode 1) and the building of the atomic bomb (mode 2) as typical of all sciences. Such a gap would not be apparent had they tracked the history of science through chemistry. Thus, while generally celebrating the transition to less academically oriented mode 2 research, the modists remain academic snobs in their historiographic bias.

21. Inkster 1991, 133–38.

of List's "mental capital," or Moltke's "intellectual infrastructure," was how and where natural-science courses should be introduced into the curriculum. Mach and Planck both realized that in the balance hung the survival of not only Germany but also science itself. Generally speaking, educational reform moved on two orthogonal fronts: on the one hand, educational opportunities were expanded for the populace, while, on the other hand, educational credentials were introduced to differentiate and stratify occupations.[22] Today, movement on both fronts is taken as a defining feature of modernizing societies. Indeed, in the writings of such systems theorists as German sociologist Niklas Luhmann, "growth through functional differentiation" enjoys axiomatic status, bolstered by a strong organismic analogy (and little empirical argument). However, the relationship between the expansion and specialization of education was much more controversial at the end of the last century.

When relatively few people received formal schooling, education served as a vehicle for both "vertical" and "horizontal" social mobility. On the one hand, the lower classes could improve their social standing; on the other, the upper classes could move more easily among careers. (Of course, this generalization applied only to men, not women.)[23] Not surprisingly, democratic theorists called for universal formal education as a means of completely equalizing employment opportunities. Yet many educational reformers feared that, left unchecked, the mobility of the newly educated

22. Mueller, Ringer, and Simon 1987.

23. This point is worth stressing because, despite their political differences, German scientists of the period were alarmingly uniform in their belief that women were unsuitable for at least the most rigorous (and prestigious) forms of scientific work. Of Planck and Mach, Planck's conservative views on this matter are the less surprising. Thinking much like a latter-day Lamarckian, Planck held that while the occasional woman may be fit to be trained in physics, a large influx of women into the field would upset the balance of nature, leading to sterility in the species. Mach's official views are more elusive but only slightly more encouraging. Mach's fellow traveler, Wilhelm Ostwald, was sufficiently liberal to allow women to train in all subjects except those requiring laboratory experience. Like Planck and other contemporaries, he claimed that women lacked the willpower needed to sustain highly focused inquiry. However, somewhat counterintuitively but in keeping with Mach's demystified attitude to scientific specialization, Ostwald argued that this female incapacity was actually a mark of moral strength, since a purer sense of satisfaction could be achieved through motherhood than scientific discovery. Planck's and Ostwald's views on women are taken from a survey conducted by the theologian Arthur Kirchhoff in 1896 on the educability of women. The survey is discussed in Proctor 1991, 112 ff.

A sense of Mach's personal attitudes toward women can be gotten from the bons mots that his sister, Marie, preserved in her memoirs. One such Machian saying is, "The only dignified female profession is to keep out of the limelight." Quoted in Blackmore 1992, 16. Here I must thank Sujatha Raman for forcing me to look into what turned out to be a rather seamy side of early twentieth-century German scientific thought. Since the academics surveyed in Kirchhoff's study drew the line on subjects fit and unfit for female study at different places, it would be interesting to investigate the extent to which they corresponded to a given academic's sense of the *Geisteswissenschaft-Naturwissenschaft* distinction.

masses would lead to social disorder. They wanted to reinvent within the "liberalized" educational system the sorts of discriminations that had traditionally restricted access to education. In practice, this meant predicating employment on credentials, the acquisition of which required the student to undergo a course of study whose content would be controlled by the relevant academic specialists. In Germany, much of the public debate over these matters centered on the implications of introducing the natural sciences into the secondary schools, the *Gymnasien*, the curricula of which had remained uniformly humanistic throughout the nineteenth century.[24] Most parties agreed that some form of (natural) science education should be made available in at least some of these schools—but in what form, and to what end? Answers to these questions turned on what was taken to be the distinctive epistemic contribution of the natural sciences, and its relevance to the "modern" German citizen who may not pursue scientific research as a career, but whose continued support would be needed for science to continue at its current pace.

Against this backdrop, Mach and Planck engaged in a sustained and highly personalized debate about "the ends of science" that attracted considerable public attention from 1908 to 1913.[25] Broadly speaking, Mach's instrumentalism drove him to see mass empowerment and scientific credentialism as incompatible goals for education, whereas Planck's realism led him to endorse credentialism as a necessary complement to a rapidly expanding educational system. Mach's vision of "spreading science" throughout the school system could not be more different from Planck's. Where Mach envisioned a more *mobile* citizenry, Planck saw one that was more *mobilizable*. In Mach's accounting scheme, science *saved labor*, whereas in Planck's it *added value*. The cases of Mach and Planck are especially illuminating because they *lived* their respective ideologies. Mach championed the cause of adult education in the Austrian Parliament and authored a series of middle-school science textbooks and college science textbooks for nonscience majors.[26] Through these texts, many of this century's leading thinkers on scientific matters—including Einstein, Heisenberg, Carnap, Popper, and Wittgenstein—were introduced to Machian instrumentalism.[27] Planck, on the other hand, administered many of the institutions responsible for shaping the professional identity and public voice of the German natural-science community both before and after World War I. These

---

24. On the general character of this debate, see Mueller 1987 and Albisetti 1987. As the debate concerned science matters, see Blackmore 1973, 135–37.
25. The texts of the major rounds are provided in Blackmore 1992, 127–50.
26. Blackmore 1973, 235.
27. Blackmore 1973, 141.

TABLE 2. *A Playsheet of the Mach-Planck Debate, 1908–1913*

	MACH	PLANCK
philosophy of science	instrumentalist	realist
"autonomous science"	pathology (means turned into ends)	purity (end in itself)
value of science	labor-saving	value-adding
source of unity in science	translation of science into everyday experience (psychophysics)	reduction of everyday experience to ultimate physical constituents
preferred basic science	phenomenological chemistry	atomic physics
science in general education	technology	problem-solving mentality
history in science education?	yes (future opened up by recovering past)	no (future confused by recovering past)
politics of science	liberal democratic	state corporatist
"citizen-scientist"	empowered	mobilizable
hero of science	Galileo (breaker of tradition)	Newton (founder of tradition)

included the German Physics Society, Berlin University, and the Kaiser Wilhelm Institutes, as well as several international scientific unions.[28] The multiple levels on which the Mach-Planck debate was registered are depicted in table 2.

4. Should We Expect More or Less Science in the Future?

Mach famously located the value of science in its ability to economize on thought: Phenomena that once could be handled only with great mental and physical effort—because they were thought to be disparate in nature—now could be epitomized in a single mathematical equation or set of formulas. In most general terms, science is an abstract labor-saving device that facilitates the satisfaction of human needs, thereby freeing up time for people to pursue other things. Mach did not hold the practice of academic science to be itself an especially interesting or ennobling pursuit. He went so far as to ridicule the practicing scientist's taste for the odd and the exceptional that was often pedagogically dignified under the rubric of "curiosity." Not surprisingly, Mach concluded that secondary schools could do

28. Heilbron 1986, 47–86.

justice to science without requiring an inordinate amount of specialized science education. Indeed, beyond the point of showing science's historical tendency toward economizing on effort, science education may even become self-defeating. The last thing Mach wanted was to replace the time students wasted on mastering humanistic arcana with time spent mastering scientific arcana. In fact, he wanted to reduce the amount of time students spent in school generally.[29]

Notice that Mach's economy of *thought* is diametrically opposed to the economy of *nature*, that is, the idea that everything in nature has been put there for a reason, such that if we could read God's mind, we would immediately see why things have been arranged as they are. If the economy of nature adapts human expectations to the complexity of nature (by rationalization), the economy of thought works in reverse as the means by which humanity focuses its finite energies strategically to reduce that complexity.[30] Mach held that human survival was the ends in terms of which the economic value of various intellectual projects, including the natural sciences, should be judged. He was unique among late nineteenth-century physicists in regarding evolutionary biology—and not some ultimate branch of physics—as the background constraint on all inquiry. Moreover, Mach's joint commitment to psychologism and liberalism distinguished him from some of his fellow evolutionists, for he believed that "human survival" was determined not by natural forces beyond human recognition or control, but by the collective interests of humanity at the times when accounts are taken. (In fact, for Mach, survival was not "brute" at all, but closer to the modern concept of welfare.) Even if our biological natures and environmental conditions determine the times in which accounts are taken and decisions made, Mach held that they do not dictate the content of those accounts and decisions.[31]

Planck focused his public battle with Mach on whether "economy of thought" should be included among the ends of science. Symptomatic of what bothered Planck was that when Mach cited examples of economized thought from the past, he treated, say, the multiplication tables and Snell's law on a par. Both save human labor by providing for the efficient storage, transmission, and retrieval of large amounts of information.[32] As a result, people do not need to work so hard to repeat the past and each other, and can move on instead to more personally satisfying and socially enriching projects. Mach observed that this tendency in human history probably has

29. Blackmore 1973, 133–36.
30. Blackmore 1973, 174.
31. Blackmore 1992, 137–38.
32. See Georgescu-Roegen 1971, 27 n. 14.

had more to do with the introduction of writing as a prosthesis for memory than with the actual development of scientific thought. In fact, the sciences are only one among many social practices whose development has been fostered by the spread of writing and its technological descendants, printing and — to update Mach — computerization.[33]

Continuing Mach's line of reasoning, to think that there is something epistemically remarkable about the uniformity, stability, and closure of scientific theories is to be overimpressed by the role that fairly ordinary forms of social and technical control play in scientific practice. Admittedly, science may start to look remarkable if one focuses exclusively on the fact that scientists in different laboratories, working under different paradigms, can reliably generate the same experimental effects. For precisely this reason, sociologists have found it important to investigate hidden uniformities in the professional training of these scientists, as well as various gatekeepers who, behind the scenes, shape and coordinate the responses that scientists make to their respective lab environments.[34] Ultimately, only the finest of lines separates rational consensus from groupthink, a point that both philosophers and scientists — though not Planck — have recognized from time to time.[35]

Nevertheless, the Machian critique was meant to apply only to the

33. While Mach would have no problems accepting most of the intellectual economies performed by computers today (e.g., storage capacity, retrieval speed), he would probably struggle with the idea that computers could become prosthetic extensions of the human body, i.e., the *cyborg* concept. For Mach, as for Leibniz before him, the ultimate economy of thought was ideographic script, as practiced in China and Japan, whereby one could literally read an idea off a character. This suggests that even in its most advanced and stylized form, the cognitive process is irreducibly grounded in ordinary human sensations: in the final analysis, ideas are abstract pictures. Thus, computer languages, as mechanized versions of mathematical notation, while undoubtedly useful in many tasks, would always appear "arbitrary" in their design and hence not fully assimilable to natural modes of human thought. At least, this is how I imagine Mach would regard the matter. In any case, this speculation turns out to have some bearing on alternative lessons drawn by Mach's followers on how to construct a universal language of thought. For example, as we shall see below, Neurath's Isotype notation took Mach much more literally than Popper or Carnap were willing to do. On Mach's own pronouncements, see Mach 1960, 577–78. Mach may well have been influenced by the rapid rise of Japan on the world scientific scene in the last quarter of the nineteenth century. An important source of that rise was the translation of European scientific texts into ideographic script, which enabled Japanese students to grasp scientific concepts much more easily than European students who had to disentangle the concepts from their obscure Greco-Latin names. See Fuller 1997d, 121–34.

34. Graeme Gooday has studied the concerns that eminent Victorian scientists (including T. H. Huxley) had about routinized laboratory training, which suggested that similar methods must lead to similar results, thereby squelching any creative impulse. See Gooday 1990. The sketch of a philosophical model for explaining scientific consensus that takes account of the Machian considerations raised in the text can be found in Fuller 1996c.

35. Of course, this was Popper's main concern about Kuhn's valorization of normal science. On the groupthink mentality, whereby group members adjust their beliefs to match those of a forceful member, see Janis 1982.

promotion of stability and closure as *epistemic* virtues worth instilling in a community of inquirers. Stability and closure clearly have other, more industrially oriented virtues. Many experimental effects can be generated with such mechanical precision that they might as well be produced by machines, thereby allowing laboratory scientists to do something else with their time. The same even applies to so-called closed theories, such as the core of Newtonian mechanics, which can be effectively black-boxed as computational devices. As with manual labor, once intellectual labor can be automated in this fashion, it is unclear what exactly is to be gained — other than some vague sense of aesthetic satisfaction — by encouraging humans to continue performing the labor under the rubric of "epistemic virtue." On the contrary, the failure to dissociate science-based technologies from the scientific community not only perpetuated groupthink among scientists but also placed nonscientists in a perpetual state of epistemic dependency. For a vivid sense of Mach's point, imagine how elite an activity driving a car would be, if one had to possess knowledge of auto mechanics before being granted a driver's license.[36]

In essence, Mach envisioned scientific careers as self-sacrificing, not self-perpetuating: The labors of one scientist today should enable the next generation of humans to apply less effort to reach the same ends. From this criterion follows a revisionist reading of salient developments in the history of science that marginalizes such characteristically "professional" activities of scientists as theorizing about unobservable entities. As we have already seen, for Mach, a scientific theory is little more than a prototype for an information technology. Indeed, his desire to detach the stability of scientific practices from the expertise of scientific communities has a rather contemporary ring. Nowadays, such detachability is central to many current multicultural claims to epistemic parity with the West. Consider the Incas, who designed an agricultural station at Machu Picchu the sophistication

36. Mach was roundly criticized by philosophers in his day — most notably Edmund Husserl — for regarding it as a virtue that technology enables us to do things unreflectively (and hence easily) that in the past had required great thought (and hence effort). For Husserl and his various twentieth-century followers, ranging from Martin Heidegger with his existential phenomenology to the Frankfurt School with its critique of instrumental reason, Mach has been portrayed as an enemy of critical reflection. This impression was not relieved in the Marxist world by Lenin's *Materialism and Empirio-Criticism* (1909). However, while understandable in terms of the various ideological struggles in which Machism figured, the charge does not do justice to Mach himself, as his systematic rejection of Planck's scientific realism should make clear. For Mach, reflective criticism of the sort advanced by his opponents as propaedeutic to large-scale social change presupposes that people have already reached a level of economic security that provides them the leisure to think through their futures more deeply. The labor-saving character of technology is essential to the promotion of this end. On Mach's response to Husserl, see Mach 1960, 594–95. A recent proposal that (despite its title) seems to be in the Machian spirit is Gorz 1988.

of which would be matched by European agronomists only in the nineteenth century.[37] A claim like this makes sense only if the Incas did not have to rely on a class of professionally trained scientists to stabilize their agricultural knowledge. The "universal" component of scientific knowledge lay precisely in that part that is naturally available to all cultural backgrounds and does not require any prior "civilizing" for its assimilation.[38] In a sense, this was another application of the economy of thought principle, specifically for students to learn the most by applying the least effort, which is to say, by capitalizing on what they already know.

Mach contributed to the advancement of multicultural science in both his historical scholarship and political activism. In matters of antiquity, Mach rated more highly the intelligence displayed by "savages" than by the classical Greek philosophers, especially Aristotle. Mach traced the origins of his own science, mechanics, to primitive handicraft, arguing that proficiency at the javelin, slingshot, and bow and arrow implied that the "savages" had implicitly mastered the principle of inertia. However, once the Greeks relegated manual labor to slaves, the remaining free citizens lost any intimate acquaintance with the principle. This is how Mach explains Aristotle's botched account of moving bodies, which in turn retarded the advance of science in the West for several hundred years, since the technology of writing enabled the Greeks to spread their mistaken ideas much more effectively than could illiterate "savages," however correct their tacit understanding of motion may have been. Mach justified this harsh judgment of the Greeks on the grounds that if the essence of science is technology, and the Greeks contributed little to that and merely superimposed bad accounts of how technology worked, then what is there to praise?[39] In terms of political activism, Mach's multiculturalist sympathies are reflected in a joint venture with Wilhelm Ostwald to provide moral and financial support to Buddhists in Ceylon (now Sri Lanka) who resisted the overhaul of native educational practices that accompanied British imperial rule. Here Mach willingly indulged Ostwald's "energeticist" metaphysical maxim, "Always conserve energy!" as grounds for the Ceylonese not having to replace their traditional epistemic practices.[40]

Looking back at scientific debates in the early twentieth century, we may

37. Weatherford 1993.
38. Lewis Pyenson has done the most to highlight the "civilizing mission" of the spread of the exact sciences to colonized peoples during the imperial period, even though most of the science taught was ornamental and did not contribute to appreciably improving the welfare of those taught. A good point of entry to Pyenson's voluminous work is Pyenson 1990.
39. Mach 1960, 3, 62. Note that Mach's very interesting historiographical sensibility resists the label of "Whig," "Prig," or even "Tory."
40. Blackmore 1992, 185.

get the impression that words like "economy," "simplicity," and "unity" are used interchangeably and mean the same things to all parties. Upon closer inspection, however, this is not the case at all. For example, Planck repeatedly asserted science's mission to arrive at the unified theory of physical reality, but refused to interpret this goal as "economic" in Mach's sense. For Mach, science is one—perhaps heretofore the most prominent—means of helping people save labor as they go about the business of living. For Planck, science is taken to be an end to be pursued in its own right, independently of its ability to ameliorate the human condition. Planck would treat any convergence between the ends of humanity and the ends of science as a lucky coincidence.[41]

Given his construal of "the ends of science," Mach held that the natural sciences' historical role in the economy of thought should not be presumed to extend indefinitely into the future. Contemporary society was not obliged to complete some ongoing scientific project that began (as Planck maintained) with Newton's *Principia Mathematica*. In other words, the past track record of natural science must be treated as a "sunk cost" when determining the direction that tomorrow's science policy should take. As suggested above, Mach was specifically concerned with the prospect that increasingly specialized scientific research would exhibit diminishing returns on investment. The chief indicators here included the accumulation and persistence of anomalous phenomena, as well as the appeal to arcane mathematics and unobservable entities. To pursue research under these conditions would clearly require more expensive and customized laboratory equipment. So far heavy capital investment in scientific research had been mainly limited to state-corporate sponsorship of chemistry, where the practical dividends were very apparent. But is the extension of this policy to all the natural sciences warranted, especially if public benefits are not likely to be forthcoming?

Planck recognized this problem as well, but his gaze was firmly fixed on the future of the physics community. However much Mach's views on science education might lead students to respect the accomplishments of science, they were not designed to encourage students to enter science or even to adopt a scientific mind-set—at least one that bore the stamp of the physics community. Planck's own strategy was to enroll the entire citizenry into the mission of mature science via science education. Planck held that science imparted an increasingly coherent world-picture (*Weltbild*) whose strictures could deepen the understanding and formalize the practice of virtually any field. For example, engineers became better engineers by

---

41. Blackmore 1992, 142–43.

mastering some of the problem-set of physics. By learning how scientists construct and solve well-formed problems, students would acquire what Kuhn has taught us to call "exemplars" and "disciplinary matrices" that can be used for shaping their own practices. This was more than a matter of applying theory-neutral equations as one pleased. It required that students be sufficiently committed to some overarching model of physical reality — such as atomism — to think through its implications for some experimentally testable cases.[42] Presumably, students would then be persuaded that they cannot fully grasp the formulas and techniques without internalizing some of the scientist's professional orientation. If such a strategy does not actually entice students to enter scientific careers, the students' efforts at mastering some of the theory will at least enable them to sympathize with the labors of future cutting-edge scientists, even when they are not producing anything of immediate practical benefit.

With this much of Planck's rationale revealed, a standard realist objection to instrumentalism becomes vivid: Without a theoretical framework to suggest objects beyond phenomena that have already been economically saved, what motivation would there be for continuing to do science? In practical terms: If you do not convey some of the research orientation of professional scientists in general science education, how do you then expect to recruit the next generation of scientists and to sustain public support for cutting-edge scientific research? For Mach science was a fit subject for general education as long as students could easily assimilate it into their normal lives, whereas Planck wanted students to become acquainted with the more demanding qualities of science that contributed to its distinctive place in modern culture. Moreover, contrary to the labor-saving image of science that Mach promoted, Planck believed that as physics approached unification, each additional increment of knowledge would require increased effort without necessarily issuing in any direct practical benefits. A public accustomed to seeing science as an economizing tool could well become discouraged by such prospects.

For his part, Planck argued that the epistemic distinctiveness of physics lay in its ability to reach closure on an ever wider body of observations by a larger number of observers, all encompassed under a single unifying theory. This theory may not make sense to someone without the proper training, but then the validity of such a theory would not be checked by such anthropocentric means, but by its ability to deliver the same results to any inquirer anywhere, even on Mars.[43] Any critique that failed to

42. Heilbron 1986, 42, 55.
43. Blackmore 1992, 128–29.

respect this fundamental aspiration did not deserve the title of "science." Planck's appeal to the "independent" and "invariant" character of ultimate reality was certainly a familiar one, yet his argument implicitly conceded to Mach that universal convergence on this reality presupposed the elimination of individuality from the creative process by the enforcement of a uniform research orientation.[44] We see here a vivid admission of the paradoxical interdependence of scientific realism and the theory-ladenness of observation.

The unified theory that Planck desired was "simple" in that it explained the most diverse phenomena by the fewest common principles. He cited Boltzmann's statistical unification of thermodynamics and mechanics as evidence that this sense of unity was imminent in physics.[45] Such a theory, however, would not be "simple" in Mach's economical sense because it would require that people reinterpret their experience of reality in terms of the esoteric language of physical theory. What would ordinarily appear as qualitatively different experiences—such as heat and magnetism—would now be discussed as the product of the statistical motions of unobservable entities, namely, atoms. Planck portrayed scientific unity in terms of the exhaustive explanation of one type of phenomena in terms of a "deeper" type to which only cutting-edge physicists had epistemic access. Mach's own vision of unity corresponded to his sense of simplicity, the prototype of which was Fechner's laws of psychophysics. Here the mathematical formulas were said to correlate sets of experiences—physical stimuli and sensory responses—that were regarded as ontological equals.[46] If Planck's principle of unity was *reduction*, Mach's was *translation*. Not surprisingly, Mach found many allies across the sciences, but they were nearly all *outside* physics.

5. Contesting the Roles of Common Sense and History in Science Education

From a pedagogical standpoint, Planck's view implied that ordinary people would either have to learn how to rearticulate their experiences in terms of the new physics or simply to let the physicists speak on their behalf. Mach found neither option appetizing: the former added to the labor that students would need to expend in schools, whereas the latter subtracted from the power that people would have to do what they want. Here it is

44. Blackmore 1992, 130–31.
45. Blackmore 1973, 213–20.
46. Blackmore 1973, 29–30.

important to appreciate the relationship that Mach saw between scientific research and scientific education, as that has been subject to considerable misunderstanding by such self-professed Machians as the logical positivist Rudolf Carnap (1891–1970).[47]

Mach's pedagogy primarily aimed to rid students of fixed ideas that prevented them from making intelligent judgments about their experiences.[48] These fixed ideas included not only the esoteric categories of contemporary physics but also the implicit ontology of common sense, which Mach believed owed more to medieval Scholasticism than to phenomenological authenticity. Thus, unlike the American pragmatists, whose phenomenalism easily lapsed into commonsensical modes of experience, Mach maintained that teachers needed to make students conscious of the metaphysical baggage imported even in expressions like "Heat causes objects to feel hot." In this respect, Mach advocated a kind of language therapy in the classroom.[49] However, unlike Carnap of the *Aufbau*, Mach did not believe that this therapy had to be followed by an axiomatic reconstruction of the world in terms of a formal language of sensory elements. Indeed, Mach made it very clear that Euclid's *more geometrico*, the model for Carnap's project, was at bottom responsible for making both scientists and laypeople less receptive to their own experience, as they are forced to reason through the implications of artificially chosen first principles.[50]

Mach objected that the Euclidean method turned what is essentially a tool for facilitating thought into a model for thought itself. The formal presentation of Newton's laws and their consequences under particular boundary conditions may enable people to locate problematic propositions more precisely without saying a thing about the cognitive contexts in which such searches can and should be made. A more generalized version of this critique appears in Mach's aversion to mechanistic explanations: To promote the use of labor-saving machines is quite different from claiming that reality is mechanical or that we think like machines.[51] In fact, from Mach's standpoint, to the extent that we can populate the world with machines, we can allow the nonmechanical aspects of ourselves to flourish. Among Carnap's colleagues in the Vienna Circle, Otto Neurath (1882–1945) was more in

---

47. An informed and sympathetic account of the development of Carnap's philosophy is provided by perhaps the youngest student in attendance at the Vienna Circle meetings: see Naess 1965, 3–66.

48. A sympathetic account of Machian pedagogy that stresses its relevance to contemporary science education is Matthews 1994, 95–100. Matthews focuses primarily on Mach's use of "thought experiments," an "economical" use of the mind to simulate the already artificial conditions of the laboratory.

49. Blackmore 1973, 137–38.

50. Blackmore 1973, 303–4.

51. Blackmore 1973, 192–93.

line with Machian thinking about scientific research and education. Neurath followed Mach's lead in supporting adult education, extending this to the idea of "social museums," which workers could visit to learn about their socioeconomic conditions and political opportunities.[52] Neurath was preoccupied with developing a theory-neutral language of experience that would allow the workers direct understanding of their situation. However, Neurath's frequent allusions to mass advertising should make us wary of overintellectualizing "understanding." Neurath's interest was more in enabling collective action than in creating a common mind-set. Much as Mach suspected the theoretical neologisms of physicists, Neurath wanted to avoid relying on the introduction of a new *verbal* language, as that would bias understanding in favor of those with prior exposure to the lexicon from which the words were chosen. (Thus, even logical syntax would be biased toward the algebraically adept.) In other words, a universal language must be universally *accessible*, especially if the information conveyed in that language is likely to upset preconceptions and focus action.[53]

But Mach's objections to Planck's pedagogical strategy were not limited to the cost-effectiveness of nurturing diseconomic forms of inquiry and forcing students to learn artificial modes of thought. Mach's reservations also extended to the ideological content of natural-scientific knowledge itself, given its combination of high societal prestige and intense theoretical contestation. Mach's worries here marked a new chapter in the history of academic freedom. When debates over the place of science in the secondary schools first erupted after the Franco-Prussian War—and the principal antagonists were Ernst Haeckel and Rudolf Virchow—the issue was whether socialist supporters of evolutionary theory might jeopardize the gains that the natural sciences had recently made as university subjects.[54]

---

52. Dvorak 1991.
53. In this spirit, Neurath invented a language of visual icons called "Isotype," which was used at the Vienna social museum. The Isotype was influenced by three pictorial conventions: (a) artistic renditions of historical battles in which the size of the rival forces represents their relative strength; (b) Egyptian art, which marked social rank by clothing and body image; (c) children's art, which tends to abstract only important data and ignore the rest. See Sassower 1985, 115–19. For the Marxist context, "Red Vienna," in which Neurath developed his proposals, see Cartwright et al. 1996, 56–82, 85–86. The influence of children's art was especially significant as an effort to capture a distinction Mach drew between "presentational" and "representational" forms of experience. Mach wanted this distinction to enter pedagogical practice, so that students learned to separate their actual experience from whatever inferences may be unconsciously attached to it. The leading academic psychologist in Neurath's Vienna, Karl Buehler, had argued that children's art constituted a "prerepresentational" form of experience focused entirely on whatever grabs the child's attention. Buehler's project, then, was to see how the child comes to learn to draw in the ways that adults find "representational." However, Karl Popper, who did his doctoral dissertation under Buehler, contested whether Neurath's Isotype was truly in the spirit of Mach and Buehler; see Bartley 1974, 322 ff.
54. Baker 1961.

Now, it would seem, the tables had turned. The question became whether natural-scientific research could continue to flourish *without* the ideologies associated with these fields dominating the secondary school classrooms. Planck was convinced that it could not.

Mach's caution about overextending the authority of science was characteristic of the Austro-German brand of "positivism" that would come to dominate Anglo-American philosophy of science. Its political cues were taken more from Mill than Comte. In the following generation, a more familiar expression of this concern would be found in Max Weber's call for "value-free" science, which is often misconstrued as simply a defense of pure research. At least as important was the need to persuade educators not to preempt their students' value choices by passing off their own speculations and prejudices as though they had the status of empirical truths. Indeed, there were two faces to the German conception of academic freedom: the freedom (of faculty) to inquire as one pleases and the freedom (of students) to learn as one pleases.[55] A contemporary descendant of this sensibility may be found in Feyerabend's insistence that Creationism be taught alongside evolutionary theory so as not to rob from students the right to believe what they want, especially when the truth (concerning the origins of life, in this case) is far from clear.[56] Indeed, Feyerabend explicitly draws on Mach as the originator of this sensibility.[57] However, it is worth noting that whereas Feyerabend held that "freedom" came from letting a thousand theoretical flowers bloom in the classroom, Mach was of the opposite opinion that freedom required the breathing space afforded by a sparsely populated theoretical landscape.

Conversely, such ideological self-restraint would also make scientists "value-free" in the sense of being less prone to confuse empirical corroboration with popular support. After all, science educators may be tempted to pass off as fact theories held in high esteem by the public, but whose ultimate validity has yet to be decided by the scientific community. This practice would then subtly put pressure on scientists to adapt their research

---

55. Weber's classic statement of value-neutrality is his 1918 address to graduate students in sociology at Munich University: see Weber 1958. However, Weber periodically vented his frustration at colleagues who seemed to violate this self-restraint in the classroom. Notable in the context of the issues raised in this chapter was his critical omnibus review of books on "energeticist culture theory" that attempted to raise energy to a moral principle and worldview. For a discussion, see Rabinbach 1990, 194 ff. On the relationship between the doctrine of academic freedom and the professionalization of German sociology, see Proctor 1991, 134–54, which stresses the self-policing practices associated with the doctrine. The classic history of academic freedom (written in the United States during the McCarthy era) remains Hofstadter and Metzger 1955, which gives extensive coverage to the German background.
56. Feyerabend 1979, 73 ff.
57. Feyerabend 1979, 195–205.

programs to the latest fashion. This worry would be realized in the Weimar Republic, as quantum physicists adapted to the irrationalist sentiments already present in general education by plumping for an indeterminist interpretation of microphysical reality.

Finally, Mach's pedagogical policy would serve to discipline the imaginations of students attracted to the metaphorical resonances of scientific theories like atomism or Darwinism. Later, upon entering public life, these students might wish to draw upon such theories in order to lend extra rhetorical weight to their political pronouncements. Mach had in mind the tendency of jingoistic German politicians to use Newton's laws—suitably refracted through the philosophical lens of German idealism—to justify the inevitability of international conflict and realpolitik. Thus, colliding inertial masses easily metamorphosed into egos with opposing wills.[58] Here Mach's view was simply that if a theory such as atomism is true, then its hypotheses will yield to equations with specifiable applications, which, in turn, will render the atomic theory itself redundant for classroom instruction. Until these results are forthcoming, the theory has the status of a heuristic device for research, but one without purchase outside that arena. Here we have the germs of the idea of language therapy that would later preoccupy the early Wittgenstein and his logical positivist followers.

So far it seems that Mach has a completely negative attitude toward the advancement of science. Is his approach truly "critical" and "instrumentalist" or merely skeptical, as Planck liked to think? Actually, to quote the subtitle of Mach's celebrated 1883 work, *The Science of Mechanics*, his approach is "critical-historical," an expression taken from a brand of Enlightenment theology championed, in the nineteenth century, by the people whom Marx called the "Young Hegelians" in *The German Ideology*. The theme that runs through this tradition is that Christianity has been mystified by ecclesiastical attempts to suppress the historicity of Jesus. A properly historicized Jesus would, among other things, show that Jesus was imperfect (but hence humanly approachable), that his teachings should be seen as universalizable (and hence not the property of a particular religion), and that the historical record has been repeatedly compromised in order to maintain Church authority. Mach's innovation was to model the history of science on this vision of the history of Christianity.[59]

58. Blackmore 1973, 234–35.
59. Inspired by Hegel's youthful work, *The Positivity of the Christian Religion* (1795), the most celebrated work of critical-historical theology was David Friedrich Strauss's *The Life of Jesus* (1838). A good account of the philosophical and political fallout of this book's notoriety is Massey 1983. The Oxford idealist philosopher F. H. Bradley (1846–1924) began his career with an epistemological critique of the critical-historical approach, which he linked to the emerging human sciences of his day associated with Wilhelm Dilthey, thereby providing a

Despite its long-term role of removing history from the arena of science criticism, the Mach-Planck controversy was, in its own day, interpreted as isomorphic to a common debate in critical-historical theology: Does one grasp Christ's message by looking at what is distinctive about Christianity or what Christianity shares with the other world religions? Similarly, does one grasp the ends of scientific inquiry by looking at what is distinctive about contemporary natural science or what it shares with other knowledge-producing practices throughout the world? The "universal message" of science that Mach believed has always faced institutional resistance is that science progresses only as an economical response to outstanding human needs.

In this respect, Mach was party to the "science versus religion" wars of the late nineteenth century when he endorsed the prevailing image of Galileo as someone who succeeded by not letting theology obscure his ability to confront nature directly. Galileo's studies of the heavens pointed toward the amelioration of the human condition, even as it meant the destabilization of the Roman Catholic Church. However, Mach was provocative in continuing this line of argument into the present, suggesting that the institutional structure of physics itself—especially its insistence on a uniform theoretical orientation to research—impeded scientific progress. As evidence, Mach highlighted fundamental objections to Newtonian mechanics that remained just as potent as when they were first made nearly two centuries earlier, but had been suppressed from the professional training of physicists. The most famous of these objections pertained to the existence of absolute space and time, the ether, atoms, and even mass itself. Indeed, Einstein credited Mach with keeping them alive long enough so as to suggest the need for what became relativity theory.

But Planck was especially incensed by the fact that Mach did more than

---

British port of entry for what became known as the "philosophy of history," whose most brilliant practitioner was Robin Collingwood. See Bradley 1968. For more on the relationship between critical-historical theology and turn-of-the-century critiques of science, see Gregory 1992.

One point that becomes especially clear from reading Bradley is that critical-historical theology constituted the first organized application of the "scientific method" to university-based scholarship. Indeed, in the German context, the theology faculties were instrumental in assimilating the natural sciences to academic culture. (Here it is worth recalling that, in Germany, 1898 was the first year in which the number of university matriculants in all of the natural sciences surpassed those in theology. See Inkster 1991, 97.) Indeed, in 1890 the Kaiser Wilhelm Institutes came into being through the sponsorship of the German research minister, the liberal theologian, Adolf von Harnack, who saw "Big Science" as a natural correlate to Big Government and Big Industry. In each case the work was led by one person (the director) by performed by many. See Johnson 1990, 43. The German experience of having a leading theologian act as midwife for the institutionalization of the natural sciences was not unique, as we saw in chapter 1, section 6, when considering the career of William Whewell, the Anglican priest who coined the word "scientist."

give voice in his historical writings to scientific dissidents. Mach also promoted (often in his own students) the revitalization of defunct research programs that attempted to develop scientific principles based on regularities in the experience of ordinary people. This sometimes involved experimentally examining "unschooled" Czech and Austrian craftsmen whose folk knowledge was based on highly developed perceptual capacities.[60] In one sense, these projects in, say, phenomenological optics, acoustics, and chemistry could be understood as attempts to resurrect early nineteenth-century *Naturphilosophie*, a German competitor to Newtonian mechanics associated with the poet Goethe, which was superseded once the rest of the physical sciences started to match the mathematical and experimental sophistication of classical mechanics. However, from a broader perspective, these Mach-inspired projects aimed at constructing "folk sciences" as alternative sources of epistemic authority that would enable the enlarged student body of the modern school system to recover some of the cognitive ground that was rapidly, albeit implicitly, being ceded to the emerging class of scientific experts.

6. Historiographical Implications

The Mach-Planck exchanges are often taken to be a cornerstone debate in the philosophy of science, but in a much more limited sense than has been presented here. Usually, the focus is on the substance of their arguments, which were informed by the changing fortunes of atomism as a research program in physics, and more importantly, the success of reductionism as a metascientific research program, given that the success of atomism implied the reducibility of chemical principles to physical ones. Moreover, the overall positions staked out by Mach and Planck do indeed resemble those of "instrumentalists" and "realists," respectively, as they appear today in the philosophy of science. However, if these positions are taken purely in their technical philosophical sense, the motivation for holding them can be easily lost, especially in the case of instrumentalism. For example, philosophers frequently argue that science's great leaps forward have required a realist perspective that transcends conventional ways of knowing, while instrumentalism, because of its stress on the application of existing knowledge and its agnostic metaphysical posture, has actually retarded scientific progress.[61] This is certainly the stereotyped image of the two positions that Planck used to define the terms of the dispute to his advantage. Circum-

---

60. Blackmore 1973, 59, 89, 206–7, 222.
61. For a critique of this widely held view, see Kragh 1987, 40.

stantial evidence supports the image as well, since many illustrious instrumentalists—notably Mach and Duhem—were among the forces of reaction in what turned out to be revolutionary moments in the history of science of their day.

Nevertheless, we need to bear in mind that Mach and other instrumentalists appear reactionary only against the backdrop of a vision of the history of science in which science is punctuated by revolutionary breakthroughs in insight—in other words, the realist vision they strenuously opposed. What someone like Planck takes as signs of instrumentalism's tendency to inhibit a specifically scientific form of progress appears to someone like Mach as signs of instrumentalism's promotion of more general human progress, which he sees Planck's scientific realism as inhibiting. Had Mach's side come to define the terms of the debate, we would now be thinking about realism and instrumentalism as affirmative and negative positions on whether science can progress independently of general human progress, with the realist bearing the burden of proving that scientists are something more than self-serving in their claim to a unique sense of progress.

I believe that the exchanges between Mach and Planck are at least as significant as the ones between Thomas Hobbes and Robert Boyle presented in *Leviathan and the Air-Pump*, the most important historical work influenced by the sociology of scientific knowledge.[62] "Hobbes versus Boyle" represents a turning point in the history of science in which social and epistemological problems are solved simultaneously. As the contestants came into the debate, the fork in the road toward which each pointed seemed equally viable. In the mid–seventeenth century, serious questions had been raised about the possibility of gaining knowledge exclusively through verbal reasoning (Hobbes's path), yet the experimental mode of knowledge production (Boyle's path) had not proven itself in a wide of range of settings. Which way to go? Likewise, "Mach versus Planck" represents a turning point, one of greater contemporary relevance given today's increasingly Mach-like calls to downsize Big Science and to relate scientific research more directly to the educational needs of ordinary citizens. But is a turn away from Planck and back to Mach possible—or even desirable?

The suppression of Mach's critical-historical approach in recent discussions of the role of history in the philosophy of science is yet another indication of the long-term triumph of Planck's vision of science over Mach's. Planck's attitude toward history is rehearsed whenever we appear forced to choose between a "science textbook" history that depicts the past purely in

---

62. Shapin and Schaffer 1985.

terms of its anticipation of the present and a "professional historian's" history that depicts the past purely in its own terms without any reference to the present. Kuhn canonized this false dilemma—between Whiggism and relativism—when he argued, on the one hand, that normal science cannot proceed without scientists' having an Orwellian sense of their own history and, on the other, that the past cannot be properly understood if the historian has a stake in how the events turn out.

Missing from this convenient division of scientific and historical labor is the idea that the very consignment of certain events to "the past" may reveal the implicit limits of contemporary scientific modes of thought. Mach certainly recognized this when Planck treated as "merely historical" Berkeley's and Leibniz's criticisms of Newton's assumption of absolute space and time, which was itself a holdover from when the two physical dimensions defined God's interface with his or her Creation. Mach's revival of these two-century-old criticisms turned out to be instrumental in Einstein's revolution in physics. In other words, *tertium datur: History can be used not only to legitimate the present and to recover the past, but also to alter the future by reintroducing silenced voices from the past into present-day concerns.* This is more than simply telling the stories of scientists "warts and all," but rather an attempt to renegotiate the disciplinary boundary between "history of science" and "science proper." As I observed in the introduction, such critical-historical work regularly occurs in the social sciences. Not surprisingly, the major historians of these fields are themselves typically regarded as field practitioners. By contrast, a historian of physics is not usually said to make a direct contribution to physics. And this is exactly how Planck and Kuhn would have wanted it, for such a division of labor enables the pursuit of normal science.

Metaphysically speaking, the differences between Mach and Planck reduce to a matter of temporal orientation, specifically their contrasting attitudes toward history. Both are clearly situated in "the present." But is the present continuous or discontinuous with things that might be called "the past" and "the future"? Analytic philosophers following the lead of Charles Sanders Peirce, Bertrand Russell, and Hans Reichenbach have considered the logic of the words people use to identify themselves in time and space.[63] Among these words are such "token-reflexive" indexical expressions as "here," "now," "there," "then." A common thread in this literature is that the scope of an indexical expression may vary quite substantially, depending on the context of utterance. For example, "now" may mean "right this

---

63. Still an excellent survey is Gale 1967.

second" or "the twentieth century" or "the modern era," according to context. Thus, in certain contexts, "the past" and "the present," or "the present" and "the future," may significantly overlap with each other. Given the token-reflexive character of temporal attributions, charges of "anachronism" have bite only if one presupposes that the historical episode in question is excluded from what is deemed "the present." And in that case, the bite goes both ways: Not only is the present not to be used to judge the past, but also the past is not to be used to judge the present. Thus, we have the complementarity of Kuhn's and Planck's respective historiographical injunctions.[64]

Planck draws a sharp boundary between the scope of "then" and "now," and aligns the present with the future. For him, objections raised two centuries ago are irrelevant to contemporary concerns. The past is dead, and best left to historians, who, in turn, should stay away from the education of practicing scientists. Interestingly, Kuhn sides with Planck over Mach, but whereas Planck protected the interests of contemporary scientists from historians of science, Kuhn protected in reverse. The result is the same, however: Neither Kuhn nor Planck enables history to intervene critically in contemporary science. Moreover, Planck's future-oriented present enables him to challenge the recoverability of lost opportunity costs in the history of science. Even if Mach were correct that Goethe's science was given shabby treatment one hundred years earlier, it is now too late to repair the damage. The Newtonian paradigm has displayed further strengths since Goethe's day, and scientific culture has remade society in its image. Thus, for Planck, past opportunities are not permanent possibilities for action.

As against Planck and Kuhn, Mach treats the past as continuous with the present and open to the future. His critical-historical sensibility is "philosophical" in the deep sense of refusing to discount the significance of a research tradition's nonempirical (i.e., conceptual, methodological, metaphysical) problems in light of its empirical successes.[65] In other words, Mach regards objections raised at all moments in history, no matter how far from the present, as always relevant (if not adequately answered) sub specie aeternitatis. This enables him to interpret the failure of contempo-

64. In an uncharacteristic feat of reflexive consistency, Kuhn exemplified his own Planckian perspective on history when writing about Planck's formulation of classical quantum mechanics. The monograph works through Planck's original assumptions and arguments without introducing the subsequent revolutions in the field: see Kuhn 1978. Although this work proved disappointing to *Structure*'s admirers, professional historians have acknowledged its significance in the logic of Kuhn's own thought. In that sense, the reception of the Planck book reflects just how much Kuhn has been misunderstood. See Buchwald and Smith 1997, esp. 370.

65. I am making a veiled critical reference to the criterion of scientific progressiveness put forward in Laudan 1977.

rary physicists to answer two-hundred-year-old objections to Newtonian mechanics as the suppression of voices in an ongoing conversation. Mach is also thereby able to appeal to the lost "opportunity costs" (to mass empowerment) involved in silencing more phenomenologically based approaches to science. By making the past continuous with the present, Mach can convey the impression that those opportunity costs are still recoverable, if enough scientists follow the example of his students and see how experimental approaches may be used to develop the insights of Naturphilosophie and craft knowledge. At the same time, Mach's past-oriented present is discontinuous with the future. This explains the ease with which he can regard the successful history of Newtonian mechanics as a sunk cost that should not bias our judgment as to the kind of science that requires support in tomorrow's world. Straight rule induction is no ready guide here.

Nevertheless Mach lost. One good way to see this is in the widespread acceptance of Kuhn's oft-repeated claim that professional scientists have a self-serving sense of their discipline's history that sharply diverges from that of the professional historian. While Kuhn gives the impression that this divergence in sensibilities has been inherent in the conduct of scientific inquiry at least since the physical sciences entered a "puzzle solving" phase in the late seventeenth century, the fact remains that historical objections to contemporary scientific theories were routinely lodged in the physical sciences until Mach's defeat at the hands of Planck less than a century ago, and they continue to be lodged in the biological sciences by the most senior practitioners, usually naturalists such as Ernst Mayr. In other words, scientists not only discount historically based arguments but they seem unaware that this very neglect is itself a relatively recent phenomenon and, contra Kuhn, *not* endemic to their enterprise. This point becomes all the more striking when considering the practice of a scientist who clearly knows the history but then proceeds to disregard it. A good recent example is physics Nobelist Steven Weinberg's admission that Werner Heisenberg, "one of the great physicists of the twentieth century," had some proto-postmodernist ideas about the inextricability of subjective and objective factors in physical measurement, which only reflected that "he could not always be counted on to think carefully."[66] The confidence with which Weinberg separates self-serving wheat from inconvenient chaff marks the openness with which the sort of historical philistinism licensed by Planck and Kuhn is practiced by scientists today.

66. Weinberg 1996, 12. Weinberg continues by observing the technical mistakes Heisenberg committed when he led the German atomic bomb project during World War II. Weinberg is taken to task by a variety of historically informed correspondents in the 3 October 1996 issue of *The New York Review of Books*.

7. Re-Platonizing the Politics of Science: Implications for Education and Research

A common assumption of the realism-instrumentalism debate as it is conducted today is that science is, in some significant sense, autonomous from the rest of society. At the very least, the success of science can be explained without referring to the societies that have supported it. However, once RID is fought in the arena of educational policy, such autonomy becomes difficult to maintain: How can science justify its autonomy, while at the same time meriting inclusion, if not privilege, in the key processes of social reproduction? From the Mach-Planck debate, two kinds of answers can be discerned that, over the course of this century, have proved as influential as they are unsatisfactory. The realist appeals to science as an exceptionally rational world picture whose adoption promises to add disciplined thought to any line of work. The instrumentalist portrays science as an economizing tool that can assist everyone in pursuing her ends without imposing any value orientation of its own. In practice, these alternative visions of "autonomy" have amounted to the dilemma: *Use or be used.*

When positions are developed dialectically, it is common for each side to scrupulously avoid the interlocutor's shortcomings, yet to remain unwittingly blind to one's own deficiencies. For Mach, the biggest threat posed by Planck's realist educational policy was clearly indoctrination. This fear led Mach to conceptualize the cultural significance of the natural sciences in terms that closely conformed to the liberal doctrine of "academic freedom" as applied to both researcher and student. Unfortunately, this doctrine was originally designed with the humanities as the center of the educational system. By the first decade of the twentieth century, the range of applications of natural-scientific knowledge had become wider and potentially more dangerous than those of humanistic knowledge. Thus, to deny science its own value orientation was to license indirectly the appropriation of scientific knowledge for any purpose, including destructive ones, as World War I would ultimately demonstrate. For every Mach who resolutely refused to involve his scientific expertise in the war effort, there were plenty of Machians, especially among the chemists (including the "pacifist" Wilhelm Ostwald), who "freely" enrolled in the kaiser's cause.[67]

Unlike Mach, Planck did not believe that the problem of scientific autonomy would be solved by a sharp separation of science and values, for that would only make science captive to those who have the power to impose their values on it. Science had to be socially recognized as its own

---

67. Johnson 1990, 180–83.

value orientation, alongside yet noncompetitive with the state, religion, and industry. An elite functionary for most of his career, Planck was alive to corporatist tendencies in the modern nation-state that eluded Mach's democratic liberalism. Thus, Planck organized scientists in ways that enabled them to take collective control of the direction and application of their work, a strategy that included insinuating a distinctive natural-scientific perspective throughout the educational system. Nevertheless, as the religious rhetoric of the Mach-Planck debate brings out, the realist strategy placed science—especially an advanced science like physics—in an awkward political position of its own. For if the ends of science are not merely distinct, but increasingly divergent, from other societal ends, then students will need to be given early exposure to the scientific world-picture, in order to be attracted, or at least rendered sympathetic, to scientific careers. Thus, increased control of the curriculum would seem to be necessary for continued control of the research agenda.

As suggested earlier, a crucial factor in Mach's long-term loss to Planck was his strict, perhaps even anachronistic, adherence to the old Socratic ideal of dialectical inquiry, admittedly the very one promoted by Wilhelm von Humboldt under the guise of "Enlightenment" when he reinvented the German university in Berlin at the dawn of the nineteenth century. Most closely associated with the classical humanistic mission of the university as citizen education, it was a resolutely praxis-oriented view in which the ends of science were regularly realized together by professor and student in the classroom. All students were, as the social psychologist Jean Lave now puts it, "legitimate peripheral participants" in the knowledge production process, active inquirers in their own right, not passive recipients of knowledge.[68] In Mach's idiosyncratic version of this vision, the trac-

68. On learning as legitimate peripheral participation, see Lave and Wenger 1991. Lest Mach appear completely atavistic, I should say that the fall from Humboldtian grace had become obvious only in the generation prior to the Mach-Planck exchanges. Even in the late nineteenth century, students in the natural sciences were treated as researchers in that they were encouraged to try out experimental possibilities to see what worked. In that sense, there was no discrete periodization of "training" and "research." See Olesko 1993, esp. 22. A brief history of the decline of the Humboldtian ideal in the German university system may be found in Schnaedelbach 1984, 12–32.

For German academics who lived through the final stages of the decline, such as the sociologist Max Weber, Friedrich Nietzsche (1844–1900) symbolized the "man of science" who lost his faith after the Humboldtian ideal had died. Nietzsche was a precocious philologist who fell into a permanent depression after his first book was severely criticized on "scholarly" grounds. (The rest is history.) His *The Birth of Tragedy in the Spirit of Music* (1872) invited readers to use the contemporary music of Richard Wagner to re-create in their own times the context that had enabled Greek tragedy to flourish over two thousand years earlier. Encapsulated in Nietzsche's fate is a profound difference in sensibility between what "teaching" and "research" cultures in the academy have wished to conserve as the collective memory of their societies. Nietzsche represented the teachers, who stressed the recovery of the

tability of science to common modes of experience should constrain the development of science nearly as much as science should revise and discipline common modes of experience. Translation, not reduction, was his principle of scientific unification. Consequently, the research and education functions of the university were never sharply distinguished in his mind. Moreover, Mach imagined that the absorption of the natural sciences into the universities (and the *Gymnasien*) would serve to divest these institutions of their residual elitism.

However, ends need to be tied to institutions, in which case the desirability of the ends turns out to rest on the feasibility of the institutions. And here Mach's vision came up short. Notwithstanding its Humboldtian provenance, the image of students in active inquiry with their professors was quickly molded by the institution of the *seminar* in the historical disciplines, which subsumed teaching under research, thereby converting the classroom into a market for recruiting the next generation of the discipline's practitioners. This mentality was transferred to the laboratory, once the experimentally based sciences were brought into the universities, with psychology providing an especially instructive case in point, since most of the early labs were housed in philosophy departments, the core of the humanities.[69] Gradually, an alternative vision emerged, namely of a community of scientific researchers whose real work occurs in settings quite separate from the lecture halls where students are encountered en masse. Whatever remained of the old praxis-based ideal in what became Planck's vision, it lay, as we shall see below, in the role that science plays in reproducing the social order, not in the conduct of science itself, which increasingly lost any sense of a realizable end.[70]

---

contexts of the past that could continue to enliven the present, whereas the emerging tendency was to stress the recovery of past *contents*, even if that meant stressing the "otherness" of the Greek conceptual universe to our own. Implicit in the latter, now dominant, viewpoint is that the past is very much a "foreign country" best left as a preserve to specialists. The Priggishness of contemporary relativism discussed in sections 4–5 of the introduction should be seen as coming from *this* perspective, rather than the one Nietzsche represented, which would erase any clear boundary between the past and the present. On the trade-off between context and content as goals of translation more generally, see Fuller 1988, chap. 6; Fuller 1998b.

69. Danziger 1990, 17–67.

70. Given the current fashion of speaking of the work done in a modern scientific laboratory as consisting of "practices," it is worth stressing that a citizen of the Athenian polis would find little in this usage that cohered with his understanding of *praxis*. For a thoroughgoing critique of the appeal to practices in social science, see Turner 1994, which was subject to a symposium review in *Human Studies*, July 1997, including responses by two of the leading practice theorists in contemporary sociology of science, Andrew Pickering and Michael Lynch. In what follows, I draw partly on the French Marxist anthropologist Godelier (1986, 130–37).

For the Athenian, a *praxis* was an activity that was done "for its own sake" in a special sense of that expression. The activity had to have a natural trajectory and ending that would

It is beyond the scope of this chapter to explain exactly how Planck's vision came to triumph, but the signs of its success are relatively easy to mark. In the first place, the unity of research and teaching of the classical humanist university underwent a subtle metamorphosis, one that can already be detected in the reactions to Max Weber's famous "Science as a

---

be recognized by the person engaged in the *praxis*. In addition, a *praxis* could define the trajectory and ending of other activities that did not have such self-defining qualities built into them. For example, before the introduction of slavery, agriculture was the paradigm case of a *praxis*. The cultivation and harvest of food were not interminable affairs, but ones pursued only as long as was needed to sustain a household. The emphasis was placed not on transforming nature but on participating in an activity pursued by many at once. Such praxis was the primary means by which a sense of civility was instilled, as each household tilled the soil in ways that enabled others to sustain themselves. However, after slaves started doing the agriculture, praxis was limited to free speech, that is, speaking one's mind (and *only* one's mind).

Two related notions need to be introduced here: *techne* and *poiesis*. The former designates any of a variety of client-oriented trades that require specialized (usually esoteric) training, while the latter refers to the characteristic products of such a trade. The client engages in praxis by directing the tradesman to make something to the client's specifications. The tradesman's job is to realize the client's idea in the medium of his trade. The client might oversee the tradesman's work, just to make sure all is going to plan. If the final product is judged good, then the *client*, not the tradesman, would receive credit for its successful execution. (The tradesman would receive payment for his labor, of course.) While such a merit system may seem odd by modern standards, consider the case of today's teacher who typically receives credit for the excellent performance of her students on exams. In both the ancient and modern case, merit lies more in the design of realizable standards of performance than in the sheer ability to realize those standards in performance. If techne enables a tradesman to infuse matter with form, praxis enables the client to infuse the tradesman with direction.

Various complications can be added to this story. For example, what if some enterprising people try to turn speech itself into a techne? Would undergoing the relevant training enhance or diminish the status of speech as praxis? Such were the worries that animated Plato's and Aristotle's philosophical response to the Sophists. Nevertheless, our Athenian forebears would be in agreement that crucial features of today's "scientific practice" disqualify it from counting as praxis. First, consider the claims made for the open-ended, indeed interminable, nature of scientific inquiry, as well as the stress placed on the essentially unintended ("serendipitous") character of major scientific achievements. But perhaps even more telling is the image of a mounting body of scientific literature that seemingly diminishes in its ability to capture its objects of inquiry. More and more is written about less and less. In short, the sorts of reasons that have often been given for why the Greeks would never have countenanced the unbounded market mentality of capitalism can be used to show why they would refuse to dignify contemporary natural science with the title of praxis. Someone endlessly driven to make money may be quite skilled, in the sense of possessing a techne, but the lack of purposeful closure to his activities would mark it as pathological.

At this point, Alasdair MacIntyre's objection to this characterization of praxis is worth facing. His list of praxes includes scientific inquiry, as well as musical performance. See MacIntyre 1984, 187 ff. However, the difference between the two kinds of cases are important. A musical performance is properly regarded as a praxis because each moment of a competent performance recognizably contributes to an overall goal, the execution of a piece of music, during which both the performer and the client-audience experience pleasure. With the notable exception of improvisational performances, one's musicianship typically depends on the ability to execute a piece of music that reaches a scripted end. There is no such clear connection between technical proficiency and success in scientific practice. Given the long hours of unproductive labor often involved in laboratory science, scientists are taught *not* to tie their efforts too closely to the likelihood of success. Rather, they are taught to see themselves as

Vocation" address of 1918.[71] Few academics any longer talked about students' intellectually maturing as they participated in the professor's scholarly interests. Rather, the professor herself is the one who now undergoes self-transformation through the process of scholarly discipline. Weber's particular way of characterizing this process—which involved subsuming one's ego to an endless and largely vicariously realized quest—was roundly criticized for its obvious reliance on the image of natural-scientific inquiry fostered by Planck and the physics community. Thus, humanists familiar with the classical pedigree of praxis, such as Ernst Robert Curtius, found Weber's vision a monstrous perversion of the ends of inquiry.[72] However, in its place, Curtius could only invoke the scholar's solitary participation in the thought of past minds, and the intuitive sense of private closure that it brings. Yet this too was certainly a far cry from the dialectical participation of professor and students in the Humboldtian classroom. It had become clear to all that academics were no longer in the business of bringing their students to intellectual maturity, let alone to personal empowerment.[73]

In short, Planck did not forsake Plato but shifted the Platonic imagery from Socrates to the philosopher-king. Planck sharply divided between the research and educational functions of the university. In practice, the state and the scientific community struck a deal. The state heavily invested in cutting-edge research as "bets" on the curriculum that will be needed to credential the next generation of citizens in jobs that will ensure the smooth operation of the social system.[74] In return, the scientific community offered its services in designing new principles of social stratification, ones based not on "status" or "class," but on "general intelligence" and

---

potential players in a story that will eventuate in a complete world picture: see Kuhn 1970b, 38. For example, Max Planck was quite explicit about the *vicarious* character of the pleasure that most scientists would receive from their activities. And if Planck is right, that a mark of mature science is that increasing effort is needed to achieve comparable results, it would seem that we have on our hands less a praxis and more an *addiction*—with hallucinatory elements thrown in to capture the virtual status of any given scientist in his paradigm's narrative of progress!

71. See Weber 1958.

72. Curtius (1989) catches Weber's implicit reliance on a physics-driven model of inquiry, but he neglects Weber's explicit tendency to accept the convertibility of thought styles common to modern academic and corporate culture: entrepreneurship requires inspiration, and an efficiently organized scientific enterprise conforms to the same principles as govern a successful company. Interestingly, Weber sees this convertibility as not only inevitable but also of American provenance, when in fact the Americans refashioned it from the Germans. In chapter 4, section 3, this history comes to the fore, as Conant is largely responsible for importing the German model of academic research to the natural sciences after World War I. The model, in turn, was forged in the midst of heavy state and industrial involvement. See note 20 above. My thanks to Nigel Pleasants for helping me crystallize these connections.

73. Ringer 1979, 19.

74. Stinchcombe 1990, 312–13.

"problem-solving ability." These qualities supposedly pertained to all kinds of jobs, but ultimately they were modeled on one's facility with the kind of "work problems" found in classical mechanics textbooks. Indicative of this new insertion of science into the mechanisms of social reproduction and control was the attitude of Gestalt psychologist Wolfgang Koehler toward the widespread use of intelligence tests in schools.

Not only had Koehler originally studied physics with Planck in Berlin, but he also credited Planck—especially in a speech made during his exchanges with Mach—with having discovered that the mind strives toward the kind of closure that would come to characterize Gestalt experimental findings.[75] Planck had described the solution to a problem in physics as a part needed to complete a preexistent whole, or "field." Thus, the Gestaltists generalized across individuals and sensory modalities a process that physicists more abstractly and self-consciously enacted whenever they constructed equations with a unique set of solutions. That physicists should have beaten the psychologists to this insight was explained by the electromagnetic "isomorphism" of psychological and physical fields.[76] All of this was a far cry from the simple capacity for precise observation and critical judgment that Mach wished to carry over from physics to general education.

The influence of this physics-centered view of thinking and intelligence was felt throughout Gestalt psychology. In their textbooks, Gestaltists were at pains to deny the Machian-popular view that physics was an inappropriate model for studying normal psychological processes because of its preoccupation with unobservable entities in artificial settings.[77] The earliest Gestalt-oriented experiments on problem solving were based on accounts of the electrical discoveries made by Benjamin Franklin and Michael Faraday.[78] These experiments, conducted by Otto Selz, highlighted the ability of subjects to transform their environments—essentially turning them into thought laboratories—and thereby remove the obstacles in the way of their problem-solving tasks. Moreover, while the subjects typically could not recount their thought patterns as a train of images, they nevertheless felt an

---

75. Koehler 1971, 112–13. On the vexed sociohistorical situation of the Gestaltists, see Ash 1991. The view that Koehler derived from Planck—that physical reasoning is ordinary reasoning rendered self-conscious—was popularized in late nineteenth century Berlin by Hermann von Helmholtz. See Hatfield 1990b, chap. 5.

76. Koehler 1971, 237–51.

77. Koffka 1935, 57.

78. See Humphrey 1951, 142–43. I say "Gestalt-oriented" because Selz, a member of the Wuerzburg school of psychology, held to a sharp split between the character of thought and perception, the former being "imageless." The classical Gestalt psychologists typically held to a monistic view of the mind. On the significance of this difference for epistemology, see Berkson and Wetterstein 1984, 8–10, 106 ff.

unconscious "determining tendency" that kept them motivated until a solution was reached. The report of such experiences suggested that even ordinary people normally partook of some of the committed and slightly mysterious character of physical inquiry as Planck portrayed it. So perhaps Mach had overstated the case for the exotic nature of physics. At least, this was the verdict of the instigators of the "second cognitive revolution" of the late 1950s.[79]

Nevertheless, younger adherents to Mach's critical approach to science, such as Karl Popper (1902–94), remained concerned. Popper's doctoral dissertation was partly devoted to showing that these early cognitive psychology experiments did not clearly distinguish between thought that was simply driven to fit the facts into preexisting patterns — as is often connoted in the idea of a "mental set" — and a genuine scientific breakthrough, which addressed the facts in ways that substantially reconfigured preexisting patterns. We see here a psychological basis for Popper's famous falsifiability criterion.[80] Given this lineage, it is hardly surprising that Kuhn found his studies of paradigm acquisition in the physical sciences moving him toward more general considerations of how children learn concepts.[81] The relevant developmental psychology experiments, which built upon the early Gestalt work, presuppose that children are inchoate physicists before they are anything else.

On the specific topic of intelligence tests, Koehler revealed his Planckian sensibilities most clearly.[82] He sharply separated the administrative role of the tests from their proper scientific merit. In fact, Koehler held that the primary function of the tests was to reduce uncertainty in how teachers classified students when tracking them through the educational system. Such bureaucratic efficiency was worth whatever interpretive confusions the tests' coarse-grained measures might breed in the public — as in encouraging the idea that "intelligence" is a univocal substance that people have in varying amounts. Psychological researchers would not be so fooled, since (so Koehler believed) intelligence tests are clearly artifacts of the cul-

79. A good historical source on this revolution (and its learned despisers) is the series of interviews presented in Baars 1986, esp. 365–66 (interview with Herbert Simon on the influence of Otto Selz and the early Gestaltists on his own work). An indication of the continued closeness between psychology and physics is the story recounted by early the cognitive revolutionary Jerome Bruner concerning the controversial atomic physicist J. Robert Oppenheimer (in his later capacity as director of the Princeton Institute for Advanced Studies), who asked of Bruner's own neo-Gestalt experiments: "Perception as you psychologists study it can't, after all, be different from observation in physics, can it?" From Bruner 1983, 95–96.

80. A good source on Popper's resonances with the history of psychology is Berkson and Wetterstein 1984.

81. Kuhn 1977a, 308–19.

82. Koehler 1971, 187–88.

tures that construct them and not the royal road to cognitive capacity. The ease with which Koehler could harbor such vastly different attitudes toward the educational value and the research value of these tests is reminiscent of Plato's tolerance for "noble lies" that serve to stabilize the social order. While Planck and Mach would disagree on the import of this resemblance, both would agree that only practices that appear to have the warrant of the natural sciences are in a position to manufacture such myths in the twentieth century.

Planck's view was effectively "naturalized" with the publication of *The General Theory of Knowledge* in 1925, written by Moritz Schlick, who acceded to Mach's chair in Vienna. Explicitly recalling the Mach-Planck exchanges of the previous decade, Schlick once again sought the source of "the value of knowledge."[83] At first glance, Schlick seemed to give a Machian answer that addressed the question within an evolutionary naturalist framework. However, on closer inspection, it became clear that he neutralized Mach's arguments against Planck's perspective. Schlick initially granted that knowledge originally was sought to maximize pleasure while minimizing pain. However, after a point, people realized that the pursuit of knowledge itself brought pleasure and thus started to pursue it as an end in itself. Indeed, knowledge turns out to be such an exquisite pursuit that other needs in life become subordinated to it. This is the mark of civilization.

Such an explanatory narrative had been previously used by the psychologist Wilhelm Wundt in order to show that teleological and deontological ethics were earlier and later phases in the moral development of humanity. Schlick had now purported to show an evolution in the conception of economy, with Mach representing the earlier phase of "minimal effort" and Planck the later phase of "minimal principles," the pursuit of which may actually involve much (pleasurable) effort. The crucial ambiguity in Schlick's account, which carries over into contemporary philosophy of science and science policy, is whether he is talking about the development of the individual inquirer or that of a community of inquirers. The ambiguity matters when interpreting the idea that other life needs come to be subordinated to scientific ones: Is Schlick talking about the self-sacrificing scientist or the society that increasingly adapts its other functions to the needs of scientific institutions? The elision of these two interpretations clearly worked to promote Planck's vision of science in society.

The social historian Fritz Ringer has examined the long-term effects of basing access to secondary and tertiary education on entry and exit exami-

83. See Schlick 1974, 94–101.

nations, the so-called merit-based system that has prevailed in Europe since the end of the Franco-Prussian War. Ringer observes that the merit system has reduced the differences in social advantage between average members of the old class and status groups, but at the cost of increasing the differences *within* those groups, as each of the old groups has roughly a normal distribution of people in the various categories of academic merit.[84] In other words, criteria of academic merit have gradually collapsed traditional ways of categorizing people to become *the* basis of social stratification. And rather than being a vehicle of democratization, the examination basis of academic merit has enabled degrees of discrimination—percentile rankings—unprecedented in their precision. It may even be that whatever original appeal mass education had as a mechanism for increasing social mobility has been offset by a decline in *alternative* paths to upward mobility. Thus, when academic credentials are required even for low-level management positions, opportunities for working one's way up from the stockroom disappear.

As the employment opportunities of academically credentialed people have increased over the last century, they have also displayed a distinct pattern, which Daniel Bell originally took to mean "the coming of postindustrial society."[85] The pattern is one of *intermediation*, that is, the increasing need for academically credentialed people to survey, digest, and translate the work of other academically credentialed people for a third group of academically credentialed people. In some recent popular works, this labor process has been described as "symbolic analysis" and "knowledge brokerage."[86] For science policy analysts, intermediation is the institutional correlate of cognitive complexity; for critical political theorists, it marks the corporatist sublimation of democratic impulses.[87] For still others, especially the German systems theorist Niklas Luhmann, these are two ways of talking about the same positive development, namely, an increase in social integration brought about by the continual redistribution of uncertainty across the various "intermediators."[88]

In contrast to these diagnoses, intermediation can equally be seen as symptomatic of the growing disparity between the *content* and *function* of

84. Ringer 1979, 27–29.
85. Bell 1973
86. See esp. Reich 1990. On the strength of his analysis, Reich became President Clinton's first secretary of labor in 1993.
87. See, respectively, Pavitt 1991 and Held 1987, esp. 143–220.
88. Luhmann 1983. Academic intermediation is the intellectual wing of the state's assumption of welfare and security functions that shelter the citizenry from the advances of capitalism. On how the need for such intermediation arises, see O'Connor 1973. Fuller 1999b, chap. 5, offers a severe critique of academic intermediation.

scientific knowledge, a tolerance for which we can already see in Planck and especially in his students Koehler and Schlick. If the ends of science are not merely distinct, but increasingly divergent, from other societal ends, then greater efforts need to be taken to render the two compatible with each other. The problem is exacerbated if the ultimate *Weltbild* seems to recede into the indefinite future and scientists are left with sheer "productivity" as their metric of progress, especially if the productivity of Big Science follows the familiar trajectory of product life cycles in Big Business. Later research is enabled by the rapid obsolescence of earlier research. Indeed, according to standard science indicators, the "harder" sciences are also the more "brittle," as measured by the rate at which research is superseded.[89]

However, the plethora of fads and jargons associated with such volatility is quite unlike the stabilizing function that science continues to serve as the premier mechanism of social reproduction. Indeed, all the more reason — as we shall see James Bryant Conant argue in the next two chapters — for a segment of the scientific community to take a special interest in stabilizing background social conditions so that whatever changes do occur in the research agenda are the result of specifically scientific concerns. Yet the need to make the diverging ends of science and society meet increasingly require "intermediators" who rationalize bits of the difference to each other, but none who can make it all the way around from content to function, showing, say, how the continued administration of aptitude tests squares with our best theories of how human cognition develops. Given such a disparity, one might suspect that *both* science and society have lost their ends.[90]

To conclude, we still live with the mutual suspicions that made the original rounds of the Planck-Mach debate so memorably rancorous. What is to be feared more: a closed science that has been reduced to an off-the-shelf technology (Planck's fear of Mach's policy implications) or the mass indoctrination of a scientific theory that affords it an extrascientific significance that it would not otherwise deserve (Mach's fear of Planck's policy implications)? *The Technological Menace* or *The Ideological Menace?* This

89. De Mey 1982, 111–31.
90. One of the most celebrated expressions of the disparity between the content and function of science is the *reflexive modernization* thesis associated with the work of the sociologists Ulrich Beck and Anthony Giddens. See Beck 1992; Giddens 1990. Although usually attributed to the emergence of global environmental hazards, reflexive modernization may equally be the result of our increasing reliance on experts whose opinions change sufficiently often (*perhaps* because of their sensitivity to evidence) that the sheer publicity of these changes may be more responsible for the general perception of risk than any putatively "objective" source of risk.

polarization reflects the inherent instability of science as objective knowledge. Importantly, it is an instability that emerges more from the classroom than the laboratory. Contrary to Kuhn's Planck-like view of the history of science as the circulation of research elites—the cutting-edge innovators versus the gatekeeping traditionalists—the Mach-Planck debate reorients us to the struggle within the educational community to represent those features of science that are worth reproducing in society at large.

On the one hand, if science is portrayed as the search for the most reliable means of adapting the material world to human ends, as Mach thought, then the success of science could be measured by the ease with which anyone can appropriate those means without having to rely on the presence of scientists. On the other hand, if science is portrayed as the search for the most comprehensive picture of reality, as Planck thought, then someone whose perspective is reduced or, in some sense, marginalized by science's current trajectory will interpret it as that imperial form of cognitive relativism known as "hegemony," the imposition of one special interest group's view of the world on everyone else. On the one view, the scientists are invisible; on the other, they are overbearing. A closely related tension obtains between saying that science is both "universal" and "expert" knowledge at the same time. If it is universal, then presumably it is codifiable and perhaps even automatable, which would number the days of an elite scientific community. But if science is expert knowledge, then presumably it is esoteric and of limited inherent relevance unless scientific standards are attached to some mode of domination.[91]

8. Life after the Fall: Polanyi's Escapist Strategy for Sanctifying Science

If anyone epitomized Whewell's vision of the mechanic remade as high humanist for the generation who came of age in the wake of the Mach-Planck debate, it was Michael Polanyi (1891–1976), autonomous science's most eloquent twentieth-century champion. Although Polanyi had received his training in Mach's favored field, chemistry, at the peak of German imperial expansion just before the First World War, he had no doubts that Planck was in large part responsible for ending the acrimony in

---

91. The trade-off between universal and expert forms of knowledge is pursued in Fuller 1991, which is followed by several comments and a response by Fuller. We can also think about the choice between Planck and Mach in terms of the science education policy issues that helped motivate their debate: What best reveals that students have made progress in their knowledge of arithmetic—that they no longer need to memorize the multiplication tables because they have access to calculators (a Machian criterion) or that they have mastered the relevant parts of number theory needed to explain how multiplication works (a Planckian criterion)?

German science and raising its moral tone.[92] Polanyi ended the German phase of his professional career at the Kaiser Wilhelm Institute for Physical Chemistry under the directorship of Fritz Haber (1868–1934), the 1918 Nobel Prize winner (for his work on ammonia synthesis) who had developed poison gas warfare during World War I, while struggling to keep his institute's military commitments separate from those relating to "pure research."[93] However, Hitler's rise to power in 1933 forced both Haber and Polanyi to emigrate, with Polanyi settling in the United Kingdom, where he became professor of physical chemistry at Manchester University. Symbolic of his pioneering work in polymer chemistry, Polanyi's Manchester laboratory was located near a textile factory that specialized in the manufacture of synthetic fibers. From the 1930s to the 1960s, Polanyi was a fixture in British science policy debate. During that time he did much to conjure up in the public's mind the presence of a "scientific community" dedicated to cultivating "practices" whose "tacit dimension" could not be fully understood unless one were "personally committed" to pursuing "the life of the mind." These en-quoted expressions highlight Polanyi's rhetorical legacy to contemporary science policy debate, not least including Kuhn's own manner of expression. However, his science policy vision was crystallized only upon an invitation by University of Chicago sociologist Edward Shils (1911–95) to contribute a piece to the inaugural issue of the journal *Minerva* in 1962.[94]

Polanyi wrapped himself in the rhetoric of nineteenth-century humanism to offer a passionate defense of the culture of science in the face of barbarous bureaucrats and creeping socialists. But by virtue of his participation in the quotidian battles that science fought in the public arena, Polanyi could not manifest the sort of detached "scientific" rhetoric that pervades most of *The Structure of Scientific Revolutions*, a point in no small measure responsible for many of Polanyi's original insights subsequently being attributed to Kuhn. Nevertheless, it is not hard to see that Kuhn owed more to Polanyi than the appreciative footnote to his magnum opus, *Personal Knowledge*, would suggest. Consider what Kuhn says here:

> If authority alone, and particularly if non-professional authority, were the arbiter of paradigm debates, the outcome of those debates might still be revo-

---

92. Polanyi's opinion is reported in Shils 1997, 253–54. The only other name Polanyi mentioned was Einstein's.
93. Johnson 1990, 195–98. Polanyi had first encountered Haber as a student when the latter was a lecturer at the Karlsruhe College of Technology. It is worth noting that one possible source of Polanyi's humanistic tendencies was his original training as a medical doctor in his native Hungary.
94. Polanyi 1962, which was followed by responses in the journal over the next five years, including one by Stephen Toulmin, which itself generated significant response (see chapter 6, section 8).

lution, but it would not be *scientific* revolution. The very existence of science depends upon vesting the power to choose between paradigms in the members of a special kind of community. Just how special that community must be if science is to survive and grow may be indicated by the very tenuousness of humanity's hold on the scientific enterprise.[95]

As Kuhn proceeds to discuss the special character of the scientific community, it becomes clear that he trades on the Polanyiesque trope of converting cognitive virtues to moral ones, qualities of *understanding* to ones of *trust*. Thus, in place of scientists holding beliefs on the basis of defeasible evidence, Kuhn finds scientists committed to a vision of reality on the basis of intuitive judgment. The sort of radical criticism that is the philosopher's stock-in-trade appears, in the Kuhn-Polanyi lexicon, as a failure to respect the difference between one's own social station and that of another inquirer. It challenges the soundness of trusting expert judgment during times of disagreement over the relevant standards of judgment.[96] Of course, Polanyi did not believe that all experts deserved to have their discretion so respected. In particular, he believed that social scientists could not be left to their own devices because of their tendency to extend some philosophical caricature of scientific practice—a "methodology"—to aspects of everyday life where it played no constructive role and, indeed, where it could be quite destructive if used to challenge established scientific judgment, the grounds for which may not be easily explained to the public.[97]

Polanyi's dissatisfaction with social-scientific appeals to methodology

95. Kuhn 1970b, 167; Polanyi 1958, based on his Gifford lectures of 1951–52. For Kuhn's acknowledgment of Polanyi, see Kuhn 1970b, 44 n. 1.

96. A leading sociology of scientific knowledge practitioner, Harry Collins, sides with Polanyi in a way that reveals the conservative implications that can be drawn from the day-to-day openness of scientific inquiry: "Even among the experts themselves, who have been trained to many levels above what can be expected of the public's understanding, radically different opinions are to be found. . . . It is dangerously misleading to pretend that the citizen can judge between the competing views of technical experts when even the experts cannot agree." See Collins 1987, 691. Ironically, then, while sociologists can step into the breach when philosophers cannot decide among themselves which methodology best explains a certain historical episode of scientific theory choice, the sociologists' lack of scientific expertise prevents them from intervening in contemporary disputes among scientists trying to resolve their own theory choices! The underdetermination of theory choice by the evidence thus licenses, for the SSK practitioner, not the introduction of specifically sociological variables, but rather the discretionary judgments of the local experts.

97. Aside from the atomic bomb, the public image of the natural sciences was tarnished by the attempts of leaders of the scientific community to suppress without testing Immanuel Velikovsky's best-selling *Worlds in Collision* (1950), which challenged the received wisdom of physics and chemistry by psychoanalyzing the creation myths of several cultures. When it turned out that some of Velikovsky's astronomical predictions were correct, several social scientists took the opportunity to chastise natural scientists for failing to live up to their own avowed standards. This was the episode that specifically caused Polanyi to retrench his own position about the necessarily inexplicit character of true scientific expertise. See Polanyi 1967.

reflects an important but neglected point about the terms on which he advocated an "autonomous" scientific enterprise. Polanyi was no positivist. In fact, his low opinion of social science can be traced to its positivistically inspired doctrine of "value-free science," which effectively established a division of labor between sociologist and policy maker: The policy maker dictates the ends she wishes to be pursued, and the sociologist designs the most efficient means for achieving them. This view, which Polanyi associated with Max Weber, enables the sociologist to live a Janus-faced existence as both policy instrument and pure inquirer.[98] The Weberian sociologist refuses to participate in deciding the external ends to which her expertise is put because her vocation dictates ends of its own. Thus, the sociologist appears to be the unholy hybrid of Cassandra and Cyclops, a seer who is at the same time professionally trained to turn a blind eye. While this position still haunts sociology today, not least sociology of science (as we shall see in chapter 7, sections 5–6), Polanyi would have recognized its origins in his own discipline, specifically as the stance adopted by his fellow chemists in the years leading up to the First World War. Arguably, this attitude was responsible for the wholesale recruitment of chemists to the kaiser's cause, the ultimate failure of which eventuated in the massive public reaction against science that characterized the Weimar Republic. As the Mach-Planck debates demonstrated, the triumph of the physicist's worldview (as championed by Planck) over the chemist's worldview (as championed by Mach) has been fundamental for understanding the character of the philosophy of science for the rest of this century, including Kuhn's distinctive position. Here it is worth recalling the features of the chemist's worldview that inspired a value-free standpoint, in response to which Polanyi's philosophy was constructed.

At the turn of the last century, it was common for chemists to claim that they were engaged in "basic research" but in a sense that was intimately tied to practical ends. To today's ears, this sounds contradictory, but an examination of the rhetoric of such leaders of the German chemistry community as Emil Fischer (1852–1919) and Wilhelm Ostwald—the respective winners of the 1902 and 1909 Nobel Prizes in chemistry—reveals that when they spoke of pursuing science as an "end in itself," they meant to include the human intellect's penetration into what had been traditionally regarded as "natural barriers," especially the scarcity of land and raw materials that many Germans felt had given the Americans and Russians an advantage in the race to global domination.[99] Since the advent of polymer chemistry in

98. Polanyi 1974.
99. Johnson 1990, esp. 1, 9 n. 26, 73, 202.

the mid-nineteenth century, an important research strategy of the German chemistry community was to replace natural resources with human-made ones by analyzing naturally occurring materials into their essential elements and then introducing catalysts to rearrange the molecules into new materials according to their durability, resilience, and other humanly desirable qualities. Among the enduring fruits of this strategy were plastics and synthetic fibers.[100] Ostwald himself was famously converted to this ideology once Germany's ill-fated support of the Dutch in the Boer War of 1899–1902 led to a restriction on German imports of fertilizer nitrates from South Africa.[101]

Most chemists espoused an "energeticist" philosophy that resembled Aristotelianism in many respects, especially its denial of any ultimate atomic constituents to matter coupled with an image of "nature" as a realm of pure potentiality that is given a specific form only through the application of human intelligence. The idea that nature might have an inherent character that scientific theory aims to represent was associated with the contemplative stance of the physics community, which (from the chemists' standpoint) seemed to draw an artificial distinction between the "human" and the "natural." Following Max Planck, physicists gave direction to their inquiries by sharing a common world picture in terms of which their particular research projects made sense, regardless of their remoteness from practical affairs. And as Kuhn would later do, most famously with his concept of "paradigm," Polanyi appropriated this last aspect of the physicist's worldview in order to prevent science from being further compromised in the public sphere.

However, considering Polanyi's own scientific work in chemistry, the emphasis he placed on the "passion" and "commitment" of scientists to their object of inquiry must be understood in a special light that cuts against some recent attempts to make him out to be a latent feminist, if not a postmodernist more generally. Nowadays Polanyi is often read as having claimed that scientists need to empathize with what they study, that is, to adopt the object's standpoint to the greatest possible extent—to "know it from the inside," as it were.[102] However, it is not at all clear that Polanyi the chemist believed that objects were *by nature* so sharply defined as to have the sort of intrinsic qualities associated with a distinct "standpoint" or

100. Bernal 1971, 825–27.
101. Johnson 1990, 39.
102. A major case in point is the celebrated feminist biography of the Nobel Prize–winning geneticist Barbara McClintock, who spoke of her method of understanding maize chromosomes in terms of becoming one with them. For the biographer, Evelyn Fox Keller, McClintock's manner of expression pointed to the characteristically holistic way in which women, by virtue of their early rearing, relate to the world. See Fox Keller 1983.

"subjectivity." Rather, Polanyi's assimilation of scientific inquiry to "personal knowledge" is better understood as a matter of scientists coming to realize the Promethean power their work contains and hence to take responsibility for defining the terms in which that work is used, that is, the contribution they make to materially constituting the objects of their inquiry. What can be too easily interpreted as the scientist's receptiveness to nature is in fact the very opposite, a manifestation of the scientist's will to power that is constrained only by a sense of moral responsibility, which in turn requires community reinforcement.

In short, Polanyi drew a very sharp ontological boundary between the human and nonhuman aspects of nature, *but not between humanity and nature as such.*[103] The problem with the "kaiser's chemists" was that they believed their responsibilities as scientists ended with a demonstration of technical proficiency and usable results. Thus, regardless of the rigor of their training, they were not properly constituted as a moral community, a flaw that is all the more glaring precisely because the objects of inquiry by themselves did not exert sufficient constraint on the ends to which they may be put. In this respect, despite his Platonist sensibilities about the politics of science, Polanyi — like Ostwald before him — comes close to an Aristotelian metaphysical perspective that regards "artifice" positively as the mark of completion that humans impose on an otherwise unformed nature when they operate by the principle of intelligent design.

At the same time, the elitist undercurrent in Polanyi's account of science is unmistakable, especially the unflattering analogy he frequently drew between, on the one hand, medieval self-governing guilds and indentured servants and, on the other, basic and applied researchers. (He seemed to think that all social scientists could be fitted into the latter category, perhaps in light of his Weberian interpretation of what they do.) This point has been typically missed by practitioners of science and technology studies who have attempted to contribute to the public understanding of science. When Polanyi alluded to science as a craft, he was envisaging it as an

---

103. This distinction can be understood in terms of the history of German idealism in the period between Kant and Hegel. Johann Gottlieb von Fichte (1762–1814) and Friedrich Wilhelm Schelling (1775–1854), two ideologues of the modern German university, represent the intended contrast. Schelling held that "Nature" has a will of its own that can be in either conflict or harmony with that of humanity. This view, integral to many an ecological sensibility, was *not* Polanyi's, though it is presupposed in the sharp distinction between *Geisteswissenschaft* and *Naturwissenschaft* that had begun to dominate German academic life in the mid-nineteenth century. Rather, like Fichte, Polanyi stressed the primacy of the human will, with nature characterized in largely negative terms as resistance to the will's strivings, but without any positive characterization of its own: pure passivity to the will's pure activity. On this period of the history of German idealism, see Beiser 1987, 1992.

aristocratic leisure activity comparable to a sport or game in that its value lay in its intrinsic pursuit, not in some specifiable consequences. What Polanyi most certainly did not intend was that science was like more proletarian forms of labor, such as auto mechanics.[104]

However, Polanyi was not above mixing his metaphors, even when it made a mockery of the history of economics. He likened the mutual adaptability of basic researchers to a market in which the knowledge producers treat each other as the primary consumers of each other's products, thereby yielding a "spontaneous coordination of individual initiatives." Unfortunately, had the medieval guilds oriented themselves in such an exclusive fashion, it is unlikely that enough wealth would have been generated to enable the transition to capitalism.[105] Furthermore, Polanyi believed that the published record of scientists would function as the analogue of prices in the marketplace, especially insofar as the scientific literature indicated the researchers on whose shoulders their authors were standing. However, here Polanyi seemed not to distinguish between current and future prices, and hence failed to account for the fact that unproven products may appear more attractive to speculative investors than proven ones — and, more to the point, that these speculations may be right just often enough to motivate investors to seek a speculative big gain over an assured modest one. This phenomenon, amply in display in the boom-and-bust cycles of the world's stock exchanges under advanced capitalism, was the raison d'être for John Maynard Keynes's *General Theory of Employment, Interest, and Money* and the subsequent financial safety nets introduced by governments

104. As portrayed in, e.g., Shapin 1992b, 1992c. This perspective was expanded in the first STS book explicitly aimed at a popular market: Collins and Pinch 1993. The book consists of some well-known STS case studies, shorn of their controversial philosophical implications, presented in the spirit of opening the door of the laboratory and letting the reader judge for herself what she sees. Collins and Pinch advised scientists that their own interests would be served by dropping inflated talk of "rationality," "objectivity," and "truth," and by promoting instead the more ordinary image of science as "craft" described in the case studies. In that way, the public would learn to have more reasonable expectations of science, and scientists would not feel a need to promise what they cannot deliver. Despite the authors' intent of providing friendly advice to scientists, the proletarian overtones of their craft rhetoric seriously backfired and only served to fuel the ongoing "Science Wars" discussed in chapter 7, section 5.

105. Polanyi 1962. A comprehensive critique of Polanyi's pseudoeconomics of science is provided in Philip Mirowski, "On Playing the Economics Card in the Philosophy of Science: Why It Didn't Work for Michael Polanyi" (paper delivered to the 1996 Philosophy of Science Association biennial meetings). The main reason it didn't work, according to Mirowski, is that Polanyi's "free" pursuit of science would have required coercing the rest of society to produce enough surplus to allow scientists to pursue potentially fruitless avenues of research with impunity. In other words, society would have to be adapted to the trajectory of science. For a similar diagnosis, see Fuller 1993b, 283 ff. We shall later see that Kuhn's account of scientific change helped instill precisely this attitude in policy makers who defined ours as a "knowledge society."

under Keynes's name. This would seem to justify a science policy of protected risk seeking over that of enforced risk aversion.[106]

One plausible explanation for the metaphorical malfeasances in Polanyi's economics of science is that ultimately his frame of reference was neither medieval guilds nor modern markets, but rather the fiduciary exchanges of nonliterate societies, especially two Sudanese tribes—the Nuer and the Azande—as described by Edward Evans-Pritchard (1902–73), professor of social anthropology at Oxford in the second half of Britain's imperial period in Africa.[107] The spontaneously reciprocal exchanges of information and insight that enable the political economy of science to work so smoothly may be traced to the remarkable level of trust that scientists place in the integrity of those exchanges. In short, an unquestioning faith in the overall system allows for the identification and often tolerance of discrepant beliefs in particular cases. A precedent for this extraordinary behavior may be found in nonliterate societies, which typically lack objectified traces of their history, a fact that extends beyond the sheer lack of written records to the absence of any institution designed to preserve and interpret artifacts as vestiges of the tribe's past.[108] Individual accountability is thus based on

---

106. The disanalogies between Polanyi's normative political economy of science and advanced capitalism run even deeper, for Polanyi suggested that it is better to suppress novel results than allow foolish error to infect the body of scientific knowledge and perhaps provide a pretext for alarming the general public. Yet capital-intensive science has caused modern societies to endure exactly the opposite, such that novel but untested scientific results, when produced in a major research facility, are often not only publicized but also used as the basis for policy, so as to thereby instantly prove their worth. One would probably expect Polanyi to disapprove of these developments, were it not that the public's belief in the efficacy of science rests largely on just this instant applicability to policy matters.

Interestingly, the contrary case—that research should be restricted to that which can be communicated freely—has been forcefully argued by the Oxford metaphysician Michael Dummett (1981). Unlike Polanyi, Dummett believes that even if research has unintentionally bad consequences, these can nevertheless often be anticipated and hence prevented before they happen. This is a point that had been made by Karl Popper in the course of arguing that the social responsibility of scientists grows as their ability to predict the unintended consequences of their activities increases. (The year was 1968 and the context was the participation of academic scientists in military research.) See Popper 1994, 107–10. One reason why Polanyi may have failed to acknowledge this point is its own rather negative consequences for the history of physics. As Dummett himself puts it, "What sane man, magically given the ability, in 1900, to foresee the nuclear weapons which it would make possible, would not have opted, if given the power, to prohibit all future research in physics?" (291). My thanks to David Gorman for having alerted me to the Dummett piece many years ago.

107. Polanyi 1958, 287–94, draws explicitly on Evans-Pritchard. A sympathetic introduction to his work by one of his former students is Douglas 1980.

108. See Douglas 1980, 88–89, where such an orientation to the past is explicitly likened to Kuhn's point about science lacking any interest in museums and other repositories of history. Again drawing on Kuhn's conception of scientists' sense of history, Douglas 1986, 69–80, subsequently developed this point in terms of a society's "structural amnesia." Given the rootedness of its communication practices in writing, science can forget only by design, whereas forgetting is literally second nature in nonliterate societies.

qualities that tribespersons manifest at the moment of questioning that reveal their affirmation of tribal values.

The viability of such a system obviously depends on the exercise of considerable control over the initial acculturation of tribal members, thereby alleviating any lingering doubts that the requisite virtue could be easily simulated in performance. Likewise, while publications are very much constitutive of the fabric of scientific knowledge, nevertheless they are not treated as a historian or lawyer would, namely as evidence of something that underlies and potentially undermines their surface pronouncements. Rather, they are treated as acts of "virtual witnessing," in Steven Shapin's terms, as if the reader were encountering the author face-to-face. It is this sense of epistemic transparency—rather than efficiency or precision—that should be seen as crucial to Polanyi's analogy between the scientific publication and the price mechanism. The complexity of scientific and tribal exchanges may be unfathomable to all but the initiated, but it must be fathomable *at least* to the initiated. In both cases, the security attached to the knowledge that insiders possess is a measure of the security of their collective grasp on what lies outside themselves.

Polanyi was only among the first of a half-century's worth of philosophical thinkers who have turned to Evans-Pritchard's ethnographic fieldwork to test their intuitions concerning the possibility of cross-culturally valid standards of reasoning.[109] Nowadays the only vestiges of this late imperialist legacy are puzzles involving Nuer claims to have descended from birds and Zande adherence to oracles even in the face of substantial contradictory evidence. The work that generated these puzzles was conducted in the southern Sudan in the 1930s and 1940s as part of the Colonial Service in the context of worries about the role that Africa might play in a second Great War with Germany. Britain wanted to ensure, at the very least, that the natives did not become so restless that they diverted strategic resources from British resistance to the German forces. Moreover, with a little luck, the Sudanese tribes might actually contribute to the British resistance, should the theater of war move to Africa. Colonial rule without force or condescension was thus required for the maintenance of tribal allegiances.[110]

---

109. This debate's more celebrated participants include Ernest Gellner, Peter Winch, Alasdair MacIntyre, Charles Taylor, Steven Lukes, Martin Hollis, and Ian Jarvie. See Wilson 1970, along with its follow-up volume, which incorporates the sociology of scientific knowledge as a more sophisticated development of Evans-Pritchard's position: see Hollis and Lukes 1982. Though often billed in such global terms as "rationalism versus relativism" or "the foundational debate in the social sciences," parties to it have been largely confined to Britain and its spheres of influence, which only serves to underscore the late imperialist origins of the concern.

110. On Evans-Pritchard's keenness to mobilize the natives in Britain's war effort, see Goody 1995, 63–66.

This meant operating within the tribal cosmologies. It was not enough to do as Evans-Pritchard's teacher Bronislaw Malinowski (1884–1942) had done, to reveal the socially functional character of native beliefs and actions in Darwinian and Freudian terms. That would be to provide a sense of rationality that went above (and, in the case of Freudian explanations, "below") the heads of the natives. Rather, Evans-Pritchard relied as much as possible on the natives' own accounts of why they believed or acted as they did. Besides advancing the philosophical fortunes of cultural relativism, this strategy enabled the British to maintain the confidence of the natives by showing that they could deal in their terms. Indeed, Evans-Pritchard was unique among his contemporaries in insisting on the need for anthropologists to eschew interpreters and master the native languages themselves.[111]

From the standpoint of the history of the sociology of knowledge, Polanyi's deployment of Evans-Pritchard to fuse the moral and cognitive orders in science marks Emile Durkheim's revenge on Max Weber's largely accepted account of rationalization as the mark of progress in complex societies. For Weber, rationalization incurs greater levels of potential scrutiny and accountability than ever before, as people have less firsthand knowledge of things that bear on their lives; hence, the increasing importance of such bureaucratic procedures as examinations and audits — not to mention recorded trials and experiments — where the adjudicative significance of written records presupposes a distrust of the inconstancies of personal observation, memory, and testimony. In contrast, Polanyi's Durkheimian vision upgrades these cognitive deficits to moral virtues by inserting a curious form of what economists call *time-discounting*.[112]

Time-discounting is normally presented as a problem in rationality. People often favor short-term gain even if it means forgoing a much larger gain in the long term. In that sense, people discount the future in favor of the present. The processes of rationalization described by Weber arguably fall in this category, as the meticulous keeping of records ensures accountability only in the short run, since in the long run the number of stored records undermines the ease with which they can be accessed in future accounting exercises.[113] From this perspective, the Durkheimian discounts

---

111. On the need for anthropologists to master native language and customs, see Evans-Pritchard 1964, 79–80.

112. For an examination of time-discounting from an economic and psychological standpoint, see Price 1993; Ainslie 1992. I apply these ideas to the historiography of science in Fuller 1997d, 95–101.

113. Even when records can be translated into computer files, a cost is incurred, and indeed the specific mode of computerization may impede access after several generations of improvements in computer technology.

in reverse, preferring long-term goals to short-term advantage. In the long term, the members of a tribe or a scientific community must be able to live with each other, given the indefinite duration of their collective projects. Thus, a premium is placed on social mechanisms designed to avoid the introduction of friction into its members' interactions, unless some clear collective gain is in sight. The burden of proof is shifted so that one trusts unless there is reason to doubt. In more individualistically competitive societies, the requisite "reason" may have just as much to do with personal as collective gain, but in both cases, the challenge to the status quo will succeed only if it serves the collective: i.e., I alone cannot stand to gain by doubting a colleague's word, others must as well. It follows, in the scientific context, that the period of doubt is likely to lead to a still stronger commitment to the collective enterprise. This attitude is evidenced in not only Kuhn's account of scientific change, but also the quotidian practices of scientific journal editors who refuse publication of negative results or theoretical critiques, unless the path ahead is as clear as the one being rejected.[114]

114. This position should be contrasted with the provocative approach of the French anthropologist Dan Sperber, who has promoted the concept of "quasi-beliefs" to capture native epistemic states. Whereas Durkheim and Evans-Pritchard (as well as Polanyi and Kuhn) assume that native knowledge claims imply a deep, almost religious, commitment to the associated propositions, Sperber argues that if alien cultures incorporate metaphor and irony the way we do, then perceptions of radical difference may simply reflect a misplaced literalism on our part, an interpretive failure born of an invalid methodology. In other words, relativism may be itself a quasi belief shared by Western anthropologists, which the natives realize and use as a basis for telling them what they want to hear and otherwise playing inside jokes on earnest Western inquiries. (The case of Margaret Mead amongst the sexually active youth of Samoa comes to mind here.) See Sperber 1982.

III
The Politics of the Scientific Image in the Age of Conant

1. The Old Public Image of Little Science

James Bryant Conant subtitled his 1970 autobiography "Memoirs of a Social Inventor," referring to the many opportunities he had from the 1920s to the 1960s to fashion institutional solutions to the outstanding social problems of his day. These ranged from nuclear weaponry and military service to the future of university research and inner-city public schools. To be sure, he was not the only "action-intellectual," as the celebrated journalist Theodore White dubbed Conant and the other scientists-turned-administrators who shaped America's Cold War vision.[1] Indeed, Conant sometimes played only a supporting role in the era's defining dramas. For example, in the establishment of a federal agency devoted exclusively to basic scientific research (the National Science Foundation), the name of MIT vice-president Vannevar Bush (1890–1974) looms larger than Conant's.[2] However, Conant's interests ranged more widely than Bush's and displayed common patterns of thought that, especially given his sponsorship of Kuhn, makes him the rightful American heir of the Whewellian task of normalizing the place of the natural sciences in society.[3]

---

1. White 1992, 411–13.
2. For an interesting attempt to politically rehabilitate Conant and Bush — that is, to portray them as having been *disappointed* by the direction of American science policy after the Second World War — see Reingold 1991, 284–333. The key to Reingold's analysis is that Conant and Bush opposed the Manichaean political universe of "Big Government versus Big Business" presupposed by New Deal social analysts of science and harked back to a more individualistic (Yankee Republican?) sense of enterprise, which was the implicit background in terms of which they defended, for example, the patenting of federally financed scientific innovations. Unfortunately, this charitable reading neglects to mention that Conant and Bush also sat on the directorates of several major corporations. For an acute account of the interlocking interests between the scientific elite and major American corporations in the founding of the National Science Foundation, see Kleinman 1995. On Bush more generally, see Owens 1990.
3. There is at least one other heir apparent to Whewell in the United States and his represents the road not taken. As president of the University of Chicago at roughly the same

If you asked the average American before 1945 about her image of a "scientist," she would have probably pointed to either an independent inventor like Thomas Edison or a white-coated male chemist working in an industrial laboratory. She would not have pointed to someone who worked in a university.[4] This was even the impression of senior government officials up to the start of the Second World War. Conant recalled that in the First World War, President Woodrow Wilson appointed Edison, who received no formal training in science, to head a scientific consulting board attached to the Navy. Edison, in turn, selected only one physicist to be on the committee "in case we have to calculate something out."[5] However, the striking success of the atomic bomb project placed *academic* scientists, especially physicists, in the forefront of the public imagination, while creating a public relations problem of unprecedented magnitude. Depending on the viewer's latent disposition to science, the visage of a fluffy-haired Albert Einstein seemed to be crowned with a saintly halo or a monstrous electrical field.

Conant sold the public value of "basic research" (as academic science came to be known) by providing contrasting explanations for the success of the atomic bomb project and the failures of Lysenkoism in Russia and eugenics in Nazi Germany. The atomic bomb was built because the relevant aspects of atomic physics were already known. This knowledge was the natural outgrowth of the physics research agenda, not the agenda of some government planning board.[6] Indeed, Vannevar Bush went so far as to define basic research by its autonomous character: "Basic research is a long term process—it ceases to be basic if immediate results are expected on short term support."[7] By contrast, the Russians and Germans wanted science on demand to suit their ideological goals. This resulted not only in political failure, but also in the perversion of science.[8] Thus, instead of

---

time as Conant presided over Harvard, Robert Maynard Hutchins, a legal philosopher of Thomistic leanings and patron of Mortimer Adler's Great Books crusade, attempted in true Aristotelian style to cast *biology* as the core discipline in the liberal arts curriculum because of its multiple footings in issues common to humanists and professional scientists. Interestingly, as in the Harvard case, this presidential initiative was met with great resistance by professional scientists whose interests—even in the 1940s—were oriented more toward their colleagues than the larger community of inquirers. See Buck and Rosenkrantz 1981, esp. 377–81. My thanks to Harry Marks and Skuli Sigurdsson for directing my attention to this highly informative piece.

4. LaFollette 1990, 45–65.
5. Conant 1952b, 8–9.
6. Conant 1961, 296–99.
7. Bush's 1945 congressional testimony, as quoted in Greenberg 1967, 147. For this reason, it is fair to claim that the "basic/applied" research distinction depends more on the accountability framework than on the content of the science itself. See Fuller 1993b, 237–51.
8. This familiar argument about the elective affinity between scientific knowledge and democratic politics omits some inconvenient facts about the controversies surrounding

philosophers, it is now scientists who must undergo an extended period of disciplined incubation before being given the reins of power. In *Structure*, Kuhn adds an air of facticity to Conant's account. He shows that scientists in their role as inquirers never *intend* to be socially useful—they just try to solve their paradigmatic puzzles the best they can—but that the overall result of this activity makes science's utility evident to society at large.[9] Presumably, utility, like nobility, can be recognized when publicly presented. The products of science sell the power of science without the scientists themselves ever having to deliver a pitch. At least, Kuhn's silence on the source of science's utility suggests as much.

Before Conant and his colleagues were on the scene, American industrial firms and affiliated foundations had managed to lure promising researchers away from the excessive teaching burdens that characterized even the best universities.[10] To many research-oriented scientists, the call to "pure science" that Henry Rowland had voiced as early as 1883 at the annual meeting of the American Association for the Advancement of Science sounded like a coded proposal for a new academic elite that would be

---

genetics in the middle third of this century. Whereas the disasters of Soviet food production are customarily attributed to the fondness of Lysenko (Stalin's agricultural minister) for a genetic theory (Lamarckianism) that fitted Marxist theory better than the facts of plant reproduction, little is made of the fact that the Nazi eugenicists operated within a strongly Darwinian-Mendelian perspective. Given the inability to intervene directly at the genomic level, if genetic traits cannot be altered by altering the organism's interaction with its environment, then it would seem that the only way to eliminate certain traits is to eliminate the individuals that carry them, or at least prevent them from producing offspring. Moreover, Mendelianism was supported by minute laboratory experiments (often on insect reproduction), the results of which were only beginning to be systematically related to the natural-historical data still favored by most biologists not involved in agricultural research. (This integration is what is known as the "neo-Darwinian synthesis" forged in the 1930s by the likes of Theodosius Dobzhansky and George Gaylord Simpson.) Thus, it was not unreasonable for the British Lysenkoists, such as J. D. Bernal, to claim that the Soviet commissar offered the prospect of a less alienated, more holistic biological science.

Finally, it is important to keep in mind the faulty causal inferences to which both supporters and opponents of Lysenko were prone. On the one hand, it was common for opponents to take the Soviet persecution of Mendelian scientists as evidence against the validity of Lysenkoism, whereas one could distinguish the scientific from the political issue by opposing the persecution of Mendelians while still supporting Lysenkoism. (To his credit, Michael Polanyi was someone who managed to keep Stalin's persecutions out of his critique of Lysenkoism.) On the other hand, Lysenko's supporters would frequently claim that the improvements in crop yields experienced by Soviet agriculture early in Lysenko's regime were due to the biological soundness of his Lamarckianism, whereas it could be just as attributable to the economic soundness of collectivized farming. See Jones 1988, 16–37; Werskey 1988, 205–10, 293–313; Levins and Lewontin 1985, 163–96.

9. Kuhn 1970b, 38.

10. Robert Kohler has perhaps done the most to illuminate the "progressive" role of the private sector in providing a basis for interdisciplinary innovation. I use the word "progressive" advisedly because I see the role of industry and its affiliated foundations more as a dialectical foil to the ossificatory tendencies of universities than as a good in itself. See Kohler 1982, 1991.

just as unproductive as their humanist forebears. It was therefore not difficult for agents of the Rockefeller and Carnegie Foundations to persuade ambitious young scientists that their disciplinary affiliations were little more than honorific titles bestowed on their university departments — bureaucracy masquerading as ontology. By "corrupting" the purity of science, corporate sponsors provided the less formal, more interdisciplinary space needed for the development of the biomedical and biochemical sciences, and were behind the move to reform medical education in a more "scientific" direction. Admittedly, industry's call for mathematically and physically trained scientific personnel suited their own commodity- and control-oriented imperatives, which eventually turned the American university into an enormous credentials mill. Nevertheless, at the same time, it set a standard of scientific performance that went well beyond the universities' meager mandate of training the scions of the aristocracy.[11]

2. Conant's Mission: To Design a New Future — but Not a New Deal — for Science

But at the close of the Second World War, both business and government leaders wondered aloud about the benefits that might be reaped from harnessing university-based science to economic interests. In order to tap into this proscience sentiment without having to turn the direction of science over to nonscientists, Conant, Bush, and other wartime science administrators lobbied for a National Science Foundation by arguing that science could, indeed, yield these benefits, but only as by-products of the autonomous pursuit of research. For political reasons tangential to the merits of this argument (including a Republican majority in Congress after the 1946 elections), Conant and his allies got their way, which continues to this

---

11. Thus, we find inscribed in this moment in the history of higher education the Marxist trade-off between the unproductive but (potentially) humane rentiers in the college classrooms and the productive but (potentially) inhumane bourgeoisie in the industrial labs. An especially interesting aspect of the analogy lies in what the two classes take to be the overall goal of economic production and higher education, respectively. In the economic sphere, the rentiers retained the old Aristotelian goal of *oikonomikos*, or householding, that is, producing enough for those under one's charge but not production for its own sheer sake. The bourgeoisie, of course, saw no principled end to production, the only issue for them being the efficiency of production, or "productivity." The classic discussion of this difference in sensibility, and of the latter's political ascendancy over the former, is Polanyi 1944. As for higher education, the academics proposed an Aristotelian goal of "completing" or "actualizing" the person by providing the knowledge he or she would need to assume a position of authority in society. While this vision typically presupposed a restricted sense of the number of people who would ever qualify for higher education, the idea was to minister to the total person, not merely one aspect of his or her character. The idea that higher education was for the maximization of particular techniques and skills relied on the universities' adapting to the imperatives of industries. See Dennis 1987.

day—though increasingly challenged—in the federal government's largely decentralized, peer-reviewed support of scientific research.[12]

In drawing the distinction between the contexts of discovery and justification, Whewell had been trying to find a place for the universities (and, by extension, other civil agencies such as the Patent Office) in certifying and disseminating knowledge, but not necessarily in producing it. Certainly, none of Whewell's most famous interlocutors—Faraday, Darwin, and Mill—ever called a university his home. In contrast, Conant's job was to show that universities were integral to the very production of knowledge. Again, there was little evidence that such had been the case in the past. Newton and Maxwell were exceptional in having done pathbreaking science while holding a professorial chair.[13] In both cases, the science was low-tech, indeed, largely mathematical—with its technological applications left to future generations outside the university to discover (or invent, as the case may be). Even Conant's favorite scientists—the chemists Boyle and Lavoisier—were, on the one hand, gentlemen of leisure and, on the other, glorified civil servants. It was perhaps understandable, then, that Conant's Democratic opponents in Congress believed that academic scientists needed some friendly federal guidance in adapting to their new role of frontline knowledge producers. Given that Conant, Bush, and like-minded scientists sat on the directorates of many of America's largest corporations, can the great science populist Harley Kilgore, Democratic Senator of West Virginia, be blamed for thinking that when the scientific elite called for greater "autonomy," they were simply submitting themselves to the "invisible hand" of capitalism?

While it is true that the people who designed the atomic bomb were also the ones who had developed the relevant background theories in atomic physics, it was only through the mediation of politicians that these scientists managed to sequester the large amounts of public space and money needed to build the bomb. Consequently, it was a genuinely open question whether the bomb's success should be attributed to factors "internal" or "external" to science. To those, like Kilgore, who supported an externalist perspective, the sudden empowerment of science was to be seen as resulting from the concentration of financial investment during the war effort, not the release of some magical powers hidden in the physical equations of Einstein, Heisenberg, and Bohr. Thus, the question for Kilgore was

12. Kevles 1977.
13. The important institutional exception to the claim that research has traditionally come from outside the universities is physiology, which historically developed out of the medieval medical faculties. See Joseph Ben-David, "Scientific Productivity and Academic Organization in Nineteenth Century Medicine" (1960) reprinted in Ben-David 1991, 103–24.

whether academic science was to be subordinated to the state or Big Business. There was no third way of "autonomy," as far as he was concerned. Kilgore supported the public finance of scientific research—and, indeed, coined the phrase "National Science Foundation"—as a continuation of Franklin D. Roosevelt's New Deal policies. This meant that science policy would become another tool for redistributing wealth, in the name of facilitating a fully functioning democratic electorate.

For example, Edwin Land, the inventor of the Polaroid camera's self-developing film, persuaded Kilgore to think about scholarships for science students as seed money for potentially lucrative inventions.[14] Others went so far as to suggest an "affirmative action" program for traditionally disadvantaged regions of the country (especially the South and the West), whereby they would receive a disproportionately larger fraction of the funds than regions that had benefited from private endowments. Another proposal in this vein was a "national service" scheme that would have set aside scholarships for clever students from impoverished backgrounds to matriculate at the major universities, on the condition that they would then staff, if not start up, technical training and research facilities in the economically underdeveloped regions of the nation.[15] Interestingly, such schemes invoked the universalist idea that the scientific mentality was available to everyone, and that all differences in scientific performance could be attributed to differences in the quality of the training and research facilities at one's disposal. They presupposed the fast-paced or cutting-edge research that characterized World War II as an aberration that would, in peacetime, be remedied by a science policy that was sensitive to the rate at which scientific innovation can be assimilated by the society at large.[16]

While a policy of deliberately slowing the pace of scientific change might seem perverse today, it had been taken seriously at least from the onset of the Great Depression in 1929.[17] A wide array of scientific, religious, political, and industrial constituencies held that chronic unemployment resulted from firms' developing new forms of automation in the hope of undercutting the competition's production costs. If the amount and type of innovations that industry could introduce into the workplace were restricted, high levels of joblessness could be curbed. Indeed, some went so far as to suggest—on the model of farmers being paid not to raise crops—

14. Kevles 1977, 19.
15. Mills 1948.
16. In fact, federal science funding has been subject to recurring cycles of mobilization and conversion. The phases of the cycle since 1917, the start of America's involvement in World War I, are charted in Reingold 1994, 367.
17. A good survey of this "revolt against science" is Kevles 1995, 236–51.

that scientists be paid *not* to do research![18] Perhaps the most visionary of the New Deal prophets of democratic science was Thurman Arnold, a Yale law professor who was part of the frontline of FDR's Brain Trust. Arnold believed that the technical potential of America just needed to be "organized" with the right legislation and incentive schemes, so that people no longer felt that they had to *choose* between working for themselves and the government.[19] Unfortunately, Arnold's vision of the plasticity of human nature fell afoul of a Cold War in the wake of the Allied victory that forced rapid deployment of already able personnel.

The New Dealers believed that collusion with Big Business rendered science a regulatory monster heading in two equally unpalatable directions. On the one hand, scientists fueled business speculation with the prospect of new inventions that often failed to live up to their hype but nevertheless created volatile financial markets, like the ones that precipitated the Great Depression.[20] On the other hand, there were concerns that one firm might gain a monopoly over the production of a certain form of highly useful knowledge and thereby hold the rest of society in its grip. To preempt this possibility, in his guise as FDR's main trust-busting attorney, Arnold proposed that science funding in the private sector be monitored for potential violations of antitrust laws.[21] A related concern, one that defined the terms of disagreement between the New Dealers and the scientific elite during the National Science Foundation hearings, was whether federally funded scientists should be prohibited from obtaining exclusive patents for the products of their research.[22] Vannevar Bush's allies managed

18. Proctor 1991, 238. A measure of just how far science policy over the last fifty years has departed from the original New Deal sensibility is that when members of Congress today call for a more equitable distribution of science funding across regions, they are immediately accused of "pork barreling." On the other hand, had the New Dealers won, we would probably treat the special pleading of elite scientists for larger and better facilities as "quark barreling."

19. See esp. Arnold 1966. Here we should be sensitive to the difference between regarding "organize" as a reflexive and as a transitive verb. The former has a long history, but the latter was a neologism of the period. Management guru Peter Drucker has made much of this semantic shift—such that disparate entities can be somehow rendered into an organism by an act of intelligent design. For him this was one of founding moments of the "knowledge society." See Drucker 1993, 49 ff. Nowadays, the transitive sense of "organize" has so come to dominate our usage that self-styled innovators in social and biological thought have had to resort to the pleonasm, "self-organization" (a.k.a. "autopoiesis") to name their theories.

20. In his final published interview, Kuhn says that his father, an industrial engineer, worked for a bank providing just this sort of investment advice before joining the New Deal's National Reconstruction Administration. See Kuhn et al. 1997, 146.

21. Kevles 1977, 6–7.

22. Arnold's trust-busting instincts came closest to an idea that was otherwise conspicuous by its absence in the founding debates of the National Science Foundation: Universities engaging in research that led to commercially viable products—the reason for wanting to have exclusive patent rights in the first place—should be taxed for those products, if not lose their

to fend off the calls for such prohibitions by brandishing none other than the U.S. Constitution, which stipulates that exclusive patents provide an incentive to innovate that might otherwise not exist.[23]

3. The Weight of Theory against Big Science

There were international precedents for the New Dealers' unwillingness to carry over science's recently scaled-up status into peacetime, ones that envisioned dangers for *both* science and society. The most prescient theorist of the adverse effects of a war economy on the peacetime development of science was Arthur Pigou (1877–1953), the "missing link" of welfare economics at Cambridge University between Alfred Marshall and John Maynard Keynes. World War I had demonstrated to the rest of the major European powers (as well as the United States) what Germany (and Japan, for that matter) had known for almost a half-century, namely the mutually enriching consequences of coupling military and industrial policies. However, Pigou was singular in arguing that such a marriage constituted a Faustian bargain that could stifle scientific innovation if continued indefinitely. A military-industrial complex is innovative in wartime because scientific talent is concentrated in competition against a similarly outfitted pool of researchers from the opposing nations. But once the war is over and the competitors vanquished, the same concentration of talent amounts to an effective domestic monopoly that can discourage any further inno-

---

tax-exempt status altogether. Admittedly, this was an era that had yet to hear of "intellectual property," but even then someone should have wondered whether the university might have begun to jeopardize the spirit of tax-exemption—whose paradigm case, after all, is a church—when it declared utility as one of its ultimate values. For a sophisticated justification (by a professional lawyer) of how universities economically benefit their communities by their tax-exempt status, see Bok 1982, 217 ff. However, Bok, then president of Harvard University, was writing just before the establishment of intellectual property officers on American university campuses.

23. On Bush's congressional testimony, see U.S. Congress, Senate Subcommittee of the Committee on Military Affairs, *Hearings on Science Legislation*, 79th Cong., 1st sess. (November 1945), 225–27. The U.S. Constitution declares patents to be a stimulus to invention in article 1, section 8, clause 8. In fact, the Patent Office was one of the first branches of the American civil service to be established (1836), predating Britain's by nearly twenty years. The interesting point here is *not* that the U.S. Founding Fathers made special provisions to encourage invention, but rather that patents were chosen as the incentive scheme. After all, the desire for fame alone had largely motivated scientific inquiry up to that point; hence, the traditional preference for prizes over royalties. In section 4, we consider how salaries and grants came to supersede prizes as rewards for new knowledge. However, the idea that royalties could motivate inquiry requires explanation, especially since the patent process typically arrests the publicity function that is crucial to scientific knowledge being, at least in principle, universally available. One possible explanation is that the Founding Fathers presumed that people would find it more economical to stick with what they already know, unless they are provided an incentive to stake their time and effort on the prospect of a better form of knowledge.

vation. Thus, Pigou called for the decoupling of military and industrial interests after World War I, and the disaggregation of the large research teams that had been assembled as part of the war effort. In that way Britain would regain the market conditions that enabled scientific innovation to flourish throughout the relatively peaceful nineteenth century.[24]

Shortly after the start of the Great Depression, another Cambridge man, John Desmond Bernal (1901–71), the distinguished X-ray crystallographer and Marxist historian of science, offered a related analysis of the emerging military-industrial complex of post–World War I Britain.[25] According to Bernal, as industrial capitalism evolved in the nineteenth century, "scientists" (understood broadly to include technologists and inventors) could no longer rely exclusively on the benefaction of philanthropists to finance the research and development of their projects. Increasingly, scientists gravitated to big business, which in turn constrained the limits within which acceptable innovation could occur. For example, the impulse of chemists to produce indestructible synthetic materials had to be curbed, lest it lead to the saturation of markets for such materials, as consumers would never need to purchase replacements for them. In addition, left unchecked, the mercurial spirit of scientific creativity could destabilize the monopoly control that large manufacturers enjoyed over certain markets.

As Bernal saw it, the only way that science could be ensured not to upset the capitalist order was by having its creative energies harnessed to the design of ever more sophisticated forms of warfare and surveillance. These would then be deployed to protect a nation's economic empire from threats both at home and abroad. If scientists were to escape captivity to the mili-

---

24. Pigou 1921. According to most nineteenth-century political economists, war would be a complete economic liability, were it not for the fact that it reversed people's natural priorities from preserving the past to inventing more efficient ways of maintaining one's current circumstances in the face of attack. But these inventions were unlikely to take root in society at large without the stability offered by a peacetime economy. Yet despite their generally negative attitude toward warfare, virtually every political economist from Adam Smith to John Stuart Mill admitted that increased prosperity forced nations to spend a proportionally larger portion of their income on defense, ideally to the point of discouraging poorer nations from thinking it in their interest to attack them. This served to undercut their general belief that greater wealth opened more possibilities for action. In practice, this conceptual tension was negotiated via a conception of progress that involved a sufficiently large investment in one course of action that it was no longer economically feasible to question its continued pursuit. See Goodwin 1991. My thanks to Sujatha Raman for bringing the Pigou and Goodwin pieces to my attention.

25. Bernal's vision of science's military-industrial complicities is most comprehensively presented in Bernal 1939. For a critical but sympathetic assessment of the legacy of this work, see Freeman 1992, 3–30. According to Freeman, the Achilles heel of Bernal's analysis was his taken-for-granted belief that science was the ultimate expression of human capabilities, which led him to undervalue ordinary democratic political processes. We shall return to this point in chapter 7, section 2, as it contributed to the radical science movement's failure to generate strong public support outside the scientific community.

tary-industrial complex and regain control of their creative potential, they would have to organize into a class that spoke in one voice aligned with those whose interests lay with the majority of humanity, that is, the socialists.[26] Since the majority of recruits to the scientific ranks were still from nonelite backgrounds, the polytechnics and factory shop floors, Bernal's call for solidarity appeared less apocalyptic then than it might today.[27]

As it turns out, neither Pigou nor even Bernal had realized the irreversible character of the Great War's effects on scientific research. Neither, for that matter, had Joseph Schumpeter (1883–1950), the economist who most consistently promoted the role of scientific innovation in wealth production. In fact, Schumpeter, always too clever by half, rationalized what he should have seen as a deviation from the historical norm. Schumpeter believed that corporate capital would naturally grow to the point of making it too risky for big companies to invest in entrepreneurs whose innovations promised to "creatively destroy" market conditions that worked to the past advantage of these companies. He saw this process as endemic to the dynamic of capitalism, not simply the local result of carrying over wartime military-industrial relations into peacetime; hence he did not entertain the reversibility of the process by legislation, as Pigou had suggested. And so, once again, the Cunning of Reason has the last word on attempts to predict the course of history. Pigou turned out to have a better understanding of the relationship between science and the economy, but because his policy advice was not followed, a world much more like the one Schumpeter envisioned has resulted, though not necessarily for the reasons put forward by his theory.[28]

26. As further testimony to the complexity of science's status in the military-industrial order, Bernal, like most political economists who gave thought to these matters, failed to anticipate the ways in which science would transform the character of warfare and military service. This transformation historically corresponds to the ascendancy of the air force over the other branches of the military, and the consequent shift in officer training to emphasize adeptness in computer simulations over on-site combat experience. Science was enlisted less to ensure the efficient production of conventional weapons than to intensify the military's intelligence-gathering operations and solidify its logic of nuclear deterrence. These shifts served to make defense contract research appear more intellectually challenging and morally neutral than even Bernal et al. had imagined. Moreover, it was by no means clear that science was rendered completely subservient to military ends; indeed, there is evidence that the reverse was the case, at least insofar as the military came to see technical sophistication as an end in itself, regardless of whether it made for greater military effectiveness. See Kaldor 1982.

27. Bernal's view of science policy as a form of labor-management relations was influenced by Alfred Mond, head of the Imperial Chemical Industries during the First World War. See Jones 1988, 6–14. By publishing in mass-circulation American magazines, Bernal's impact extended across the Atlantic. See, e.g., Bernal 1935.

28. Like all great rationalizers, Schumpeter perceived other ironic twists in the wake of science's scaled-up peacetime status. In particular, intellectuals, whose numbers had grown with the nationalistic expansion of higher education over the previous 150 years, spilled over into the civil service, the employment sector that grew the most as the state assumed increasing

4. Interlude: Peer Review as the Cornerstone of Conant's Brave New Science Policy

Even Schumpeter tended to underestimate the extent to which the critical disposition of intellectuals would be institutionally circumscribed once they entered the civil service, specifically, through "academic freedom," a guild right of academic communities whose jargons and techniques are safely insulated from the concerns and comprehension of the wider public.[29] The so-called peer review process was designed precisely to operate in such an environment. Instead of making direct appeals on their own behalf to segments of the public, academic intellectuals appealed to each other, leading to a mutual cancellation of more "dogmatic" or "ideological" elements (typically enabling third parties, especially the state and industry, to appropriate the purified remains as "objective knowledge"). In the National Science Foundation proposal favored by Conant and Bush, peer review was presented as the omnibus regulatory mechanism for science, when in fact it had traditionally been mainly used to evaluate research once it had been completed, as in the case of a professional journal publication.[30] The idea that peer review should extend to decisions concerning

---

responsibility for public welfare. Thus, whereas in Marxist theory, the intelligentsia were entrusted with heightening working-class opposition to the dehumanizing effects of capitalism, in welfare state theory, they came to be in charge of muting that opposition. Nevertheless, so Schumpeter claimed, the final result would be the same: Capitalism would die — however, not with a bang but a whimper, through the routinization of innovation. See Schumpeter 1950, 154–55. For a more elaborate version of this diagnosis that invests somewhat more faith on the emancipatory capacity of civil service intellectuals, see Gouldner 1979.

29. Schumpeter regarded the overproduction of idle intellectuals as the biggest threat to stability in advanced capitalist societies. His potted "sociology of the intellectuals" begins by noting that rulers originally extended patronage to intellectuals as a form of bribery to stop them from spreading ill-will about them. In the Enlightenment, intellectuals (Voltaire is his archetype) regained their critical independence by circulating among several patrons, playing one off against the other in the name of "cosmopolitanism." See Schumpeter 1950, 148–49. The nation-state recaptured the intellectuals in the nineteenth century by placing them in institutions of higher learning that instilled a common national consciousness, while they inoculated an ever-expanding credential-seeking public against radical perspectives they might encounter once they left school. Schumpeter neglected the long-term conservative effects of this institutionalization because he could not envisage higher education growing to such an extent that it absorbed or contained virtually all the unruly elements.

30. There is much disagreement as to the overall efficacy of peer review in the scientific publication process. Some general findings, however, include the following: Peer review more often takes the form of circumscribing the author's knowledge claim (for what exactly should she be given credit, if we take her results at face value?) than actually validating the claim. Articles that receive high peer agreement for publication are not generally the articles most highly cited in the field. A highly cited article is often controversial, with public commentary often raising issues that, had the original referees been different, could have easily led to the peer rejection of the article. There is little systematic information about the effects of peer review on the utilization of knowledge by people outside a given field. Two good syntheses of studies on peer review are Chubin and Hackett 1990 and Daniel 1994.

who is entitled to pursue research in the first place was novel and unapologetically elitist. As Conant put it, science differed sufficiently from other public policy initiatives to warrant judgments more of the "man" than the project proposal itself.[31]

Once again, the precedent for expecting scientists to converge on resource allocation issues was the atomic bomb project, which was shaped by the fact that all the relevant scientists fought on the same side against the Axis powers. Prior to the alignment of such formidable foes, scientists were not known to have regularly agreed on who should receive research money, regardless of the consensus reached by journal editors on what to count as creditable research in a given field. This point is masked over by the fact that scientists have traditionally turned to a variety of sources—some public but mostly private—to support their research, so that the "scientific community" as a corporate body has been rarely called upon to deliver final funding judgments. Indeed, until the nineteenth century, *prizes*, not grants, had been the dominant form of payment in science. In a prize-based culture, the scientist would be paid for achieving results, typically being the first person to solve a publicly announced problem within specific constraints, but regardless of that person's credentials or track record. In such a culture, it is assumed that knowledge of reality is, in principle, available to all, and so credit should be assigned simply on the basis of being the first to plumb its depths according to the contest rules. The overall shift from prizes to grants in nineteenth-century science coincided with the Whewellian professionalization of the "scientist," for whom science was not a leisure activity but a full-time job the remuneration of which would be based on one's credentials and previous work—more so than even the promise of the project per se, which had been the traditional grounds on which inventors had been awarded grants from industry and private foundations.[32]

One very American way of avoiding peer review's blatant strategy of "the rich get richer, the poor get poorer"—what Robert Merton has gingerly called the "principle of cumulative advantage"[33]—would have been to institute a system of checks and balances into science funding policy. Such a scheme was proposed to Senator Kilgore by Chauncey Leake, provost of the University of Texas. He argued that the National Science Foundation

31. Conant 1961, 323.
32. See MacLeod 1971a, 1971b. However, as the status of science as an academic guild has been increasingly challenged from various quarters, more economic arguments are now being heard for reverting to the old prize-based system, which would effectively roll back the reach of the peer review process. See Hanson 1995.
33. Merton 1973, 439–59.

was worth supporting only if its board did not consist of people who were on the editorial boards of the major scientific journals or the governing boards of the major scientific societies. This would bring a wider cross-section of scientists into the governance of their fields, creating a broader sense of scientific "peerage." Leake feared that if these checks were not built into the science funding system, federal funding would go the way of corporate funding and be concentrated in precisely the private universities that already attract large industrial investment. While Leake was wrong to suggest that only private universities would be included among the beneficiaries of a peer-reviewed National Science Foundation (for example, he failed to see that his own public university would eventually enter the orbit of the elites), nevertheless the spirit of his concerns continue to be valid, given that 50 percent of federal funding of science today goes to a mere *thirty-three* (of a possible twenty-five hundred) institutions of higher education in the United States.[34]

It is worth recalling the imaginative range of alternative futures for science that were voiced in the public forum in 1945 because we may now be at a crossroads of comparable significance. Equally important is the fact that "orthodox" spokespersons could be originally found for views that would now appear beyond the pale of feasibility. For example, Frank Jewett, president of the U.S. National Academy of Sciences and director of Bell Telephone Laboratories, opposed the establishment of a National Science Foundation, mainly because he feared it would lead to precisely the concentration of resources that would discourage scientific creativity. Jewett suggested that the state should roll back its research investment to pre–World War II levels. Instead, the public should be encouraged to regard scientific research programs as charities, contributions to which may be the source of tax relief. This would certainly provide an incentive for the public — at least those wealthy enough to have incurred a significant tax burden — to acquaint itself with competing research programs.[35]

5. The Harvard Strategy for Resisting a New Deal for Science

When it came to aligning scientists with the labor movement, Bernal had his work cut out. Scientists were routinely blamed for the Depression by

---

34. Senate Subcommittee of the Committee on Military Affairs, *Hearings on Science Legislation*, 967–68; Office of Technology Assessment 1991, 263–65.

35. U.S. Congress, House of Representatives, Committee on Interstate and Foreign Commerce, *National Science Foundation*, 80th Cong., 1st sess. (March 1947), 73–76.

inventing labor-saving techniques that displaced workers. Moreover, the New Deal diagnosis of the Depression as resulting from overproduction squarely laid the blame on "scientifically managed" factories devoted to manufacturing as many goods as fast as possible. Indeed, the characteristic difference between social-scientifically based economic reforms such as those associated with FDR's Brain Trust and the natural-scientifically based ones associated with Bernal's admirers was that the New Dealers actually wanted to shrink the economy by shifting labor from machines back to humans (as in the New Deal's famed public works projects), whereas the Bernalists continued to encourage the design of technologies that relieved the drudgery involved in most forms of work, but insisted that their collective ownership would be a safeguard against mass unemployment in the future.[36] It was becoming clear, even in the relatively serene precincts of Cambridge, Massachusetts, that a crossroads had been reached on the role of science in modern society. American critics of capitalized science such as Thorstein Veblen, Charles Beard, Stuart Chase, and Lewis Mumford were required reading for MIT student agitators.[37] By the time Bernal did a public lecture tour of the United States to promote his 1939 book, *The Social Function of Science,* he was drawing big crowds of disenchanted scientists, most of whom were drawn from the nonelite backgrounds that, in Bernal's Marxist eyes, made them ripe for unionization.[38] However, efforts were made shortly after the start of Conant's presidency to ensure not only that Harvard's own scientists did not go down that route but more importantly that the rhetoric of the broadly leftist agenda linking the New

36. Kuznick 1987, esp. 9–70. The most politically prominent U.S. advocate of economic shrinkage as an antidote to the Depression was the Iowan populist, Henry Wallace, who was initially FDR's secretary of agriculture, then vice-president during his third term (1941–45), but finally dropped in favor of Harry Truman just before FDR's fatal fourth term, when Wallace himself started to envision the Soviet Union as a role model for America.

37. Kuznick 1987, 60–61.

38. Kuznick 1987, 237. Bernal found support at not only MIT but also Harvard, though Conant did all he could to contain its spread. Merton's deradicalization of Bernal's Marxism (see next note) was probably most effective in the long term, though some scientists persisted in their Bernalism in the face of local indignities. A case in point is Bart Bok, an astronomer who in the early 1930s helped organize a Harvard chapter of the Bernal-influenced American Association of Scientific Workers to study the effects of science on society. See Kuznick 1987, 231–32. Fifteen years later, when Conant launched his General Education in Science initiative, Bok proposed one of the first courses, indeed, the one with the most explicit sociological content. Social Relations of Science revolved around the writings of Bernal and other British Marxist scientists, as well as the founding documents of UNESCO and the transcripts of the debates over the remit of the U.S. National Science Foundation. See *Minutes,* 2 December 1946. When interviewed by Conant's biographer, Bok recalled that in the early Cold War era he was repeatedly pressured to accept U.S. military money and stop supporting the left-wing causes promoted by his colleague Harlow Shapley. See Hershberg 1993, 573–75, 617–18.

Dealers and the Bernalists was neutralized and ultimately co-opted by the old elites. This patrician reaction is summed up in the Pareto Circle.[39]

Harvard's Pareto Circle, which was convened from 1932 to 1942, spawned a theory of revolution that stressed, true to the original Latin, a "turning back," a restoration of natural order. This sense of revolution has perhaps left its most lasting influence on today's intellectual scene through Kuhn's theory of scientific change. The circle's convener was the practicing biochemist, certified physician, and amateur sociologist Lawrence J. Henderson (1878–1942).[40] Among the circle's members were such future luminaries of American sociology as Talcott Parsons, George Homans, and Robert Merton, as well as the intellectual historian Crane Brinton, and occasionally Joseph Schumpeter, who had taken up a professorship in Harvard in 1932 (after having failed as finance minister of his native Austria).[41] Henderson, as it happened, was also Conant's uncle-in-law and an influential member of the committee that made Conant Harvard's president and later

---

39. An initial sense of the Pareto Circle's subtle but pervasive influence can be gleaned from Robert Merton's successful deradicalization of Bernal's Marxist sociology of science. A junior member of the circle, Merton was instrumental in watering down Bernal's message for the American audience, especially after 1943, when Merton perceived that the social sciences were themselves just as susceptible to political control as the natural sciences. Merton's strategy was to neuter the distinction between factors that were "internal" and "external" to the development of science. He replaced Bernal's original Marxist sense of internalism—that is, scientists' developing class consciousness—with an "internalist" perspective that was closer to Max Weber's sense of scientists pursuing their vocation for its own sake. Merton's sense of "externalism" similarly muted any specific reference to capitalism as the source of science's compromised utilitarianism. In this way, Merton's famed "four norms of science" were born. See Merton 1942; Mendelsohn 1989; also Kuznick 1987, 85.

40. Henderson's sociological works are compiled in Barber 1970. It is worth remarking that Henderson cotaught a general history of science course at Harvard with George Sarton that attracted such founding American members of the field as I. B. Cohen. Cohen 1984 provides reminiscences. My thanks to Stephen Turner for originally alerting me to Henderson's significance. A good place to begin following the leads initiated here is Buxton and Turner 1992.

41. Heyl 1968. It was through Henderson that Parsons, then a young assistant professor, became acquainted with Pareto's organismic conception of "social system," which was to leave a permanent imprint on Parsons's style of "grand theorizing," starting with his first major work, Parsons 1937. Alvin Gouldner was probably the first major sociologist to recognize the significance of the Pareto Circle for the subsequent history of American social thought. See Gouldner 1970b, 149 ff. However, his take on Pareto's influence at Harvard is somewhat different from mine. Whereas Gouldner stressed the entitlement that the Harvard faculty felt they had to determine the course of social thought, I would put it more in terms of noblesse oblige, that is, the belief that those fortunate enough to have made it to Harvard incur an obligation to conserve what has been of value, lest it otherwise be lost in the normal course of events. (I shall consider this point in more detail, with special reference to Kuhn's social position, in chapter 8.) Not only do I believe this explains the constrained sense of innovation pursued by members of the Pareto Circle (as well as Kuhn), but also the grain of truth contained in the uncharitable observation that Harvard is a large cage for gilded birds. On the Pareto Circle's influence on Conant's educational ideas—to provide freedom for the elites and structure for the masses—see Lagemann 1989, 194–95.

established the Society of Fellows as a pool for nurturing the future academic elite. It was here that Conant recruited Kuhn and much of the staff for the General Education in Science program.[42]

The official charge of Henderson's group was to discuss the ideas of the Italian founder of modern sociology, the political economist Vilfredo Pareto (1848–1923), an independently wealthy scholar of encyclopedic ambitions and Machiavellian dispositions. Benito Mussolini regarded Pareto as such a formative influence on his own Fascist ideology that he made him a peer in Italy's House of Lords (though, like all good intellectuals, Pareto refused to be co-opted for the few months he survived his title). The middle third of the twentieth century witnessed the peak of Pareto's popularity in the United States. This "Marx of the Master Class" was routinely listed as one of the founding fathers of sociology, along with Marx, Emile Durkheim, and Max Weber.[43] His ideas had made their mark on the American scene shortly after the onset of the Great Depression, when they were used to explain the emergence of a managerial elite on whose judgment corporate shareholders were increasingly dependent, mainly as a result of the increasing complexity and volatility of the market in advanced capitalist societies. Thus, when the Depression made it clear that many managers had steered their shareholders in financially unsound directions, the shareholders sought new managers whom they felt they could trust. But, as Pareto would have it, the shareholders did not themselves regain control of the company.[44] In the political sphere, Pareto depicted a class of professional politicians competing for votes. This image proved attractive not only for Fascists (most of whom have been democratically elected) but also

42. Conant 1970, 87–90, 108.
43. This point is most evident in the book that first crystallized the canon of sociological theory, Parsons 1937, on the strength of which Parsons received tenure at Harvard. The other major national tradition in sociology that has been, and continues to be, so influenced by Pareto is the French. See esp. Aron 1970. Pareto's influence on recent French sociology of science is discussed in chapter 7, section 6.
44. Moreover, rather than reversing this tendency, FDR's corporatist New Deal policies merely extended the forfeiture of democratic control across all segments of society, a point that became apparent only with America's entry into World War II. See Hooks 1991. The most popular text from this period to make this point is Burnham 1941, written by a Trotskyite-turned-neo-Machiavellian (compare the metamorphosis of Daniel Bell). A latter-day successor to this line of thinking is the Viennese-turned-New Yorker, Peter Drucker (1909–), the inventor of the genre of managerial self-help literature that has reinvented the market niche for Machiavelli's *Prince* in airport bookstalls across the world today. An interesting analysis of the moral asymmetry implied in the Paretian approach to management—namely, between the concentration of power in an elite and the distribution of responsibility to the masses—with specific reference to Conant and Parsons is MacCannell 1984. MacCannell draws attention to the fact that, on the one hand, Conant strongly supported vocational training for the poor, yet on the other, he equally supported the concentration of these people in ghettos so as to enable a fully functioning state, even if a substantial percentage of the American population were targeted in a nuclear war.

for Max Weber and Schumpeter, as well as the most influential American journalist of the period, Walter Lippmann. They all held that in large, heterogeneous democracies, the public is best invited to the ballot box only for purposes of symbolically ratifying policy measures already in progress or making choices between candidates whose differences are unlikely to upset the status quo.[45]

Pareto was inspired by Machiavelli's distinction between two types of human instincts, one oriented toward tradition and one toward innovation. To these corresponded the two sorts of elites that Machiavelli observed in history, which was firmly anchored in Thucydides' *Peloponnesian Wars*: traditional "lions," symbolized by the disciplined Spartans, and innovative "foxes," symbolized by the crafty Athenians. Societies are more secure when the lions are in power. But being ingrained in their ways, the lions find it increasingly difficult to adapt to changing circumstances (in Kuhn-speak: "anomalies"), thereby causing the society to stagnate. Here the foxes make their mark, albeit usually for only a brief spell, after which a new den of lions is installed. Machiavelli first presented his image of political history in the late fifteenth century as a cyclical account familiar from Greco-Roman historiography.

Four centuries later, Pareto embedded this account in a view of society as an organism in search of equilibrium, which is what first grabbed Henderson's attention.[46] As Pareto saw it, the nineteenth century belonged to the innovators, but these foxes created such a volatile political climate that people in the twentieth century were destined to seek strong, stable leaders. The new lions would get their way, not by providing rational justifications for their policies but by appealing to people's hopes and fears. However, these nonrational appeals would prove to be the most efficient means of restoring social equilibrium, and hence rational at a systemic level. Perhaps the best indicator that equilibrium has indeed been restored is that the pursuit of politics is successfully insulated from potentially disruptive interventions by the public. (Compare Kuhn's requirement that a discipline is not scientific unless its research agenda is autonomous from external

---

45. An excellent discussion of this thesis, especially in terms of its role in the rise of political science in the United States, is Purcell 1973, esp. 95–113, which culminates with the reception of Walter Lippmann's *Public Opinion* (1922).

46. Henderson's views were aligned to those of his colleague, Walter Cannon (1871–1945), concerning what Cannon called "homeostasis" (the Greek translation of the Latin "equilibrium") as the fundamental principle of organic survival. Henderson attempted to wed the principle of homeostasis to Darwinian accounts of natural selection in *The Fitness of the Environment* (1913). His own experimental work in biochemistry led to the standard explanation for acid-base neutrality in bodily fluids by reference to physiological buffers that inhibited any extreme tendencies in either direction.

interests.) Pareto believed that demagogues, whose own political survival depends on cultivating the goodwill of the masses, had a better working understanding of the human condition than conscientious social democrats whose advocacy of "rational social planning" was an unholy hybrid of theory and practice. Theorists of the social order, according to Pareto, best function as clinicians and maybe even coroners, but not planners and founders.[47]

From the Great Depression of the 1930s to the student revolts of the 1960s, Pareto was widely seen in the United States as steering a middle course between liberalism's naively optimistic views of spontaneous individual advancement through free trade and socialism's excessive trust in a "people's revolution" as a remedy to current inequalities of wealth. As it turns out, Henderson's access to Rockefeller Foundation money enabled him to apply the Paretian framework to study worker fatigue and other industrial hazards, especially as these led to work stoppages.[48] Henderson believed that addressing worker dissatisfaction at the moment of its surface expression would prevent a full-scale overturn of the status quo later on, as that dissatisfaction deepened. A similar strategy had worked in Germany for Bismarck, who introduced the world's first social insurance scheme as a sop to the socialists. These conservative adjustments to resistance are based on the hope that the resisters will be flattered by *any* concession to their claims.[49]

This strategy provides the clearest precedent for Kuhn's ideas about the "essential tension" in science between tradition and innovation, which later came to be articulated in the distinction between normal and revolutionary science. Tellingly, these ideas were first presented in a conference devoted to "the identification of scientific talent" at the University of Utah.[50] As Kuhn made clear in his introductory remarks, several of the

---

47. This clinical approach to the study of society enabled several members of the Pareto Circle, especially Talcott Parsons, to function as informants to the FBI and CIA. See Heims 1991, 183–84.

48. Buxton and Turner 1992, 386–93. See Purcell 1973, 101 ff. for Lippmann's influence in legitimating Elton Mayo's "human factors" approach to the psychology of industrial relations, which Henderson helped promote at Harvard with Conant's blessing. The Machiavellian character of Mayo's research lay in his belief that workforce pacification was necessary for steady productivity and that this could be brought about only by indirect means—that is, the manipulation of work conditions.

49. The ease with which radicals can be co-opted to the establishment was a regular theme in political sociology at the dawn of the twentieth century. See esp. the "fallacy of optimism," as discussed in the introduction to *Reflections on Violence* (1908) by Georges Sorel (1847–1922) and *Political Parties* (1915) by Roberto Michels (1876–1936). These works are discussed in Horowitz 1968.

50. Kuhn, "The Essential Tension: Tradition and Innovation in Scientific Research" (1959), in Kuhn 1977a, 225–39. Both the year of this conference (1959) and the declared search

participants were interested in finding ways of making scientists more open-minded, flexible, and risk seeking. Kuhn devoted his contribution to scuppering this aim by arguing that radical change is licensed in science only as the last resort once a series of incremental changes of the established tradition have proven inadequate.[51] A few years later, in *Structure*, Kuhn would openly declare, in Paretian fashion, the first sign that a scientific paradigm needs to be replaced is that the *elites* are dissatisfied with it.[52]

To be sure, Kuhn never cited Pareto. But given that Paretian ideas were taken for granted in the intellectual environment of Kuhn's formative years and even informed the institutions of which he was a part, it is natural to conclude that the concept of revolution available to him owed more to Pareto than, say, Marx. The clearest witness to this point is the popular intellectual historian and doyen of Harvard humanists, Crane Brinton (1898–1968). In explicitly Paretian terms, Brinton argued that belief in democracy is contradicted by the evidence of human history. Do we therefore treat democracy as an ultimate but inaccessible ideal—the secular version of the Christian salvation story—or do we resort to cynicism toward the ideal altogether, as the Fascists had and (Brinton suspected) the Communists did? Brinton's third way was a kind of "political realism" recognizable to Conant and other Harvard Paretians as predicated on the idea that whatever progress has been made in human history has come from understanding and adapting to our failures in achieving the democratic ideal.[53] (In the next section, we shall examine the conceptual framework informing this third way.) In his classic *Anatomy of Revolution*, Brinton exploited the metaphorical implications of this approach, especially the idea that social

---

for "talent" betray the influence of Sputnik-driven initiatives to have the United States keep pace with the Soviet Union in the "science race" that had come to be the main arena in which the Cold War was fought. In *The Endless Frontier*, Vannevar Bush had already written of the need to maintain and enhance the nation's "scientific capital" by cultivating "talent" in science at the secondary-school level. This charge was extended to the National Science Foundation—previously an exclusively research-oriented agency—through the National Defense Education Act of 1958, which helped fund conferences such as the one in which Kuhn first presented the "essential tension" thesis. On the politics surrounding this newfound interest in scientific talent, see Montgomery 1994, 208–15. Conant's General Education in Science program met its end in the midst of this consolidation of national educational priorities. See chapter 4, section 7.

51. The examples of political revolution that Kuhn uses to introduce the idea of scientific revolutions are drawn from the Balkan Wars that helped precipitate World War I. These are usually attributed to the failure of the Austro-Hungarian and Ottoman Empires to maintain order among internally divisive ethnic and cultural groups, much like a mature paradigm's inability to contain the spread of anomalies. The resolution of a paradigmatic crisis by redistributing scientific labor to more specified domains of inquiry may be likened to the devolution of the old eastern empires into independent nation-states. See Kuhn 1970b, 93.

52. Kuhn 1970b, 92.
53. Brinton 1950, esp. 259–61.

science is a form of clinical practice that diagnoses without predicting.[54] Here revolution appears as integral to the body politic's immune system — its natural response to illness, starting with a crisis akin to a fever, which peaks as delirium, followed by a period of convalescence, after which the body returns to a state of equilibrium with its environment.[55]

Nowadays Brinton's theory occupies a distinct place in the gamut of theories of revolution.[56] It provides the paradigmatic explanation of revolution in terms of the frustration experienced by semiprivileged individuals (as opposed to, say, the mobilization of the disgruntled masses). The frustration comes from the increasing gap between economic and political power. Put more psychologically, desire outstrips realization. Seen from the outside, which is the most natural vantage point from which to appreciate Brinton's theory, the revolutionaries represent an emerging elite whose position in society has improved — but not as quickly as these upstarts would like — just as the remaining obstacles to its ascendancy have been clearly identified. Precisely because they are upstarts, they are more concerned about the dimming prospects for progress in the future than about the strides that have been already made to conserve worthwhile innovations within traditional institutions. In Kuhnian terms, Brinton's revolutionaries occupy the position of the younger generation of scientific elites in a mature paradigm that appears to be accumulating anomalies faster than it can solve them. That the paradigm had solved significant problems in the past does not inspire respect in the younger generation; on the contrary, it magnifies their expectations and impatience when these are not fulfilled. We shall return to this feature of the revolutionary change in science, the so-called Planck effect, in chapter 6, section 4.

6. The Teacher's Teacher: A Portrait in Political Realism

James Bryant Conant was probably most responsible for redefining the mission of American academic life in a century that has witnessed two "hot" world wars and at least one "cold" war, all in a world that had become capable of self-destruction, largely through the ingenuity of some of the best academically trained minds. We can begin to understand how Conant's political vision was translated into Kuhn's academic practice by considering the general mentality bred by the Cold War, which Conant and

---

54. Brinton 1952 generalizes from the English, American, French, and Russian Revolutions.
55. Brinton 1952, 17.
56. See, e.g., Coleman 1990, 472–79.

Kuhn expressed in their respective spheres of action: *political realism*. Political realism consists of the following four propositions, which together define a world where the absence of violence passes for peace:[57]

1. that the political arena is limited to a few "superpowers" whose interests are, in principle at least, not jointly realizable;

2. that these superpowers are supposed to have roughly the same destructive capabilities;

3. that the combination of such mutual antagonisms and destructive capabilities make it likely that any open conflict would quickly escalate to total war; and

4. that the superpowers distrust each other sufficiently to make it unlikely that their differences can be resolved by diplomatic negotiation or any other relatively straightforward means of communication.

Political realists are quite clear in portraying the interests of the superpowers as distinct and incommensurable. On the one hand, this renders impossible any reconciliation of differences; on the other hand, because the political realist acknowledges that each superpower is prone to act in ways that promote its own ends but also have morally offensive consequences, there is a de facto tolerance of superpower actions. Thus, a card-carrying political realist such as Henry Kissinger would advise President Richard Nixon on how America's interests might be best advanced in the face of Soviet actions; yet he might also be the first to acknowledge that human rights violations and the murder of millions would be in the USSR's interests and, in that sense, completely understandable.[58] One way of gauging the effect that this sort of judgmental relativism has had on histories of the recent past is to note the greater reluctance that historians have had in condemning Stalin vis-à-vis Hitler, even though, strictly speaking, it is very likely that more people died under Stalin than under Hitler. Were political realists as dominant in the historiography of World War II as they are in Cold War historiography, Hitler would probably not appear in such an unequivocally negative light.[59]

Conant's personal version of the Cold War vision can be observed in four mental tendencies, each of which left a residue in Kuhn, who in his last interview called Conant "the brightest person I had ever met."[60] First, the Cold War vision licensed mild forms of irrationality—say, the threat of holding one prospect hostage to another—to create an immunity against still greater irrationality. The idea was to normalize the destructive

---

57. On the history and philosophy of political realism, see Smith 1986; Rosenthal 1991.
58. On Kissinger's application of political realism, see Smith 1986, 192–217.
59. Gaddis 1993. My thanks to Robert Newman for drawing this article to my attention.
60. Kuhn et al. 1997, 146.

capabilities of American science so that the public would be prepared to stave off alarmist arguments from both the Communists abroad and the pacifists at home.⁶¹ Consider Conant's first major defense of public support for science after the explosion of the atomic bomb:

> If, following Emerson, we think of the potential power of destruction of the atomic bomb as the price we must pay for health and comfort and aids to learning in this scientific age, we can perhaps more coolly face the task of making the best of an inevitable bargain, however hard.⁶²

We are never told why we must have all of science or none of it—why, say, medical research must be held hostage to research into subatomic particles. But in the Cold War vision, trade-offs are ineluctably made: once again, a noble lie for a still nobler truth. There is a general metaphysical belief in the inseparability of things good and bad, such that everything good comes at a cost. Moreover, the cost one is willing to absorb is often the best measure of the value one places on something. The range of legitimations for this intriguing mental tendency run the gamut from Reinhold Niebuhr's theological invocation of original sin to Hans Morgenthau's more scientifically inspired view that the recognition of what Jean-Paul Sartre called "dirty hands"—the coincidence of good and bad consequences of the same act—is a natural result of an increased understanding of the causal structure of reality. For both Morgenthau and Niebuhr, as well as Conant, to act decisively under such a state of enhanced knowledge is the true standard of moral courage.⁶³ Kuhn's commitment to the double-truth doctrine is plausibly seen as born of this environment, especially given Conant's remarks to him about the multiple purposes served by presenting the history of science in various lights.⁶⁴

Conant's second tendency may be seen as an abstraction from the previous two. Because the Cold War constructed the world as an "either/or" rather than a "both/and" moral sphere, Conant could not accept the compatibility of a stable social order and regularly occurring social criticism. If

---

61. During this period, social psychologists at Yale University did a brisk business providing experimental evidence for the efficacy of this strategy. The charge of this research was to make Americans resistant first to Fascist and later to Communist propaganda. The initial findings suggested that propaganda was more easily resisted if subjects had already been exposed to a weakened version that could be either easily assimilated or refuted. The earliest study in this vein was Hovland, Lumsdaine, and Sheffield 1949. However, the classic statement of inoculation theory is McGuire and Papageorgis 1961.

62. Conant 1947, xiii.

63. On this Platonic debt of Cold War political theory, see Rosenthal 1991, 121–50. For a philosophical analysis of the problem of dirty hands, see Williams 1981, chap. 4. Perhaps the most learned consideration of these issues is Aron 1966, chaps. 19–20.

64. Letter to Kuhn dated 11 October 1950, in *Conant Presidential Papers*. See chapter 4, note 76, for the relevant context.

one is already committed to a given regime, then it would constitute nothing less than treason to encourage criticism, unless the critic were prepared to provide an appropriate alternative to the status quo. To do otherwise would be to expose the society to external threat. Part of the point of having political decisions absorbed by a class of expert civil servants in modern nation-states was to ensure that potential sources of social criticism were treated as problems of administration — that is, until the professionals themselves were overwhelmed, in which case a significant break in policy was licensed, one inviting public participation. Translated into the realm of science we find Kuhn's exaggerated fear of institutionalized criticism, or "permanent revolution," in Popper's adaptation of Trotsky.[65] Conant said virtually the same thing fifteen years earlier: "We can put it down as one of the principles learned from the history of science that a theory is only overthrown by a better theory, never merely by contradictory facts."[66]

As is to be expected of someone who addressed many different problems at once, Conant was given to conceptualizing several domains in largely the same terms; hence his third mental tendency. Conant repeatedly informed the public that by virtue of their training, scientists organized their research into a "conceptual scheme," in terms of which appropriate "strategies and tactics" for tackling specific problems were defined.[67] Unsaid in this otherwise unremarkable observation is that the tendency to organize one's thinking in a given domain may enhance its transferability across domains, as is already apparent from Conant's characterization of an orientation to scientific research in terms of a militaristic metaphor. Kuhn illustrated the same tendency in his fashioning of an expression from his Air Force experience in the Second World War: "mopping up operations," the

---

65. Kuhn 1970b, 77 ff. This is a rare occasion in which Kuhn is in fundamental disagreement with Michael Polanyi, a point that came out when Polanyi commented on a sketch of *Structure*'s argument at a conference in Oxford in 1961. While Polanyi agreed that normal science indeed requires the suspension of "universal doubt," he felt that Kuhn neglected the role of "original minds" who purposefully reintroduce such doubt, typically against the tide of collegial opinion. Whereas Kuhn saw a radical change in research orientation as purely a reaction to accumulated anomalies, Polanyi offered a more proactive, perhaps even heroic, image of the scientific revolutionary. See Polanyi 1963, 379–80.

On the other hand, when treated as a doctrine for interpreting experimental results, Quine's underdetermination thesis can be read as an abstract expression of the Conant-Kuhn strictures against regular public criticism: "a single theoretical hypothesis cannot be conclusively falsified, so any statement can be held to be true come what may if we make drastic enough adjustments elsewhere in the system" (Quine 1953, 43). One can easily imagine the "drastic enough adjustments" in the background assumptions of the tested theory as akin to an Orwellian Ministry of Information that continually rewrites the theory's logical structure to preserve its truth.

66. Conant 1947, 36.

67. A good philosophical analysis of Conant's appeal to conceptual schemes is D'Amico 1989, 32–51.

bombing that follows an initial strike to ensure that all the targets have been hit. Thanks to Kuhn, this expression came to mean the puzzle solving that is needed to complete a paradigm's mapping of some domain.[68] Indeed, given the technical status that "mopping-up operations" now has in the science studies disciplines, Kuhn may have actually succeeded in laundering the expression of its militaristic origins.

One conceptual scheme that Conant shared with other Cold Warriors was the idea that, with the end of World War II, the torch of democracy had been passed from Europe to the United States—this against the backdrop of an increasingly menacing Communist bloc. However, Conant's distinct spin was to convert this conceptual scheme into a rubric for thinking about the needs of the postwar university curriculum. Conant identified the future of science with one particular democracy, namely the United States. The two rose and fell together in his mind. From the tenor of Conant's remarks in the foreword to Kuhn's first book, *The Copernican Revolution*, it is clear that Conant linked the decline of Europe as a world power with the persistence of an outdated humanities-based university curriculum. Russia, in Conant's thinking, stood for the educational system that had come to recognize the power of science, but at the same time undermined the integrity of science by tying it too closely to technological applications and political aims. The United States potentially stood alone not simply in the increasing importance it gave to science in the university curriculum, but especially in the respect nonscientists gave to the integrity of the scientific enterprise.[69] Thus, the American educational system, at least in Conant's mind, offered a unique resolution of C. P. Snow's "two cultures" problem of the "arts" versus "sciences" that was by the late 1950s very much in the air. In chapter 7, section 2, we shall consider the educational resolution instituted in Snow's Britain, which laid the groundwork for science and technology studies.

68. Kuhn 1970b, 24.
69. James Bryant Conant, foreword to Kuhn 1957, xiii–xviii. This point was reinforced by Harvard-trained sociologists of the period, especially Robert Merton and Bernard Barber. See Novick 1988, 296–97. Barber followed up Conant's contrast between American and Soviet science policy initiatives in his (unsuccessful) attempts to persuade the National Science Foundation to provide support for sociology of science within its recently (1953) conceived history and philosophy of science division. Interestingly, Barber ran afoul of Herbert Dingle of University College London, then head of the oldest and largest history and philosophy of science program. In 1950, Conant had dispatched Kuhn to check out Dingle's program as a possible source of ideas for Harvard's own General Education in Science courses. (See letter from T. S. Kuhn to J. B. Conant, 9 May 1950, in *Conant Presidential Papers*.) Dingle was concerned that sociologists might notice that most science had so far been done in Western countries and thereby conclude that only those of European stock could do science. Thus, Barber's eagerness to play the Cold War card was trumped by Dingle's recollection of Nazi precedents. See Rossiter 1984.

As for the fourth and final mental tendency, Conant never looked back once a decision had been made. If all decisions, however good, are tainted with bad consequences (per the first tendency), then not much is to be gained by replaying the decision and imagining an alternative course of action. Indeed, the good one wishes to achieve with a decision is directly tied to the steadfastness with which one follows through on the decision. Consequently, the trajectory of Conant's policy thinking followed a predictable course. His first considered view on some public issue was typically a "liberal" one: a generous attitude toward the Communists, say, a dislike for universal military service, or an openness to congressional involvement in science policy. However, Conant's threshold for being converted to the opposite position was quite low. One major violation of this trust was enough to have Conant call for a *permanently* suspicious policy toward the violators. He showed little interest in negotiating intermediate positions; rather, he preferred to contain any possible future violations.[70]

This habit of mind nurtured a particular view of democratic participation in the policy process. On the one hand, Conant frequently called for public debate about national military or educational policy; on the other hand, once the relevant policy decision had been taken, Conant took the lead in ensuring that the decision was publicly perceived as the only one that could have been made. Like most Cold Warriors, Conant was convinced that acknowledged policy reversals were signs of moral and political weakness. The most vivid embodiment of this conviction was what is now recognized to have been Conant's crucial role in convincing U.S. secretary of war Henry Stimson to publicly portray the dropping of the atomic bomb on Japan as "inevitable"—a moment in Orwellian historiography, if there ever was one.[71]

It is here that political realism left its strongest stamp on Kuhn's thought. For example, a Kuhnian paradigm emerges from the successful disentanglement of science from ideology, where "ideology" implies irresoluble disagreement over principles of theory and method.[72] Once some substantial

---

70. In theoretical discussions of strategic action, the need to match force with counterforce is often called "tit-for-tat." However, there are two styles of tit-for-tat. One style (championed by Conant, George F. Kennan, and most Cold Warriors) called for a permanent state of suspicion in response to one episode of violated trust. Another style, championed by one of the founders of peace research, Anatol Rapaport, would limit retaliatory action to the round in which trust was violated. According to a computer simulation run at the University of Michigan by political scientist Robert Axelrod, Rapaport's strategy was found to be more likely to foster long-term survival. See Rapaport 1989, 225–43.

71. Bernstein 1993. My thanks to Robert Newman for drawing this article to my attention.

72. Kuhn 1970b, chap. 2. The locus classicus for this conception of ideology is Bell 1960. Kuhn is explicitly invoked to legitimate the next phase of international relations theory, which surrenders the idea of one world paradigm for the incommensurability of multiple

points of agreement have been found, they are then intensified and ramified through the disciplining processes of "normal science," which—as we shall see in chapter 4, section 3—seems to encompass all possible arenas from the classroom to the laboratory. In the terms of political realism, "world order" is operationalized as the containment of conflict to an arena where disputes can be efficiently settled, if not preempted altogether. The agreement of scientists on experimental standards for the assessment of competing theories is, therefore, much like the agreement of political leaders on standards by which to assess the balance of power.[73] Without such binding agreements in either science or politics, chaos would presumably soon return—especially in our nuclear age. Whereas political force in the past had been defined in terms of direct geographic control, the prospect of nuclear war shifted the standard to a nation's capacity for destruction at a distance. But "luckily," an Orwellian sense of history discourages scientists and politicians from ever looking back at the sacrifices that had to be made to secure this "advance" in geopolitical order, thereby facilitating an "end of ideology."

During a Kuhnian crisis, candidates for the next paradigm impose incompatible cognitive and material demands. Under the circumstances, the actual record of the history of science can be envisaged as an ambiguous duck-rabbit Gestalt that may be "completed" in a variety of ways, depending on which future policy initiatives are projected, as was suggested in our discus-

---

superpower paradigms. See Jervis 1976, 143–202. International relations is perhaps the one field in which Kuhn has been used to model not the field's research trajectory *but its very subject matter*, which suggests that perhaps Jervis and his colleagues have excavated the unconscious origins of Kuhn's image of incommensurability. Ironically, while we have been arguing that Kuhn's image of paradigms and incommensurability partook of the Cold War worldview, Jervis (p. 5) justifies the relevance of Kuhn on the grounds that the thought patterns of decision makers are probably no different from—and certainly no better than—those of scientists. Here we see a case in which ignorance of the sociological roots of one's own knowledge claims can make two opinions of common ancestry—Kuhn's and the average Cold Warrior's—seem like independent corroboration.

73. Max Weber (1864–1920), the sociological founder of political realism whose thinking about the relationship between politics and science most resembles Daniel Bell's in our own time, publicly fulminated against "dilettantes" who could not decide whether they wanted to play by the rules of politics or the rules of science, and so typically operated by a pseudo-scientific conception of politics, such as racialism or revolutionary socialism. Weber vented much of his spleen against Lenin and the Bolsheviks, who invoked Marx's authority to claim that the time was ripe for overturning the Russian monarchy. The mark of Weber's political realism lay in the literalism of his belief in politics as a "profession" with its own specialized knowledge and standards of competence. Indeed, for Weber, the advent of such professionalism was one of the few senses in which politics could be said to have progressed over the last two thousand years. See Anderson 1992, 182–206. Although Weber's is not an especially robust sense of progress, it is nevertheless one we have already encountered in his phenomenological follower, Alfred Schutz, who in turn informs the sociological assumptions of much of contemporary STS (see chapter 4, section 8). It is a sense that sees "knowing one's place" to be an advance over boundary-spanning criticism. It is clearly a sense with which I disagree.

sion of the realism-instrumentalism debate in chapter 2.[74] Moreover, since there is no principled way of resolving this ambiguity, whatever resolution is reached will require continual reinforcement, lest the previous state of uncertainty and conflict return. The political-realist subtext of this general picture may be seen by considering the genealogy of the images promoted by two philosophical theses broadly associated with a paradigm in conflict: the *underdetermination of theory by data* and the *incommensurability of worldviews*. According to the former, the same facts can be represented in many different ways; according to the latter, there are various ways of representing the world, each projecting a different set of possible facts.[75]

In chapter 1, we saw that underdetermination has been a problem for philosophers following Whewell's lead in trying to find a unique route to scientific truth, while incommensurability is the natural outcome of cultures, languages, artistic traditions—and scientific research programs—appearing to possess an organic, specieslike unity. Although both theses can be found in Kuhn's corpus, generally speaking underdetermination looks most compelling in the context of different theories competing to occupy the same conceptual space at a given point in time. The single-tracked model of scientific change presented in *The Structure of Scientific Revolutions* fits this description much better than the essays collected in *The Essential Tension*, which are more focused on the historian's ability to reconstruct the conceptual framework of past science. Regarded in terms of temporal orientation, underdetermination follows history in order of occurrence, namely, from the past to the future, while incommensurability reverses this order, as the historian tries to recover the past from the standpoint of the present and—as we saw in chapter 2's discussion of Mach's excavation of repressed scientific traditions—a desired future.

But most significant to a genealogy of political realism is the alternative geopolitical imagery inspired by the two theses. On the one hand, underdetermination is comparable to imperialism, whereby several nations aspire to colonize a vast terrain that lacks its own sovereignty according to international law. On the other hand, incommensurability is akin to sovereign nation-states that restrict trade among themselves by the imposition of high tariffs. Thus, the "scramble for Africa" among European powers starting in the 1880s provides a pointed example of underdetermination projected on an entire continent, whereas incommensurability is evidenced

---

74. This interpretation of the nature of the conflict between alternative paradigms is developed in Taylor 1982. We shall return to this interpretation of incommensurability in chapter 6, section 7.

75. On the complementarity of underdetermination (understood as an application of Quine's indeterminacy of translation thesis) and incommensurability, see Fuller 1988, chap. 6.

in the distinct tribal visions of the colonized African peoples, who became the objects of inquiry of colonial anthropologists, such as Edward Evans-Pritchard, an inspiration for contemporary relativism, as we saw in the final section of chapter 2. It is worth stressing here that the relationship between the political and scientific worldviews runs deeper than mere analogy, given that both underdetermination and incommensurability were first formulated as theses about *science* in the same period, with the competing research programs typically identified by their national origins and distinctive styles of thought.[76]

Kuhn's commitment to the paradigmatic containment of inquiry is perhaps most apparent from the pronounced absence of *criticism* as a regular feature of science, a point to which we shall return in section 6 of the next chapter. According to Kuhn, unrestricted criticism is the mark of preparadigmatic inquiry; from the political standpoint of the Cold Warrior, it constitutes sedition and verges on treason. Even Kuhn's biologistic view of "progress," ostensibly modeled on Darwinian evolution, reveals his profound aversion to conflict. On this view, science evolves, as paradigms continually divide and subdivide without end, simply in response to their selective adaptation to specific environments, much like how new species are supposed to come into being.[77] However, Kuhn presumed without argument that (after a scientific revolution) these new species inhabit niches that peacefully coexist in the same environment. There is no perpetual struggle for survival in Kuhn's evolutionary epistemology, caused by, say, the scarcity of funding or even overlapping claims to cognitive authority. Consider Kuhn's recent clarification of this point:

> Some characteristics of the various practices [of science] entered early in this evolutionary development and are shared by all human practices. I take power, authority, interest and other "political" characteristics to be in this early set. With respect to these, scientists are no more immune than anyone else, a fact that need not have occasioned surprise. Other characteristics enter later, at some developmental branching point, and they are thereafter characteristic only of the group of practices formed by further episodes of proliferation among descendents from that branch.[78]

---

76. According to Ian Hacking, Heinrich Hertz's *Principles of Mechanics* (1894) was the first book to argue that the same facts can be represented in many different ways. See Hacking 1983, 143. Hertz presented the problem of theory choice as a matter of competing imperial systems (British, French, and German), each operating with its own set of fundamental principles, but all designed to colonize an ontologically undifferentiated realm of phenomena. For the cultural context of Hertz's mechanics, see Janik and Toulmin 1973, chaps. 5 and 6.

77. Kuhn 1970b, 205–6. The most thorough study of Kuhn as an evolutionary epistemologist is Callebaut 1996.

78. Kuhn 1992, 17.

Given the absence of sustained conflict in Kuhn's account (or for that matter, any clear sense of an external environment that acts as a selection mechanism), he should have perhaps turned to embryology—the internally driven processes of growth through functional differentiation—rather than evolution as his source of biological metaphors: more mitosis than speciation.[79] His participation in the "end of ideology" motif would have been made even more apparent. Nevertheless, the postideological sensibility that Kuhn inherited from political realism has been passed on to the moral indifference—or "symmetry"—that characterizes the sociologists of scientific knowledge (SSKers), who claim Kuhn as important part of their pedigree.[80] We shall return to this theme in chapter 7.

79. A careful reader of *Structure* should not find Kuhn's deafness to biological matters surprising. For example, it is clear that he did not have Darwin in mind when he conceptualized scientific revolutions as requiring a "crisis" resulting from the accumulation of unsolved empirical problems. A popular attempt to rejig Kuhn's model to enable Darwin to pass as a revolutionary is Ruse 1979.

An interesting perspective on this matter is provided by the venerable biological taxonomist, Ernst Mayr (b. 1904), who distinguished between Kuhn's account of scientific change and Darwinian evolution, arguing that the latter's "anti-essentialism" provides a better model of the historical trajectory of biology. By "essentialism," Mayr means the tendency to see the world in terms of intrinsic natures that a given thing either has or not. Examples would include the pre-Darwinian conception of species, in which each species is specially created by God, as well as Kuhn's conception of paradigms, between which change occurs only through such radical breaks as personal conversions or collective revolutions. Mayr 1994 diagnoses Kuhn's residual essentialism as a product of his training in theoretical physics, which disabled him from seeing that only professional theorists feel the need to "choose between" theories and hence are inclined to see the intellectual world in the black-and-white terms of the essentialist. Here Mayr reinforces a point I made in chapter 1, section 7. A similar critique is lodged in Toulmin 1972. However, neither Mayr nor Toulmin relates essentialism to political realism.

80. Indeed, it has been argued that the sociologists have greater affinities with the political realists than Kuhn himself did (especially if Kuhn is perceived as a defender of the moral authority of science). See Proctor 1991, 224–31, an extended critique of the SSK classic, Barnes 1975.

IV
From Conant's Education Strategy to Kuhn's Research Strategy

1. Conant's Clientele: The Postwar Guardians

The main constituency for the General Education courses that Conant designed for Kuhn and other disaffected scientists of the World War II generation was the war veterans, whose education at Harvard was being funded by the Servicemen's Readjustment Act of 1944, otherwise known as the G.I. Bill of Rights.[1] These students were destined for managerial positions in a world that had been radically transformed by science while they were on active duty. When the bill was first proposed, just as the tide had turned in favor of the Allied Powers, President Franklin Roosevelt believed that the economic drain on the nation that would result from the bill's generous scholarship provisions would offset any increases to what is now called the nation's "stock of human capital." But FDR's misgivings were assuaged once he was told that America's political stability would be jeopardized if the returning soldiers did not seem partake of the anticipated postwar economic boom.[2] In that case, it was feared that they would be susceptible to Communist infiltration.

Be it on political or economic grounds, the G.I. Bill was passed by Congress, in spite of the opposition of most leading academicians, including Conant himself. However, once the act was passed, Conant warmed to it, as the veterans' military discipline rendered them better prepared for the rigors of university instruction than the average college matriculant from

---

1. For a more detailed account of the causes and effects of the G.I. Bill, see Olson 1974.
2. In fact, the justification for the G.I. Bill was more economic than Roosevelt had realized, for if the veterans had not become scientifically literate, their subsequent unemployment would have struck a much bigger blow at the postwar American economy than the cost of their training. See Galbraith 1952.

an upper-class background.³ Reflecting his newfound faith in military training, Conant famously titled Natural Sciences 4, the course that Kuhn eventually took over, "Strategy and Tactics of Science."

Such was the testing ground for both Conant's and Kuhn's ideas about the history of science. It is well documented by the generous acknowledgments that they make to each other's pedagogically inspired writings during the 1950s and early 1960s. A comparison of Conant's 1951 "popular" work, *Science and Common Sense*, and Kuhn's 1962 "scholarly" work, *The Structure of Scientific Revolutions*, reveals many similarities of conception. Perhaps the deepest yet least remarked upon of these is, ironically, an asymmetric view of the relationship between the dynamics of *research* and *teaching*. Both portray research as a naturally self-organized activity that proceeds at its own pace. In contrast, education can be explicitly designed to convey the fruits of that research, and especially its attendant mind-set, to those who dwell beyond the "cutting edge." Accordingly, while there can be no discovery before its time, a discovery once made can be disseminated more or less immediately, if the educational system is properly planned.

I raise this asymmetry between the spontaneity of research and the planning of education because it is shared by most historians of science today, who continue to focus on the vicissitudes of the research trajectory to the exclusion of the means by which its products are distributed and incorporated into other practices, thereby providing science its legitimacy in society at large. Telling in this regard is that while Kuhn accepted Quine's doctrines of the underdetermination theory by data and the indeterminacy of translation, he denied Quine's metatheoretic doctrine that science is an extension of common sense. Whereas Quine here followed the general pragmatist line that science is a consistently applied encounter with empirical reality, which is the natural tendency of human beings unless impeded by what Charles Sanders Peirce called "fixed ideas" (a.k.a. religion), Kuhn regarded the adoption of a scientific worldview—especially a new paradigm—as something that does not happen naturally, but rather requires disciplined effort.

In addition, there are two important but related differences between Conant and Kuhn. First, missing from Kuhn are Conant's explicit Cold War references, especially the idea that basic scientific research is the ultimate bulwark of democracy against the Communist menace. In contrast, all of

---

3. Had Conant and his colleagues recalled the examples of the Jesuits being grounded in military discipline and the Inquisition in Scholastic examination, they would probably not have been so surprised at the veterans' adaptiveness to academic standards of performance.

Kuhn's "politics" are to be found in the metaphors he uses to characterize the internal workings of science, "revolution" being only the most striking example. Indeed, contrary to his popular reputation, Kuhn follows traditional history and philosophy of science in portraying the larger society as only minimally affecting the conduct of the scientific community. Science is permeable to politics only during a "crisis," a state that is to be avoided whenever possible and terminated quickly whenever unavoidable.

The second difference is one that Conant himself observed and criticized in Kuhn's work. Conant portrayed science as subject to multiple viewpoints that collectively advance knowledge by a process akin to trial and error. Each experimental scientist operated with her own "conceptual framework," yet that fact alone did not preclude communication or agreement on experimental outcomes. Kuhn, by contrast, offered a view of the scientific community—under the rubric of "paradigm"—that was so homogeneous and entrenched that it could be overturned only at the risk of revolution. Moreover, failure to maintain the uniformity of the paradigm—say, by allowing anomalies to accumulate without resolution—can easily lead to communication breakdown and the disorganization of inquiry.[4]

What is the relationship between these two points of divergence between Conant and Kuhn? Kuhn basically conflates two claims about the governance of science that Conant tried to keep separate in the various rhetorical contexts in which he operated. One of Conant's self-appointed charges as "action-intellectual" was to make the direction of scientific research largely impermeable by the state and industry. In this context, it was easy to suggest that the reason science should not be externally regulated is that it is already internally regulated by something like a Kuhnian paradigm. However, Conant never explicitly made this argument, and probably did not believe that such an argument could be made, given his rather individualistic sense of how science is done. Indeed, Conant resembles the free-market ideologue who presumes that whatever economic order issues from the removal of government planning will be "natural." No mechanism is ever offered for how this putative free market of science governs itself.

---

4. Quoted in Hershberg 1993, 860 n. 84. To appreciate Kuhn's transformation of Conant's position, simply compare their accounts of a standard case history used in Natural Sciences 4: Priestley versus Lavoisier on the discovery of oxygen, which launched the chemical revolution. See Conant 1950; Kuhn 1970b, 53 ff. Kuhn shifted the account of the discovery of oxygen from Conant's emphasis on the advantage of Lavoisier's mental preparation over Priestley's to the incommensurability of the two conceptual frameworks, which originally led to their mutual miscommunication and now makes it difficult for historians to tell when exactly oxygen was discovered. It is worth emphasizing that Kuhn's reinterpretation did not involve the introduction of any new archival material.

Enter Kuhn, who postulates such a mechanism for the self-regulation of science and, in the process, renders Conant's public strategy for demarcating science from the rest of society constitutive of scientific practice itself. Had Conant gotten his way, the success of his arguments for the autonomy of science would have enabled scientists to go about their business without worrying about the politics of demarcation. In contrast, Kuhn seemed to presume that scientists are not only in the business of advancing theories and conducting experiments, but also—and even in the very same activities—policing the boundaries of science. Thus, every Kuhnian normal scientist already has a bit of Conant's demarcationist impulses in her own soul. In essence, Kuhn did to Conant what Freud did to Plato. Plato drew a sharp distinction between the guardian and philosophical castes in the *Republic*, comparable to the difference Conant saw between his own activities and those of the practicing scientist: the former protects the latter from corruption. Kuhn's Freudian twist is to internalize the protector in the manner of a superego.

2. Kuhn's First Assignment: Cut and Paste the Sources

Conant's vision of post–World War II America was one in which scientific research would be making greater demands on the public coffers as the pace of progress quickened. There would be greater disagreement among scientific experts about the path that science should take. Consequently, the public (and Conant had in mind the typical Harvard-trained corporate executive or government official) had to become "expert in judging experts."[5] Here, in his characteristically no-nonsense manner, is how Conant put the matter:

> Whether we like it or not, we are all immersed in an age in which the products of scientific inquiries confront us at every turn. We may hate them, shudder at the thought of them, embrace them when they bring relief from pain or snatch from death a person whom we love, but the one thing no one can do is banish them. Therefore every American citizen in the second half of this century would be well advised to try to understand both science and the scientists as best he can.[6]

Conant's key presumption that survives in Kuhn is that the bridge between science and society is best constructed by the public growing more accustomed to science—rather than by scientists growing more accus-

---

5. Conant 1952a, xiii.
6. Conant 1961, 3.

tomed to the public.⁷ Conant's vision was operationalized by the chemist Leonard Nash, with whom Kuhn taught Natural Sciences 4 and to whom Kuhn dedicated *The Copernican Revolution*, his first book:

> The goal of our course [is] the development of a healthy and informed attitude toward, an appreciation and understanding of, science ... In discussing the earlier and the more recent activities of science it is carefully pointed out to our students that the urgency of the desire for the solution of a particular scientific problem constitutes no guarantee that it will be or even can be solved ... That great strides will be made in science cannot be doubted, but that they will lead immediately to a desired end ... can be legitimately questioned. This unbounded optimism can be tempered by the suggestion that unlimited scientific progress in a specified direction cannot be bought or otherwise induced. Favorable conditions for an advance can be created, and the advance may be accelerated; but the advance itself cannot be created.⁸

When Nash proceeds to diagnose "two delusions" that cause students to ask of science more than it can reasonably give, he arrives at suspiciously philosophical disorders: "the cult of the fact" and "the cult of the method." The remedy that Nash proposed as "the case approach" involved presenting students with carefully edited versions of classic experimentalist writings that will enable them at least to follow, and perhaps even to reenact, the experiment in question.⁹ Using a line of argument that

---

7. I do not mean to deny that Conant was concerned that scientists be properly "socialized." However, he responded by offering courses that exposed scientists to the old liberal arts culture that dominated the elite colleges prior to World War II. See Freeland 1992, 79.

8. Nash 1952, 116–17.

9. Nowadays an appeal to "cases" in studying science suggests an approach that draws, at least by analogy, on ethnographic fieldwork. In fact this is quite different in spirit from the Harvard approach, which was modeled on effective teaching techniques developed in the law and business schools, whereby stylized episodes would be presented to students as the basis for deriving relevant, generalizable principles. Conant was instrumental in spreading the case study method across the university by giving free rein to the business dean, Wallace Donham, to develop a school of public administration that would apply principles of management to government. As Conant observes in his autobiography, the advantage of cases was that they simulated the experience of, say, passing a verdict or managing a firm, in which a set of particulars are given and then one must decide which principles apply. This was in contrast to the more deductively oriented training that students received in economics. See Conant 1970, 438–40. Students who acquired the mind-set (paradigm?) suggested by the cases were presented as capable of managing organizations with comparable characteristics, regardless of the goods or services they were designed to provide. The great strength of this acquired mind-set was its ability to discern simple patterns amid complex data.

The most striking career made with this mentality was probably that of Robert McNamara, who, upon graduating from the Harvard Business School in 1940, went from the accounting department to the presidency of the Ford Motor Company, and then became U.S. Secretary of Defense (at the height of the Vietnam War) and ultimately retiring as president of the World Bank (in charge of development aid for Third World countries). A good biography of

Kuhn would later adopt in explaining his Gestalt-based notion of paradigm, Nash defended the suitability of the case approach on the grounds that to understand science is to understand how patterns can be extracted from ambiguous data, a cognitive process that has basically remained unchanged since the rise of experimentation.[10] Thus, Nash had no qualms about selecting only experiments from before the mid–nineteenth century. Indeed, such experiments had the additional advantage that their accompanying texts presupposed relatively little esoteric background knowledge. Even so, the amount of pedagogical guidance involved in this process — "preparation of the materials," in Nash's own clinical terms — must have been considerable, as Nash anticipates and meets charges of "spoon-feeding" the students.[11]

A comparison of the original texts with Nash's "prepared" versions reveals that left on the cutting room floor was any trace of the social production of knowledge, especially as expressed in appeals to disciplinary audiences and other potential readers. Of course, such excisions become more difficult to perform on later scientific writings, in which the "logic" of the text is determined almost entirely by the writing conventions of a given discipline, conventions that themselves cannot be understood without delving into the background interests involved in the formation and maintenance of the discipline's boundaries.[12] However, the course that Nash and Kuhn taught was designed to introduce such larger social factors only *after* the students had grasped the "technical detail" and "tradition of work" that are highlighted in the edited texts. Thus, a strong internal/external history of science distinction is manufactured in the student's classroom experience, in a way that would not be apparent if the students were simply taught the texts in their original form. As Kuhn might put it, the incommensurability that exists between today's scientists and students is bridged by having students observe historical cases of science as simulations of how science is performed today.

But were matters quite so simple? As it turns out, the difference between internal and external histories of science even needs to be manufactured in the minds of professional scientists. At least, Kuhn was forced to

---

McNamara that focuses on the formative role of his education is Shapley 1993, esp. 20–25. For an illuminating discussion of the political context from which the case study method emerged, see Buxton and Turner 1992. I draw some implications for normative philosophy of science from this lineage in Fuller 1993b, 215–17.

10. Nash 1952, 116; Kuhn 1977a, 293–319.
11. Nash 1952, 115.
12. For a cross-disciplinary treatment of the style manual as contested terrain, see Bazerman 1988.

acknowledge diplomatically this methodological lesson a decade later, when in charge of compiling source materials for the history of quantum mechanics.[13] It seemed that the people his team interviewed—the surviving founders of the twentieth-century revolution in physics—had a difficult time recalling the technical details of their revolutionary work, though they had no trouble recounting its social setting and larger cultural, often philosophical, significance.[14] At one point, Kuhn even confessed that the well-prepared interviewer often found himself prompting, if not instructing, the likes of Bohr and Heisenberg in the details of the physics that made them famous.[15]

It may be tempting to interpret the curious behavior of Kuhn's scientific interlocutors as symptomatic of the selective memories that mark "great men" in their dotage. (Indeed, Niels Bohr died at the age of seventy-seven the day after Kuhn interviewed him in Copenhagen.) However, this is a natural interpretation only if one is already committed to a normative vision of scientific inquiry in which the technical work should both be separate from and take precedence over the various contexts that make the work meaningful to those who did it. Failure to live up to this vision, then, would constitute a source of error. Of course, a more straightforward explanation—one that does not require attributing mental infirmity to Bohr, Heisenberg, et al.—is that, because at the time of their training physics had not

13. Kuhn et al. 1967.
14. The anachronistic tendency to marginalize the larger philosophical and cultural significance that the originators of quantum mechanics invested in their work continues to the present day. Witness Weinberg 1996, an attempt to debunk Heisenberg's and Bohr's philosophizing. An interesting question to pose to Weinberg, given the constructed character of the internal/external history of science distinction, is whether he observed such a sharp distinction *before* the work for which he himself received the 1979 Nobel Prize in Physics was recognized as a solid achievement. In the period since the end of World War II, scientists on the cutting edge of research have been less likely to register the internal/external distinction than either professional historians or scientists laboring in the research hinterlands. The psychology that informs this difference is not too difficult to fathom. Someone whose work has not yet been canonized will be more alive to the factors that potentially impede its canonization than someone who no longer suffers from that problem.
15. Kuhn et al. 1967, 4–5. Kuhn's observation casts in an ironic light his injunction that the historian attend to the technicalities of the science under investigation—since the scientists themselves might not be so attentive! See Kuhn 1977a. 157–58. I believe that a historian like Kuhn can so neatly distinguish the "technical" from the "social" aspects of scientific work, not because he was trained in the technicalities of the science (though that might explain how the historian can give the distinction some content), but because the historian was not existentially involved in the original production of the science, and hence lacks the holistic vision needed for understanding what it was like to have lived as a scientist at a given place and time. Significantly, one of the physics-trained graduate students working under Kuhn, Paul Forman, has brilliantly recaptured this vision in his various histories of twentieth-century physics, especially Forman 1971.

yet become so differentiated from larger philosophical concerns, Bohr and Heisenberg did not share Kuhn's normative vision and hence had no reason to devote a "part" of their memory (assuming such a locution makes psychological sense) to remembering the lines of reasoning that would strike the internal historian of science as important.[16]

From the standpoint of the branch of cognitive psychology that deals with the systematic elicitation of memory — "protocol analysis" — the type of structured recollection of the history of quantum mechanics that Kuhn and his colleagues attempted to assemble in their interviews is best seen as a series of narratives jointly constructed by interviewer and interviewee, as constrained by specific dates, places, and collateral evidence.[17] If the reader concludes that this "narrative" is a piece of fiction, comparable in its epistemic warrant to a historical novel, then so be it. However, the point I wish to emphasize is that Kuhn was clearly interested in fitting the scientists' recollections within the genre of internal history of science.

Consequently, once Kuhn discovered that scientists naturally spoke in autobiographical terms about their careers, he designed questionnaires for his interviewers that proceeded chronologically, forcing the interviewed scientists to disentangle the specifically "technical" bits from their account and address issues that could serve to fill gaps in extant histories of quantum mechanics.[18] In courtroom parlance, Kuhn trained his interviewers to "lead the witness." This is all the more striking, considering that Kuhn was adamant that one could not do a proper history of a period in which one had a vested interest in the outcome. For, while Kuhn personally abided by his own stricture, his amassment of archival materials was done in the spirit of enabling the writing of a specific kind of history in the future. In this respect, it is telling that in an otherwise adulatory review of Kuhn's career, his former student Jed Buchwald felt compelled to remark upon — yet leave tactfully unanalyzed — Kuhn's inability to study unprocessed archival materials.[19]

---

16. A long line of research in the psychology of artistic creativity questions the idea of losing one's mental powers with age. Originally, this research was focused on mental illness (especially schizophrenia), but interest has increasingly shifted to such degenerative disorders as Alzheimer's, which potentially offers an alternative interpretation of the physicists whom Kuhn interviewed. A theoretically astute critique of an allegation of mental infirmity in a contemporary artist is Fraser 1998.

17. On the methodological perils associated with protocol analysis, see Ericsson and Simon 1984. On the epistemological implications of protocol analysis for the historiography of science, see Fuller 1993a, 155–62.

18. Sample questionnaires appear in Kuhn et al. 1967, 150–55.

19. Buchwald 1996.

3. Kuhn's Second Assignment: Normalize "Normal Science"

If Conant did not actually invent normal science, he was certainly America's leading importer of the concept from the German laboratories where it was honed in the years leading up to the First World War. Readers of *Structure* will find this is a strange thing to say, given that, at one point, Kuhn traces the practice of normal science as far back as Aristotle's *Physics*.[20] But on the face of it, the idea that "normal science" was being conducted in antiquity seems profoundly anachronistic. It would mean that the *Physics* served as a blueprint from which a team of researchers were then able to carve out well-defined projects for the next two millennia. Given that that paradigm case of paradigms, Newtonian mechanics, was able to hold the collective attention of cutting-edge researchers for barely two centuries, these works of antiquity would have had to cast quite a spell over inquirers. Can this literally be what Kuhn meant when he called a work like the *Physics* a "paradigm" at first and a "disciplinary matrix" later? If so, it would reinvigorate all the old stereotypes of the premodern era as a Dark Age characterized by blind obedience to authority.

But fortunately, we can lay those old ghosts to rest, if we attend more closely to Kuhn's historical examples. For, like Ptolemy's *Almagest*, the great astronomical treatise of antiquity, Aristotle's *Physics* does not figure in Kuhn's text as an exemplar for research until the epistemic authority of antiquity is challenged in the sixteenth and especially seventeenth centuries. Of course, I am referring to that hyper-Copernican, Galileo Galilei (1564–1642). When Galileo argued that reality was not big enough to accommodate the claims of both Copernicus and Aristotle-cum-Ptolemy, he was thinking of their two bodies of knowledge as operating in the sort of scarce epistemic environment presupposed by the existence of paradigms, only one of which can dominate a field at any given time; hence, the title of Galileo's most popular work, *Dialogue Concerning the Two Chief World Systems*, and his talk of having founded a "new science." However, as the famous exchanges between Galileo and Cardinal Bellarmine made clear, this way of framing the relationship between Copernicanism and the epistemic orthodoxy was a much more difficult pill to swallow than the actual Copernican beliefs themselves.[21] Bellarmine's adherence to a doctrine of multiple truths (some sacred, others secular; some for the masses, others for the elite) did not commit him to Galileo's agonistic view of knowledge.

---

20. Kuhn 1970b, 10.
21. The classic analysis of this exchange is Duhem 1969.

Nowadays, it is easy to imagine that Bellarmine let his scientific scruples be compromised by the realpolitik of his Holy Office. He simply should have admitted that there *is* a contradiction between the two "world systems." But regardless of the ultimate soundness of this point, it is worth recalling just how difficult it is, even in our own self-consciously scientific age, to acknowledge and articulate meaningful conflicts between bodies of knowledge. For example, the social sciences overlap significantly with each other—and often with biology—in the phenomena they attempt to explain, yet the patent incompatibility of their various explanatory frameworks is normally regarded, in Kuhnspeak, as "incommensurable," and therefore not in urgent need of resolution. Perhaps even physics and biology are incommensurable in this sense.[22] Bellarmine would have appreciated the attitude.

The one important kind of exception to this pluralism arises when specific public policies hang in the balance. In those circumstances—such as whether intelligence tests "really" measure cognitive ability and whether such ability, in turn, corresponds to a particular configuration of genes—more aggressive knowledge claims are subject to severe methodological critiques that typically inhibit policy implementation, yet without precluding further development of the lines of research associated with the odious policy proposals.[23] Galileo's achievement becomes all the more striking, then, since he managed to motivate the perceived incompatibility between two bodies of knowledge *in a pure research context*.[24]

The Galilean encounters were significant in causing both pro- and anti-Copernicans to reconstruct their histories, such that the earth-centered and sun-centered universes began to appear as if they had *always* been rivals for the same paradigmatic space. In this sense, Galileo was responsible more for a *metascientific revolution*—that is, a radical change in the ends and means assigned to scientific inquiry—than a "scientific revolution," in Kuhn's sense of the term. Galileo's metascientific revolution was twofold: astronomical phenomena had to be explainable by physical principles, and theological claims about the natural world had to be consistent with those principles.

22. For an elaboration of this point, see Fuller 1997d, chap. 2, which includes a discussion of Stephen Jay Gould's efforts to prevent bioevolutionary explanations from being judged by the standards of physics.
23. The increasingly important class of exceptions to such epistemic tolerance involves research programs competing for scarce resources, often resulting in what I have dubbed *cognitive euthanasia*. See Fuller 1997b.
24. To be sure, fears that Galileo's rhetoric might play into Europe's ongoing religious wars ultimately figured in his censorship. But it is equally clear that such concerns were foreign to Galileo himself, who never renounced allegiance to the Church of Rome.

To appreciate the advance on Copernicus himself, consider the subtle difference between saying—as Copernicus did in his magnum opus, *De Revolutionibus* (1543)—that theologians should leave astronomy to the mathematicians and saying—as Galileo did on numerous occasions—that the results of mathematical astronomy (and physics) can be used to contradict the claims of theologians. The former argued for the methodological autonomy of astronomy from theology, while the latter took the further step to claim that the two disciplines compete in a common epistemic space. No doubt, Copernicus's move made Galileo's possible. But they are not the same move—and, more importantly, Copernicus could not have been regarded as a Kuhnian revolutionary until Galileo had reinterpreted the significance of his work.

Here it is tempting to argue that Copernicus was more modest than Galileo in his claims—avoiding all talk of "world systems" and "new science"—because he feared persecution by the Church, especially given his employment as a cathedral canon. But this would be misleading, since from his early *Commentariolus* (1513), Copernicus saw himself as critiquing and correcting Ptolemy, but he stuck rather close to mathematical arguments based on a Pythagorean preference for uniform circular motion, which could be easily evaluated without raising theological considerations. This, in turn, enabled his work—such as had circulated in his lifetime—to be generally well received by Catholic scientists.[25]

However, harking back to the problems raised in the issues that opened chapter 1, I would say the difference is that Galileo presupposed that there was one, publicly accessible truth, whereas Copernicus believed in multiple truths to which people had varying degrees of access, depending on their mental preparation. This is more than simply the difference between Galileo being a "realist" and Copernicus an "instrumentalist" about astronomical theories in modern philosophical parlance. "Instrumentalism" is especially misleading as a characterization of Copernicus because it implies that he either suspected that the heliocentric hypothesis was false or was unconcerned about its ultimate truth-value. To be sure, Ptolemaic astronomy was effectively an off-the-shelf technology for predicting the location of the heavenly bodies for navigational and astrological purposes. It did this quite well, and the Copernican alternative certainly did it no better. This would seem to undermine the instrumentalist reading of his work that enabled *De Revolutionibus* to be read throughout Europe until Galileo managed to get it on the Church's *Index of Forbidden Books* in 1616.

25. Gjertsen 1984, 98–99. This book is an excellent source for the histories of the receptions of the major scientific works from Euclid to Darwin.

Indeed, Copernicus was not much of a practicing astronomer in the modern sense, in that he made virtually no new observations of the heavens. Instead, Copernicus saw himself as a mathematically inspired natural philosopher walking in the footsteps of Pythagoras and Plato; hence, his desire to find the smallest number of circles to explain the orbits of the planets, even if it meant dislodging the earth from its central position. Thus, at the end of his life, Copernicus could claim, in good conscience, that his main achievement had been, not the heliocentric hypothesis, but the elimination of one of Ptolemy's calculational gimmicks, the equant, which brought astronomy closer to the Pythagorean ideal.[26] These Platonic scruples, I maintain, extended to an adherence to the double-truth doctrine, which disabled scientific advancement from being a vehicle of revolutionary change.

Often Copernicus's self-imposed "esotericism" is taken as evidence of his ability to foresee the "full import" of his own theory—how it could provide the basis for a radically new cosmology—and hence his fear of its socially destabilizing consequences. This is meant to suggest that Copernicus had revolutionary impulses but did not believe the time was ripe for action. However, it may well be that Copernicus's concerns went in quite the opposite direction, namely that someone might get the *mistaken* impression that his theory had implications that went far beyond the precincts of astronomy—to include, for example, theology. In other words, Copernicus may have become more guarded as he grew older because he saw more clearly the perverted directions in which a full-blown version of his theory could be taken. In fact, Galileo may have been precisely the sort of person Copernicus was worried about—someone who was inclined to place various bodies of authoritative knowledge in conflict with each other and try to settle the differences between them by conjuring up striking empirical demonstrations that can sway the impressionable (as opposed to esoteric mathematical calculations that only the adept can follow).[27] If this speculation is correct, then Copernicus did not have a revolutionary bone in his body.

The point of conjuring up all these counterfactuals is to suggest that there is reason to believe that Copernicus would not have approved of the role he has been made to play—in the hands of Galileo and others—as the founder of what we now call the Scientific Revolution. No doubt, similar

---

26. I. B. Cohen 1985, 113–14, quoting Owen Gingerich, who in turn was referring a statement Copernicus made to his follower Erasmus Reinhold.

27. For a diagnosis of Galileo's troubles that turns on his specific challenge to theology (as opposed to his challenges to Ptolemy, Aristotle, etc.) because of the Protestant Reformation, see McMullin 1997.

things could be said about others who have been assigned roles in the major metanarratives of the history of science. The point may be pressed more strongly. If Copernicus would have disowned Galileo, what would *we* make of Newton or Darwin were they to pay us a visit, courtesy of some time-traveling technology? Newton would probably be much too fond of the theological implications of contemporary physics and Darwin of the socio-biological implications of evolutionary theory to sit comfortably with the standards of "political correctness" in the scientific community, one that restricts discussion of the value implications of research to the precincts of science popularization. I would wager that Newton and Darwin would find their soulmates in, respectively, Paul Davies and E. O. Wilson rather than, say, Steven Weinberg and Richard Lewontin.

In other words, an unprompted visit by our revolutionary founders could prove to be a public relations disaster for the scientific profession, comparable to the depiction of the Second Coming of Jesus in Dostoevsky's "The Grand Inquisitor." Notwithstanding Richard Rorty's catchy talk of "the conversation of mankind," the fact remains that the civility of that conversation depends on the likes of Newton and Darwin sitting in respectful silence as we talk about the implications of their views. The history of science is not just an account of scientists standing on the shoulders of their predecessors trying to see a bit further ahead. In addition, the later scientists reconstitute the ground on which the earlier ones stood, which means that the perspective attributed to their predecessors is altered. This is the difference between a *linear* and a *reflexive* reading of the history of science, which I shall discuss in section 5 below.

Only if one already holds that astronomy must be accountable to physics does Ptolemy's theory display any significant "anomalies" in Kuhn's sense of outstanding unsolved puzzles. That this belief was not yet widely held may be witnessed by the thirteen centuries of scholars who tolerated the metaphysical discrepancies between Ptolemaic astronomy and Aristotelian cosmology, the one treating the apparent motions of the planets as imperfect manifestations of Platonic circular forms and the other treating them as real but qualitatively different from earthly motions. According to Kuhn, there are no grounds for promoting an alternative paradigm until the anomalies are acknowledged. Thus, Galileo famously had to resort to all sorts of experiments, telescopic displays, and rhetorical tricks to get the scholars of his day to take seriously that the models of Ptolemy and Copernicus could not simply sit side by side as alternative abstract instruments of stargazing.

As Galileo performed his metascientific revolution, he did much to elaborate and correct Copernicus's original theory. Yet it would be a mistake to call Galileo's efforts "normal science" because these activities—which occurred more than a half-century after Copernicus's death—were precisely the ones that eventuated in Copernican astronomy replacing Ptolemy's. It might seem, then, that Galileo was engaged in Kuhnian "revolutionary science." Unfortunately, the story does not quite end here because even after Copernicus was regarded as having the superior astronomical theory, there continued to be more a selective appropriation of the theory than its systematic development. Reading the history from past to future, rather than in reverse, this would be the best way to characterize the work of Kepler and Newton, unless one wished to entertain seriously the idea that they too were "normal scientists" in the Copernican paradigm.[28]

When marking the transition between two paradigms, the first image that comes to mind is the "gestalt switch" between Ptolemaic and Copernican astronomy, which was already a standard philosophical example of radical conceptual change in science by the time Kuhn came on the scene.[29] The basic idea is that supporters of Ptolemy and Copernicus would see the same data—the changing location of the heavenly bodies over time—as evidence for diametrically opposed conclusions, the former holding that these bodies circled the earth whereas the latter held that the earth itself was moving. If we think of a paradigm as an interpretive scheme and overall worldview, then Ptolemy and Copernicus may be said to have operated with radically different "paradigms." But were they paradigms in the sense of providing the background assumptions for the conduct of normal science? Everything said so far suggests no, since we can in principle acknowledge two radically alternative ways of seeing the world without feeling the need to choose one. For example, Kuhn regarded two paradigms as a bilingual person regards the two languages she speaks, which makes perfect sense unless one is already in the grip of a picture of science that

---

28. This point comes out nicely in the account of Copernicus's reception given in Gjertsen 1984, esp. 113–20.

29. Kuhn 1970b, 85. The most influential account of the gestalt switch between the two astronomies appears in Hanson 1958, 4–8. Hanson traces the example he shares with Kuhn to William James. Hanson's version specifically compares what two sixteenth-century contemporaries—the geocentrist Tycho Brahe and the heliocentrist Johannes Kepler—see when they look at the Sun. As Kuhn was to do later, Hanson drew upon the Wittgensteinian image of the reversible "bird-antelope" gestalt to epitomize the alternative meanings that can be read off the same sensory data. The first depiction of Copernicus as having effected the sort of change in perspective nowadays associated with a Kuhnian gestalt switch is often thought to have occurred in the preface to Kant's *Critique of Pure Reason* (1781). However, for an interesting analysis of Kant's text that disavows any such intent, see I. B. Cohen 1985, 237–53.

demands one to make *irreversible* decisions, not mere back-and-forth switches. Thus, Kuhn was dogged by philosophers for failing to explain how one paradigm would and why it should replace another.[30]

In any case, for us now to reconstruct the history of science in paradigm-driven terms is quite different from assuming that scientists in the past either conducted research in those terms or felt that their history had to be continually reconstructed to appear progressive. The former issue is dealt with below, but the latter captures the points made in chapters 1 and 2 of this book. Whewell made the first step toward Kuhn by arguing on normative grounds that the history of science had to be captured under one encompassing theory-driven rubric. Planck then triumphed over Mach by claiming that Whewell's narrative rubric should retain its unitary focus while suppressing the historical character of that focus—specifically, facts relating to why a particular focus is being promoted at the particular time, at the expense of viable alternatives.

But even granting those points, the spread of Orwellian historiography has been surprisingly slow in the twentieth century. One problem is that, in the social and even the biological sciences, historical investigation often remains the best source for uncovering deep conceptual objections that refuse to go away, such as the status of physical laws in open systems or the ontological status of biological species.[31] Another problem is that professional scientists still read histories of their field, so that they turn out to be less ignoramuses than philistines when it comes to appreciating the role of the past in informing their future endeavors. In other words, scientists continue to acquire enough knowledge of their field's history in nonprofessional settings that their current research is effectively immunized from the professional historian's knowledge.[32]

As far as scientists' conducting paradigm-driven research is concerned, here too I would argue that before the dawn of the twentieth century,

---

30. See Kuhn 1970a, 207.

31. It is rare to find a scientist born after 1920 who follows Mach in using historical arguments to criticize her field of inquiry. Any number of books by the thermodynamicist Ilya Prigogine and the natural historian Ernst Mayr exemplify this "old school" approach to these matters.

32. This specific sense of historical philistinism is illustrated by scientists' tendency to compare the history of science to military history, in terms of their allegedly clear sense of purpose and achievement. In part, such philistinism is a residue of Whewell's attempts to represent the advance of science by cartographic metaphors, as discussed in chapter 1, section 6. A good recent example is Weinberg 1996, 15, esp. the concluding paragraphs. However, historians of science have sometimes mimicked the same tendency to characterize their own pursuits—to be sure, much to the consternation of their more discerning colleagues. See Holmes 1997. The article was critiqued in Stuart Leslie's review of Soederqvist 1997, which circulated in the H-SCI-TECH-MED listserv in June 1998.

science was only sporadically conducted in canonical Kuhnian fashion, with the most striking instances of normal science to be found described in the philosophical utopias of Thomas More, Francis Bacon, Auguste Comte, and Henri de Saint-Simon. Indeed, only toward the end of the nineteenth century does science begin to approximate the dimensions of the modern industrial enterprise either presaged or promoted in the utopias. But even at this late date, things are not entirely Kuhnian. Discussing the organization of the physics lab of Hermann von Helmholtz (1821–94) in 1870s Berlin, Jed Buchwald makes a valuable distinction between a *blueprint for calculating* and a *design for experiment*. Helmholtz designed his research program to maximize the generation of novel results that would challenge the Newton paradigm. The spirit of this enterprise is much more Popperian than Kuhnian. It is also strikingly different from the pattern found in twentieth-century particle physics, whereby large-scale research is planned to calculate the results of equations whose validity is presumed, unless the values veer wildly from what the background theories predict. That is more in the Kuhnian vein of normal science research.[33] This is the aspect that Conant exploited in the 1920s to convert Harvard into a frontline producer of scientific research. In this context, it is worth observing just what a historiographical mirage that most paradigmatic of paradigms, Newtonian mechanics, turns out to be.

Although Newton's achievement in *Principia Mathematica* was hailed in many quarters, ranging from British natural theologians to French Enlightenment wits, and Newton himself held a chair in mathematics at Cambridge, there was nevertheless no Newtonian "school" dedicated to solving the remaining puzzles in the master's paradigm. To be sure, Newton had several notable scientific correspondents and critics — much more so on the European continent than in Britain itself — who helped define the strengths and weaknesses of Newtonian mechanics. But Newton's interlocutors generally made these contributions in the context of promoting their own research agendas, not Newton's. The establishment of the Ecole Polytechnique in the aftermath of the French Revolution, one hundred years after Newton, seemed to provide the basis for a disciplined pursuit of mathematical physics in the Newtonian vein, as inspired by Laplace. However, with the restoration of the French monarchy, this pursuit declined, as mathematics retreated from its alliance with experimental science to its traditional liberal arts base.

It was only with the expansion of the German university system in the third quarter of the nineteenth century that a continuous tradition of

33. See Buchwald 1993, esp. 184 and 202–3 n. 16.

normal science is established for a mathematical physics that treats Newton as its source. However, the immediate cause of this development was less a desire to emulate Newton's achievement than to ensure common measurement standards so that an error in one laboratory did not contaminate the entire national scientific effort. Interestingly, these historiographical corrections to the Kuhnian thesis were made by Kuhn himself, partly in response to those who took his account too much at face value.[34] The point that Kuhn failed to observe, of course, is the correspondence between a full-fledged sense of paradigm-driven science and the modern nation-state in the nineteenth century, at the start with Napoleon and at the end with Bismarck. This point was not lost on Conant.

As we consider the competing images of normal science that figure in Kuhn's book, it will become clear that the concept is less *anachronistic* than *syncretistic*. It has been less a throwback to inquiry's idyllic past than a superimposition of perspectives from different moments in the history of science.[35] Therefore, we need an "iconography" of Kuhn's mythical image of normal science, which would involve stripping the various superimposed historical layers. Like the components of all good myths, these layers represent tendencies that have occurred to varying degrees in various fields at various times and places. *But they have never all occurred in one place at one time.* Ironically, this is the feature of myths that has led anthropologists in the past to regard "primitive" peoples as lacking a historical sensibility. Someone of a more charitable disposition may wish to think of normal science as an "ideal type," in Max Weber's sense of a hypothetical construct that sensitizes the historically minded social scientist to salient features of an unwieldy object of inquiry. However, were normal science meant to play such a heuristic role, it should not exhibit the internally contradictory features that I shall have occasion to observe below.[36] Our point of entry will be the "division of labor" that is so crucial for Kuhn's conception of normal science.

34. See Kuhn, "Mathematical vs. Experimental Traditions in the Development of Physical Science," in Kuhn 1977a, 60–65. Kuhn also develops these points in response to Joseph Ben-David, who, partly on Kuhn's authority, locates the origins of the nexus of mathematical and experimental traditions that defines modern science in the seventeenth-century Royal Society, culminating in Newton: see Ben-David 1972. For his considered views, see Ben-David 1984, 75–87.

35. In linguistics, *syncretism* refers to the fusion of different roots or inflections into one form. See Hjelmslev 1961, 88–93.

36. One reason why the syncretistic character of normal science does not seem to bother historians of science is that they interpret Kuhn's promiscuous borrowings from the history of science as a nod to the venerable "comparative method." For example: "[Kuhn's] habit of seeking patterns through the consideration of scientific ideas separated by time and place, a method not unknown to historians of science but more natural to colleagues in philosophy, emboldened others to think more comparatively about the past" (Servos 1993, 9).

According to Kuhn, a key feature of normal science puzzle solving is the appearance of a clear division of labor in science, at the very least to determine the skills needed to solve a particular puzzle.[37] But if normal science is defined primarily in terms of the possession of the relevant puzzle-solving skills, then it would be no more than a century old, since only at the end of the nineteenth century did scientific skills come to be distinguished in so clear a fashion as to call for specific forms of technical training. Even in a field like experimental physics, where one would expect the uniform training characteristic of normal science to appear first, there is considerable debate as late as the third quarter of the last century over so basic a methodological issue as "curve fitting," that is, whether one should include or exclude data that lie outside an emergent pattern of results. The problem basically gets resolved and the paradigm consolidated, Kuhn-style, by the publication of a widely adopted textbook.[38]

Given that a paradigm needs a univocal sense of measurement error in order to determine whether or not a series of experiments has produced anomalous results, it is surprising to learn that astronomy is the only branch of physics that has such a sense until the 1890s.[39] Kuhn himself claims that the precision and accuracy of physical measurements start to be systematically addressed only around 1840, and formal agreement is reached on the values of the fundamental physical constants only in the 1920s.[40] Thus, it was only in the last quarter of the last century that a form of inquiry bearing most of the defining features of Kuhnian normal science begins to be institutionalized in the physical sciences in Germany. Indeed, the idiosyncratic character of this development is signified by the expression "German science," which is coined to describe the look of the laboratory as a cross between a monastery and a factory, in which devoted bench workers never

---

37. For a comprehensive sociological analysis of this feature of normal science, see Fuchs 1992, 143–92.
38. Olesko 1993, esp. 24–28.
39. Hacking 1983, 234.
40. Kuhn 1977a, 220. For the backroom politics behind the negotiation of the Birge ratio, the formula used to set these values, see Mirowski 1996. Since histories of the natural and social sciences are still conducted largely independently of each other, no one has yet explored the possibility that a science concerned with human beings, such as experimental psychology, may have beaten physics in the calibration of the accuracy and precision of its measurements. Indeed, the original data for experimental psychology were the discretionary errors—what astronomers called the "personal equation" that dogged all physical observation. See Boring 1950, 134–56. (A good place to investigate the possible cross-fertilization on these matters would be the period of overlap in Leipzig [1887–1905], when those scientific polymaths, Wilhelm Wundt, experimental psychology's canonical founder, and Wilhelm Ostwald both held chairs.) Historians have now made it clear that statistical inference was developed to deal with the uncertainties of the human condition, and that many physical scientists as recently as a hundred years ago found it unsuited for their concerns. See Porter 1986; also Krueger, Daston, and Heidelberger 1987.

questioned the point of repeatedly taking the same measurements.[41] The name most popularly associated with this style of organizing research was Wilhelm Ostwald, whose physical chemistry laboratory in Leipzig trained Conant's predecessor to the chemistry chair at Harvard, Theodore Richards, the first American to receive the Nobel Prize in chemistry.[42]

The industrial organization of German science deeply impressed Conant in a 1924 visit to several German university chemistry departments. What struck Conant's eye was not just the uniform textbooks, but the fact that the Germans had relinquished the old apprenticeship model of inquiry, whereby graduate students would learn the full range of research skills—theorizing, experimenting, evaluating—from a master professor.[43] In its place had emerged the master professor as research manager, one who constructed a "program" divided into discrete projects, each of which would be delegated to a freshly minted Ph.D. specializing in that area. In turn, these postdoctoral fellows would train graduate students in the technical skills relevant to that part of the research program. At this point, a more complete characterization of German science may be welcome, this time written in a less-than-sympathetic, but no less revealing, manner by Pierre Duhem in the heat of Franco-German hostilities during the First World War:

> When, in a dream of the future, Professor Ostwald catches sight of the Europe he desires, Europe organized by a German triumph, he configures it entirely like one of those vast chemistry laboratories on which the universities beyond the Rhine pride themselves. There, each student punctually, scrupulously, carries out a small bit of work which the chief has entrusted to him. He does not discuss the task which he has received. He does not criticize the thought that dictated this task. He does not get tired of always doing

---

41. Duhem 1991.
42. Not surprisingly, Ostwald lent enthusiastic support to Taylorist schemes for the "scientific management" of all labor. See Rabinbach 1990, 254. An interesting contrast between Ostwald and Kuhn is over the significance of the "simultaneous" or "multiple" discovery of the energy conservation principle by scientists of several nations over a twenty-five-year period in the mid–nineteenth century. Reflecting his Taylorist scruples, Ostwald argued not only that this convergence pointed to the relatively equal distribution of scientific talent across humanity but also (and as a result) that scientists may be more efficiently organized into large research teams with minimal overlap of effort, so as to avoid any future incidents of multiple discovery. See Ostwald 1910, 185. In contrast, Kuhn was focused more on the role of the historian in arranging these disparate scientists in a common narrative. Characteristically, Kuhn eschewed policy recommendations, except to note that scientific textbooks rely heavily on such an account of discoveries to give contemporary inquiries a long-term historical focus. Ostwald would have effectively removed the main historiographic marker of scientific objectivity, namely, the "independent corroboration" of a finding. See Kuhn 1977a.
43. Conant 1970, 69–72. One of the last "pre-industrial" physical scientists was Heinrich Hertz (1857–94), whose training is described in Buchwald 1993. On the attitudes of the German physics community as they began to witness the consequences of this division of labor, see Hiebert 1990. My thanks to Skuli Sigurdsson for bringing this article to my attention.

the same measurement with the same instrument. He does not feel any desire to put some variety into his work, to exchange his habitual task for that done by some other student nearby. A toothed gear exactly meshed into a precise mechanism, he is happy to turn as the rule says he should turn, and he has no concern for the finished product produced by the machine. By virtue of his natural tendencies, he lives in the laboratory to which he is attached in the same fashion as, by virtue of his vows, the Benedictine or the Carthusian lives in his monastery.[44]

This description is remarkable in capturing the defining tensions of Kuhnian normal science that we shall explore below. Thus, we witness here the slide between mature scientific researchers and science students, between industrial and communitarian models of organization, between forced and voluntary silence, and so on. But before launching into a deeper exploration of these tensions, it may be worth looking at Kuhn's own words on the matter:

Few people who are not practitioners of a mature science realize how much mop-up work of this sort a paradigm leaves to be done or quite how fascinating such work can prove in the execution. And these points need to be understood. Mopping-up operations are what engage most scientists throughout their careers. They constitute what I am here calling normal science. Closely examined, whether historically or in the contemporary laboratory, that enterprise seems an attempt to force nature into the preformed and relatively inflexible box that the paradigm supplies. No part of the aim of normal science is to call forth new sorts of phenomena; indeed those that will not fit the box are often not seen at all. Nor do scientists normally aim to invent new theories, and they are often intolerant of those invented by others. Instead, normal-scientific research is directed to the articulation of those phenomena and theories that the paradigm already supplies.[45]

By his own account, Conant's attempt to import the German model met resistance from many of America's leading scientists, notably Robert Millikan, who, like Conant, had worked with Theodore Richards.[46] Nevertheless, Millikan and his colleagues felt that the model stunted the creative development of the "postdocs," whose best years may be spent as glorified indentured servants in some master professor's lab. In contrast, Conant was impressed by the overall productivity of the German scientists and especially the ease with which the research director could move from his laboratory to other administrative functions. The free-floating manager was

---

44. Duhem 1991, 122. Notwithstanding Duhem's nationalistically inspired disparagement of German science, his own philosophical attitudes toward the physical sciences were rather close to Ostwald's, as discussed in chapter 2, section 1.
45. Kuhn 1970b, 24.
46. Conant 1970, 71–77.

then able to bring the scientific touch to situations that would otherwise dissolve into endless battles of competing interests. Of course, this turns out to be the story of Conant's own career and, eventually, the signature attitude of the action-intellectuals after World War II.[47]

The imprint of the industrial model of normal science is felt whenever Kuhn and other contemporary philosophers of science — including most of his foes — speak of science as an activity whose success can be measured in terms of sheer output, either by the overall number of problems solved or by the efficiency with which they are solved, especially against the backdrop of a constant state of competition that drives researchers to try to establish the dominance of their respective paradigms.[48] However, unlike most of his interlocutors, Kuhn does not limit his account of normal science to that of a knowledge-producing factory. In addition, he clings to the older apprenticeship model that Conant had apparently superseded. In this vein, Kuhn speaks of the puzzle-solving collective as a "scientific community" whose members share a species of "tacit knowledge" that distinguishes them from the members of other communities.

As our discussion of Polanyi in chapters 1 and 2 already suggested, Kuhn was not alone in conjuring up the enquoted expressions. In particular, the expression "scientific community" acquired a currency among cultural commentators of the 1950s that it had previously lacked. The usage did not reflect nostalgia for the simpler times of the medieval craft guilds; rather, the image of scientists jointly and deeply committed to free inquiry was a defensive posture against the ever-present threat of government intervention.[49] This posture is readily seen in Conant's popular writings, in which he switches back and forth between an *industrial* and a *communal* rhetoric of science, depending on whether he is displaying science's strengths (industrial mode) or camouflaging its weaknesses (communal mode). However, because Kuhn's *Structure* was not written as a public defense of science, these two rhetorics are allowed to sit side by side as two halves of the same putatively empirical account of science.

---

47. Daniel Bell seems to have shared Conant's view of the well-organized university rendering its professors "dispensable" and thereby free to pursue affairs of state. See Bell 1966, 94–95. Bell derives this positive spin on dispensability from Max Weber's explanation for the predominance of practicing attorneys in national legislatures (namely, their regular jobs do not require constant attendance and, if necessary, can be done by others with relative ease).

48. I especially have in mind Laudan 1977. However, in many respects, the recently fashionable philosophical fascination with experimentation as the motor of scientific progress is even more beholden to the industrial mentality, especially when the distinguishing feature of the natural sciences turns out to be the manufacture of robust phenomena in the laboratory — a criterion that would have met with the approval of Ostwald and Mach, but not Planck and Einstein. See Hacking 1983.

49. Hollinger 1990.

Perhaps the most enduring legacy of the conflated models of normal science is a fundamental ambiguity as to whether Kuhn intends the normal scientist's habits of mind to belong primarily to the seasoned researcher, the hack technician, the college student—or to some combination of the three.[50] Generally speaking, those who approve of Kuhn's depiction of normal science tend to assimilate its mind-set to the seasoned researcher, whereas those of a more negative disposition tend to imagine that Kuhn is talking about bovine undergraduates. About the latter, Popper memorably remarked:

> Kuhn's description of the "normal" scientist vividly reminds me of a conversation I had with my late friend, Philipp Frank, in 1933 or thereabouts. Frank at that time bitterly complained about the uncritical approach to science of the majority of his Engineering students. They merely wanted to "know the facts." Theories or hypotheses which were not "generally accepted" but problematic, were unwanted: they made the students uneasy. These students wanted to know only those things, those facts, which they might apply with a good conscience, and without heart-searching.[51]

The ambiguity is maintained in Kuhn's account by two crosscutting trends in the history of science that together produce the syncretism that constitutes normal science. On the one hand, the distinction between the seasoned researcher and hack technician has broken down considerably in the twentieth century with the introduction of an industrial division of labor into scientific training. Highly specialized, instrument-driven research, however academically creditable, looks increasingly like hack work. In the context of other lines of work, this subtle form of deprofessionalization is often called "deskilling."[52] On the other hand, Kuhn generally insisted on portraying mature scientists as self-consciously refining and extending their skills at the same time that they solve paradigmatic puzzles. This craftlike feature of science confers some measure of plausibility on the image of scientists as "perpetual students" laboring on the shoulders of their forebears. It is the stuff of which more iconographic histories of science are made.

---

50. Credit for clearly articulating this point goes to Joseph Agassi, in a review of Lakatos and Musgrave's *Criticism and the Growth of Knowledge*, reprinted in Agassi 1988, 324–26.

51. Popper 1970, 53.

52. The locus classicus for deskilling is Braverman 1974. The reason that specialized training in science is not usually taken to be an instance of deskilling is that the scientists in question often move into research and management posts, especially in the private sector, where their status and incomes are higher, not lower, than that of their academic colleagues. My thanks to Sujatha Raman for first suggesting this connection in her unpublished paper, "Progress in the Sociology of Skill" 1995.

4. Kuhn's Third Assignment (Extra Credit): Translate Pedagogy into Historiography

Conant wanted students to see contemporary Big Science through the lens of Little Science because he wanted to maintain the integrity of science as an institution. Kuhn's own rationale was somewhat different, though born of the same environment and molded of largely the same sensibility. It can be understood as Nash's pedagogical technique pumped up with ontological gas. In his final years, Kuhn signaled this point by drawing a sharp distinction between *knowledge* and *science*. The former stands for the internal character of inquiry that can be abstracted from its historically varied institutional arrangements, while the latter stands for just these changing external arrangements.[53] Accordingly, one would want to study several centuries of scientific experiments to observe the perennial patterns of the scientific mind—that is, its knowledge-oriented qualities—at work.

But what if the external character of science changes so much that its knowledge-oriented qualities become virtually unrecognizable? What if, say, the pursuit of research is determined more by utilitarian considerations than by the logic of normal science? After all, Kuhn was himself someone who went into physics to probe deep philosophical problems, only to be drafted into a U.S. Air Force project designed to jam German radar signals during World War II.[54] That experience led him to rethink his commitment to physics. Nevertheless, Kuhn consistently maintained throughout his historical and philosophical career that his interest in science extends only to its providing an unusually good basis for studying the nature of what he now calls "knowledge." Though true to his roots in Natural Sciences 4, Kuhn's judgment neatly sidesteps the delicate question of whether contemporary science is a fit source for epistemological inquiry at all. The example he set by his own research practice would suggest not—a point to which we shall return in chapter 8.

It is worth dwelling for a moment on the fact that Kuhn's ideas about the incommensurability of paradigms arose from the demands of teaching, not research. The appreciation of incommensurability is typically associ-

53. Sigurdsson 1990, esp. 19, 24. For my own related distinction, see Fuller 1994d.
54. Kuhn's involvement in this form of military research was not accidental, but reflected the interests of his doctoral dissertation director, John van Vleck (1899–1980), who headed Harvard's Theory Group at its Radio Research Laboratory. Van Vleck, who won the 1977 Nobel Prize in physics for his work on the magnetic properties of atoms, had been a member of the committee of U.S. physicists chaired by J. Robert Oppenheimer, who concluded in 1942 that a bomb based on atomic fission was indeed possible. My thanks to Stephen Brush for bringing this point to my attention.

ated with the need to understand cultures whose thought patterns are radically different from our own. Here historians and anthropologists have sought methodological safeguards against the hasty assimilation of native categories of thought to our own. Kuhn's insight was to suggest to them that incommensurability is overcome, not by a sentence-to-sentence translation between the languages of the two cultures, but by a kind of cultural bilingualism that enables the interpreter to switch between the worldviews represented by the two languages.[55]

To be sure, this interpretive problem also figured in Kuhn's pedagogical charge. As we saw in chapter 3, section 3, the New Dealers who first raised the National Science Foundation as a political possibility after World War II found the assimilation of science policy to other public policy issues quite a natural move to make. Therefore, the only way to forestall this assimilation would be to teach future policy makers to dwell on the distinctiveness of a life of scientific inquiry from other ways of engaging with the world, and hence not to apply the same cost-benefit criteria as they would to other matters of public welfare. By invoking incommensurability to instill in students a sense of science's authority, Kuhn came to realize that interpretive charity does indeed begin at home!

We can go further, given Kuhn's own admission that he probably would not have had these insights into incommensurability had he not been teaching in Conant's program; that is, before he identified himself as a historian of science.[56] Take Kuhn's famous example of trying to make sense of Aristotle's seemingly self-contradictory conception of velocity, which combines elements of instantaneous and average velocity. Kuhn originally found himself in this predicament because he was trying to make sense of Galileo's experiments in the course of constructing case histories for Natural Sciences 4.[57] In other words, Kuhn was really concerned with understanding how Galileo's mind worked, and to do that, he would have to come to terms with the fact that Galileo's experiments were relatively direct responses to Aristotelian doctrines of motion.[58]

---

55. Kuhn 1970a, 207. For a concise account of the development of Kuhn's thinking about incommensurability, see Buchwald and Smith 1997, esp. 370 ff.
56. Perhaps the fullest account of this formative period in Kuhn's career is Kuhn et al. 1997, 159 ff.
57. The pedagogical origins of Kuhn's historiographical interests are often overlooked because Kuhn's most famous account of such matters occurs in the course of discussing the differences between how (sympathetic) historians and (unsympathetic) philosophers read past scientific texts. By then (the mid- to late 1960s) Kuhn was reflecting on his experience with specialist graduate students at Princeton, rather than generalist undergraduates at Harvard. See Kuhn, 1977a, 3–20.
58. Sigurdsson 1990, 20.

But that does not quite capture all of Kuhn's interpretive task. Galileo had long been portrayed—indeed, celebrated—as someone who had utter contempt for the dogmatic promotion of Aristotelian physics by the Catholic Church. And certainly Galileo's flamboyant rhetoric conveyed that this was his view. Moreover, Galileo's apparent contempt for Aristotle could be easily explained by the surface incoherence of the latter's doctrines. Why, then, should one suppose that Galileo had believed that something deeper was transpiring in Aristotle's texts? After all, we saw in chapter 1, section 4, that even Kuhn's historiographical mentor, Alexandre Koyré, had gone so far as to credit Galileo with initiating the Scientific Revolution by scrapping Aristotle and renovating Plato.

The answer to this puzzle lies, I submit, in the seriousness with which Kuhn took the idea that the scientific mind-set has remained unchanged throughout the ages, regardless of its subject matter or institutional setting. However, this premise of Conant's course had to be itself institutionalized in some form of historiographical practice. To be persuasive, this practice would have to show that scientists down through the ages have treated each other with the same circumspection as Conant would have science policy makers treat scientists today. In that case, were Galileo portrayed as someone who dismissed Aristotle out of hand for having patently false views, then students might well conclude that the need for science to pass through successive paradigms simply reveals how scientists can accumulate and rationalize errors almost indefinitely—that is, until a heroic genius comes along and wipes the slate clean.

While a bit of this image remains in Kuhn's idea of scientific revolutions, it is given quite a different spin, one designed *not* to embolden future policy makers into thinking that the course of science can be improved simply if one has the will and the nerve to do so. Another way to see what Kuhn was trying to avoid is to imagine, as the philosopher Hilary Putnam has, the epistemological consequences of observing that most of what scientists believed a hundred years ago is regarded as false by today's scientists.[59] On what basis, then, should we believe that today's scientists are arriving at lasting truths of the natural world? To avoid the prospect of such inductive

---

59. Putnam 1984, esp. 146. Inductive skepticism toward modern science goes back to Montaigne, who in 1580 reflected on the futility of progress, given that an astronomy that had been serviceable for over a millennium (Ptolemy's) should yield, in rapid succession, major challenges from Copernicus and Tycho Brahe. Clearly Montaigne did not have the sanguine revolutionary attitude of the logical positivists when they witnessed relativity and quantum theories overturn the hegemony of Newtonian mechanics. But perhaps Newton's mere fifth of a millennium of dominance can account for the difference? (I think not.) On the context of Montaigne's pessimism, see Gjertsen 1989, 229–30.

skepticism, while retaining the idea of a continuous scientific mind set, Kuhn had to suppose that Galileo himself possessed the same kind of "bilingual" skills as historians and anthropologists do today. But is it plausible to think that Galileo, excellent experimenter as he may have been, also had the talent, patience, and specialized knowledge to practice the sort of depth hermeneutics on Aristotle that Kuhn suggests he did? The answer is probably yes, but with qualifications that cut against Kuhn's charitable image of the revolutionary paradigm spanner.

To be sure, for all his caricatured dismissals of Aristotle, Galileo had been steeped in Aristotelianism as a medical student in Padua. So the issue is not how much he knew, but rather how sympathetically he should have presented his opponents' arguments before passing judgment. Galileo had more respect for the Aristotelians who followed the spirit rather than the letter of Aristotle: the former looked to Nature (as Aristotle himself did) without trying to reconcile Aristotle's original pronouncements. The Jesuit scientists were generally of the "spirit," not the "letter" mode. As is clear from his *Dialogue Concerning the Two Chief World Systems*, Galileo had much more time for them than for the text-bound Aristotelian literalists.[60]

From this standpoint, Galileo's mind-set looks more like that of a contemporary scientist (or philosopher of science) interested in pointing out contradictory expectations and moving toward empirical demonstrations and conceptual clarifications—not someone with Kuhn's gestalt-switching hermeneutical sensitivities, interested in identifying the hidden background assumptions that render the incoherent coherent. Indeed, because many years had passed since Galileo was part of the Aristotelian scene, he may have been especially insensitive to the degree of offense his caricatures caused in more orthodox circles.[61] If that is the case, then a good part of Galileo's success may be attributed to the relatively high tolerance that his audience had for ridicule, especially when heaped upon the more conservative Aristotelians.

However, this conclusion would hardly set a good precedent for Kuhn's pedagogical practice in Conant's course. On the contrary, I argue that Kuhn wanted to portray Galileo as able and willing to understand his

60. See Wallace 1983.
61. Biagioli 1993, 232–42. Biagioli's discussion occurs as part of an argument for the incommensurability of Galileo and his interlocutors, mainly on the basis of Galileo's rhetorical misfirings (which are then traced to his awkward social position). However, my point is to examine the basis for attributing the metalevel sense of incommensurability proposed by Kuhn, which implies that *Galileo himself* could recognize the two worldviews as such and switch accordingly between them. Thus, what Kuhn glosses as Galileo's bilingual adeptness, Biagioli recasts as his tripping over linguistic barriers.

opponents sympathetically because he needed to demonstrate the workings of the scientific mind so as to make it appear worthy of respect and emulation. This was especially relevant to the students in Kuhn's course, who were likely to be, as managers or bureaucrats, making decisions on the often counterintuitive and possibly frightening schemes that scientists would propose for research funding. To be sure, Kuhn's students were to think like revolutionary, not normal scientists. For as Kuhn himself has noted, most scientists will refuse to perform the requisite "gestalt switch," even if it only means trying out how the other paradigm thinks. Moreover, this strategy is usually justified, given the effort scientists invest working within the current paradigm. Where the material stakes are high, to interpret charitably a competitor operating with radically different assumptions is to place at risk the future of one's own research program. But if students were taught this way of "scientific understanding," then once placed in a policy setting they might be prone to reject scientific proposals that seem alien to the "paradigm" that currently defines the public interest.

Here we see the difference in context that separates Kuhn from a pure Platonic cultist like Koyré. Koyré's depth interpretation of Galileo could reveal a scientific hero who manipulated his audience with fake experiments and talked metaphysically over the heads of his Aristotelian opponents because, unlike Kuhn, Koyré did not then have to turn his interpretation into a model for how students of either science or science policy should understand scientists in today's world. From Paris to Princeton, Koyré conveyed his teaching to elites who had already been instilled in the ways of esoteric historical and philosophical scholarship. While Kuhn himself could have been a member of Koyré's seminar, his own pedagogical predicament was precisely the one that a Platonic cultist always hopes to avoid, namely, a mixed audience of elites and nonelites. Kuhn's creative resolution lay in a generalized incommensurability thesis and the interpretive charity it encourages.

In application, this thesis sent two messages at once: to students of science and science policy, scientists appeared open and trustworthy; to students of history and philosophy, scientists appeared closed and elusive, and hence fit subjects for hermeneutical inquiry. Failure to appreciate the difference in social contexts in which Koyré and Kuhn operated has led to some of Kuhn's most prominent admirers to regard this interpretive charity as a superimposition of the historian's second-order awareness on the first-order awareness of the scientists they write about. Thus, Kuhn's authority has emboldened Richard Rorty to claim that revolutionary scientists are arch-hermeneuticians fully cognizant of the traditions they have transformed in making their mark on history.[62] After conflating these levels of awareness, Rorty then chides Kuhn for inconsistency, since Kuhn still seems to want to explain "the success of science," a phenomenon that a hermeneutically sensitive account of science would seek to dispel, not resolve.[63] However, as I have argued repeatedly in this book, Kuhn's point is the Orwellian one that the success of science is largely predicated on the historian's second-order awareness of the past *not* becoming part of the scientist's first-order awareness of her current activity.

5. Interlude: How Kuhn Failed Hegel

In the introduction, I observed that Kuhn lacked a sense of his own place in history, or *historicity*, for short. In other words, he never reflected on how the past may have already laid the groundwork for the acceptance of his vision of scientific change as "natural," by not only natural scientists but also (as we shall see in chapter 5) social scientists who read *The Structure of Scientific Revolutions* as a blueprint for presenting their own activities in a normatively acceptable light. For example, Kuhn never commented on the range of issues raised in I. B. Cohen's *Revolution in Science*, which highlights the contested and remarkable recency of claims to revolutions in science. Although Kuhn placed great store on how scientists reconstitute their history in postrevolutionary textbooks, he did not treat his own periodization of the normal and revolutionary phases as itself a product of post facto reconstructions of the history that have become common lore.

That modern master of philosophical history, G. W. F. Hegel (1770–1831), would not have been pleased by Kuhn's failure of world-historic awareness. Hegel would interpret the compelling character of *Structure* in terms of its being the final stage in a sequence of moments, each of which had limited the range of possible moves its successor could make. Over the course of what we now regard as "the history of science," each major feature of Kuhn's account was incorporated as part of the scientific self-image, such that by the time Kuhn's theory was actually proposed, it was quickly seen as articulating what scientists had already taken for granted. In that sense, Kuhn rendered the history of science "self-conscious" by making plain to scientists (and their emulators) the sense in which they have had a history of their own. The point of this section is to highlight the difference between the history of science as recounted in Kuhn's own account and this Hegelian understanding of Kuhn's historical project. To see the differ-

62. Rorty 1979, 322–23.
63. Rorty 1979, 324–25.

ence, I shall refer to the two senses of temporality first introduced in chapter 1, section 7: *mythos* and *kairos*, corresponding respectively to Kuhn's and what I imagine would be Hegel's sense of the plot of *Structure*.

The narrative of scientific change presented in *Structure* may be read as either a reconstruction of the past done with the benefit of hindsight from the standpoint of a normatively acceptable present *(mythos)* or a reconstruction of the past done at each critical juncture in history with an eye to providing normative constraint on the prospects for future development *(kairos)*. We may call these readings of *Structure*, respectively, *linear* and *reflexive*. This gives us two ways of telling the story contained in *Structure*, which are epitomized in the following one-sentence plot summaries:[64]

> *Linear reading*: Revolutionary science is made possible only after a crisis has been precipitated by the accumulation of unsolved problems in a paradigm.
> *Reflexive reading*: The creation of a crisis simultaneously reveals the existence of a paradigm that has been suffering from unsolved problems and makes possible the revolutionary science that will overcome them.

The word "creation" in the reflexive reading signals a greater scope for the expression of will, novelty, and "construction" in the philosophically charged sense that already appeared in chapter 2, section 2, and will be associated with "social constructivism" in chapter 7. To appreciate this point, imagine the simplest version of the Hegelian dialectic as a sequence of thesis-antithesis-synthesis. Just because "thesis" precedes "antithesis" in the logic of narration, it does not follow that *consciousness* of this logic proceeds in the same order.[65] To put it in characteristically Hegelian terms, it may be only once the antithesis has become "self-conscious" that a prior thesis and a prospective synthesis are identifiable.[66] This would certainly capture Galileo's status as someone whose world-historic intervention turned Copernicus into a rival of Aristotle and Ptolemy, when Copernican astronomy had been previously regarded as little more than an incommensurable alternative to those ancient authorities.

---

64. If the reader finds the linear/reflexive dichotomy vaguely familiar, it may be because it corresponds to a distinction in what linguists call *aspect*, as illustrated in the difference between the perfect and imperfect tenses: e.g., "She has come" versus "She is coming," The aspect in each case refers to the relationship between the time an utterance is made and the time of the action mentioned in the utterance. If the action is completed at the time of the utterance, the utterance is spoken in the perfect tense *(mythos)*; if the action is not yet completed, then it is spoken in the imperfect tense *(kairos)*. A full introduction of this subtle but essential feature of linguistic expression is Ducrot and Todorov 1979, 304–13.
65. This point should resonate with the Scholastic distinction between *ordo essendi* and *ordo cognoscendi*—the order of being and order of knowing—except that in Hegel's constructivist picture, the order of knowing *precedes* the order of being.
66. Compare my remarks on the in medias res character of rhetoric: Fuller 1993a, 17–24.

Therefore, in his role as antithesis, Galileo identified Copernicus as the precursor whose work opened the space for Newton to enter the world-historic stage. The import of Galileo's intervention was to show that a plurality of research agendas was intolerable, and that a choice between logically conflicting programs had to be made. This move implicitly undermined the authority of the Catholic Church, which had licensed many agendas without privileging any, so as to reinforce the image of a divinely inspired order that encompassed many partial empirical understandings of it. Accordingly, the greater the disparity of the visions tolerated, the stronger the evidence for God's cognitive elusiveness; hence, the theological basis for the empirical underdetermination of scientific theory choice, championed most famously by the Catholic physicist, Pierre Duhem.[67]

Pursuant to Galileo's ground clearing, Newtonian mechanics became the first "paradigm" in the sense of an "exemplar" that self-consciously set out to replace the entire apparatus of ancient cosmology. Before Galileo, it would have been unclear why anyone would have wanted to do such a thing. However, at the metalevel, constructing the need for a successor to replace a predecessor was only the first catalytic moment in the path that led to the acceptance of *Structure* as a "natural" presentation of the history of science. The sense of direction implicit in the succession of paradigms also had to transpire by an internal logic. William Whewell's work in distinguishing the logics of discovery and justification, as elaborated in chapter 1, section 6, turned out to be the next crucial juncture. His interest in generalizing the character of Newton's achievement (i.e., applying to backward disciplines the "method" by which Newton had sorted the wheat from the chaff in earlier traditions and combined them into a program for future research) came from the pressures that industry had begun to exert on the university as the training ground for future elites.

Whewell's mission, then, was to exclude technology from determining the course of science, just as Galileo had excluded religion two centuries earlier. Thus, in response to the evident tendency for financially rewarding inventions to come from nonacademic quarters, Whewell argued that the scope and significance of those inventions could not be understood without the "scientific" training provided by universities. Such training enabled students to explain the success of inventions as hypothesized deductions of previously established theories rather than simply trial-and-error and native wit. Moreover, Whewell suggested that a theoretical understanding of technological success would foster its continuation by means of principled

---

67. For more on the connection between underdetermination and Catholicism, see chapter 6, section 2.

improvements. In this way, the part of technology that could be readily captured by abstract scientific principles was endogenized to science as technology's "essence," whereas the part that could not came to be exogenized from science as the sphere of fortunate amateurs, people who stumble upon breakthroughs by accident because they cannot articulate the theories that underwrite their success.

But once again turning to the metalevel, giving a focus and direction to change are necessary but not sufficient to establish the progressive trajectory that characterizes the history of science. Whewell's rhetorical success first had to be institutionalized as the standard-issue science textbook that converts Newtonian mechanics into a "paradigm" in the sense of a "disciplinary matrix" that prescribes an explicit problem-solving strategy (or "heuristic," another of Whewell's coinages) in place of what had been previously attributed to Newton's unique genius. The pedagogy attached to the Berlin laboratory of Hermann von Helmholtz in the 1870s symbolizes this transition from genius to routine, which, as we saw in section 3 above, was inspired by both a Popperian desire to beat Newton at his own game and the need to consolidate the German national scientific effort. The latter reason would have been familiar to Hegel himself, who claimed that only states were the proper subjects of history.

However, the next transition, epitomized by Max Planck's efforts against Ernst Mach, erased the historical character of these earlier achievements from the collective memory of science, such that scientists had only to refer to a paradigm's empirical consequences, rather than its conceptual origins, to justify its continued pursuit. In this way, the direction of change in the history of science appeared irreversible: science never looks back. Newton's achievement may have been incomplete, but it was not wrong. Thus, Helmholtz's residual falsificationist impulse was purged. Not only did scientific pedagogy come to be presented ahistorically but history of science itself was precluded from the scientific curriculum. Finally, Kuhn himself enters the story to ratify this exclusion of historicity by suggesting that the points at which history raises its ugly head in scientific pedagogy — so-called revolutionary episodes — are themselves relatively short-lived and routinized into the sequence of stages of his narrative logic. And so *Structure* itself became a "paradigm" in the sense of a free-floating narrative logic that can be applied to any set of practices that aspires to become a science, especially the social sciences, at the cost of suppressing alternative developments from the past that do not conform to the Kuhnian *mythos*.

The preceding Hegelian reinterpretation of Kuhn is encapsulated in tables 3 and 4, which trace the successive stages by which Kuhn's picture of scientific change came to be seen as the only plausible one. The figures

TABLE 3. *The History of Science as a Hegelian Dialectic Leading to Kuhn*

	THESIS	ANTITHESIS	SYNTHESIS
first moment	Copernicus	Galileo	Newton
second moment	Newton	Whewell	Helmholtz
third moment	Helmholtz	Planck	Kuhn

TABLE 4. *The Dialectical History of Kuhn's Account of Science*

	"PARADIGM" MEANS	WHAT SCIENCE GAINS	WHAT SCIENCE LOSES
first moment	exemplary achievement	specific theoretical focus	piety (religion)
second moment	disciplinary matrix	internally driven change	wit (technology)
third moment	narrative logic	irreversible direction	dissent (history)

marked as "antithesis" in each moment of the history of science identified in table 3 are the ones responsible not only for facilitating the transition between the figures marked as "thesis" and "synthesis" in the narrative logic of the history of science, but also for instituting the successive moments in table 4 that define the nature of paradigm.[68]

6. The Legacy of a Bad Lesson Well Learned

Even when some, like Popper and his allies, have harbored suspicions about the syncretistic character of normal science, they have not probed the possible effects of the historical incoherence of Kuhn's account. Kuhn's internally divided conception of normal science enabled philosophers and sociologists to choose sides during the founding debates of science studies that took place in the 1970s and early 1980s. Over the course of many acrimonious exchanges, the sociologists who pioneered science studies were able to retrieve from Kuhn's text the science-as-community model in order to undermine the science-as-industry model championed by their philosophical foes. However, in practice, it was often difficult to keep straight the difference between the two models, especially when they were viewed in light of the historical record.

A striking case in point is Michael Polanyi (previously discussed in the

---

68. This section was prompted by an invitation by Salim Kemal to contribute to a lecture series on the contemporary relevance of Hegel at Dundee University, Scotland, February 1998.

final section of chapter 2), who is normally read—and certainly saw himself—as one of the premier supporters of science-as-community. Yet Polanyi's emphasis on the apprentice scientist's submission to authority does not actually represent the experience of the most successful scientific schools in the nineteenth century, which tended to encourage independence in their students' research. These arrangements have been described as constituting schools in the "institutional" but not the "cognitive" sense, since they enable the school to be relatively autonomous of external pressures without, at the same time, demanding intellectual conformity within the group.[69] However, submission became more characteristic as mounting enrollments of students preparing for careers in industry forced research directors to be more authoritarian.[70] This tendency had become exceptionally pronounced by World War I, which was when Polanyi himself had turned away from medicine to pursue an advanced degree in chemistry. Needless to say, Kuhn and Polanyi have not been alone in their failure to keep the two models of scientific organization straight.[71]

The differences between a social practice that is organized as a "community" and one organized as an "industry" are certainly clear enough as Weberian ideal types. Direct experience of these differences have fed the sociological imagination from its inception, as comparable distinctions recur in the classic writings of Max Weber, Emile Durkheim, and Ferdinand Toennies to explain one or another transition from "traditional" (*gemeinschaftlich*) to "modern" (*gesellschaftlich*) society. However, it is worth highlighting the relevant differences in the case of science, in order to understand how they could coexist so peacefully in one account.

---

69. Rocke 1993. An extended comparison of industrial (or "quasi-military") and communitarian schools of research is Fruton 1990. Fruton argues that scientists associated with schools operating in the industrial mode tended to be politically more liberal, and less in need of the holistic understanding that characterized the "high humanism" of university culture. Their students were also more likely than their communitarian peers to pursue careers in nation building, as opposed to knowledge building.

70. Rocke 1993 documents this transition in academic leadership style over the professorial career (1851–84) of Hermann Kolbe, the founder of synthetic organic chemistry.

71. Two examples from the interwar years come to mind. First is Kuhn's avowed precursor, the physician and amateur sociologist Ludwik Fleck; second is his alleged antagonist, the logical positivist Rudolf Carnap. In Fleck's case, professional sociologists who reviewed the original edition of his book were disturbed by his use of the term "collective," with its strong industrial socialist overtones, over "community," when referring to scientific groups. Although Fleck often stressed the egalitarian, if not quite interchangeable, character of research team members, he also periodically pointed out (certainly much more than Kuhn) the value of team members having unique identities and perspectives on science. Here the members' mutual trust made sustained collaboration possible. See Fleck 1979, 163–64. In Carnap's case, his industrial vision of positivist philosophers as coworkers in well-defined puzzle-solving tasks is overlain with a more medieval image of masons adding stones to a slowly emerging edifice. See Carnap 1967, xvi-xvii.

In the communal vision of science, no strong distinction is drawn between the conception and execution of scientific labor. Scientific knowledge is simply embodied in the practices of people who share deeply and equally a commitment — a worldview, if you will — to science as a way of life. Typically, this commitment has a strong subjective component, which makes it difficult for scientists to distinguish a commitment to science per se from a commitment to their particular science. Moreover, the commitment is ritually reinforced in the places where scientists work together. Thus, the laboratory becomes the symbol of science as "local knowledge." Under these circumstances, something called "the scientific method" is at best a philosophical caricature of what science is really about.

In contrast, the industrial vision of science posits a strong distinction between the conception and execution of scientific labor. The research director is endowed with managerial skills that enable her to divide the research agenda into soluble problems to which the appropriately trained people are then assigned. These managerial skills constitute "the scientific method" and are transferable from the laboratory to other sectors of society in need of rational administration. The subordinate members of the research team are typically focused on their assigned problems. They are not expected to contribute to the overall direction of the agenda. On the contrary, it is presumed that they will ultimately migrate to other laboratories to work on problems that are suited to their specialized skills.

Conant had his own characteristically clever way of conflating the *gemein-* and *gesellschaftlich* aspects of science, reminiscent of Whewell and Planck, which denied the existence of an all-purpose "scientific method" while enabling action-intellectuals like himself to apply some univocal sense of "science" to problems of public administration.[72] Significantly, Conant identified the objectionable sense of a unitary scientific method with pragmatism, not positivism.[73] For example, in John Dewey's formulation, the scientific method is a problem-driven enterprise by which

---

72. Conant 1952b, esp. 19–25.

73. Indeed, Conant was himself a member of the Unity of Science circle that developed at Harvard around Philipp Frank, once he left Austria. Yet Conant criticized Frank's reluctance to extend the unificationist impulse from science to society at large. See *Conant Presidential Papers*, letter from J. B. Conant to G. E. Owen, 25 July 1949. His comments were made upon reading Frank 1950, a collection of his lectures and essays, esp. chaps. 13–14, which concern the role of philosophy of science in the physics curriculum. Interestingly, Frank says (p. 232) that studying the original challenges to the acceptance of Copernicanism can give physics students an "inside track" in understanding current social and political problems. However, Frank operated with an Enlightenment model of science explicitly inspired by Mach, which was oriented more toward citizen education than the training of policy elites. On the transformation of the Vienna Circle's Unity of Science project after it reached the United States, see Galison 1998.

intelligence is deployed to satisfy human needs, typically by sequences of trial and error that become increasingly sophisticated as the technology of experimentation is developed. For a pragmatist like Dewey, then, science is, as it was for Ernst Mach, little more than technically mediated common sense. It is clear from Conant's presentation that he mainly objected to the catholicity of this definition, specifically the range of nonacademics—"inventors" and "craftsmen"—who would thereby count as scientists.

At a deeper level, Conant objected to the idea that science had to continually prove itself through public displays of trial and error. Rather, a mark of progress in science was its lowering "the degree of empiricism" in its methodology. Here Conant meant the capacity of scientists equipped with the right mathematical formulae to transform the world according to plan. To be sure, the knowledge encapsulated in such formulae may itself have resulted from trial and error, but at some point the need for further testing ends, and one can simply bring about the changes one wants. In that case, the *gemeinschaftlich* aspect of science comes not from the intimacy with the phenomena that comes from repeatedly testing one's hypotheses, but from the special training required to know which formulae will work the first time around.

In one crucial respect, the industrial and communal visions of science need not be finessed because they actually overlap. Both greatly diminish the role of critical discourse in scientific practice. Neither vision adheres to Popper's philosophical conceit that science is Socratic inquiry continued by other means. However, the two visions diverge markedly in their accounts of this "tacit dimension" of scientific knowledge. What the communal vision represents as an incontrovertible personal commitment to a specialist community, the industrial vision portrays as a narrowed career horizon that renders matters of commitment irrelevant. Does this conspicuous silence, then, signify *devotion* or *detachment?* Be it by existential choice or employment circumstance, neither vision provides a space for airing disagreements about the ends of science. Kuhn, of course, regarded extended debates over the fundamental questions in science as signs of instability, symptomatic of a paradigm in "crisis." The genius of the scientific enterprise, according to Kuhn, lies in its ability to resolve these crises and to erase from its institutional memory the fact that they had ever occurred.

Kuhn's experience in Conant's General Education course may explain how the industrial and communal images of science fused in his own mind, as well as how this fusion may have played a strategic role in the understanding of science that Conant wanted America's future leaders to have. When justifying courses like Kuhn's as part of general education,

Conant was quite clear that students need to "understand science" in the sense of appreciating its distinctive forms of reasoning and ultimate benefits, without becoming distracted by the more hyperbolic claims that were popularly made both for and against science at the dawn of the "nuclear age." Since the popular hysteria over science was grounded in vulgarized versions of the industrial vision—"science made to order," so to speak—it was important that these delusions not be indulged in the General Education courses. Science may well be the salvation of the new age, but its mission would only be perverted—as it had been in Nazi Germany and Soviet Russia—if public expectations and scrutiny of science were raised too high.

The pedagogical remedy was to promote a more matter-of-fact image of science, but one that kept the future managers in the course at a respectful distance from the day-to-day work of practicing scientists. Thus, Conant restricted the content of the course to case studies in which scientific thinking could be displayed in fairly concrete form, so that students could experience vicariously the uniqueness of scientific work. For the most part, this meant reconstructing great tabletop experiments from before the late nineteenth century. Conant gave pride of place to cases from chemistry and electricity, as these could be persuasively presented as having had practical payoffs. Kuhn's input into the course appears to have been twofold. Much to the eventual dismay of the students who would query their practical relevance, Kuhn prevailed upon Conant to include cases from the more philosophically influential fields of mechanics and astronomy.[74] More significantly, Kuhn curbed Conant's tendency to read back into the distant past a strong sense of the basic/applied science distinction, when in fact it required the existence of autonomous scientific academies and academic departments of natural science.[75] Anachronism on this point had wreaked havoc on the first set of course lectures for Natural Sciences 4. For example, Conant declared that scientists get their ideas from the practical arts, even though they may not have an interest in improving them. Thus, students were made to believe that Galileo was fascinated by the trajectory of cannonballs without wanting to improve the artillery capabilities of his patrons.[76]

However, Conant's attempt to recover the communal vision of science

---

74. *Conant Presidential Papers*, letter from G. E. Owen to J. B. Conant, 8 August 1949.
75. The locus classicus for this point has been Ben-David 1984.
76. *Conant Presidential Papers*, Natural Sciences 4, academic year 1949–50, 3 October lecture. Kuhn's response may be found in letter, dated 8 October 1950. Conant's response to Kuhn's advice is itself interesting as it shows Conant teaching Kuhn the ways of Orwellian historiography. In other words, Conant defends the idea that the validity conditions for science, history of science, and public understanding of science are not necessarily the same. See letter to Kuhn dated 11 October 1950.

was at odds with the pedagogical framework within which General Education was taught. By hopscotching across the centuries and crisscrossing the disciplines, the courses created the impression that it was, indeed, possible to understand the scientific turn of mind, regardless of the time and place in which science is practiced. Moreover, not only did this fit Conant's self-image as the itinerant scientific manager and social inventor, but it also captured his sense of his students' likely career trajectories, which would demand more science connoisseurship than actual scientific knowledge.

In effect, Conant wanted to shape managers who would be "patrons of the sciences" but not try to second-guess what scientists ought to be doing. This goal reinforced the one pursued by the founder of iconography, the art collector Aby Warburg, whom we identified in chapter 1, section 3, as an important independent source for Kuhn's orientation to intellectual history. The added challenge in the impending atomic age was to enable students to see the same mental processes that enabled Galileo to discover the law of falling bodies in projects whose salient feature would seem to be their cost, be it measured in dollars (i.e., the cost of construction) or lives (i.e., the cost of deployment). However, even this case had precedents in the annals of aesthetic appreciation, as connoisseurship in so-called non-representational art—whereby one could justifiably recognize both a Rembrandt and a Jackson Pollock as art—typically required extended training.

A good example of this connoisseurship is provided by the hypothetical situation on which students were tested both before and after they took Natural Sciences 4. The student is asked to offer advice about investing in a scheme for converting glucose into wood: What sorts of considerations are relevant in determining the likelihood that such a scheme would be commercially viable in the next few years? A list of arguments is then presented, and the student is asked to judge each one's relevance.[77] This is an interesting way to see whether students have a discriminating eye for good and bad research prospects. But it is by no means clear how close acquaintance with the experimental work of Boyle, Lavoisier, and Pasteur—staples in Conant's course—would help students develop such an eye. Perhaps it comes as no surprise that the course did not seem to improve the students' abilities to deal with the hypothetical situation. The students themselves wondered (in their year-end course evaluations) why the political, economic, and technological contexts of science were not more explicitly treated in the course.[78]

77. *Conant Presidential Papers*, questionnaire designed by E. P. Gross for Natural Sciences 4, academic year 1948–49.
78. *Conant Presidential Papers*, final course evaluation for Natural Sciences 4, academic year 1948–49.

A similar question might be posed of Kuhn's treatment of scientific change in *The Structure of Scientific Revolutions*. Normal science, as Kuhn would have it, is a puzzle-solving process dictated by the internal logic of the paradigm under which the scientific community is working. Members of this community are said to embody a specialized craft knowledge, due to the kinds of instruments they use and the problems to which they are applied. However, one would never suspect, just from reading Kuhn, that scientific instruments cost money to design and manufacture and that scientific problems require institutional protection and incentives for their sustained pursuit. Moreover, Kuhn would hardly give one reason to believe that this pursuit might itself have consequences—both intended and unintended—on the society in which it was housed.[79] As we shall see at the end of this chapter, and in much more detail in chapter 7, these blind spots have come back to haunt the Kuhn-inspired research in science and technology studies, which has had great conceptual difficulty integrating its findings into more general theories of social, political, and economic change—especially theories that are critical, or at least observant, of science's impact on the rest of society.

These omissions testify to *Structure*'s overall success in repressing the industrial vision of normal science. At the same time, they help explain the book's allure for humanists and social scientists. Kuhn was taken to be saying that what popularly seemed most distinctive about the natural sciences—the technological power popularly associated with physics and chemistry—really did not contribute to the status of the natural sciences *as* sciences. Thus, in principle at least, it appeared that humanists and social scientists could become proper scientists, if they consolidated their efforts a bit and constituted themselves as paradigms. While Conant and Kuhn would bristle at the thought of humanists and social scientists regarding themselves as proper scientists, Conant at least would have to admit that their understanding of Kuhn's message served his own interest in normalizing the relations between the natural sciences and the rest of the academy. These relations had grown increasingly tense after the explosion of the atomic bomb engendered a miscellany of humanist harangues over natural science's cultural nihilism. However, *Structure* turned much of the academically based suspicion around by unwittingly suggesting to nonscientists, "If you can't beat 'em, join 'em!" What Kuhn had originally presented as signs of science's uniqueness were read as a strategy for converting any field into a science.[80]

79. See Kuhn 1970b, x, where Kuhn acknowledges that while he failed to take the larger social consequences of science into account, he doubts it would have altered the plot of his story.
80. Introduction to Gutting 1979, 14.

7. The Ironic Fate of General Education in Science

Conant misjudged the long-term impact of the General Education curriculum, though he was far from alone in doing so. Shortly after the General Education in Science program was instituted, Conant was contacted by the Carnegie Corporation about spreading the course to other college campuses.[81] While it failed to be adopted outside of Harvard (largely for staffing reasons), in Kuhn's hands a failed undergraduate course of study turned out to be remarkably successful as a vehicle of faculty development. At least, widespread conversion to the pedagogical perspective of Natural Sciences 4 would explain why certain tensions in Conant's original vision have remained unarticulated by those who come to their understanding of science through *Structure*. These tensions may be summarized in the following questions:

> 1. If pure science affects us only through its applications, then why do General Education curricula focus on the creative process of pure science, rather than on the real "strategy and tactics" by which the pure and the applied are networked together? (Conant reserved the military metaphor for the heuristics scientists used to channel their creativity.)[82]
> 2. Indeed, why continue focusing on individual achievement, where an understanding of institution building would seem to be in order?
> 3. If the public understanding of science has become crucial because of the investments and impacts of Big Science, then why study historical cases of "Little Science"?
> 4. Suppose it is true that "the scientific mind" has not changed over the centuries. Why not, then, simply conclude that the scientific mind is *not* the crucial factor in fathoming the workings of Big Science?
> 5. If the value of scientific work can be judged only by fellow scientists, then how exactly is the General Education curriculum supposed to contribute to decision making about science in a democracy?

The tensions reflected in these questions represent the tricky ongoing negotiations between science and power in modern democracies. The philosophical movements that most strongly identified with experimental science prior to World War I, positivism and pragmatism, grounded science's superior epistemic status on its alleged ties to socially beneficial technologies. I say "alleged" because, while science did much to explain and to legitimate the great technological advances of the nineteenth and early twentieth centuries, it contributed relatively little to their actual invention.[83] Science becomes instrumental in the construction of "progressive"

---

81. Conant 1970, 373. See *Minutes*, 14 February 1950.
82. Conant 1947, 102 ff.
83. Mulkay 1979a.

technologies around the time that it also participates in the construction of war machines. After World War I, most notably in the case of logical positivism, rationalist philosophers start to distance the theoretical trajectory of science from its real or potential applications, which were, in turn, typically presented as deflecting science from its natural trajectory.[84] Yet, this turn comes just as science is increasingly driven by military and industrial concerns, indeed, with major conceptual breakthroughs often arriving as the spin-offs of what would ordinarily be regarded as "applied" research.

Given this intaglio of interests and effects, what would the public do, once it becomes "expert in judging experts"? Conant argued that these lay metaexperts would provide a pool of funds large enough to enable all professionally approved practitioners in a field to compete among themselves, according to established rules of scientific evaluation.[85] This process, now known as "peer review," was not much different from what Polanyi advised, but it served to circumvent the problem of dirty hands. For even if Conant could not prevent the prospect of more science producing equal amounts of evil and good, he could help manufacture an asymmetry in the amount of accountability demanded of research designated "pure" (relatively low) vis-à-vis that designated "applied" (relatively high).[86] Thus, the ultimate goal of Conant's General Education curriculum was a lay public capable of discriminating between instances of these two types of research—a public conditioned to see the dirt on one set of hands but not on the other.

All of the above developments can be seen only with the range of vision that hindsight affords. Looking more myopically, at Harvard in the short term, General Education in Science met an ironic fate. In 1957, the Soviets

---

84. The two most influential German philosophical movements that were spawned in the 1920s and 1930s may be understood as desperately ingenious attempts to preserve the integrity of "science" in the face of post–World War I antirationalism. Both the Vienna Circle and the Frankfurt School were academically marginal groups that, once they gained prominence as exiles in the anglophone academic universe, began to accuse each other of abetting still more virulent strains of antirationalism that they saw afflicting the post–World War II era. See Adorno 1976. In their own way, both the Vienna and Frankfurt exiles endorsed a vision of "unified science," with the former invoking a more Platonic, mathematical sense of unity, and the latter a more Hegelian, organic sense. Both visions had a strong aesthetic dimension—roughly speaking, a high modernist style versus a late romantic one—and both hark back to a pretechnological, liberal arts sense of "science" as Wissenschaft. Whereas the positivists (and here Popper may be included) attempted to preserve the integrity of science by defining (or "demarcating") science so as to exclude politics and technology, the Frankfurt School openly blamed politics and technology for polluting science. But be it by repressing or denouncing the technopolitical dimensions of science, the two movements drove home the idea that science has its own intrinsic character.

85. Conant 1961, 321 ff.

86. I mean quite literally that the basic/applied research distinction originated in accountancy, specifically that used by the U.S. National Science Foundation. See Greenberg 1967, 31–36.

had launched the first artificial space satellite, Sputnik, and American educators were formally drafted into the Cold War. By this time, Conant had already left Harvard to become the first U.S. ambassador to West Germany, while Kuhn had been denied tenure, largely because he had failed to become a recognized specialist in any field.[87] Soon afterward, the psychologist Jerome Bruner headed a commission to investigate the place of science in general education at Harvard.[88] Bruner concluded that the courses inspired and designed by Conant imparted a dilettantish understanding of how science works. Each course seemed held together more by a familiar historical narrative than by the scientific community's own evolving practices of inquiry. Bruner proposed that the program be restructured so that non—science majors would be required to take a sustained contemporary introduction to a specific science. These new courses would not substantially differ from the elementary courses that a science major would take. Bruner's proposal passed, and in the 1960s radical educational theorists would come to hold up Bruner as a paradigm case of someone who believed that learning was a matter of seeing the world as the experts see it.[89]

A latent source of Bruner's discontent was that Conant's courses were rarely, if ever, taught by practicing scientists, whereas General Education courses in the humanities and social sciences often had luminaries such as I. A. Richards and Clyde Kluckhohn at their helm. Moreover, this asymmetry had long been a sore point in the staffing of General Education, since even the humanists made it clear that these courses should not distract from the scholarly calling, lest one be absorbed into providing "mere" service courses for the expanding natural-science faculties. Kuhn's own failure to earn tenure at Harvard was attributed at the time to his failure to observe this stricture, albeit largely due to what, by that point, was felt to be Conant's unfortunate influence.[90] This fact, coupled with the way

87. *Minutes*, 8 November 1955.
88. *Minutes*, 3 June 1958.
89. See, e.g., Postman and Weingartner 1969, 78.
90. Humanist and social-scientific concerns about General Education not being a "contribution to scholarship" appeared at the courses' inception. See, e.g., Kluckhohn's comments in *Minutes*, 30 October 1947. While in principle interested in contributing to these courses, the newly appointed B. F. Skinner wanted assurances that it would not impact significantly on his research time (*Minutes*, 8 March 1949). In addition, there were concerns that General Education would become a safe haven for dilettantes. Indeed, until finally receiving tenure in 1953, the man who would become most closely associated with the rise of Harvard's history of science department, I. B. Cohen (b. 1917), found himself dogged by such accusations. As was periodically reported in the *Minutes*, Cohen was subject to all the familiar indignities of untenured academic life: the assignment of a scientific mentor when he began teaching in the program (21 March 1946) and the importation of an outside observer on his classroom performance at tenure time (15 January 1953). While Cohen managed to survive his trials by fire, Kuhn proved less fortunate when he came up for tenure. Even Kuhn's senior colleague,

the courses were taught, made it unlikely that students would be inspired to take additional science courses, let alone switch to science majors. Thus, it did not seem that Harvard was doing its part to meet the challenge of Sputnik. Indeed, if anything, General Education in Science functioned like a course in religious tolerance that served only to immunize students from ever committing to any particular faith. True, several of the General Education instructors, like Kuhn, held Ph.D.'s in a natural science. However, at Conant's recommendation, they made very early career shifts to history and philosophy of science—and never looked back. Conant himself, of course, had closed the doors to his own chemistry lab when he became president of Harvard in 1933.

Shifting from personalities to institutions, we witness a still more significant pattern. Conant had never been interested in swelling the ranks of research scientists. Rather, he wished to retain the elitism of science but to enable a larger segment of the population to work around this elite corps, either as managers or laborers.[91] Bruner, by contrast, did not draw such a sharp distinction between science education for specialists and nonspecialists. In a world where the Soviets had shown that science was crucial to national security, nonspecialists were simply failed specialists. Bruner's vision was consonant with the emerging educational policies of the U.S. National Science Foundation. Despite NSF's origins as a research-funding agency, Sputnik added to its missions the cultivation of human capital for science. Toward that end, the metaphor of the "pipeline" was developed, which portrayed all students as potential science majors who "leak out" of the pipeline at various points. Educational policy, then, would be devoted to plugging the leaks so that more students make it through and become scientists.[92]

It is ironic that Bruner would prove to be the heavy in this story. In his autobiography, he openly admires Conant's foresight as an academic leader, and he honors Kuhn as one of the first to realize the larger implications of Bruner's own "theory-laden" view of perception, which helped give psychological plausibility to the famous "gestalt switch" account of

---

the generally supportive Leonard Nash, damned with faint praise by remarking that Kuhn's *Copernican Revolution*, while not very scholarly, was nevertheless good for teaching (8 November 1955).

91. See esp. Conant 1959, his Carnegie Corporation Report. Note the Platonic resonances of the Guardians' and Workers' being trained in the myths needed for them to accept the authority of the philosopher-kings.

92. On Bruner's contribution to Cold War science education policy, see Montgomery 1994, 213–14. My thanks to Juan Lucena, whose doctoral dissertation ("Making Scientists and Engineers for America: From Sputnik to Global Competition," Virginia Tech, 1996) first alerted me to Bruner's NSF connection.

paradigm change.⁹³ However, as the Cold War intensified, the goals of General Education in Science shifted in Bruner's mind from an appreciative understanding of science to active recruitment into the scientific ranks.⁹⁴ But more importantly, Bruner never seemed to experience the conflict of scientific visions that Conant and Kuhn tried to mitigate in their writing and teaching. It was quite easy for Bruner to assert that the best way to understand science is to "do" some of it. As a cognitive psychologist, he clung to the idea of a distinctive scientific mind-set—one common to practitioners of all the natural sciences—that remains untouched by changes in the shape and size of the scientific enterprise itself. By contrast, Conant and, to a lesser extent, Kuhn realized that the industrial vision had irreversibly transformed the character of science, but in ways that could prove detrimental to both science and society, if presented without the guise of the communal vision. Here, then, was the source of the "noble lie" that has been effectively propagated under the rubric of normal science.

The advent of the industrial model of science was Janus faced in its effects. On the one hand, it freed up the time of scientific leaders to manage extrascientific affairs; on the other, it encouraged the public to make extrascientific demands on scientific research. From his standpoint as a Cold Warrior, Conant perceived the former to be a politically stabilizing consequence, the latter a destabilizing one. Conant's General Education courses were designed to stress the former and to mask the latter by recovering the rapidly fading image of science as a community. As history has since shown, Kuhn's *Structure*, the product of his experience teaching those courses, enabled Conant's aim to be advanced on a wide front, even after Bruner recommended the discontinuation of Conant's courses.

8. Postscript: A Hands-On Approach to Science That Is Also Hands Off

Pivotal in extending the "hands-on" understanding of science officially promoted in Conant's curriculum has been the pioneering ethnographic approach to science of the sociology of scientific knowledge (SSK). Traditionally, the cognitive standpoint informing ethnography has been critical not of the people under study, but of those who would suppress their voices; hence, the field's much controverted methodological commitment to relativism. Thus, implied by SSK's critique of scientific rationality is that phi-

---

93. Bruner 1983, 85, 244–45.
94. For corroboration, see Buck and Rosenkrantz 1981, 384–85.

losophers of science have routinely misrepresented scientists—indeed, in a way that has contributed to the manic-depressive cycle with which Big Science has been received by the public: impossible expectations followed by disappointment and even betrayal.[95] This is a familiar concern from Conant and Nash, but can the connection between their history-based curriculum and SSK's ethnographic encounters be made any closer? After all, Kuhn himself has professed hostility to the ethnographic turn in SSK.[96]

Once again, the key is Conant himself, who explicitly justified the study of past experimental science as "the equivalent of a magic tour of laboratories" in which visitors would be able to observe the scientists in action, interrupting them with questions about their curious practices.[97] Of course, the currency in which "public understanding" is traded has changed over the past three decades, now that we live in what Guy Debord calls "the society of the spectacle." For what Conant originally proposed as an impossibly ideal vehicle for the public understanding of science has recently been advanced by one of SSK's leading practitioners as an idea whose time has come. As in Conant, we find here an interest in getting citizens to appreciate the "ordinariness" of normal scientific work—and hence not to pin too many hopes or fears on it. Perhaps even more now than in the 1950s, part of the appeal to ordinariness involves leaving scientists to do their work just as "you," the average citizen, would want others to leave you to do yours:

> In democratic societies it is always a sound instinct to trust the people with the truth—even if some work has also to be done to overcome institutionalized idealizations. And in this case the truth is that scientists are neither more nor less than our best current experts in their domains. Their workplaces have got all the clutter of your favorite mechanic's garage, and he is the man you trust with your car.[98]

The analogy threatens to break down upon considering that you trust the mechanic not because you see him do things to your car that you yourself would never have dreamed of doing. In fact, that alone might invite suspicion that he is incompetent or overcharging you. Rather, you trust the mechanic because your car runs better after he has worked on it. Unfortunately, the connection between workmanship and outcome is not nearly so close in the case of Big Science, a point repeatedly made by Conant, who, for that reason, stressed the need to identify science with the craft of doing

---

95. The most comprehensive treatment of this manic-depressive cycle in the popular psyche is Burnham 1987.
96. Horgan 1991, 49.
97. Conant 1961, 5.
98. Shapin 1992b.

science, which is portrayed as something quite distinct from simply being "well-informed" about the results of science, a state of mind that Conant denigrated for instilling a "merely" critical, outcomes-oriented attitude.⁹⁹ This point also explains how Conant reconciled two incompatible values that he espoused for science: *openness* and *nonaccountability*. "Openness" meant that science's operations are made available to fellow scientists and rendered visible to the general public. However, Conant's interest in science's public visibility was not accompanied by an interest in its public accountability. Indeed, Conant made it clear that when he decried the "secrecy" imposed by the military on research for purposes of "national security," he was worried that this policy would impede scientific progress while needlessly arousing public suspicions.¹⁰⁰

The appeals to hands-on experience and on-site visits to scientific workplaces common to Conant and SSK practitioners are part of what political scientist Yaron Ezrahi has called the "visual culture" that characterizes modern liberal democracies.¹⁰¹ According to Ezrahi, the "seeing is believing" brand of empiricism is perhaps the Scientific Revolution's chief continuing legacy to our understanding of democratic governance. However, the legacy is an ambiguous one that has not generally encouraged a focused critical attitude toward the objects seen. Opening the laboratories to the public certainly gives the impression that the scientists have nothing to hide. As a result, that gesture of openness may be just enough to discourage the public from further probing behind the scenes.¹⁰² Moreover, in cases where members of the public look closely at a scientist's work, they are likely to be struck by the distinctive skills on display, while failing to wonder about the overall ends that those skills serve.

One might say that the epistemological fallacy associated with spectator culture—a fallacy that has only become more trenchant with our ability to see more of the world through the "small screen" of television—is a version of affirming the consequent: that is, "seeing is believing" is fallaciously taken to imply "out of sight, out of mind." In the case of science, then, whatever escapes the spectators' visual horizon—the networks of power and channels of information that exist behind and beyond the lab environs—is simply not factored into their sense of a proper accounting for science. Indeed, given that a larger share of the scientist's time and energy is being devoted to "entrepreneurial" activities outside the lab, lab work in

---

99. Conant 1961, 315–21.
100. Conant 1952b, 14–18.
101. Ezrahi 1990, esp. 67–96.
102. This mentality corresponds to the interest in inoculation theory in social-psychological research of the period, as discussed in chapter 3, section 6.

today's world would seem to be little more than a showcase activity—perhaps, like so many tribal rituals, done *primarily* for the benefit of the spectators.[103]

Placed in the larger context of political theory, Conant and Ezrahi are articulating a conception of democracy that has become increasingly prominent in the twentieth century, namely, *pluralism*.[104] Pluralism emerged as an alternative to the pseudodemocracy of self-styled socialist regimes, whose idea of "equality under the law" was equal subordination to a central authority. The pluralist aims to diffuse such centralized power to voluntary associations, which, under ideal conditions, are sufficiently divided in their labors that the members of any such association must rely regularly on the expertise of the members of other such associations. Explicit legislation plays less of a governing role in this system than tacit relations of trust born of mutual dependency.

Pluralists see themselves as presenting a version of the "open society" suitable to democracies much larger and more heterogeneous than those modeled on the Athenian polis. Instead of questioning authority that *strays beyond* its bounds, as Karl Popper might advise, pluralists advocate tolerating authority that *stays within* its bounds. If the one aims to eliminate the prerogatives of expertise, the other aims to spread them around so that everyone ends up becoming an authority over that which she has the most direct experience. Final discretion over a small domain thus replaces fallible assertion over a larger one.[105] This is a subtle, but significant, shift in

---

103. In this discussion, I make the controversial assumption that there *is* a viewpoint from which one could comprehend a significant part of the whole scientific process, a perspective that is, at least in principle, open to anyone. Ezrahi, for one, denies it, as do other postmodernists. I explicitly defend retaining the idea of a public sphere for governing science in Fuller 1993b, 277–316.

104. Conant (1970, 366) describes when he adopted the pluralist conception of democracy. There are a plurality of pluralisms, divided mainly according to the degree of conflict between the interest groups, or voluntary associations, in the society. A good survey is Held 1987, 186–220. The model emphasized in this discussion is relatively peaceful, drawing on a coordinated division of labor, based on knowledge being distributed across the members of the society. This is also the sort of polity that SSK seems to countenance. For the clearest SSK statement of this point (albeit written without specific reference to science), see Barnes 1990.

105. Moreover, it is not quite sufficient to claim that pluralists hold that one needs to recognize the limits of one's own expertise. One must equally acknowledge the expertise of others in their respective domains. This important point—essential for a science's normative control over its practitioners—is often overlooked because it is discussed (after Robert Merton) under the rubric of "communalism," the sharing of data and credit that distinguishes true scientists from the industrial researchers who keep trade secrets in order to protect their company's profits. On the sinister side of Merton's sense of the normative structure of science, see Fuller 1997d, 63–67. Merton's formulation of communalism (originally "communism," until it resonated badly in the American context), see Merton 1973, 273–75. Despite the scent of sanctity surrounding this distinction, it is worth noting that such "sharing" amounts to paying mafia-style protection money. Scientists cannot afford *not* to share with their colleagues.

the sense of "openness" required of the open society in a world of, so to speak, Big Democracy. To be sure, it is not the only, or even the best, strategy for reinventing the open society in a scaled-up world, but it remains a popular strategy.[106] In the case of science, it marks a shift away from *criticizing theories* to what has become the signature attitude of today's science studies: *understanding practices*.[107]

In anticipation of our discussion of the politics of SSK in chapter 7, let me end by mentioning here SSK's pluralist roots in phenomenology, especially those laid down by that Viennese expert in international banking law, Alfred Schutz (1899–1959), who essentially introduced the phenomenological movement to the United States from a marginal academic base at the New School for Social Research in New York City.[108] As might be expected, pluralists of a phenomenological persuasion believe that one's epistemic authority is limited to the sphere of one's direct experience. Not surprisingly, then, Schutz anathematized the advent of mass media and

---

Otherwise, colleagues will not share with them, approve their grants, or hire their students. At the very least, it is nearly impossible to use someone else's apparatus without contacting them for background information not included in the typical scientific article, which means one needs to be in the author's good graces. This is much of the critical thrust of Collins 1985.

106. On the analogy between Big Science and Big Democracy, see Fuller 1994f. The term "big democracy" is associated with Dahl 1989, esp. 213–310.

107. For state-of-the-art discussions of this transition, see Rouse 1987, 1996. For a critique of the turn from theory to practice, not only in science studies but also in social theory more generally, see Turner 1994.

108. A genealogy of the position can be traced from Ophir and Shapin 1991, through Berger and Luckmann 1967, back to Schutz, "The Well-Informed Citizen: An Essay in the Social Distribution of Knowledge" (1932), in Schutz 1964, 120–34. Schutz, in turn, wanted to provide a phenomenological basis for the brand of Austrian economics associated with Friedrich von Hayek as part of the microfoundations for a unified social science. The pluralist polity would thus be a minimal state. An illuminating but frequently neglected fact is that, while in Vienna, Schutz studied with the legal positivist Hans Kelsen and the neoclassical economist Ludwig von Mises. These studies predated his more publicized acquaintance with Edmund Husserl, the founder of the phenomenological movement. See Prendergast 1986. See also Polanyi 1969.

One reason for dredging up this genealogy is to counter the recent tendency, partly inspired by Kuhn, to infer from the appearance of a functioning scientific community to its members' holding a large number of beliefs in common. Supporters of such "consensus" theories of validity should consult Schutz and Hayek to see how it is possible—and perhaps even necessary—for a complex polity to function with agreement over little more than whom to have adjudicate their disputes (i.e., the bare rule of law, which begins as an arbitrary uniform imposition but proves itself by the beneficial activities enabled by its constraints). See Hayek 1960. With a foot in both microeconomics and phenomenology, Schutz provides the link between Hayekian notions of distributed intelligence (where economic agents have unique abilities and needs that cannot be encapsulated more efficiently by the sort of central planner postulated by Keynesian and Marxist economics) and ethnomethodological notions of multiple realities (where social agents possess unique perspectives that cannot be reconciled without remainder by either a Parsonian or Marxist sociology). For their part, consensus theorists seem to suppose that a community's lack of explicit disagreement must imply general agreement: *tertium non datur*. An informed discussion that nevertheless concludes with an endorsement of the consensus theory is Kim 1996.

public opinion surveys as instruments for "artificially" extending the horizons of one's life world. He especially decried the tendency of media moguls and social scientists (notably his fellow Viennese–turned–New Yorker Paul Lazarsfeld) to conjure up situations that conferred a false sense of competence on ordinary citizens, whether it be the radio broadcast of a political debate or a telephone — or still worse, door-to-door — survey of citizens' opinions on subjects that they would not have otherwise thought about.[109] Indeed, Schutz and his fellow phenomenologists believed that this illusory sense of competence could help explain the volatility of the electorate in the 1920s and 1930s that resulted in the democratic election of demagogues such as Mussolini and Hitler.[110]

An important legacy of this normative phenomenological sensibility is that when Bruno Latour, Karin Knorr-Cetina, and Michael Lynch finally embarked on ethnographies of "laboratory life" in the late 1970s, they had to redescribe their observations so as to render them natural objects of humanistic understanding, lest they be accused of reporting things that were alien to *their own* life worlds; hence, subsequent generations of science studies practitioners have come to regard laboratory apparatus as "inscription devices," an expression that continues to bemuse practicing scientists.

---

109. One can imagine what he would have made of more recent mass marketing of computer simulations as "virtual reality" machines!

110. One sociologist has gone so far as to argue that Schutz included a dislike of politics as part of the "natural attitude." This includes the avoidance of both criticism and commitment, a knowledge of one's own limits, and a disdain for not only risk taking but even utility calculations. This apolitical attitude is summed up in the word "coping." See Gilliatt 1995. In light of issues raised earlier in this book, it is worth noting that Gilliatt treats Schutz's phenomenological sociology (and Harold Garfinkel's ethnomethodology) as the mirror image of classical stoicism (see chapter 1, section 2). Depending on which discipline is regarded as the metascience of the other, (phenomenological) sociologists can be seen as naturalizing a normative perspective or (stoical) philosophers as moralizing the natural attitude.

V

How Kuhn Unwittingly Saved Social Science from a Radical Future

1. Why *Structure* Provides Social Scientists Less than Meets the Eye

As we shift from what influenced the writing of *The Structure of Scientific Revolutions* to the various impacts it had on its readers, the first thing to notice is that the book was taken up in rather different circles from Conant's works. Whereas Conant persuaded U.S. government officials and much of the educated public to support the funding of pure science without requiring too much accountability in return, *Structure* occluded the tensions latent in Conant's vision for the potentially more critical audience that was brewing within the academy, namely professional social scientists who strove throughout most of the century—with decidedly mixed results—to acquire the legitimacy that had been increasingly given to the natural sciences. Yet, in response to an invitation to participate in a conference in which *Structure* was proposed as one of the major advances in twentieth-century social science, Kuhn had this to say: "I know a great deal less than I should, and in any case virtually nothing at all, about the social sciences and I will not create confusion by bluffing it."[1]

This is not false modesty. The amount of influence that Kuhn has exerted over social scientists is truly remarkable, especially considering that *Structure* displays little of what a social scientist would recognize as methodological sophistication. Even Kuhn's appeals to gestalt switches, cognitive psychology experiments, and theories of psychological development are more in the spirit of metaphor than evidence.[2] And while Kuhn's model acquires rhetorical force by his claim, early in *Structure*, to have

---

1. Letter to Karl Deutsch, quoted in Deutsch, Markovits, and Platt 1986, 278.
2. This reflexive inconsistency between Kuhn's own method and *Structure*'s status as a blueprint for science has given at least one psychologist the temerity to question the wisdom of his field's embrace of Kuhn: see O'Donohue 1993. My thanks to Art Houts for bringing this illuminating piece to my attention.

demarcated the natural from the social sciences, he never actually compares the two types of fields in any depth. By his own account, his knowledge of the workings of the social sciences was based on casual observations, while a fellow at Stanford's Center for Advanced Study in the Behavioral Sciences in 1958–59.[3]

In this context, Kuhn observed that social scientists could never agree on what counted as an exemplary piece of research, and so could never establish a common frame of reference for anchoring their disputes. Marx or Freud may be esteemed by one colleague, only to be dismissed out of hand by another. Accordingly, Kuhn concluded that the history of the social sciences has not witnessed a clear succession of paradigms because social scientists have been unable to agree on research exemplars to underwrite the activity of normal science. The amazing thing is that so many social scientists have taken Kuhn's casual observation as a deep criticism of their fields. After all, the peculiar character of the social sciences can be traced historically to social scientists' having been guided by larger, conceptually unwieldy social problems that cannot be reduced to well-defined puzzles. Consequently, historians of each of the social sciences have portrayed their disciplines' trajectory as pulled in three distinct directions that can be characterized independently of any intrinsic concern for the nature of social reality:

1. from *above* (in an administrative or managerial capacity, as the trustees of state or business);[4]

2. from *within* (in an ethical capacity, as the secular successor of pastoral theology);[5]

3. from *below* (in a rhetorical capacity, as the voice of politically disenfranchised groups).[6]

3. Kuhn 1970b, vii–viii.
4. See Wagner, Wittrock, and Whitley 1991; Rueschemyer and Skocpol 1996. Most of the very few general histories of social science follow this trajectory: e.g., Manicas 1986.
5. This is most characteristic of the histories of British and American social science, although the sociological roots of these developments are somewhat different. Roughly speaking, the British case is fueled by middle-class philanthropy, whereby support for social-science institutions appears largely to assuage the guilt for capitalist success. See Collini, Winch, and Burrow 1983. In contrast, U.S. sociology originated as a secularization of the gospel of social reform preached by the more liberal branches of Protestantism in reaction to social Darwinism. See Ross 1991; Lasch 1991. A much-underestimated precursor in this context is Fuhrman 1980.
6. Given its rarity, it is worth specifying the pressure from below that enabled the emergence of a social science discipline. Geography, typically omitted from general histories of the social sciences, is perhaps the only major academic discipline to have been built from the grassroots of sub-university-level teachers. Geography did not have a clear research agenda or even merit a professorship in the United Kingdom until elementary- and secondary-school teachers lobbied Parliament to get the subject established in the universities so that teachers could be formally trained in it. The teachers were partly reacting to the difficulty of teaching a coherent history, given that Britain, at the end of the last century, had involved itself in the

In what follows, it will become clear that the "lower" the source of social-scientific activity, the more resistant to Kuhn's charms it is likely to have been.

2. The Path Not Taken: Social Science as Critique of Natural Science

By the second term of the Eisenhower administration (1957–61), such prominent American sociologists as C. Wright Mills (1916–62) and Alvin Gouldner (1920–81) had started to overtly renounce any pretensions that the social sciences may have had to emulate the natural sciences. The goals of the social sciences were increasingly distanced from those of the natural sciences, the latter being increasingly portrayed as amoral, if not immoral, technologies of control available to the highest bidder, complicit in what Eisenhower himself had seen by the end of his presidency as the emerging "military-industrial complex."[7] An assumption that united Mills and Gouldner was that, since the world is causally too complex to afford any unmitigated benefits, the only intellectually responsible attitude to the status quo, including science itself, was reasoned skepticism. Here was a full recognition of science's dirty hands that Conant had tried to hide.

The distinction that Mills and Gouldner implicitly drew between the natural and social sciences did not aim (as it so often does today) to replay the cluster of turn-of-the-century *"Natur* versus *Geist"* debates on the nature of systematic inquiry (or *Wissenschaft,* in the original German). These typically turned on philosophical considerations of whether one's methodology tended to "generalize" or "particularize," or one's ontology consisted of "material" or "mental" entities.[8] Rather, for these critical sociologists, the natural and social sciences epitomized two types of reasoning that Karl Mannheim (1893–1947) had identified at the dawn of the Second World

---

affairs of so many different places across the globe. Geography provided a more convenient way of teaching about this imperial involvement, though the organization of its studies remained "makeshift" (i.e., classroom driven) until the universities started to certify proper "researchers." At the time, geography's conceptual space was occupied in universities by history, languages, and even geology. The main difficulty in institutionalizing geography in universities turned out to be that it was a hybrid social-natural science subject. Whether its eventual institutionalization served the interests of the geography teachers is an open question. See Goodson 1988, 160–96.

7. Medhurst 1994. See also Mills 1956 and Gouldner 1970a. Gouldner's article was the 1961 presidential address to the Society for the Study of Social Problems, and in part a reflection on the career of Mills. For a comprehensive and critical account of the effects of the Cold War on U.S. academic life, written by a politically active natural scientist, see Lewontin 1997.

8. For a critical review of these debates in the German philosophical tradition, see Apel 1984. For the more recent versions of the debates, which center on the reception of Winch 1958, see Dallmayr and McCarthy 1977. I characterize the split between the natural and social sciences as historically entrenched, institutionally enforced self-divided inquiry in Fuller 1993a.

War: a free-standing *functional* rationality that operated without consideration of the ends it served (natural sciences) and a socially embedded *substantive* rationality that extended inquiry into the ends that reason served (social sciences).[9]

Mannheim's model for this distinction between functional and substantive rationality was the relationship between foot soldier and military commander, the one following orders as efficiently as possible without question and the other capable of justifying the orders in terms of an overall strategy he helped hatch at headquarters. The distinction would probably have not seemed so compelling before it was possible for the top brass to operate at a safe physical distance from the battlefront. This was a luxury first made possible during the Franco-Prussian War by Moltke's provision of telegraph service as part of Germany's material infrastructure.[10] A comparable distinction in peacetime is between the workers on the factory shop floor who assemble goods without questioning their destination or use and the managers in the corporate skyscraper whose livelihoods depend on their ability to justify their production schedules to shareholders by referring to a global market strategy.[11] And of course, as we saw in chapter 4, section 3, the distinction surfaced in Conant's "industrial" image of scientific research as centering on a laboratory led from a distance by a master professor who spends most of his time applying the scientific touch to other matters in society.

The first lesson in science and technology studies (STS) is usually that Mannheim refused to extend the critical perspective of his sociology of knowledge to mathematics and the natural sciences, which is in turn diagnosed as a failure of nerve that provides the raison d'être for the "Strong Programme in the Sociology of Scientific Knowledge," the Edinburgh source of STS. To be sure, Mannheim never incorporated anything like a

---

9. Mannheim 1940, 51–60.

10. Mannheim may be read as providing a realpolitik understanding of the instrumentalist/realist distinction. Basically, a realist philosophy of science presupposes that there is a class of people ("theorists") who operate at a sufficiently safe distance, from both a science's practical applications and its day-to-day drudgery, to discern the field's overall trajectory. In the 1930s, this asymmetry in epistemological perspective between the theorist and practitioner was reinvented in not only Mannheim's sociology of rationality but also Jean Piaget's psychology of cognitive development. Specifically, Piaget distinguishes between the child's ability to interrelate objects in a *spatiotemporal* and *logico-mathematical* world, the former corresponding to functional and the latter to substantive rationality. In each of these cases, the source of the asymmetry ultimately boils down to a difference between *bottom-up* and *top-down* cognitive processing—in more old-fashioned terms, why induction and deduction are not reversible versions of the same reasoning process. See Fuller 1992b, 453 ff.

11. For the separation of management (understood explicitly as "power") from ownership in high capitalist economies, see Berle and Means 1932, which showed that the two hundred largest nonbanking corporations in the United States were owned by such a dispersed number of shareholders that they were effectively controlled by the people they hired as managers. This work landed Berle a position as part of FDR's Brain Trust in the New Deal.

critical STS perspective in his major published works. But this was not because he thought it could not be done. By the Second World War, Mannheim had come to believe that the sociology of knowledge applied to the natural sciences just as much as to other more explicitly social forms of knowledge. Here we need to consider Mannheim's 1944 correspondence with Michael Polanyi, another Hungarian intellectual who relocated in Britain during the war.[12]

Polanyi argued that the sociology of knowledge in its strong form—that the content of knowledge is determined by its social origins—contradicts the scientist's faith in the possibility of experiencing a reality that lay beyond the all-too-contingent events in the laboratory. Just as Planck (and later Kuhn) held that a scientist's knowledge of the history of science could be corrosive to the scientist's vocation, so too Polanyi adopted a similar posture against the sociology of knowledge. Polanyi granted only that sociology can alert the scientist to "opportunities" for discovery that are either taken or missed in the course of inquiry. Invoking the image of Galileo, Mannheim responded by observing that the sociology of knowledge may be seen as applying a "scientific" perspective to science that perhaps Polanyi would prefer to avoid as it would challenge the more specifically disciplinary character of his training as a chemist. Here we see an interesting contrast in what science can supposedly "transcend." For Polanyi, a scientist's training enables her to transcend immediate experience, whereas for Mannheim a scientist should be willing to transcend the limits of her training upon learning the conditions under which it has occurred. Had Mannheim attempted to relate these insights to his distinction in rationalities, then he might have argued that under the regime of a Kuhnian paradigm, a scientist was merely engaged in functional rationality, whereas substantive rationality corresponded to the revolutionary scientist who sets the terms of her own research agenda. In that case, Mannheim might have avoided some of the abuse heaped upon him by Karl Popper in his rampage against authoritarian thinkers.[13]

Here it is worth recalling the source of Mannheim's critical orientation, one that attracted Mills and Gouldner, but not SSK. Following the neo-Hegelian line of Marxism pioneered by his older contemporary, György Lukács (1885–1971), Mannheim conceived of the sociology of knowledge as an "oppositional science" that did not simply aim to reflect the diversity of perspectives of society, but rather to represent them as symptomatic of society's latent class conflicts. Thus, Mannheim periodically distinguished

12. For the text of the Mannheim-Polanyi correspondence, see Woldring 1986, 374–77.
13. Popper 1945, chap. 23.

between existentially *determined* and *conditioned* thought *(Seinsgebundenheit* vs. *Seinsverbundenheit)*. In the former case, agents are not conscious of the social factors influencing their thought (hence they experience their perspective on the world as the only one they could have), whereas in the latter, the agents are conscious (hence their perspective is experienced as changeable). Part of the critical mission of the sociology of knowledge, qua oppositional science, is to get people to realize that their thought stands in some systematic relationship to taken-for-granted social conditions. Indeed, that was the original sense in which the sociology of knowledge was supposed to be "reflexive," a spirit that has been perhaps more faithfully reproduced in Mills and Gouldner than in its later SSK incarnations, whereby "reflexivity" has often come to mean little more than textual self-consistency.[14]

Returning to Mannheim's original distinction, substantive rationality is no less scientific than its functional counterpart, if by "scientific" is meant systematic empirical inquiry; but substantive rationality is more comprehensively critical, "reflexive" as Gouldner would later say, by virtue of placing the status of the inquirer within the scope of inquiry. Thus, in addition to wanting to know why X is the case, the substantively rational scientist would also ask *who* wants to know, and in that way include the prospective knower among those responsible for constituting the knowledge system: How is a given piece of knowledge placed in the hands of, say, a disciplinary community or a corporate client or a government bureau likely to figure in any subsequent action they might take?[15] Max Weber's call for a "value-

14. On the sociology of knowledge as oppositional science, see Frisby 1992, 173. Mannheim's distinction between existentially determined and conditioned thought was perhaps presented most clearly in a German encyclopedia article on the sociology of knowledge that appears as an appendix to the English edition of *Ideology and Utopia*. At a philosophical level, Mannheim's point in drawing the distinction was to show that "necessity" and "contingency" were ultimately grounded not in the nature of things, but in the degree of one's consciousness of the nature of things: as consciousness is heightened, the sense of necessity is transformed into one of contingency. See also Woldring 1986, 173 ff. This sense of reflexivity may be called "constructivist" in that agents are presumed to have some control (and hence responsibility) over how much they critique the conditions of their own knowledge claims. This explains the moralistic tone that Mills and Gouldner, as followers of Mannheim, adopted in their reflexive sociologies. More recently, under the influence of the "linguistic turn" in social science, some SSKers have advocated a sense of reflexivity that is more "realist" in that self-reference is treated, not as something the agent adds to thought but something "always already" present by virtue of the linguistic nature of thought. While advertising itself as radical — and certainly it has generated "new literary forms" that straddle the fact/fiction divide — the reflexive turn in SSK has succeeded only in constituting reflexivity as a bounded field of inquiry much like any other in sociology. The summa of this approach is Ashmore 1989.

15. Gouldner 1968. This paper was written as a response to Howard Becker's 1966 presidential address to the Society for the Study of Social Problems: Becker 1967. The culmination of this line of thought was Gouldner 1970b. We shall return to Gouldner's critique in chapter 7, section 5B.

free" science, by result if not design, had enabled social researchers to suspend this kind of question as being beyond their professional expertise. Unfortunately, this meant that the knowledge they produced functioned as "wild cards," playing to the advantage of whoever happens to possess it.[16]

Mills had sympathized with Mannheim's sociology of knowledge from the start of his career.[17] He held that science's biggest contemporary threat to society was its incorporation into what Mills dubbed the "science machine," the mechanical execution of military orders, albeit by means of intellectually stimulating forms of research, such as the quest for radar-insensitive aircraft or atomic bombs (later called "neutron bombs") equipped to annihilate people but not buildings. To counteract this tendency, Mills urged that scientists air their disagreements with their employers and administrators, pointing to the emergence of defense industry whistle-blowers as exemplary catalysts to a more public science policy discourse. Moreover, he denied the public importance of the "science race" between the United States and the USSR, even before it had gotten into full swing. Mills dared to ask in 1958, "Who wants to be number one to the moon?"[18]

3. *Structure* as a Recipe for Making the Future Look Like More of the Same

However, *Structure* deflected sociology's critical gaze on the natural sciences by showing (by example, though not by intent) how even sociology could legitimate itself as a "real" science, and thereby (presumably) no longer be excluded from the ruling power structure that Mills and Gouldner were trying to fight.[19] An apt slogan for this turn of events is the subtitle of *Dr. Strangelove*, the period movie starring Peter Sellers: *How I Learned*

---

16. The devil in Gouldner's eyes would be a successful player in the Weberian game, such as Paul Lazarsfeld (1901–76), who emigrated from Austria to the United States when the Nazis came to power, and quickly established himself as sociology's leading quantitative methodologist. Lazarsfeld would overcharge his government and corporate clients for surveys that gave them what they wanted and then use the extra money to finance methodologically sounder surveys to satisfy his disciplinary community. The success of this strategy presupposed that sociology was insulated from, if not disinterested in, the uses that the state and industry made of its findings. See Turner and Turner 1990, 85–132. Lazarsfeld also comes in for some blows as the archetypal "abstracted empiricist" in Mills 1959, 60–86. Mills was a colleague of Lazarsfeld's at Columbia.

17. Mills's attraction to Mannheim first appears in his doctoral dissertation on American pragmatism's legacy to sociology. As it turns out, during this period (the late 1930s), Mannheim increasingly acknowledged the influence of pragmatism on his thought. See Nelson 1995.

18. Mills 1958. esp. 174–83. For a trenchant critique of Mills as romantic, see Bell 1960, 47–74, which is on Mills 1956, his most important sociological work.

19. A rather sanguine analysis of Kuhn's use by social scientists is given in Weingart 1986. In contrast, as Kuhn's work became more widely known, Gouldner explicitly distanced "critical intellectuals" from normal scientists in the Kuhnian mold. See Gouldner 1979, 48–49.

to Stop Worrying and Love the Bomb.[20] But if Kuhn failed to anticipate the sociological uptake of his ideas, sociologists equally failed to anticipate that by constituting themselves as one or more self-contained paradigms, they were in fact decreasing the likelihood of their having *any* impact, either critical or supportive, on that powers that be.[21] However, it must be said that through Kuhn-inspired self-discipline, the social sciences have managed to render themselves more collegial company within the academy.[22]

Kuhn's disclaimers notwithstanding, social scientists were attracted to his book precisely because it seemed to provide a blueprint for how a community of inquirers can constitute themselves as a science, *regardless* of their subject matter. However, this move created a free-floating legitimating narrative—a "myth," properly speaking—that could be used by any discipline in need of boosting its status: quite the opposite of what Kuhn would have wanted.[23] Here it is worth recalling an astute account of the social function of narrative recently suggested by literary critic J. Hillis Miller.[24] Miller argues that narratives can sublimate the potentially disruptive character of social change in two ways.

On the one hand, a narrative can script such change through the reenactment of a sequence of events that in the past had enabled others to reach a social status similar to the one currently desired. The clearly labeled plot structure of Kuhn's account of scientific change contributes to this function well, as it allows others to see that one's discipline is behaving as a science

---

20. Peter Sellers, in another of his famous roles, figured as an inspiration for this book's conception, as discussed in the preface and Fuller 1992a.

21. See esp. Scott and Shore 1979, which explicitly blames an increased disciplinary consciousness among sociologists for the field's failure to affect U.S. government policy. However, Scott and Shore argue that the solution is for sociology to adopt a more client-centered orientation, not, as Mills or Gouldner might suggest, to consider a wider range of society toward which the sociologist might orient herself. My thanks to Nico Stehr for bringing this work to my attention.

22. An especially striking example of this point appears within a few pages of a standard American textbook in law relating to science. An abridgement of *Structure* is initially presented in a section of the book devoted to how judges should decide cases concerning genetic engineering. Here Kuhn appears to testify that scientific research has its own value system, quite apart from the way its results are used. Yet, at the end of that section, students are invited to consider that law itself may be a science in just this Kuhnian sense, thereby problematizing the impulse to have the courts regulate genetics research. For if law conforms to the Kuhnian account of science, then it can restrain science only as much as it would itself; if law fails to conform, then it would seem to lack any legitimate basis for telling scientists how to conduct their affairs. See Areen et al. 1990, esp. 163.

23. The disembedded character of *Structure's* narrative was not lost on his fellow historicist philosophers, who suspected Kuhn of having interdefined his key terms, especially "paradigm" and "revolution," to such an extent that there was no way for a discipline to have a paradigm without a revolution, and vice versa. The twin specter of tautology and unfalsifiability that Hanson and Toulmin raised against Kuhn were bugaboos to the philosophical mind, but safe havens for fields in need of legitimation. See Hanson 1965; Toulmin 1972, 98–130.

24. Miller 1990.

would at a certain stage in its development. On the other hand, a narrative can effect a subtler social channeling, one that operates as a safety valve for dissipating subversion. In this case, the wording of the narrative incorporates, typically by extended metaphor, elements that could threaten the social order that underwrites the narrative's legitimacy. We may call this function *verbal co-optation*. The desired effect is that the threatening elements are contained by severing any connection that they might have to a world outside the narrative. In the Kuhnian narrative, with its talk of "crises" issuing in "revolutions," the political economy of science has been so verbally co-opted that Kuhn's sociological followers, be they Mertonian or SSK, have been able to cast several models of scientific activity in the discourse of political economy without either making links with the larger political and economic scene that sustains Big Science or even drawing much on the empirical and explanatory resources of political science and economics.[25]

As evidence for this pseudopolitical economy of science that has been sustained by Kuhn's work, I will focus on the simple locution, "Big Science," coined by the historian Derek de Solla Price (1922–83), as that forms the basis for much of the other work in this vein.[26] Price is generally rec-

---

25. Followers of the postmodern scene will recognize a pronounced "Baudrillardian" element in the legitimacy that social scientists have acquired by means of the Kuhnian narrative. Jean Baudrillard (b. 1929) has coined two useful terms in this context: *hyperreality* and *simulacrum*. Once enough research programs have persuaded enough audiences that they constitute sciences, even if it is impossible that they could all be right (because a true science has only one paradigm at any given time), and even if an increasing number of historians doubt that the Kuhnian narrative is itself right, those new "sciences" have already acquired their own *hyperreality*, which creates similar expectations in other fields desirous of becoming sciences. At the same time, social scientists lose any interest in recovering their own origins, or at least this interest becomes increasingly alien to those who regard themselves as "cutting-edge" researchers. Thus, like a Hollywood depiction that becomes reality for its audience, the Kuhnian narrative displaces the actual history of social science and, in that sense, functions as a *simulacrum*. See Baudrillard 1983.

26. Price 1986. It is worth remarking that, like Kuhn, Price trained in mathematics and physics (his experimental work was on the atomic emissions of hot metals) but failed to make the transition to Big Science after World War II. Thus, his references to the "sealing wax and string" tradition of "Little Science" should be taken as having been made in the spirit of nostalgia. My thanks to Tim Rogers and Donald Beaver for insight into this matter. A good guide to the artifice committed by a manner of speaking that is now taken for granted is to examine those who balked when it was a neologism. In the case of "Big Science," one need look no further than Crowther 1968, 318–19.

Crowther regarded the expression as a piece of Orwellian Newspeak designed in the interest of policy makers who wanted to cover up the fact that science had relinquished its role as the catalyst of social and intellectual change for becoming the bulwark of national security in a world dominated by the Cold War. Whereas Price portrayed the transition from Little to Big Science as a natural evolutionary process, Crowther regarded it as the result of planned stagnation. Not only did Crowther make the now-familiar point that no scientific revolution — even in the twentieth century — was instigated as a result of massive capital investment, but he went so far as to suggest that the increasing concentration of scientists in universities (recall

ognized as the founder of "scientometrics," the discipline that designs quantitative measures for tracking the growth of knowledge, typically with an eye to science policy concerns. Although Price's original work was independent of Kuhn's, scientometricians have subsequently focused on identifying the "life cycle" of scientific specialties, deriving the stages in this cycle from Kuhnian categories of old paradigm–anomalies–crisis–revolution–new paradigm.[27] Notwithstanding the merit of much of this work, it rests on simplifying assumptions about the international political economy of science that cannot help but contribute to the image of science as a self-contained, self-sufficient enterprise. For example, the major "product" of science is presumed to be the journal article, whose value is measured in terms of the other articles that cite it as instrumental in their own production. Indeed, at one point, Price operationally defined science policy as investment strategies for producing the largest number of highly cited articles.[28]

What makes today's science so "big," then, is the number of articles produced, and especially the exponential rate of their production. Price justified his circumscription of scientometrics by appealing to a contrast between the politically motivated accounting procedures that governments

---

that this is written during a period of unprecedented university expansion) would ensure science's continuing role in stabilizing the social order through its triple function of training, research, and legitimation—very much like Planck's Faustian bargain, discussed in chapter 2, section 7. However, with the end of the Cold War and the onset of state divestiture of science funding, we might wish to revisit Price's evolutionary scenario and suggest, in a spirit Crowther might appreciate, that the biggest creatures are rarely the best adapted, in which case before we infer that bigger science is always better, we might consider the fate of the dinosaurs.

27. De Mey 1982, 111–70, esp. 148–70. One scientometric indicator of where a field is in the Kuhnian life cycle is Price's index, which measures the obsolescence rate of journal articles in terms of diminishing citation patterns. A rapid obsolescence rate means a rapidly advancing research front, a paradigm at top puzzle-solving performance levels.

28. Price 1978. Some of those influenced by Price paint an orderly Mertonian picture whereby science's "reward structure" is governed by internally generated status markers, such as citation counts, which amount to elaborate exercises in gift giving. See Hagstrom 1965, who credits his knowledge of the history of science to Kuhn's tutelage at Berkeley. However, some French theorists of science, starting in the 1970s, projected Conant's original military metaphor of "strategy and tactics" on a social "field of play" to paint a more agonistic picture of science, yet one still captured within a broadly economic framework. Chief among these has been Pierre Bourdieu (b. 1930), whose anthropological fieldwork in Algeria in the late 1950s, just prior to its colonial revolt from France, led him to introduce the idea of *symbolic capital*, namely acquired (and sometimes inherited) cultural resources that are convertible with economic ones. Specifically, symbolic capital approximates the strictly economic sense of capital, the more competitive the field of play, such that win-win exchanges (as in gift giving) are rendered impossible. See Bourdieu 1977, 171 ff.; Bourdieu 1975. Whereas Bourdieu analogically extended the findings of his anthropological fieldwork to science, the connection was made literally, and to spectacular effect, in Latour and Woolgar 1979, esp. chap. 5. For a critique of this "post-industrial" mentality in science (which I dub *meta-industrial*), see Fuller 1997d, 67–76.

use to estimate research and development expenditures and the policy of free trade and open markets demonstrated by the publication practices of scientific journals.[29] Price seemed to be impressed with the fact that out of a myriad of national interests can emerge a global picture of science that is reducible to fairly simple and intuitive S-shaped logistic curves, which represent exponential functions whose rate of growth rises quickly until it reaches a plateau. Such curves supposedly attested to the existence of science as a spontaneously generated and self-sustaining organism.[30] Moreover, he believed that given the nature of logistic curves, there would soon be a surplus of scientists, each of whose efforts would constitute a diminishing return on national investment, if they are confined to a research environment. Rather than decreasing the number of scientists produced, Price envisaged that they would spill over into government, thereby contributing to an overall rationalization of society.[31]

Price's line of thought here was hardly idiosyncratic. We have already seen it as the mark of Conant's action-intellectual, the scientist who moved easily between the lab bench and the corridors of power. What distinguished Price was his belief that this scientization of society would emerge, without planning, as an unintended consequence of the surplus production of scientists. It would be hard to overestimate Kuhn's influence in the promotion of this picture of science as a self-organizing system in science policy thinking in the final third of the twentieth century. Indeed, a case can be made that *Structure* was used to legitimate this view of science policy even before it had caused academic disciplines to constitute themselves as paradigms.[32] The result has been to diminish the normative dimension

29. Price 1986, 137–38.

30. For all his genuflection to the idea of science as a self-organizing system, even Price seemed to realize that Big Science required the right economic background conditions to be in place. Specifically, he observed that there was a strong positive correlation (at least in the 1960s) between the nations that were the top ten per capita producers of scientific papers and the top ten per capita consumers of electricity. See Price 1978.

31. Price 1986, 95–102. Nowadays, this would be quite a lot of spillage, since holders of science degrees are as likely—about a 1 in 3 chance—to land permanent positions related to their field of study as holders of humanities and social-science degrees. However, in the case of science degree holders, the change in career trajectory positively reflects on the versatility of scientific training, whereas in the case of the others, the change negatively implies the dysfunctionality of humanistic and social-scientific training.

32. One of the most influential policy theorists of the post–World War II period, Don K. Price (b. 1910, no relation to Derek de Solla Price), discusses Kuhn's views in some detail in a 1965 book, critical acclaim from which helped propel Price into the presidency of the American Association for the Advancement of Science. Reflecting on his experience as an officer in the Ford Foundation's program of technical assistance for developing countries, Price argued that both science and politics had their own proper ends, and that effective science policy was forged by keeping communication channels open, while respecting the inherent differences between these two spheres of activity. Price appeals to Kuhn in order to draw a connection between scientists' instinctive rejection of non-paradigm-based accounting procedures, such

of policy making from *planning* to *forecasting*, which, as we shall see in the next chapter, corresponds respectively to the distinction between *prescription* and *evaluation* in the philosophy of science.

In scientometrics, the stress on forecasting over planning caters to a "clinical" picture of science policy (akin to Crane Brinton's Paretian historiography in chapter 3, section 5), whereby the policy maker attempts to diagnose and treat an objective condition of a science on the basis of key symptoms, or "indicators." The science under scrutiny is taken to have a natural trajectory, which the policy maker can then slow or hasten, or, more drastically, divert in some other direction, perhaps halting progress in the science altogether. Each more substantial intervention is seen as incurring a greater risk to the delicate balance of factors needed for the science to flourish.[33] It has become common in Western countries for science policy makers to see themselves as primarily in the business of adjusting the tempo of scientific change, and preparing the public for absorbing the intended and unintended technological consequences of that change. The burden of planning is taken to lie in reconstructing not the scientific enterprise, but the larger society, presumably because a society's ordinary institutions are safely manipulable to meet the public interest, whereas the "natural" ends of science can be easily perverted if forced to serve interests other than its own. The result is a society in which science can be more readily applied, received, and consumed—indeed, perhaps even more than if the policy makers had set out, socialist-style, to adapt science directly to society's needs. It is precisely this reversal in the direction of social shaping that

---

as a federal budget, and their ability to get at the nature of some subject matter. Defending those instincts, Price proceeds to argue that if, on the contrary, scientists were driven primarily by political imperatives, they would be engaging solely in applied research, leaving more discipline-based pursuits to grow fallow. Unfortunately, this argument obscured the delicate fact that modern governments have been just as eager (if not more so) to support research that scientists deem "pure" as that which they deem "applied." In the name of pure science, they could get high-quality information ("intelligence") that might someday serve to legitimate otherwise dubious political acts, in return for the scientists' assurances that they will not feign competence in determining the appropriate uses for their work. Once again, Conant's problem of science's dirty hands is solved by the principles of democratic pluralism, this time as midwived by Thomas Kuhn. See Price 1965, 172–77.

33. Thus, in concluding about "the future of science," Hagstrom says: "The central problem facing society with regard to all of the professions is the same: how can they be controlled without having their effectiveness destroyed?" (Hagstrom 1965, 294). It would be interesting to examine which groups would be placed at risk by a stronger sense of science planning than the forecasting approach countenances. My guess is that at risk are the social scientists, whose legitimacy rests on their having correctly followed a presumptively universal pattern of scientific development! For a survey of the social causes and effects of forecasting over the last thirty years, see Dublin 1989, which argues that, contrary to its own hype, forecasting has politically conservative, not liberating, consequences because it is based on extrapolating from current trends. The high degree of likelihood attached to the future's resembling the past artificially raises the level of risk in potential deviations.

is often meant by those who follow Daniel Bell's lead in claiming that we reached the "end of ideology" and have now entered a *knowledge society*.[34]

That most academically successful of disappointed Marxists, Daniel Bell (b. 1919), was commissioned by Columbia University in 1963 to study the disintegration of general education programs like Conant's at Harvard and even Columbia's, the oldest of its kind in America, whose equally war-related inspiration is revealed by its date of origin: 1917, the year the United States entered World War I.[35] Bell's charge was to suggest a curriculum for what was soon to become the baby boom generation of offspring of World War II veterans. Already, in response to the Cold War, the U.S. government

---

34. See Stehr 1994, 103. Perhaps the most intellectually sophisticated development of the knowledge society idea was made in the late 1960s by the distinguished social psychologist and social-science methodologist, Donald Campbell (1916–96), under the rubric of "the experimenting society." Drawing on his career as a pioneer in "quasi-experimental methods" (i.e., methods that simulate experiments outside laboratory conditions) and a participating scientist in Lyndon Johnson's project of social reform, the Great Society, Campbell envisaged the populace as a testing ground for social science, once their responses to policy interventions are systematically solicited, recorded, and fed back into hypothesis formation process. Campbell believed that the experimenting society would help consolidate the kind of "social intelligence" that John Dewey held was essential for the growth of democracy. In that way, advancement in social science would correspond to advancement in society itself. A curious feature of the experimenting society is that for all its noble aspirations, it did not include science as itself one of the aspects of society on which quasi-experimental methods should be applied. Consequently, while Campbell encouraged public participation in the testing of social-scientific hypotheses, he did not extend the encouragement to the actual hypothesis formation process. See Campbell 1988, 261–314; Fuller 1998d. For a sympathetic account of the experimenting society that situates it amid other trends in social theory and social policy in the 1960s and 1970s, see Coleman 1978.

35. Bell 1966, 14–15. Even in the late 1970s, when I was an undergraduate at Columbia College, general education at Columbia revolved around two required courses, "humanities" and "contemporary civilization." The former is a Great Books course that involves the student in reading and discussing a classic a week, typically with an instructor who is not a specialist in any of the works under study. The latter is a more issues-oriented course, centering on the principles grounding the just (Western) society, as seen through the ages. Part of the motivation for the humanities course was to develop a "lay" appreciation of Western culture that could be used to counteract the revisionist tendencies of scholarly specialists, whose incursions in the public sphere threatened to undermine the legitimacy of the social order. Just as Martin Luther retrieved Christianity from the clerics by encouraging the faithful to trust their own understanding of the Bible, so too have the likes of Mortimer Adler and other advocates of the Great Books program tried to retrieve secular culture from academic revisionists. The paradigm case of such a dreaded revisionist whose radical scholarship threatened to subvert middle class values was Charles Beard. His quasi-Marxist interpretation of the U.S. Constitution and explicit pacifism in World War I eventuated in his expulsion from a history professorship at Columbia. See Novick 1988, 206–24.

The contemporary civilization course was originally conceived as a course in "war issues" and "peace issues," operating on the assumption that students may increasingly find themselves in leadership roles that straddled these two important sets of concerns. For an excellent account of the Columbia experience, especially in relation to Harvard's attempts at instituting a general education curriculum, see Buck and Rosenkrantz 1981, esp. 374–77. For a general account of the campaign to inoculate middle-class college goers against revisionist views of Western culture, see Rubin 1992.

was providing liberal incentives for citizens wishing to acquire technical training under military auspices, most notably the Reserved Officer Training Corps, a natural complement to the original G.I. Bill of Rights. This, combined with an increasingly credentials-based social order, had swelled the college enrollments of eighteen- to twenty-one-year-olds from 14 to 46 percent between the end of World War II and the start of the Vietnam War.

Bell was noted at the time for arguing that the university had unwittingly acquired the institutional mission of empowering the traditionally powerless.[36] However, the faculty waiting to instruct these new recruits to academia had come to expect students who were sufficiently rooted—if not in Western culture per se, at least in its material maintenance—to withstand the intellectual assault of reading Marx, Freud, and Nietzsche. It was not that the new breed of students suffered from economic hardship; on the contrary, they had been raised in relative wealth and hence without having had a stake in sustaining American society in its current form. Thus, they were increasingly skeptical of the ultimate value of the "higher education" they were receiving, with some students even demanding that they have a hand in university governance. In fact, this tension only increased during the three years it took Bell to file his report, and within five years Columbia experienced the first of its student revolts against U.S. military involvement in Vietnam.

In both his report to Columbia and *The Coming of Post-industrial Society*, published in the following decade, Bell declared that the dawning of a new age of "intellectual technology"—what is now called, less mystifyingly, "information technology"—would at once rationalize and depoliticize the complex divisions that had come to characterize American society.[37] At the helm of his vision was the university, the leading manufacturer of new forms of human capital and capital goods, technical expertises, and their computer simulations. If class strife was born of industrial society, a newly differentiated and integrated workforce would issue from "post-industrial society." In hindsight, we can say that while it is certainly true that every sector of the economy has raised its academic requirements for employment and computer-based products have become indispensable aids, these developments have not added up to a harmonious social order supported by a benevolent technocracy.

Among those who regarded Bell's dream as a nightmare from the start were the intellectual leaders of the campus unrest in the late 1960s.[38] Per-

---

36. Geyer 1993, 505.
37. Bell 1966, 77–82; Bell 1973.
38. The leading radical leftist student group of the 1960s, Students for a Democratic Society (SDS), was founded on this principle as an extension of participatory democracy. The

haps because many of these critics were trained in the humanities or the "softer" social sciences, there has been a tendency to caricature their objections as antiscientific and Luddite. However, a perusal of a major manifesto of the period, *The Dissenting Academy*, reveals a more nuanced concern. The editor was the medieval political historian, Theodore Roszak (b. 1933). Roszak and his comrades, who included Noam Chomsky, feared that the postindustrial mentality sacrificed the critical mission of science—associated with the Enlightenment motto "Dare to know!"—in the name of increased productivity, which, in turn, strengthened the grip of the military-industrial complex on American society.[39] Technocracy for them was not science applied but science betrayed. While Bell did not hide the fact that the university figured so prominently because embodied information—both human and machine—was quickly becoming the leading factor of production in the economy, his critics' concerns centered on the quality of the products—again, both human and machine—that this new economy was generating. In particular, the dissenters were worried about the kind of *people* that such an economy produced.[40]

Although artificial intelligence research was still very much in its infancy in the 1960s, it was already clear that even relatively unintelligent computerized systems could perform well in environments that simulated the complex problems facing a harried bureaucrat, manager, or field commander who cannot wait for perfect information before deciding on a course of action. The advertised virtue of these machines was, in large measure, a function of their very stupidity. Precisely because these machines could not capture the full range of factors that would influence a human decision maker, they ended up saving time and resources by avoiding potentially irrelevant lines of thought. The obvious but politically abrasive point in all this was that these machines could not effectively operate with-

---

movement's founding Port Huron Statement (1962) was influenced largely by the thought of the recently deceased C. Wright Mills. See esp. Mills, "Letter to the New Left," published originally in the *New Left Review* (1960) and reprinted in Waxman 1968, 126–40.

39. Roszak 1967.

40. In terms of the institutionalization of knowledge production, the postindustrial dissenters were self-consciously revolting against the German model of the research university, which the United States had too eagerly adopted earlier in this century. Consequently, the Enlightenment mission had been artificially divided (into "service" and "scholarship," or more realistically, clientelism and scholasticism) and conquered (by a university bureaucracy that regulated the relationship between the academic's intra- and extramural activities). From the dissenters' standpoint, when the American Association of University Professors censured academics like Columbia historian Charles Beard for his opposition to U.S. entry into World War I, it had already betrayed its original mission of protecting the conscience of academics from institutional stricture. The dissenters vowed to ensure that such capitulation did not occur in the 1960s. See Theodore Roszak, "On Academic Delinquency," in Roszak 1967, 3–42.

out a hospitable human environment. In other words, the attending humans had to be just as focused on the goals pursued by the computers as the computers themselves were. Any critical interrogation or creative reconfiguration of the software would defeat the purpose of "informatization." However, postindustrialists like Bell were well aware that old human habits die hard, and so new forms of research and teaching entered academic life that soon made "simulation" seem like second nature.[41]

Regularity and reproducibility—the virtues that Bell had identified in the new information technologies—appeared, in the eyes of his critics, as the latest trained incapacities of the learned classes. Classroom simulations favored those whose conception of accountability was limited to their assigned frame of reference. The implicit message of these exercises was that a complex decision should be made at some distance removed from the issues—and perhaps even the very parties—potentially affected by the decision. So, whereas the fuzzy-minded would deal directly with the concrete lives at risk when deciding how to trade off productivity against safety in the workplace, the rigorous simulator would start by translating these people into more abstract units of analysis. From under the veil of the algorithm, then, postindustrial reasoners were to be insulated from the remote consequences of their decisions. This style of reasoning would become the signature of American military strategists in the Vietnam War, especially the Harvard-trained civilian, Defense Secretary Robert McNamara.[42]

The discontent generated by the new information technologies in the 1960s was epitomized in a word: *alienation*. However, this word was used in so many different ways that Bell and his critics ended up arguing at cross-

41. Business schools were in the vanguard of these developments, with Herbert Simon (b. 1916) at Carnegie-Mellon University playing a leading role. See Simon 1991, chap. 9 ff. In the last thirty years, computer simulations have been increasingly used as research tools in the natural and social sciences. Indeed, the ease with which scientists today confer on simulations the title of "experiments" speaks volumes about how "natural" the artificial has become. Simulations nowadays often replace costly laboratory experiments, which a couple of generations earlier would have themselves been regarded as lacking naturalness or "external validity." However, dissenters from postindustrialism were originally more concerned about the role of simulations in the *classroom*—not the actual use of computers, but the attempt to get teachers and students to think more like computers. These simulations often assumed such humanistic guises as "games" and "role-playing," but were in fact no different from what military and business strategists had begun to call "scenarios." The players of these games occupied positions in an artificially constructed situation calling for a decision. However, the abstract setting sufficiently resembled the "real world" to be of some pedagogical value. Each position came built with assumptions that constrained the player's deliberations. Adeptness at play came from making the best decision possible operating exclusively within those constraints. In fact, players would be faulted for "breaking frame" and introducing "extraneous information" that nevertheless would be pertinent in the real-world setting. Sometimes these faults would be diagnosed as "errors in formal reasoning," if not signs of outright "irrationality." See Noble 1991; Edwards 1996.

42. On McNamara as an exemplar of the Harvard curriculum, see chapter 4, note 9.

purposes. At first, the critics complained mainly about how the computerized environment disabled the skills classically associated with participatory democracy, especially critical deliberation in a public setting. Their fear was that, given the "proper training," people would come to lose any interest in collectively questioning the ends that their work serves. It would be enough for them to do the work as efficiently as possible. Bell and his defenders interpreted these complaints to be more about the character of postindustrial work itself—whether it retained enough "craft" elements to give aesthetic satisfaction to the people doing it.

At the most basic level, "alienation" refers to the process by which people come to lose what they regard as most human. But, of course, there are as many senses of "being human" as there are schools of philosophy. Addressing the alienation of *zoon politikon*—the concern of the dissenting academicians—is quite different from addressing the alienation of *homo faber*—the concern of the postindustrial prophets. However, in the confused heat of the debate, the latter concern came to dominate the former. In large part, this shift reflected a widespread belief that the social order projected by the postindustrial prophets was, indeed, an inevitability, and that the best one could do was to adapt one's sense of intellectual integrity to it. The call to craftsmanship was one such adaptation.

Bell's inspiration for introducing craft elements into technocratic work was Kuhn's attractive picture of normal science.[43] Instead of making all scientists look like Galileo and Darwin—figures that inspired Roszak to coin the word "counter-cultural"—Kuhn portrayed the average scientist as focused on technical puzzles that make sense only in the context of the "paradigm" under which she is laboring. Opposing those who had caught on to Kuhn's claims that outsiders are often the source of revolutionary science, Bell shifted the focus from the innovators to their audiences, without whom the innovators could not bring about a revolution.[44] Bell argued that revolutions have been recognized only against the background of a common core of learning shared by a paradigm's practitioners. This provided justification for requiring that undergraduates major in proper disciplines with well-established paradigms—that is to say, *not* in the emerging interdisciplinary fields of "area" (i.e., Soviet, Asian, African, Latin American) studies and women's studies. These fields, then popular with students because of their "relevance" to the contemporary world's problems, would be restricted to the graduate level.

43. Bell 1966, 109–10, 248–50.
44. On the claim that scientific revolutionaries tend to be outsiders or "new to the field," see Kuhn 1970b, 166

Kuhn's claim that scientists are averse to protracted paradigmatic crises also played into Bell's technocratic resolution of scientific revolutions, namely, the subdivision of paradigms into two peacefully coexisting specialties, a model Kuhn regarded as "evolutionary" but that easily falls under the technocratic rubric that Bell favored for converting political disputes into administrative problems. Although Kuhn himself clearly wanted to restrict his account to the natural sciences, his tendency to identify a paradigm with a self-selecting, self-governing community of inquirers helped persuade many humanists and social scientists that they too could enjoy scientific status — or at least raise their standing in the university — by doing things that look like normal science.

4. *Structure*'s Role in Socializing Sociology and Depoliticizing Political Science

Thus, by the 1970s, the rhetoric of the dissenting academy had subtly shifted from claims that the sciences were allowing the military-industrial complex to colonize the university to claims that one's own field had the right to pursue its paradigm alongside the natural sciences. Perhaps most striking in this regard was Robert Friedrichs's *A Sociology of Sociology*, to this day perhaps the most impassioned application of Kuhn's ideas to a discipline's intellectual trajectory. Nevertheless, despite winning the American Sociological Association's leading research award in 1971, the work displays the sort of self-subversive radicalism that we have by now come to associate with Kuhn. Friedrichs opens boldly with a chapter entitled, "The Structure of Social Scientific Revolutions":

> [Kuhn's] central thesis is that the communal life of science . . . demonstrates considerable affinity to the life-cycle of the political community. Indeed, Kuhn's posture is quite literally a "radical" one, for the political community that he offers as a pattern is not the constitutional community that fits the ideological proclivities of the contemporary West but rather — as the title of the volume would indicate — the *revolutionary* community.[45]

However, by the end of the book Friedrichs's revolutionary fervor has been dampened by the realpolitik of the academy, though ironically he fails to see that his solution is in keeping with the spirit, if not the letter, of Kuhn's thought: "Hard evidence in support of one paradigm to the exclusion of another is too difficult to come by to dull the social scientist's appreciation of the long-range value of 'peaceful' paradigmatic coexistence."[46]

---

45. Friedrichs 1970, 1 (italics in the original).
46. Friedrichs 1970, 325.

Although Kuhn would not accept Friedrichs's proposal that sociology as such be regarded as a "multiple paradigm science," the image is clearly inspired by the lack of unity in Kuhn's own overall conception of science.[47]

The rhetoric of "Let a thousand paradigms bloom"—or more prosaically, "If you can't beat 'em, join 'em"—worked well in this time of plenty for the universities, when ideological enmities could be resolved by the creation of separate departments underwritten by discrete paradigms. However, the fiscal contractions of the last decade have forced somewhat different conclusions to be drawn from the normal science mode of inquiry. A paradigm in the arts and sciences not only marks its practitioners as disciplined and its practices as self-contained, but also increases the likelihood that both will be dispensable from the standpoint of general education and the public mission of the university. In the case of the professional schools, the conversion of their expertises to normal science renders them tractable to the class of information technologies called "expert systems," which similarly threaten the employment prospects of therapists, medical technicians, engineers, financial planners, legal advisors, and many more.[48] Thus, traveling under Kuhn's protective coloration, Bell managed to persuade academics to think about their work as constituting, in Herbert Simon's terms, a "near-decomposable" system. Because a paradigm is essentially a module of knowledge production whose business can be conducted largely in isolation from other bodies of knowledge, the source of its autonomy—a limited domain and well-defined procedural rules of inquiry—is also the source of its replaceability, either by a successor paradigm, in the case of revolutionary science, or by an expert system, in the case of normal science. In both cases, however, the rest of the knowledge system can proceed as if nothing has happened.

Thanks partly to the unintended influence that Kuhn's *Structure* had already exerted in the late 1960s, by the early 1970s Bell and Roszak turned out to hold complementary positions that enabled each to assimilate Kuhn's oversharp distinction between revolutionary and normal science in a way that allowed no public space for science criticism. In a nutshell, as Bell increasingly celebrated a postideological technocratism in *The Coming of Post-industrial Society*, Roszak embraced a romantic aestheticism in *Where the Wasteland Ends*.[49] On the surface, it is difficult to see how to reconcile these two positions as complementary readings of the same text.

---

47. The idea that sociology is a "multiple paradigm science" is due to Ritzer 1975. For a critique of this appeal to paradigm, see Roth 1987, 124–26.
48. Fuller 1994a, 1994i.
49. Roszak 1972. Kuhn is not mentioned in Roszak's most famous, earlier work: Roszak 1969.

TABLE 5. *Kuhn as a Reversible Template for the Postindustrial Society*

	BELL	ROSZAK
revolutionary science	ideological (−)	inspired (+)
normal science	technocratic (+)	mediocre (−)

Their differences are encapsulated in the opposing personality profiles they offer of intellectuals, as depicted in table 5.

Since we have already seen how Bell used Kuhn to uphold the virtues of normality, it is worth recalling the flavor of Roszak's revolutionary interpretation:

> Now, in truth, the process of discovering knowledge — especially at the level of a Galileo, a Newton, a Darwin — requires indispensably the unpredictable flash of genius, the unaccountable spark of insight. As Thomas Kuhn has shown in his *The Structure of Scientific Revolutions*, it is precisely by such inspired leaps that the great "revolutions" occur in the several branches of science. And then, suddenly and amazingly, there is something — an idea, an image, a conjecture, a neat synthesis, a "paradigm" as Kuhn calls it — that captures the imagination and is accepted as knowledge. Often enough, it is a matter, as Yeats has put it, "of finding similarities among things thought different and differences among things thought similar." But how such perceptions occur is beyond routinization. Which means that methodology is the stone in stone soup, the most dispensable of all ingredients. This is no less true of the so-called behavioral sciences, despite the fact that the learned journals of these disciplines brim over with the lucubrations of methodological specialists. It is all nonsense plain and simple. Who, except as an afterthought or as a pinch of incense on the professional altar, ever used another person's methodology to produce a significant idea of his own? The methodologies of a Max Weber or a Sigmund Freud yield brilliant insights only in the hands of a Weber or a Freud; in the hands of lesser talents, they yield what may be less worth having than the blunders of a great mind. One might almost suspect that methodology is the preoccupation of mediocrity, the dullard's great hope of equaling the achievements of the gifted.[50]

Much of *Wasteland* reads like this — a scholar's night thoughts, Goya's *Sleep of Reason* in linear form. Normally hushed truths are expressed with

---

50. Roszak 1972, 201–2. This passage is redolent of James 1956, 299–300. In support of his interpretation of Kuhn, Roszak cites Barnes 1969. Barnes and his Edinburgh colleague David Bloor have, over the years, promoted an academically domesticated reading of the "romantic" image of Kuhn (references to Karl Mannheim on the sociology of conservative thought replacing ones to William Butler Yeats on the phenomenology of poetic genius). See Bloor 1976, chap. 4; Barnes 1994.

what Nietzsche, referring to John Stuart Mill, called "insulting clarity."[51] The book's style is self-consciously modeled on the early nineteenth-century romantic movement, and consequently reproduces the "in your face" elitism that was its calling card (as opposed to the ironically understated elitism that characterized Pareto's disciples).[52] This aspect of Roszak's rhetoric is often lost in nostalgic reminiscences about counterculturalism. For example, Roszak believed that Bell's vaunted technocracy was persuasive just as long as the intellect was understood to be a principle of social stratification. In other words, Roszak doubted that people were intellectual equals, but believed that human equality could be grounded in the similarity of people's feelings and passions.[53] Keeping with his original training in medieval philosophy, Roszak supposed that the antidote to the orderly world projected by Bell's vision of postindustrialism was not public debate, but the promise of personal salvation from the burdens of mediocrity, a "visionary commonwealth" whose strength of imagery is the measure of its political force.

What both Bell and Roszak seemed to *disallow* was the possibility of open disagreement, that dissatisfaction with the military-industrial complex might lead to organized resistance and transformation. It would seem that the only two options on offer were either to become a well-oiled gear in the administrative machinery (computer addiction) or to envisage a personally satisfying alternative reality that enables one to "drop out" of the dominant one (drug addiction). From reading just Bell and Roszak, one could easily get the impression that explicit arguments about the role of science in society are inherently dangerous: One should either accept or reject science as is, with no room for negotiation. Given this polarization, sociologists could not be blamed for explaining the sentiments spawned by

51. Readers may wish to compare Roszak's remarks with my own in chapter 7, section 3C, concerning "contextualist boilerplate" in science and technology studies.

52. The economic origins of romanticism in the vicissitudes of the publishing trade are reasonably well known but still not sufficiently integrated in general histories of European culture. Basically, authors found the appeal to "genius" attractive when book piracy caused publishers to cut the commissions given to authors in order to make up for lost profits. Up to that point (the late eighteenth century), authors were treated as potentially interchangeable hack writers, much like the emerging class of industrial wage laborers. "Genius" suggested that the author added something unique to the book production that might not be immediately evident to either publisher or reader but merited generous payment. By citing genius in this fashion, it became possible for authors to claim a living wage, regardless of how much their works sold or were plagiarized. See Woodmansee 1984. A good survey on the conversion of authorship from a material to a notional relationship to readers is Tompkins 1980. Ironically, Roszak, a history professor at California State University at Hayward, has made a good living from eloquently expressing the alienation of many self-styled geniuses who see themselves as unappreciated by their contemporaries.

53. Roszak 1972, 204.

the student revolts of the 1960s in psychoanalytic terms. In that respect, Kuhn's habit of naming paradigms after their founding fathers was taken perhaps a bit more seriously than he had intended.[54]

Those acquainted with Bell and Roszak more by reputation than direct acquaintance with their writings might conclude that Bell's measured prose exhibited more of the usual scholarly virtues than Roszak's flamboyant discourse. And while Bell's appropriation of *Structure* probably pleased Kuhn more than Roszak's, the two were equally erudite and eclectic in their sources. Indeed, Roszak was considerably more learned than Bell in the history of science. Thus, in 1973, in the midst of the U.S. Senate Watergate hearings, Roszak was featured in a four-part series run by *Time* magazine "examining America's rising discontent with entrenched intellectual ideas: liberalism, rationalism and scientism." It was designed to coincide with the five-hundredth anniversary of Copernicus's birth. Perhaps the most notable piece in this series was the final article, in which the Harvard historian Everett Mendelsohn was quoted as saying, "Science as we know it has outlived its usefulness."[55] The article subsequently makes clear that Mendelsohn was referring to the fact that all scientific research contains normative commitments that remain hidden at the peril of those on whom that research is subsequently "applied" (in some cases, "inflicted" would be a better word). Presumably, as the public became more cognizant of science's effects, they would want to participate in decisions that include a strong scientific component.

If Mendelsohn's assertion was controversial in 1973 (as the article suggests), no one in good faith can deny its truth today. However, back then a silver lining laced Mendelsohn's cloud. It was none other than Thomas Kuhn's argument "that science is not cumulative, but that it collapses and is rebuilt after each major conceptual shift."[56] *Time* fortified Kuhn with Abraham Maslow's self-actualization psychology, which supported the idea that scientific revolutionaries have radically different personalities from normal scientists (a point supported by Roszak). Kuhn himself was photographed for the article, looking slightly bewildered, but not actually quoted. The reader was then left with Kuhn's passive endorsement of the prospect that a radically new science—perhaps one aligned with parapsychology, mysticism, or the ecology movement—was in the offing, which the public

54. Berger, Berger, and Kellner 1973, 174 ff. Berger, Berger, and Kellner note (221) that depending on whether the students were portrayed as searching for a revolutionary or a normal existence, they would be interpreted as, respectively, trying to *kill* or *find* the Freudian "father." Works that exemplify these respective tendencies (Roszak's and Bell's) are Feuer 1969; Bell and Kristol 1969.
55. *Time* 1973, 83.
56. *Time* 1973, 85.

would find more responsive to its needs and interests than the dominant "paradigm" that seemed hell-bent on the destruction of human life and environmental despoliation. Needless to say, this was hardly Kuhn's intention but, of course, that did not diminish the influence of what he was alleged to have meant.

Ironically, political science turned out to be the discipline most susceptible to the faux-revolutionary character of Kuhn's image of science. As Richard Bernstein has observed, political science enjoys the somewhat dubious honor of its leaders' having openly embraced *both* the normal and revolutionary sides of Kuhn's image of science in the 1960s. Two presidential addresses to the American Political Science Association (by David Truman in 1965 and Gabriel Almond in 1966) invoked the image of normal science, whereas prominent marginals such as Sheldon Wolin advocated the revolutionary image at roughly the same time.[57] The latter appropriation is especially interesting and perhaps even worthy of pathos. One of America's most distinguished historians of political thought, Wolin was taken with Kuhn's account of scientific change. Contrary to the relativist tendencies in his own field, Wolin argued that classical political theorists from Plato to Marx saw themselves as progressively aiming toward the Good Society, and would thus have found it patronizing to have their work understood merely in its own context or as simply offering one of many alternative visions of the good society.[58] In terms of the historiographical perspectives introduced in the introduction, Wolin's instincts would have to be classed as anti-Priggish. On that basis, he inferred that the classical theorists were engaged in Kuhn-style paradigm-driven inquiry.

But what counts as a paradigm here? Only Marxism has arguably established a paradigm in the sense of a coherent research community.[59] However, reasons Wolin, maybe political theorists have been primarily interested in influencing not each other, but the larger political community. Yet this point, which Wolin endorses, would seem to concede that political theorists as a community stand in no systematic relation to political practice, thereby leaving them free to seek whatever constituency they can find in the polis. For example, the great English seventeenth-century political theorist Thomas Hobbes thought that his predecessors failed to affect actual politics because they had not reduced politics to the sort of

---

57. Bernstein 1976, 93–106. For the Cold War context of Truman's and Almond's presidencies of the APSA, see Katznelson 1997.
58. Wolin 1968.
59. See Bernstein 1981. However, in Lakatosian terms, Bernstein believes that Marxism is suffering from a "degenerating problemshift," as its theoretical elaboration has been driven largely in response to empirical anomalies in Marx's original theory.

elementary principles that politicians could put into practice. But instead of exhaustively critiquing his predecessors, Hobbes designed principles that he believed could be taught in universities to future political leaders (that is, assuming — rather implausibly in his own day — that Hobbes was respectable enough for his principles to carry authority).

Wolin's way around Hobbes's complaint was to argue that the political system governing the theorist's society is akin to a paradigm that accumulates anomalies in the pursuit of public affairs. Thus, the leading school of political science in Wolin's day, the behavioralists, proceeded from the normative assumptions governing the United States in the 1950s and 1960s, which presupposed consensus formation as the progressive tendency of democratic societies.[60] Yet, it was not long before their positivistically inspired, survey-based research stumbled upon an unexpected finding, namely, that the impact of voters on the opinions and actions of politicians is much less than one would expect of a liberal democratic society grounded in consensus politics. According to Wolin, such a discrepancy between norm and fact (or "ideology" and "reality," in more Marxist terms) should precipitate a crisis in the political system, one that would renew the period of theorizing that eventuates in a revolution. Yet, over a quarter-century later, behavioralism has yielded only modest ground to alternative scientific orientations, while American politics perversely persists in the face of its glaring anomalies. Indeed, the behavioralists have even turned the tables on Wolin and deployed Kuhn to ironic effect: Maybe the fundamental problem of political science is that every national political tradition constitutes its own paradigm, and behavioralism is valid *only* in the United States![61]

Wolin's wishful prognosis was based on the historical tendency of great political theories to have emerged from periods of political crisis. Nevertheless, a sober appraisal of this proposal would question whether it makes sense to treat contemporary political theorists as purveyors of revolutionary science, given that their personal fates are largely insulated from political events. More likely is that, caught up in the spirit of the late 1960s (he was

60. A characteristic work of this school is Dahl 1963. The sociological foundation for this view was Parsons 1951. It was in turn challenged by Gouldner 1970, which treats the consensus politics of contemporary welfare states as little more than temporarily contained conflict. The behavioralists in political science should not be confused with the "behaviorists" who dominated experimental psychology during this period. In the former case, the appeal to "behavior" was meant to contrast with classical normative approaches that focused on the qualities associated with good leadership or citizenship, but not with, say, actual voting patterns and levels of political participation.

61. See the survey of the current scene written by one of the founders of behavioralism: Easton 1991, esp. 47 ff., which is largely concerned with the implications of Kuhn and post-Kuhnian philosophy of science for political science.

writing from Berkeley at the time), Wolin overlooked some basic facts about the institutionalization of political theory. With the exception of Hegel, every major political theorist before Marx had been a nonacademic whose livelihood depended on his ability to influence political leaders or their mentors. Once political theorists became tenured civil servants earning a steady living in public universities, they started addressing primarily members of their own guild, those on whose good graces their livelihoods now depended. The difference in style and tone between Karl Marx and Max Weber vividly marks this transformation. Gradually, theorists lost the incentive, and even the ability (in terms of writing publicly accessible prose), to align with the political movements most likely to be at the leading edge of change. Under those circumstances, theorists found it deceptively easy to imagine that by diagnosing the incoherences of an orthodox political scientist in a learned journal, they were somehow undermining the legitimacy of the current liberal democratic regime.

This bit of intellectual legerdemain was abetted, in Wolin's case, by what we earlier noted as the iconographical character of Kuhn's analysis, the attractiveness of which was conditioned by Wolin's prior acquaintance with the work of Leo Strauss. According to Strauss, by directly addressing something called "The Western Political Tradition," the theorist was indirectly addressing the political problems of the day—but in a way that safely went above the heads of potentially volatile masses and contributed to a permanent body of political wisdom. That Kuhn could enable Wolin, an avowed democratic leftist, to overlook the elitist and conservative cast of Strauss's vision of the political theorist's task testifies to the evocativeness of Kuhn's pseudopolitical imagery. In the final analysis, of course, the fact that Kuhn's ultimate interest was in depoliticizing, not repoliticizing, science should have made it clear to Wolin that any revolutionary reading of *The Structure of Scientific Revolutions* was bound to backfire.[62]

5. Putting an End to It All: From Finalization to Fukuyama

If my account thus far of *Structure*'s impact on the social sciences is to be believed, then it would seem to have corrupted all of its left-inspired appropriations. The ultimate test of this thesis is the fate that befell those who saw in Kuhn's model of scientific change a kind of evolutionary eschatology

---

62. Nelson 1974. For a sensitive interpretation of Wolin's theoretical project that nevertheless reaches largely the same conclusion as I do here, see Gunnell 1986, 116–33. Gunnell argues that Cromwell, not Hobbes, is the relevant analogue to Newton. Hobbes is more like the philosopher of science, someone who operates once removed from scientific practice. For Wolin's response, see Wolin 1986.

that was reminiscent of Marx's dialectical materialism. This interpretation took firmest root in the West German scholarly community, where *Structure* was available in translation as early as 1967, before the second edition of the book started having substantial cross-disciplinary influence in the English-speaking world. It was originally read through the critical Marxism of the then up-and-coming Jürgen Habermas (b. 1929), who eventually became Germany's leading social theorist and public intellectual. At the time, Habermas was director of the newly formed Max Planck Institute in Starnberg, dedicated to "the investigation of living conditions in the industrial world." The institute acquired political notoriety in 1976 when Germany's Social Democratic chancellor Helmut Schmidt delivered a speech at Starnberg that appealed to what had become known as the "finalization" thesis on the role of science policy in the larger context of state policy.[63]

Under the influence of *Structure*, Habermas's Starnberg group had effectively attenuated the orthodox Marxist sense of state planning for science by supposing that one must wait for a paradigm to "mature" before redirecting (or "finalizing") its research from "basic" to "applied" concerns.[64] In practice, this policy would touch only the sciences that have reached an advanced stage of puzzle solving, whereby the prospect of technological spin-offs from solving additional puzzles becomes less economical than simply having the scientists address public needs directly.[65] Despite the public polemics in which the "finalizationists" engaged with Popperian defenders of a more libertarian science policy, these self-styled Marxists were a far cry from the specters of socialist science that set Polanyi on the warpath during the early years of the Cold War. Had there been any doubt in Bernal's mind, it was clear by the time of the finalizationists that science must be made *for* the people, but not necessarily *by* them.[66] Thus, in good Kuhnian fashion, the finalizationists left open the possibility that public problems will be replaced by "clarified" scientific ones, much as a paradigmatic order replaces preparadigmatic confusion. This message appealed to the technocrats in the Social Democratic administration, many of whom were trained in the natural sciences, engineering or, like Chancellor Schmidt himself, economics. It was only once finalization attracted such

---

63. This speech is discussed in Pfetsch 1979, 116–17, as part of a documentation of the rise and fall of finalization in the German press.

64. For a good sample of historical and philosophical papers from the German finalizationists, see Schaefer 1984.

65. Kuhn's influence in attenuating the science-planning impulse of avowed Marxists was by no means limited to Germany. Just as striking was the response of British scientific socialists, who invoked Kuhn to argue that science had an "inner logic" that, once "matured," could be directed toward socially beneficial ends. See Rose and Rose 1970, 240 ff.

66. For a history of this sensibility, see Elzinga 1988. For a critique, see Fischer 1992.

technocratic interest that public opposition began, mostly from humanists and other protectors of academic freedom in the face of expanding student enrollments and external research demands.[67]

Nevertheless, despite its rhetoric of public policy intervention, finalization ultimately reduced the role of public participation in science-based decision making. To see how this happened, consider the following passage by the historian and physicist Gerald Holton, who (at the time) claimed allegiance to a modified version of the finalization principle:

> [B]asic researchers in the physical and biological sciences have only rarely looked for their puzzles among the predicaments of society, even though it is not difficult to show that the lack of relevant scientific knowledge in such "pure" fields . . . is among the central causes of almost any major social problem. (For example, a better understanding of the physics, chemistry, and biology of the detailed processes of conception is still fundamental to the formulation of sounder strategies for dealing with overpopulation and family planning.)[68]

Advocates of a more democratized science policy such as Jerome Ravetz and Barry Commoner regarded such calls for more research into conception (and presumably contraception) as a diversion from the real issue, which is to determine who stands to gain and lose from diagnosing global environmental problems in terms of "overpopulation," since, technically speaking, the earth may have enough resources to support many times its current population.[69] This is not to deny the many millions who remain ill fed and ill housed, but that problem is unlikely to be solved, according to Ravetz and Commoner, by the production of new natural-scientific knowledge—though admittedly that is what natural scientists are most used to doing. Rather, the solution would lie in the design of political and economic institutions capable of efficiently and equitably distributing existing resources. Typically, these institutions would be portrayed as enhancing or replacing key functions of the state. What we have here, then, are deep problems of social—not natural—science. Indeed, to assign the

---

67. A relevant political benchmark in this discussion is that the Popperian opponents to finalization saw the thesis as supporting the steering of science by *both* state and industry, which in the German welfare state context had overlapping interests. In the 1970s, the "open society" rhetoric of the opponents was more or less exclusively rooted in the ideology of university autonomy, without the connotations of privatized "free markets" that such rhetoric often has today. In this respect, the institutionalization of STS in France in the 1980s provides an interesting point of comparison, since by that time the success of the French corporate state started to be challenged, which opened the door to the scrambling of political allegiances, so as to invite opportunistic alliances between defenders of academic and economic structures. See chapter 7, section 6.

68. Holton 1978, 229.

69. Ravetz 1971; Commoner 1971.

problems to the natural sciences is to ensure that they will not be dealt with democratically.

Because finalization remained fixated on the trajectory of the natural sciences as dictating the pace at which social policy could occur, their account of scientific change ended up shadowing the "linear model" of technical development so common to both welfare state and Marxist economic policy thinking in the Cold War period: a scaled-up research base in the natural sciences corresponded to the formation of a capital base that was deemed necessary for reliable and efficient wealth production, while social-scientific research later provided insight into how the surplus from that production may be most equitably distributed or, as the case may be, redistributed.[70] Just as Kuhn defined progress within a paradigm largely in terms of its ability to enforce a sharp separation between the producers and consumers of knowledge in a given field, economic development was similarly defined by the ability of venture capitalists to increase their profits in a relatively unimpeded fashion until enough wealth was produced to be worthy of centralized distribution by a state planning board. Since the linear model has been implicit in most modern economic theories of growth, Kuhn's model simply added a naturalistic underpinning to its legitimation.

Before going further, two points are worth making. First, Kuhn himself never contributed to the finalization discussion, for reasons that become clear once we recall his intellectual debt to Alexandre Koyré. A science that had matured to the point of being an "off-the-shelf" technology was, from Kuhn's standpoint, dead — or "closed," in Werner Heisenberg's more euphemistic phrasing, favored by the finalizationists. Thus, it no longer fell under the jurisdiction of the account outlined in *Structure*.[71] The second, and more important, point is that, in the United States, not all Marxists accepted Kuhn's model as a version of their own. Marxist thought was split over Kuhn's implicit politics. Basically, the harder the line of the Marxist, the more skeptical the response to Kuhn. Consider this rather positivistic

---

70. Steven Yearley offers an illuminating commentary on the debate over the appropriateness of this model for development aid to Third World countries in the 1960s. Defending the linear model was Nicholas Kaldor, Keynesian economist and chancellor of the exchequer in Harold Wilson's Labour government. Opposing it was E. F. Schumacher, champion of "small is beautiful" Buddhist economics and senior economist at the U.K.'s national coal board. See Yearley 1988, 167 ff.

71. Kuhn 1970b, 79 (see chapter 1, section 4). Amid the political controversy surrounding finalization, it is often forgotten that the finalizationists assigned a special place to the biological sciences as requiring contact with the complexities of real-world problems in order to reach full theoretical maturity. Their standard case was the role that an interest in improving crop yields played in the maturation of agricultural chemistry in the mid–nineteenth century. See Krohn and Schaefer 1976.

pronouncement from the official policy journal of the U.S. Communist Party:

> The reactionary aspect of Kuhnianism stems from his rejection of the objective truth of scientific knowledge. For if physical science itself can be shown to be nothing more than a succession of subjective models, then . . . social science also would have no objective content . . . Those who say capitalist oppression is a reality are just as right (or wrong) as those who deny it . . . [Kuhn] is encouraging an ideological trend which has a paralyzing effect upon millions.[72]

However, less orthodox Marxists wishfully thought that scientific revolutions were political revolutions in disguise. This led them in one of two directions. One was to see Kuhn as elaborating Lenin's view that science is the ideological reflection of the dominant interests in society. Marxists of a more economistic bent took a second path that made them bedfellows with neoclassical economists who identified science with technology as the main mode of production in advanced capitalism.[73] Both of these directions, of course, were quite contrary to the spirit of Kuhn's work. As one astute commentator of the period observed, the fact that the scientific community can reconstitute itself with relative ease after undergoing a Kuhnian revolution implies that the ambient political culture remains stable throughout the change.[74] Nevertheless, clearly Kuhn's politically charged language provided science protective coloration from the withering gaze of leftist thought in the 1960s and early 1970s, while narrowing the differences between self-professed "democrats" and "elitists" in science policy debates.

One Marxist scholar-activist stands out for his early incisive critique of Kuhn: Robert Maxwell ("Bob") Young (b. 1935), a Cambridge-trained American expatriate who in the late 1960s helped found the British Society for Social Responsibility in Science, the organization with which Jerome Ravetz was most closely affiliated.[75] Young's critique of Kuhn was triggered,

---

72. John Pappademos and Beatrice Lumpkin, "The Scientific Outlook under Attack," *Political Affairs* 53 (November 1974), quoted in Novick 1988, 422–23.
73. Among these misreadings of Kuhn, Peter Novick cites Jesse Lemisch of the University of Wisconsin, who claimed that "what we know of scientific revolutions and of the sociology of knowledge clearly indicates that meritocracy wars with truth." Lemisch, "Radical Scholarship as Scientific Method and Anti-authoritarianism, Not 'Relevance,'" *New University Conference Papers* 2 (1970), quoted in Novick 1988, 430. Novick underscores Lemisch's apparent commitment to objectivism, not relativism, in his (mis)interpretation of Kuhn.
74. King 1971.
75. Young 1975, esp. 406 ff. Young is currently professor of psychiatry at Sheffield University (United Kingdom) and editor of the journal *Science as Culture*. Many of Young's writings can be found on the World Wide Web at http://www.shef.ac.uk/uni/academic/N-Q/psysc/staff/rmyoung/index.html.

in part, by Kuhn's response to one of his essays on the Malthusian roots of Darwin's theory of evolution by natural selection.[76] Young argued that what was essentially a long-term policy proposal for population control through government inaction (Malthus's classic 1798 essay *On Population*) provided the basis for what is today regarded as the most powerful scientific explanation of organic survival. Here Kuhn argued that what distinguished Darwin's theory of evolution from all the others—many just as much inspired by Malthus's views on population—was its careful marshaling of evidence. Indeed, according to Kuhn, scientists had been previously suspicious of evolutionary theories because of the strongly ideological character of such theories, but Darwin overcame that prejudice with his mastery of the natural-history sources. Young found this response very revealing. Kuhn gave the impression that it was clear to all sides at the time who counted as a competent scientific authority and what counted as an unscientific ideological intrusion. Yet, Young argued (rightly), anyone familiar with the details of the case would realize that these issues were among the points of contestation, thereby making it difficult to sustain the claim that Darwin's success was exclusively, or even primarily, attributable to the collective wisdom of an autonomous scientific community.

It would seem that, once again, Kuhn had carried over the pedagogical lessons of Conant's courses into his research. Notwithstanding Young's clear demonstration of the role of ideology in consolidation of scientific opinion, Kuhn was quite explicit in his denial of science's relationship to ideology: "Science, when it affects socioeconomic development at all, does so through *technology*."[77] Kuhn's perspective has been reasserted with a vengeance in the notorious *The End of History and the Last Man* by the American Sovietologist and former student of Alan Bloom, Francis Fukuyama (b. 1953). Fukuyama argued that history has revealed all the shapes that the good society can take, and that 1989 turned out to be the annus mirabilis when liberal democratic capitalism proved, once and for all, that it was much better than even the strongest socialist society.[78] Based on Fukuyama's adherence to Hegelian logic, it will not be long before liberalism triumphs across the globe, bringing with it an unprecedented period of peace and prosperity. Recalling our discussion of "embushelment" in chapter 1, Fukuyama's twist in the Platonic tale is to identify capital development as the activity that requires protective coloration to ensure the long-term survival of humanity.

76. Kuhn 1971, esp. 281–82. Young 1985 collects his essays on the ideological roots of Darwinism.
77. Kuhn 1971, 284 (italics added).
78. Fukuyama 1992.

Fukuyama's critics typically limit their disagreements to the exact vision of the good society that awaits us at the end of history. However, they generally agree on the means by which we will have gotten there. The "logic of natural science," in Fukuyama's words, plots an inevitable course that both transcends and transforms even the most historically entrenched of cultural differences.[79] In that sense, science puts an end to history: Once the good society has been hit upon by one country, history then becomes simply a matter of the rest of the world catching up by repeating the steps that the first country took. Well into the 1970s, this was how both capitalists and socialists in the first two "Worlds" thought that the Third World would be "modernized." Socialists pointed to science's role in the creation of labor-saving technologies that eventually undermine the basis for any sharp distinction between the workers and their bosses. Capitalists emphasized the role of science in enhancing people's innovative capacities and hence their ability to compete more effectively in the marketplace. The roles assigned to science are different, but in both cases they are meant to have global application. And to whom does Fukuyama turn for science's crucial role in providing humanity's "mechanism of desire"? None other than Thomas Kuhn.[80]

It may seem odd that Fukuyama would cite Kuhn, and not some full-blown scientific realist, for the idea that science displays an irreversible trajectory that propels the progress of humanity. While Kuhn certainly does not see a unified theory of reality waiting for us at the end of science, he nevertheless holds a crucial piece of Fukuyama's puzzle. Like Fukuyama's own narrative, Kuhn's account of scientific change is both irreversible and Eurocentric. After the Royal Society of London was founded as the first self-determining scientific association, the advancement of science proceeded in terms of greater problem-solving effectiveness over more rigorously defined domains. Moreover, the period in the history of European physical science from which Kuhn generalizes — roughly 1620 to 1920 — coincides with the ascendancy of Western capitalism. Indeed, in this respect, Kuhn's account is *more* Eurocentric than such philosophical competitors as Carnap and Popper *precisely* because he attempted to model the history of Western science as closely as he did. The significance of this point appears most clearly if Kuhn and, say, Popper are seen as having provided alternative normative accounts of scientific change. Whereas Popper said that genuine knowledge was not *predictable*, Kuhn presents it as not *reversible*. Those are two different radically opposed claims. The former

79. Fukuyama 1992, 80–81.
80. Fukuyama 1992, 352–53 n. 2.

associates rationality with learning from mistakes, while the latter stresses getting it right the first time.[81] All considered, then, Fukuyama is rather perceptive to regard *The Structure of Scientific Revolutions* as amenable to his own special brand of what, in international relations circles, is called "modernization theory."

Fukuyama would have recognized the "shape" of Kuhn's narrative from that classic statement of Cold War modernization theory, Walt Rostow's *The Stages of Economic Growth: A Non-Communist Manifesto*.[82] Indeed, a fair measure of the extent to which policy makers since the 1960s have been willing to treat science as an autonomous policy realm is the contrast between the easy acceptance of the Kuhn-Price vision of science-as-organism and the resistance that Rostow met when he put forward a very similar vision of national economy–as–organism at roughly the same time. Rostow, an economic historian at MIT, later became director of policy and planning at the U.S. State Department in the Kennedy administration and then, notoriously, Lyndon Johnson's chief advisor on the Vietnam War.

Intellectually speaking, Kuhn and Rostow faced largely the same challenges. Both were mounting a historical argument in fields that had been dominated by static, formalist conceptions of their respective subject matters (i.e., positivistic conceptions of science and neoclassical conceptions of the economy). Both emphasized that over the long term, growth was not continuous, but subject to spurts from one sector ("take-off" in Rostow's memorable phrase) that eventually reconstituted the entire organism, be it a national economy or a field of inquiry.[83] Rostow and Kuhn even shared a similar typology of stages, with Rostow's final stage of "mass consumption" corresponding to the "finalization" of a paradigm, as discussed above. The crucial difference came once this scheme was exported into the policy arena, and Rostow promoted its use as a basis for intervening in the affairs of developing countries, especially in Latin America. The idea was for the United States and its allies to boost capital investment ("development aid") in countries exhibiting the early signs of an economic take-off, so that they will follow through on the growth pattern of capitalism and not fall into the hands of the Communists, who typically became influential as these countries first began to shake off their economic dependency on the West.

This strategy immediately drew attention to the precariousness of what was supposedly a "natural" course of economic development. By contrast,

---

81. For a good account of the differences between Popper's and Fukuyama's visions of history that brings this point out is Williams, Sullivan, and Matthews 1997, 137 ff.
82. Rostow 1960. For an appreciative nod, see Fukuyama 1992, 128.
83. The historiographical connection between a Kuhnian revolution and a Rostovian take-off is made explicit in Pocock 1973, 13 ff.

no science policy maker armed with the Kuhn-Price vision of scientific change suggested, say, that certain developing areas of the social sciences should be targeted for heavy investment so that they can acquire the stable growth patterns of a paradigmatic science. Rather, this process was thought to happen eventually (if at all). Indeed, even without the requisite and material bases, many social scientists went ahead and declared their disciplines paradigms.

However, I would go further. The ultimate vagueness of Kuhn's evolutionism—his idea that science progresses *from* something but not *toward* anything—makes the theory easier to insert into Fukuyama's teleology than a strongly realist theory of scientific change that implies a clear end in sight. The reason is that if science were portrayed as not only the engine of world history but also an institution with overarching ends of its own, then that could be (and has been) taken to mean that the ends of science must take precedence over—if not simply overtake—the ends of humanity. We would therefore not be far from the Platonic vision of scientists as philosopher-kings who clarify their own vision in order to superimpose it on everyone else. This would do more than simply inconvenience a classical liberal like Fukuyama. Granting such a prerogative to the scientific community would unwittingly expose the antinomic character of scientific realism. On the one hand, if the ends of science appear too autonomous from the concerns of the rest of society, then scientists look like a special interest group with hegemonic designs. On the other hand, if scientific techniques are shown to advance a wide range of personal and social goals, then perhaps the particular value orientations of the scientific community are dispensable. In the one case, science appears to be a totalizing ideology, in the other a high-grade tool: Planck versus Mach all over again. Thus, unless it is kept at a strategically vague level, scientific realism is likely to devolve into a species of either imperialism or instrumentalism. *Structure* succeeds in maintaining such vagueness, and so the feared devolutions never transpire.[84]

---

84. Fukuyama recently showed his true Platonist colors when an interviewer asked him whether the scientific search for the ultimate theory of everything might replace socialism's universalist mission. Fukuyama scornfully described such people as "space travel buffs" who overestimated science's world-historic significance. It seems as if the noble lie of scientific progress was being taken just a bit too seriously by people in search for a *new* End of History. See Horgan 1996, 242–44.

VI

The World Not Well Lost
Philosophy after Kuhn

1. How Far We Have Fallen: Philosophers as Underlaborers

Speaking of his own doctrine of pragmatism, William James (1842–1910) said that a theory goes through three stages: "First, you know, a new theory is attacked as absurd; then it is admitted to be true, but obvious and insignificant; finally it is seen to be so important that its adversaries claim that they themselves discovered it."[1] It would not be too difficult to fit the philosophical reception of *The Structure of Scientific Revolutions* into this mold, especially if we concentrate on the book's reception among the American progeny of the logical positivists who dominated the philosophy of science from the late 1950s to the early 1970s and, more importantly, the people trained by them who have since then come to dominate the field. The plot of this reception narrative is one whereby philosophers of science gradually accept their diminished normative status and ultimately embrace the role of *underlaborer*. John Locke's (1632–1704) classic formulation is worth recalling here:

> The commonwealth of learning is not at this time without master builders, whose mighty designs, in advancing the sciences, will leave lasting monuments to the admiration of posterity: but everyone must not hope to be a Boyle, or a Sydenham; and in an age that produces such masters as the great Huygens and the incomparable Mr Newton, with some others of that strain, it is ambitious enough to be employed as an UNDERLABOURER in clearing the ground a little, and removing some of the rubbish that lies in the way of knowledge.[2]

---

1. James 1948, 159. Many thanks to Davis Baird, Randy Harris, and Joshua Lederberg for helping me track down this quote during a discussion on the HOPOS-L electronic mail group in November 1997. Testimony to Kuhn's having cycled through James's stages over the course of a generation is provided in Hacking 1993, 275.

2. Locke 1959, 14.

Locke lacked the mathematical training needed to understand the details of Newton's *Principia Mathematica*. Nevertheless, he functioned as a "detached" but able publicist for Newton's views, once personally instructed on the philosophical implications of the arcane calculations that constitute the core of Newtonian mechanics. Before decrying such collusion as compromising philosophical independence, what should we make of the fact that "two-thirds of the more visible philosophers of biology today have spent time in the very same lab ([Richard] Lewontin's at Harvard)"?[3] In the set of interviews with Werner Callebaut from which this quote is taken, none other than *Structure* is most frequently credited with having led philosophers to apprentice with the "master builders" and immerse themselves in the details of a particular science. Interestingly, no one ever seems to have left his apprenticeship in a major laboratory *less* committed to the science in question than when he entered. Moreover, contemporary philosophers have rarely had the temerity to argue that they could come to know *too much* of a science to recover the broader perspective that first drew them to philosophy.[4]

Philosophers who spend time in such scientific apprenticeships often think they are continuing the project of the logical positivists and their Popperian cousins. The positivists are credited with good intentions in wanting to model philosophical practices on scientific ones. But they are faulted for not mastering the details of particular sciences, with the partial exception of physics. Yet how then, for all their technical deficiencies, did the positivists nevertheless manage to exert so much influence over scientific methodology and the public image of science? One hypothesis (which I happen to favor) is that the logical positivists were *not* trying to be underlaborers at all, but rather using science to promote certain philosophical ends of larger societal import. Take the symbolic function of the natural sciences in the project of "Enlightenment" promoted most consistently in our own time by Karl Popper. The idea here was *not* one of philosophers paving the way for a mounting body of esoteric knowledge; rather, it was of

---

3. Callebaut 1993, 450. The book is an imaginatively interspersed set of interviews with twenty-four philosophically minded biologists and biologically oriented philosophers that Callebaut conducted from 1985 to 1990. Although Callebaut remains diplomatic throughout, by the end of the book it becomes clear that he is disturbed by the underlaboring tendency in contemporary philosophy of science for many of the same reasons suggested here.

4. However, at least one interviewee suggested the possibility that philosophers of biology were co-opted by their host biologists. See Alexander Rosenberg's remarks in Callebaut 1993, 463. For a more complete review of Callebaut, which appeared at the dawn of the ongoing Science Wars, see Fuller 1994h. I continued this theme in terms of the philosophy *of* science's devolution into philosophy *for* science in a symposium paper in memory of Popper and Feyerabend at the 1996 American Philosophical Association, Eastern Division, meetings.

extending to all spheres of life the critical attitude that had motivated scientists to challenge traditional beliefs in the first place. As science comes to be materially committed to particular lines of inquiry, such that the costs of reversing them becomes prohibitive, this spirit of criticism is increasingly difficult to sustain.[5]

Some followers of Popper, such as Paul Feyerabend, have gone so far as to suggest that scientific research programs need to be cut down to a size that enables criticism to flourish. Thus, when Feyerabend argued that Creationism should be taught alongside evolutionary theory in the public schools, he was primarily offering an opinion not on the probative value of Creationism per se but on the social contexts in which its probative value should be determined: namely, it should rest with local educational authorities rather than the professional scientific community. This distinction between one's personal judgments and the framework within which they should be evaluated is subtle but crucial for understanding the politics of science implied by the underlaborer model. Feyerabend intervened in the Creationist controversy as someone who wanted to square the imperatives of science and democracy. This is a classically philosophical interest that requires sustained thinking about science, but without being beholden to particular scientific research programs. Indeed, it may even involve supporting decision-making conditions that would issue in judgments counter to one's own personal preferences.

In what follows, I shall regard Feyerabend's orientation as characteristic of the "prescriptive" approach to the philosophy of science, one that survives virtually intact in the social and political philosophy of John Rawls and Jürgen Habermas. In contrast, Philip Kitcher, as he recalls to Callebaut, became involved in the Creationist controversy as a partisan for the evolutionary cause. His first impulse was to forge a "new consensus" in the philosophy of science, one whose united front would keep the Creationists out of the classroom.[6] To be sure, philosophers have always been known to

5. I develop this line of inquiry in Fuller 1999b.
6. Soon after he became director of the University of Minnesota Center for the Philosophy of Science, Kitcher was keen on using it as a springboard to forge a "New Consensus in the Philosophy of Science," largely in the face of the perceived threat of Creationism. See Kitcher's remarks in Callebaut 1993, 194–99. To the diminishingly few philosophers of science who saw their inquiries as independent of scientific partisanship, the rush of philosophers to defend the integrity of evolutionary biology against Creationism in the 1980s was a perplexing state of affairs. But to most of the younger generation it was a natural course of action. For example, Larry Laudan (b. 1941) and Philip Kitcher (b. 1947) are only six years apart in age, yet the difference in their response to the Creationist "threat" represents a generational divide in terms of how they see the philosopher of science's professional obligations in public matters. See Laudan 1982; Kitcher 1982. Both Laudan and Kitcher did their doctoral work at Princeton's history and philosophy of science department under the logical positivist Carl

slip into ideology as their political ends overtake their intellectual means. However, today's philosopher-underlaborers come dangerously close to being apologists for the masters whose houses they so dutifully clean.

At a more strictly professional level, the impulse to underlabor can be detected in the recent resurgence of *naturalism* in analytic epistemology and the philosophies of science and mind. From the standpoint of the twentieth century's most eminent American naturalist, John Dewey (1859–1952), the various senses in which naturalism has been invoked in recent times seriously compromise the critical edge of his original doctrine. These recent invocations include:

> 1. Scientific findings are used to illustrate—"confirm" in a very weak sense—philosophical claims that have been reached largely through conceptual analysis and other a priori forms of revelation.[7] In this respect, contemporary naturalized epistemology adheres to the same probative standards as natural theology, especially with respect to the falsifiability conditions on offer for, on the one hand, the existence of Truth and scientific progress and, on the other, the existence of God and cosmic design.[8]
>
> 2. The history of science is used in what can only be called a "providential" manner to argue that the most materially well-endowed and technologically applied sciences happen to produce the most reliable forms of knowledge. Since there are no generally agreed-upon measures for either counting knowledge claims or assessing their reliability, the only obvious precedent for what passes as a belief in scientific progress is the Calvinist conception of Grace applied to entire disciplines, rather than individuals: in both cases, prosperity is the mark of salvation. The mysteriousness of this position is only enhanced by the refusal of so-called naturalized epistemologists

---

Hempel during Kuhn's tenure. Indeed, Kuhn periodically cotaught with Hempel. But Laudan did so in the 1960s, shortly after Kuhn had arrived, and Kitcher in the 1970s, once Kuhn had acquired an international reputation. For my own views on Creationism, which reflects the Feyerabendian critique, see Fuller 1998c.

7. See Goldman 1986. Position 1 also fairly captures the traditional account of the relationship between philosophy and science in such seventeenth-century figures as Descartes, Hobbes, and Leibniz. However, more thoroughgoing naturalistic pictures of even these philosophers have been forthcoming in recent years, e.g. Garber 1992. The locus classicus for this entire discussion is Buchdahl 1969. Given the professional insulation of epistemology from science in the United States, it is not surprising to learn that Goldman is more of an apriorist in practice than Descartes, at least in terms of their shared inclination to protect intuitions about the nature of knowledge from direct empirical scrutiny. For a criticism of this self-incapacitating tendency in contemporary epistemology, see Fuller 1993b, 70–84.

8. Although the comparison between naturalized epistemology and natural theology may appear inflammatory, the writings of Richard Swinburne, professor in the philosophy of religion at Oxford, should disabuse one of that suspicion. I would class him as an able practitioner of *both*. Another interesting attempt to take the connection between these two disciplines seriously, especially in light of recent developments in scientific realism, is Banner 1990. The argumentative strategy shared by scientific and theological realism is normally called "inference to the best explanation." For a discussion of the commonality, see Fuller 1998c.

to take seriously the empirical evidence that highlights the polymorphous perversity of human cognitive limitations.[9] Consequently, naturalists are left puzzling over how such fallible creatures have been able to combine over great expanses of space and time to produce the "reliable" forms of knowledge associated with the modern natural sciences.

3. Formal models are borrowed from science—even formalized nonnatural sciences like economics and cybernetics—to solve philosophical problems, but usually without dealing with the controversies surrounding the models in their home science and without empirically specifying the domain to which the model would apply. For example, the appeal to neoclassical economics to model the inner workings of science itself does not actually reduce science to economics; rather, it treats epistemic entities as if they were economic ones, without saying how one would then test that hypothesis, independently of showing that it enables us to say things we already believed, albeit now in a more dogmatic form.[10]

4. Philosophers cast themselves as natural historians of knowledge production who accept a "disunified" ontology for science simply because that is how normal science works: i.e., scientists manage to get their self-assigned tasks done under such an ontologically pluralist regime. There is no acknowledgment that the catalyst for reevaluating the role of knowledge production in society has typically come from discrepancies in the metaphysical commitments between sciences that have too long been allowed to conduct their business in relative isolation from one another and hence lose track of the fact that, in the end, we all inhabit the same world.[11]

9. This is the main polemical thrust of Fuller 1993a.

10. See Dretske 1981 and Kitcher 1993. Arguably precedent for this move—though less mathematical, to be sure—can be found in the uses that Adam Smith and David Hume made of the subtle forces postulated by Newton to model the ultimate social glue: i.e., the "invisible hand," natural relations of mutual sympathy, etc. However, it is worth noting that in each case the political consequences of this move were conservative.

11. Perhaps the earliest and boldest statement of contemporary disunificationism was made by Ian Hacking in a popular philosophy of science textbook, in which he updated Leibniz's view that God created the best of all possible worlds by maximizing the variety of phenomena in nature while choosing the simplest laws. Hacking's gloss on the divine plan was to say that God used an inconsistent set of laws, thereby ensuring a plethora of incommensurable object-domains. See Hacking 1983, 219. In that sense, the "Disunification Church" was born from the ashes of a panglossian view of the world order. The most articulate development of the dogma, drawing heavily on the variegated state of contemporary biology, is Dupre 1993. Disunification's proponents currently capture the center-left coalition in the philosophy of science, many with strong ties to Stanford University. Aside from Hacking and Dupre, I would count Nancy Cartwright, Helen Longino, Peter Galison, and Joseph Rouse as leading disunificationists. A selection of their work may be found in Galison and Stump 1996.

Although disunificationism is normally cast as a radical reaction to the monolithic physics-mindedness that has dominated the philosophy of science since logical positivism, from a sociological perspective the movement reveals philosophy's institutional weakness in the academy. The clearest precedent for disunificationism is late nineteenth- and early twentieth-century neo-Kantianism, which introduced the idea that all philosophy must be philosophy of a special discipline. Neo-Kantianism was originally a survival strategy in the German university system, where Fichte and Hegel's unified idealist vision of learning had been discredited with the rise of academic training in the experimental sciences. Rather than pretend to

Absent from this inventory of neonaturalisms is any open *conflict* between science and philosophy. Yet that was what characterized most naturalistic thinkers in the nineteenth and early twentieth centuries. In the seventeenth and eighteenth centuries, philosophy and science sought comfort in each other's arms because of their mutual opposition to theology as an institutionalized obstacle to critical thought. Indeed, this short description captures the spirit of Enlightenment from that period. However, as the secularized nation-state effectively dethroned theology from its position of academic privilege, naturalists became the people who made sure that neither philosophy nor science appealed to supernatural forces in order to recapture theology's throne and thereby reinvent the original problem—which was the empirical unaccountability of supernaturalism. Consequently, some naturalists aimed their fire at philosophy, others at science, depending on whose dogmatism was seen as the most threatening.

A good example of a naturalist who set his aim at philosophy was John Stuart Mill (1806–73). Mill's *System of Logic* (1843) was the most influential work on scientific methodology in the second half of the nineteenth century. Mill did not underlabor in what is nowadays called "philosophy of economics." Rather, he believed that political economy had absorbed many of the traditional substantive questions of ethics. The only remaining question for him was whether political economy had absorbed *all* of them. By the time of *Utilitarianism* (1863), Mill had said no, which set him at odds with his godfather Jeremy Bentham, who was, as we might now say,

---

synthesize and legislate for increasingly disparate fields of academic inquiry, philosophy would articulate how these fields have historically separated themselves out, clarifying their foundations, and sorting out any disciplinary boundary disputes that might arise. Perhaps most characteristic of the last generation of neo-Kantians was Ernst Cassirer, previously discussed in chapter 1, note 31, who devoted much of his later career to comparing the forms of representation ("symbolic orders") in religion, art, and science.

Given recent disunificationist developments, I may be forgiven a sense of déjà vu here: Jones and Galison 1998. It is worth noting—if only in possible anticipation of the future—that neo-Kantianism came to be rejected in the 1930s by both logical positivists and existential phenomenologists (many of whom were trained as neo-Kantians) for its studied refusal to issue an independent judgment on the ultimate value of historical developments in science and technology. What united Popper and Heidegger—to take two extreme cases—against the neo-Kantians was their willingness to criticize, if not outright reject, the actual history in the name of some specifically philosophical ideal, which invariably reached back to the pre-Socratics. On the institutional history of neo-Kantianism, see Collins 1998, chap. 13.

The ultimate test of contemporary disunificationism, which cozily challenges the philosophical metasciences of positivism, realism, and relativism without challenging the sciences themselves, is whether they would renounce the idea that Western Europe was the site of a "scientific revolution" (in some sense, at some time) that failed to occur in China or the Islamic world. On my reading of their collective position, the disunificationists should renounce this idea, and hence do away altogether with the normative conception of science. To his credit, one historian in favor with the disunificationists has already taken this step: Shapin (1996). However, for some reasons why this idea should perhaps *not* be renounced, see Fuller 1997d, esp. 137–44.

an "eliminative economist"—that is, he held that there is no abstract entity "summum bonum" that is the goal of ethics, only material "goods" that satisfy the utility functions of individuals and that the economist can calculate and manipulate in the name of welfare policy.

For a naturalist who set his aim on science, consider John Dewey's critique of experimental psychology's reliance on metaphysical individualism, be it the solitary knower (the "spectator") abstracted from its environment or the atomized reflex abstracted from its context of activity. It goes without saying that the significance of Dewey's critique has not diminished over time.[12] For, if we truly wish to recover this lost legacy of naturalism, philosophy and science will need to be *more*, not less, critical of each other—contrary to the tendency of what nowadays passes as "the naturalistic turn" in analytic philosophy. At the very least, this will mean treating scientific theories as testable hypotheses, not virtually unconditional commitments. In that respect, the use that philosophers made of Kuhn to rediscover naturalism has marked a step back from the Popperians and even the logical positivists. To appreciate just how backward a step Kuhn has taken us, it will be helpful to consider a "missing link" between Dewey and Kuhn: Clarence Irving Lewis (1883–1964), the chair of the Harvard philosophy department in Kuhn's student days.

2. C. I. Lewis as the Missing Link to Kuhn

> *Words* such as 'life', 'matter', 'cause' and so on have been used since thought began, but the *meanings* of them have continuously altered. There is hardly a category or principle of explanation which survives from Aristotle or the science of the Middle Ages. Quite literally, men of those days lived in a different world because their instruments of intellectual interpretation were so different. To be sure, the telescope and microscope and the scientific laboratory have played an important part. As time goes on, the body of familiar experience widens. But that hardly accounts for *all* the changed interpretation which history reveals. Not sense observation alone, but accord with human bent and need must be considered. The motive to control external nature and direct our own destiny was always there. Old principles have been abandoned not only when they disagreed with newly discovered fact, but

---

12. The most developed version of this critique is Dewey 1920. My application of Dewey's critique is best seen in Fuller 1992b. Although different in detail, Ernst Mach's rejection of explanatory unification as a goal of physics, as elaborated in chapter 2, provides another instance of a naturalist who wanted to preempt science's transcendental pretensions. His phenomenalism was motivated largely by a fear that physics was turning into a secular theology, underpinned by a metaphysics of atoms, that disempowered people from using science as an economizing tool to satisfy their personal ends, as they were led to believe that their understanding of science was insufficient, unless they had mastered the leading theory of the day.

when they proved unnecessarily complex and bungling, or when they failed to emphasize distinctions which men felt to be important'.

Readers familiar with *The Structure of Scientific Revolutions* may think that this excerpt is from an article on which that book was based or a speech given subsequent to *Structure*'s publication, perhaps as republished in *The Essential Tension*. The telltale sign, of course, is the suggestion that scientists dwelling in different paradigms live in different worlds, a source of much of what was distinctive and controversial in *Structure*. The only hesitation the reader might have in attributing this passage to Kuhn is that there is little sense of science as a specialized community; rather, we see here an implicit equation of science and humanity that belongs to an older Enlightenment mentality that is alien to Kuhn's thinking.

As it turns out, this quote is taken from a talk that C. I. Lewis gave at Berkeley in 1926.[13] Lewis regularly vacationed and spent sabbaticals at the flagship campus of the University of California, where he received his first regular appointment after completing a philosophy Ph.D. at Harvard under Josiah Royce in 1910. Upon his return to Harvard in 1920, Lewis became the premier consolidator of the American philosophical tradition, not least by supervising two graduate students, Charles Hartshorne and Paul Weiss, in the compilation of the papers of Charles Sanders Peirce (1839–1914), the erstwhile founder of pragmatism who hovered at the edges of the Harvard establishment throughout most of his career.[14] Although Alfred North Whitehead was the leading luminary in the Harvard philosophy department in the 1920s and 1930s, Lewis was the chief institutional presence, a leading agent in the professionalization of philosophy in the United States through his critical assimilation of the émigré logical positivists into an American tradition that still tended to see the educated public as its natural audience.

The irony of history being what it is, professional philosophy has failed to repay Lewis in kind. As a systematic thinker whose legacy was left to an age that deplores systematicity, Lewis survives in fragments in the collective memory of his profession. He is forgotten (most unfortunately, I believe) for his rather bold claim that philosophy is primarily a normative discipline, with logic and epistemology ultimately grounded in ethics.[15] How-

---

13. Lewis 1970, 253.
14. Lewis, "Autobiography," in Schilpp 1968, 16 ff.
15. On the boldness of this claim in its own day, see Purcell 1973, 56–59. In our own time, the most distinguished promoter of this vision of philosophy has been Nicholas Rescher, who has even adopted Lewis's "conceptual pragmatism" as the name of his own metaphysical position.

ever, he continues to be assigned some modest credit for having extended symbolic logic to cover relationships of necessity, contingency, possibility, and impossibility among propositions in ways that capture our metaphysical intuitions of these modal terms, which run deeper than what is permitted by the constraints of deductive validity.

But ultimately, and least complimentarily, Lewis is cited as one of the last defenders of a strong Kantian distinction between "the given" that we experience of reality and the "conceptual scheme" that makes sense of it — or, in Kant's own terms, the difference between "synthetic" and "analytic" statements. Indeed, Wilfrid Sellars, a significant metaphysician in his own right and the son of one of Lewis's archest rivals (Roy Wood Sellars), dubbed Lewis's persistent belief in this distinction "the myth of the given," which Richard Rorty has since turned into a mantra of postmodernism.[16] To postmodernist ears, Lewis's distinction sounds naive, as it suggests that we are somehow "external" to the reality we supposedly inhabit and hence must be reacquainted with it by something like a conceptual scheme that reaches out to whatever predates our existence. (The appropriate postmodern thing to say at this point is that reality and our understanding of it are "coproduced.") On first blush, then, Lewis looks like a dinosaur who tried unsuccessfully to use pragmatism to update Kantianism for the twentieth century.

Yet, Lewis also says things that suggest that he anticipated the Kuhnian turn and deliberately recoiled from it. These are worth investigating. Consider the following reply that Lewis made to critics at the end of his life:

> Concepts which we successfully apply with consequent understanding survive, and those which pragmatically fail become obsolete. They take the initial form of hypotheses of a theory. They become "confirmed" by substantiations of such theory as correct, applicable, contributory to right knowing. "Phlogiston" does not alter its meaning; it becomes forgotten. "Quantum" (with its exact physical sense) will retain precisely that meaning which it has unaltered, but the fate of it as a physical concept will be experimentally determined. But we must forever keep separate the analytic explication of a concept, as a statement whose truth or falsity is necessarily antecedent to any attempted application of it, and the truths of objective fact which are forever at the mercy of empirical findings, and theoretically never established as better than probable — as probable, let us say, as the presently determined facts of physics.[17]

In the context where his views are usually placed, Lewis seems to entertain a Kuhn-like incommensurability between paradigms (note the remarks

16. Rorty 1979, 101 ff.
17. Lewis, "The Philosopher Replies," in Schilpp 1968, 662.

on the fate of phlogiston) but then, observing its odious consequences (i.e., the impossibility of overall progress in science), pulls back, reaffirming his belief in the analytic/synthetic distinction. However, on closer inspection, Lewis is challenging contemporary sensibilities in a more profound way. I urge that his argument be read as follows: Suppose we specify "paradigm" to mean a scientific theory treated as a closed deductive system, a move familiar to the logical positivists and their analytic offspring. We can then draw the sharp distinction between "analytic" scientific statements that act as the hypothetical premises of our inquiry (and hence are relatively immune to empirical check) and "synthetic" statements that are the logically deducible but empirically testable consequences of those premises.[18] The sharpness of this distinction effectively licenses the incommensurability thesis, as two theories governed by mutually exclusive sets of premises define distinct conceptual schemes. Nevertheless, that result is not grounds for discarding the distinction. On the contrary, the advancement of science depends on our ability to keep track of our reaction to the difference between what we originally postulate and nature's response to those postulations.[19] In other words, our sense of progress, perhaps the most distinctive feature of science, presupposes a sharp analytic/synthetic distinction. In the rest of this section, I shall explore the philosophically interesting, and generally neglected, consequences of this argument, since it helps to illuminate how Kuhn's successful blurring of this distinction has retarded the critical edge of both philosophy and science.

The first point that needs to be made is that, like the logical positivists generally and Popper always, Lewis gave a "conventionalist" spin to Kant by portraying the selection of the analytic starting point to inquiry as, more or less, a free choice. This means that whether a statement counts as analytic or synthetic may radically differ across conceptual schemes. "$2 + 2 = 4$" may be philosophers' favorite example of an analytic statement, but it could also be treated as a synthetic statement put to the test of repeated instances of counting, in which the failure of a counter to produce an outcome of "4" from "$2 + 2$" constitutes a strike against the equation's validity. Would such a practice still be mathematics? If so, would the practice be sufficiently stable to permit the pursuit of the natural sciences as we know them? However these difficult questions are answered, they draw attention to the fact that the analytic status of a statement is maintained by discounting, or at least suspending, the experience of people who attempt to use the statement for their own purposes. Be that as it may, the

18. See, e.g., Carnap 1942.
19. For Lewis's views on the analytic/synthetic distinction, see Thayer 1968, 212–31, 511–19.

conceivability of the statement falling on either side of the analytic/synthetic divide is not itself grounds for discarding the distinction. All it shows is that the distinction cannot be drawn on the basis of the statement's content, but rather its logical form, which is prescribed by the language of the conceptual scheme chosen, until we choose to do otherwise.

Indeed, precisely because we are in full control of the starting point, we can hold it constant in our encounters with empirical reality. In other words, the artificiality of our premises ensures that they can simulate one of the main properties that pre-Kantian philosophers had attributed to reality itself, namely, *invariance*. Instead of presupposing what philosophers after Hume have uncritically called the "uniformity of nature," Lewis presupposed the invariance of our conceptual scheme.[20] In its rigidity, therefore, lay its diagnostic strength. If conceptual schemes could not be held constant in their encounters with reality, then there would be no way of comparing their ability to capture reality, and hence we would not be afforded the basis for making another free choice between schemes in the future. However, in order for conceptual schemes to appear to the inquirer as distinct choices, it is necessary that they *not* be intertranslatable; otherwise, the freedom of choice would dissolve into a notational illusion: that is, saying the same thing in different ways. For example, our belief that there is a genuine decision to be made between a chemistry based on "phlogiston" and one based on "oxygen" rests on our anticipating the consequences of that decision to be significantly different in the long term. But were there overlap in the alternative possible consequences of that decision, such that today's chemistry could have just as easily started from phlogiston as from oxygen, then doubt would be cast on the control we exert over inquiry.[21]

---

20. See esp. Lewis 1929, 348 ff.

21. To be sure, if our choices at the analytic level did not make a difference at the synthetic level, it would not necessarily follow that reality was drawing us inexorably toward its ultimate nature; rather, it could equally imply that our own mental, linguistic, or conceptual dispositions are so entrenched that, no matter how hard we try, we keep ending up with the same picture of reality. A sophisticated version of the entrenchment idea that draws inspiration from chaos and complexity theory, rather than traditional notions of teleology and innatism, is the "natural attraction" theory of cultural evolution popular among anthropologists who take their marching orders from evolutionary psychology. For a very interesting presentation of this viewpoint, see Sperber 1996.

I have used the term *overdetermination* to describe the general tendency of history converging on the same end, no matter the starting point. The overdeterminationist holds that history is heading somewhere, though most of the particular events along the way need not have happened as they did. Practicing scientists are generally overdeterminationists about the history of their fields, and therefore have no problem believing that the same truth can be arrived at by many different cultural paths, and that beyond a certain point, historical detail shades into minutiae. In contrast, the *underdeterminationist* turns this picture on its head, arguing that history acquired whatever direction it has only through a series of contingent

The question that begs to be asked is *how* one chooses between incommensurable starting points for inquiry. At this point, Lewis situated the pursuit of science in a larger framework of social values that defines pragmatic criteria for judging any form of inquiry as a human endeavor. These include the usual criteria that have been accepted by most twentieth-century philosophers of science, not least Kuhn: theoretical simplicity, developmental fruitfulness, empirical adequacy, computational efficiency, predictive accuracy, explanatory breadth, conceptual depth, and so forth. However, Lewis differed from Kuhn over the judgments he reached when applying these criteria to specific cases.

A good case in point is the transition from Ptolemaic to Copernican astronomy. Kuhn and Lewis are in agreement that the transition was not merely a matter of a false theory being replaced by a true one. Kuhn is well known for arguing that although we now find Ptolemy's model of the universe much more complicated than Copernicus's, sixteenth-century astronomers already trained in Ptolemaic astronomy found Copernicus's innovations artificial and hence difficult to assimilate—that is, absent any clear physical proof, which would come only over a century later with Newton.[22] In other words, for Kuhn, the theoretical simplicity of Copernicus's model was not sufficient to overcome the cognitive inertia of the existing community of inquirers. Only once Copernicanism was shown to excel in another virtue they upheld did they change their basic orientation. In contrast, in Lewis's version (allegedly based on an example that William James periodically used in his lectures), Copernicus was accepted over Ptolemy because his was indeed the simpler model of the universe, as revealed by the various contexts in which Copernicanism, but not Ptolemaism, enabled humans to extend their control over their environment.[23]

Nowadays, we say that Kuhn's is the historically more accurate account, but only because we tend to presuppose a specific framework for the explanation of historical episodes. However, both Kuhn and Lewis can be shown correct, if we take into account the difference in the class of decision makers and the breadth of temporal perspective in terms of which the choice between the two theories is to be made. Kuhn was clearly trying to explain

---

"turning points," the outcomes of which narrowed the number of paths that history could subsequently take. When historians stress the need to write history from the standpoint of the past and not the present, they are making the underdeterminationist point that the original historical agents saw their future as open to many more possibilities than we can easily imagine, since our knowledge is colored by later events. See Fuller 1993b, 210–13; Fuller 1997d, 84 ff. (It should be clear that my use of "underdetermination" here bears no clear relation to Quine's thesis, discussed intermittently in this book.)

22. This is one of the principal themes of Kuhn 1957.
23. Lewis 1970, 251–52.

why astronomers who were roughly contemporary with Copernicus did not immediately warm to his model, whereas Lewis wanted to understand how a wider range of inquirers came to accept the model in the long run. Not surprisingly, they gave different answers to these rather different questions.

Moreover, behind these different framings of the question were substantially different normative commitments. Kuhn takes as his starting point the self-selecting and self-organizing character of scientific inquiry. Thus, he describes the theory choice episode from the standpoint of a scientific practitioner of the period he is trying to capture. In contrast, Lewis follows the pragmatist penchant for adopting the standpoint of an idealized version of what might be called "the complete human being," for whom science is primarily a means for satisfying basic human needs and not necessarily an activity worth pursuing indefinitely for its own sake. From this larger perspective, the natural inclinations of the scientific community may become an impediment to inquiry's ability to contribute to the interests of humanity. Here it is worth recalling that fears of "professionals" constituting themselves as a secular priesthood abounded early in the twentieth century, when pragmatism was coming of age. Lewis followed James and Dewey in admitting of a down-to-earth "Yankee" suspicion of any tendency that implied that the average citizen should delegate her powers of judgment to "experts" more interested in perpetuating their specialty than addressing the public's problems.[24]

This important point of difference between Lewis and Kuhn is worth mentioning because it helps to convert one of Lewis's seemingly dinosaur-like qualities into a critique of science as an increasingly closed society. Given Richard Rorty's inestimable role in making pragmatism palatable to postmodern sensibilities, we might do well here to start by distinguishing between *pre-* and *post-Rortian* interpretations of Lewis versus Kuhn. The post-Rortian view is familiar enough: Kuhn understands science from the spatiotemporal horizon of the practicing scientist, whereas Lewis approximates the transcendental, "godlike" standpoint of the traditional philosopher who aims for an "ultimate" sense of science's grip on reality. This explains, at least to the post-Rortian's satisfaction, why the philosophy of science has divided between followers of Kuhn, who tend toward historical and sociological studies of science, and followers of Lewis (such as they are), who continue to engage in largely metaphysical discussions of the

---

24. This point is made in Thayer 1968, 221. A good cultural history of learned antiexpertism is White 1957, which considers Dewey, Thorstein Veblen, Oliver Wendell Holmes Jr., Charles Beard, and James Harvey Robinson. The parallels between pragmatist and Machian concerns about the professionalization of science raised in chapter 2 of this book should be evident.

nature of "Science" that rarely touch base with the empirical character of actual sciences. In social-psychological terms, Kuhn captures science from the "inside," Lewis from the "outside."[25]

The persuasiveness of the post-Rortian reading appears in the ease with which philosophers accept the impossibility of understanding something except from within one's own conceptual scheme, as if our ability to make sense of an opposing viewpoint is predicated on our ability to have already entertained that viewpoint. For that reason, some philosophers, notably Donald Davidson, have concluded that the very idea of radically different conceptual schemes is incoherent; rather, there are simply alternative, but

25. The difference between the insider and outsider approach to conceptual schemes that we find in Lewis and Kuhn can be conveniently captured in terms of Gestalt psychology. Interestingly, Gestalt principles also captured the imagination of those two world-historic Viennese thinkers, Wittgenstein and Popper. As it turns out, they, like Kuhn, first came to grips with the idea of conceptual schemes as teachers interested in how students come to "see" what they are shown in the classroom. See Bartley 1974. Consider the "gestalt switch" phenomenon, whereby, depending on the context, the subject can be made to see, say, either a duck or a rabbit in an ambiguous sketch.

Popper stressed the need for conventions of representation in fixing perception, here defined by the experimenter's prompt to the subject to contextualize the sketch one way rather than another, since the sketch itself has no "natural" interpretation. Lewis's appeal to a neat analytic/synthetic distinction should be seen in this light, namely, from the standpoint of someone who begins inquiry without the burdens of past biases, not because she transcends those biases but because she anticipates and hence controls for them in what remains a genuinely open interpretive situation.

Wittgenstein started close to this position but gradually shifted the focus from the experimenter to the subject, observing that the subject must supply a dimension when seeing the duck or rabbit that is not present in the sketch itself, thereby revealing the tacit character of the requisite background knowledge. See Wittgenstein 1953, pt. II, xi. Not surprisingly, given the detachment that is afforded to those lucky enough to create their own situations for action, Popper and Lewis (and we might wish to include here the logical positivists) were more sanguine than the later Wittgenstein about the ease with which one's background assumptions could be identified, criticized, and altered.

Kuhn himself somewhat reluctantly accepted the gestalt switch as a model for understanding the ability of revolutionary scientists to switch between paradigms. However, his interpretation of the switch is closer to Wittgenstein's subject-student than Popper's experimenter-teacher. Indeed, Kuhn actually deepened Wittgenstein's sense of the intractability of background knowledge in that he denied the possibility of giving a relatively neutral description of the sketch that subjects perceive differently: "Scientists do not see something *as* something else; instead, they simply see it. We have already examined some of the problems created by saying that Priestley saw oxygen as dephlogisticated air. In addition, the scientist does not preserve the gestalt subject's freedom to switch back and forth between ways of seeing" (Kuhn 1970b, 85).

Notice that Kuhn conceptualizes the scientist as a completely innocent (or is it self-involved?) subject who cannot imagine what it would be like to be the experimenter, and hence cannot switch back and forth between two alternate interpretations of the ambiguous duck/rabbit sketch. Also noteworthy is that Kuhn assumes (wrongly) that when "see X as Y" is used to characterize perception, "X" must refer to the putative reality that underlies "Y," whereas "X" may also be a minimal account of the phenomenon that is not biased toward any of its possible interpretations. Thus, we might say—as would the Gestalt psychologists themselves—that Priestley interpreted a particular gaseous substance emitted by his experiments *as* dephlogisticated air (rather than oxygen).

no less intertranslatable, perspectives on the same reality. Thus, comparing conceptual schemes is no more difficult than comparing maps of the same space.[26] (Interestingly, when the map metaphor was originally used to characterize theories, it was to highlight the fact that *even though* several maps may capture the same space, it was impossible to tell which was better without first specifying the reason one wanted the map in the first place.)[27] To someone not immersed in our postmodern times, this position can easily look like a transcendental justification for closed-mindedness: If your belief is intelligible, I should be able to have it without suspending too many of my current beliefs. Closed-mindedness is precisely how the original pragmatists would have diagnosed this situation.[28]

This is a good point at which to turn to the pre-Rortian interpretation of Lewis versus Kuhn. Here Kuhn's "insider" understanding of science looks parochial, as if scientists were *nothing but* scientists—and not full-fledged human beings for whom scientific work plays an important but not absolute role in their judgments about which course of action is in their own and their fellow humans' best interests. In that case, Lewis's "outsider" perspective looks less like a feeble attempt to get into God's mind than an abstract expression of how one might understand the relative merits of alternative scientific theories for these larger normative concerns. In other words, the pragmatist wants scientists to stand back, not from their earthly existence, but from their narrow professional interests. The fact that post-Rortians are prone to so see no intermediate standpoint between that of professional interests and "the mind of God" would appear—at least to the pre-Rortian—as symptomatic of a science profoundly alienated from its social conditions.

A good historical indicator of this assessment is James's and Dewey's general distrust of the growing enthusiasm for formal logic in the twentieth century. Their suspicions were fueled by the prospect that if the criteria of scientific acceptability were reduced to purely formal logical criteria—such as avoidance of contradiction—then professional scientists would be licensed to persist in their pet theories with impunity, since contradictions can nearly always be overcome by some conservative adjustment to one's original theoretical commitments. Thus, a dilution of science's revolutionary spirit could result from an overreliance on logic for normative direction. Of course, Lewis, himself a distinguished contributor to the development of modern formal logic, did not share such views, but as a result, he

---

26. Davidson 1982, esp. 69.
27. This pragmatist point was originally presented as an insight of Wittgenstein's later work. See Toulmin 1951, chap. 4.
28. See also Popper 1994, 33–64.

may have failed to see how the growing prestige of logic within philosophy would contaminate thinking throughout the field. For, while Lewis clearly accepted the general pragmatist point that logic was literally only a "tool for thought," most twentieth-century logicians have tended to simply equate logic and normatively appropriate thinking.

To bring closure to this discussion, let us return to where Lewis first anticipated Kuhn in his endorsement of the idea that scientists from different ages lived in different worlds. In a quite un-Kuhnian manner, Lewis concluded his argument with a plea to recognize the logical distinctness of our conceptual scheme from the features of reality that it represents. Given the dynamics of the recent debates between "realists" and "antirealists" in the philosophy of science, it is easy to suppose that only a realist would want to draw such a distinction. However, Lewis justified this distinction on *pragmatic*, not transcendental, grounds. He did not advance a positive argument for the independent existence of reality. Rather, he tried to underwrite humanity's ability to improve its capacity for action, given the patent divergence of worldviews throughout the ages. The degree of freedom we have in our engagements with reality depends on our ability to track the consequences of our conceptual commitments. In the Hegelian phrase, freedom is the recognition of necessity. Logic then turns out to be the premier device for structuring our historical intuitions, so that we retain a clear sense of what follows from what, and then can decide accordingly. Our rationality lies, not in the logic itself, but in the decision that it enables us to make.[29]

Lewis's normative orientation to logic has not been completely transparent to his successors in epistemology and the philosophy of science over the course of this century—and here I mean more the orientation than its provenance. Textbook accounts of scientific inference continually run together two accounts that privilege deductive logic: the exemplars are Karl Popper's "hypothetico-deductive" (H-D) method and Carl Hempel's "deductive-nomological" (D-N) model of explanation. In the realpolitik of classroom practice, Popper's method, which presupposes a view of logic close to Lewis's, is reduced to Hempel's model. There are at least two reasons for this. One is the very strong institutional inroads made by logical positivism in Anglo-American philosophy during World War II, which occluded Popper's distinctive pragmatist revision of positivism. The other is the simple fact that Hempel and Popper were part of the same cohort of Central European philosophers who were marked as the "old guard" by

---

29. Lewis 1970, 257. This view is perhaps most radically and systematically developed in Dewey 1938.

the generation of English-speaking philosophers who came of age during World War II and began to assume major professorships around 1960. These philosophers, especially Toulmin and Hanson, both of whom were ultimately overshadowed by Kuhn, positioned their critiques so as to sweep away both Hempel and Popper in single rhetorical gesture.[30]

Both H-D and D-N can be understood as a syllogism expressed in first-order predicate logic, e.g.: "All swans are white; this is a swan; therefore, it must be white." Putting aside the scientific simple-mindedness of this stock example (proper scientific examples can be easily multiplied, especially when expressed as a set of formulae containing open variables), the syllogism is typically read as a scheme for justifying scientific findings by showing how they fall under a generalization that is tentatively regarded as a law of nature. Under that interpretation, the scheme provides a guide for what you ought to predict in new situations comparable to those that have confirmed the generalization in the past. However, Popper vehemently opposed this interpretation as a dogmatic use of logic to bolster entrenched scientific prejudices. Drawing on beliefs about both the essential creativity of human beings and the ultimately indeterminate character of physical reality, Popper argued that the quest for "laws" (the "nomological" side of Hempel's model) in science was a philosophical chimera.[31] The real reason for insisting on generalizations in our scientific hypotheses, according to Popper, was to make their test against reality as sharp as possible. If a favored theory insists that a certain event will happen, and it does not happen, then the scientist is faced with a clear choice: either abandon the theory or abandon the observation.

Perhaps because Popper supposed that the refutation of well-confirmed theories by an indeterminate reality was an inevitability and that scientists were sufficiently self-possessed to allow their own observations to trump the authority of the past (after all, for Popper, theories are no more than

---

30. Hanson's conflated account of H-D and D-N (whereby Hempel's position is given Popper's name) is still cited as an authoritative source: see Hanson 1958, 70–73, 82–86. Toulmin's complicity in the conflation appears in his review of Popper's *Logic of Scientific Discovery*, which explicitly (and misleadingly) distinguishes himself and Hanson from a positivistic interpretation of Popper. See Toulmin 1959. The conflation of H-D and D-N continues in Harre 1986, 4, where Hempel and Popper are dubbed "logical essentialists."

31. Here it is worth recalling the roots of Popper's idea of the "open society" of science in Henri Bergson's account of the origins of religious dynamism (see chapter 1, note 136). Popper's commitment to the ultimate indeterminacy of reality was a metaphysical position he shared with Bergson that comes out most clearly in his philosophy of quantum mechanics. For a critique that links Bergson and Popper, see Bricmont 1997, 147 ff. Bergson was to Popper as James to Lewis. Precisely because reality offers no clear signs of its own nature, logic was of paramount importance to both Popper and Lewis as the "rigid rod" against which epistemic claims could be judged. For their part, Bergson and James believed that metaphysical indeterminacy was inescapable except through an act of pure will.

formalized presumptions), he regarded his H-D method as a vehicle for forcing scientists, on a regular basis, to decide whether they accepted or rejected their basic working assumptions, and not simply to drift along in their current allegiances as the path of least resistance (essentially the course of action recommended by Kuhn in the guise of "normal science").[32] In his attitude toward logic, Popper is Lewis's surest fellow traveler.

Although Lewis did not explicitly raise it, the specter of world-historic self-deception clearly lay in the background of his thinking about the importance of logic as a diagnostic tool. Pragmatism was vulnerable to self-deception from two opposing directions, both of which can be read as glosses on one of William James's most vivid moments of philosophical expression, his widely read essay of 1897, "The Will to Believe," which is now often regarded as the classic defense of people's entitlement to believe whatever works for them. On the one hand, if all of science is seen as "analytic," then we are likely to overestimate our capacity to change the world, perhaps thinking that every difference in verbal formulation marks a difference in real-world consequences. This is the fallacy of *voluntarism*, to which pragmatism's more Nietzschean fellow travelers, such as the British humanist F. C. S. Schiller, were prone. On the other hand, if all of science is seen as "synthetic," then we may underestimate our capacity for change, as we suppose that the practical outcomes of our conceptual commitments that do not kill us must make us stronger and therefore should be tolerated and continued. This is the fallacy of *adaptationism*, which was likely to afflict pragmatism's French theistic adherents such as Édouard Le Roy, who interpreted the benefits we accrue from our beliefs (be they religious or scientific) as orienting us to ultimate reality. Nowadays, the Divine Light has been replaced by the Principle of Natural Selection as the beacon providing metaphysical direction, but little of the dogmatism has changed, except that religion has yielded completely to science as its source.[33]

My last comment is designed as an oblique reference to the principal

---

32. Here I draw on the interpretation of Popper's approach to logic offered in Notturno 1999, chap. 5. Notturno goes so far as to argue that, for Popper, the logic of discovery is the logic of discovering our errors (i.e., such that we choose a path of inquiry other than the one previously pursued). The following epigraph to Popper's magnum opus, *The Logic of Scientific Discovery* (1959) provides the basis to Notturno's reading: "There is nothing more necessary to the man of science than its history, and the logic of discovery . . . The way of error is detected, the use of hypothesis, of imagination, the mode of testing" (Lord Acton). This strikes me as true to Popper's spirit and should give pause to those who still believe that the English translation of *Forschung* as "scientific discovery" was a complete mistake.

33. On the British and French extremes of pragmatism, see Thayer 1968, 283–92 and 314–20, respectively. A good example of Natural Selection functioning as the Divine Light did in the past is Dennett 1995.

heir of pragmatism's adaptationist strain, Willard Van Orman Quine, whose 1951 essay, "Two Dogmas of Empiricism," is the vehicle through which most analytic philosophers influenced by logical positivism have been forced to confront the pragmatist legacy. The link from Le Roy to Quine is provided by one of Le Roy's scientific sympathizers, Pierre Duhem, who generally backgrounded his Catholic commitments when discussing the nature of scientific inquiry. This was especially true in the case of physics research, which, by the end of the nineteenth century, had become so dependent on mathematical expression and laboratory experiments that it effectively constituted its own artificial world detached from any direct experience of nature through which God might provide illumination. Given the overriding importance of the so-called Duhem-Quine thesis for recent work in both the philosophy and sociology of science, it is worth recalling that Duhem originally justified the claim that theory choice in science is always underdetermined by the available evidence by noting how physical inquiry had become severed from the forms of intuition and reasoning characteristic of our experience of nature. Unlike Quine, who subsequently turned this into a general scientific principle, Duhem regarded the situation in physics as unique and, without theological guidance, probably inimical to a comprehensive understanding of nature.[34]

A sense of how Quine constructed the legacies of pragmatism in general and Duhem specifically may be seen in his concluding remarks:

> The issue over there being classes seems more a question of convenient conceptual scheme; the issue over there being centaurs, or brick houses on Elm Street, seems more a question of fact. But I have been urging that this difference is only one of degree, and that it turns upon our vaguely pragmatic inclination to adjust one strand of the fabric of science rather than another in accommodating some particular recalcitrant experience. Conservatism figures in such choices, and so does the quest for simplicity.
>
> Carnap, Lewis, and others take a pragmatic stand on the question of choosing between language forms, scientific frameworks; but their pragmatism leaves off at the imagined boundary between the analytic and synthetic. In repudiating such a boundary I espouse a more thorough pragmatism. Each man is given a scientific heritage plus a continuing barrage of sensory stimulation; and the considerations which guide him in warping his scientific heritage to fit his continuing sensory prompting are, where rational, pragmatic.[35]

---

34. An early perceptive account of the relationship among Duhem, Quine, and Kuhn may be found in Hesse 1970a. Duhem is discussed passim in chapters 1 and 2, above.

35. W. V. Quine, "Two Dogmas of Empiricism" 1951, in Quine 1953, 45–46.

On my reading of Lewis, Quine's account of our "vaguely pragmatic inclination" is simply off base. However, an initial difficulty in making this case comes from Quine's peculiar tendency to reduce scientific inquiry to an extended exercise in set theory. In arguing that the failure to observe one member of a class (e.g., the discovery that there are no brick houses on Elm Street) does not necessarily invalidate the existence of that class (i.e., that there are brick houses), Quine is alluding to how we instinctively resolve the underdetermination of theory by data. Quine characterizes this response as "conservative," a term that he generally invests with positive value. However, he refuses to elevate our instinctive conservatism to one of Lewis's analytically definable norms that binds our action in a principled way. Yet, as Lewis showed, understanding the conditions under which we might switch to a radically different conceptual scheme—pending the inadequacy of our current scheme to our interests—requires a strict distinction between the analytic status of a theory and the synthetic status of its consequences. That is the only way we can determine whether the same theory has indeed performed as desired over time.

In contrast, if we take Quine at his word, we are simply licensed to continually reassign truth-values across a network of propositions so that equilibrium is retained between theory and experience. We are given no indication of the conditions under which it would make sense to compare alternative theoretical premises, let alone decide to frame one's inquiry in terms of one those alternatives. Indeed, given Quine's fondness for Otto Neurath's metaphor of the boat that is rebuilt at sea, it would seem that he was keen on blurring any clear sense of choice between conceptual schemes. Thus, by incremental changes, one scheme may simply metamorphose into another without notice.[36]

When James originally declared that people should believe whatever works for them, he was crucially ambiguous about whether his policy had implications at the metalevel, namely, that people should endorse whatever evaluative standards allow their current beliefs to be deemed successful, which would then create a presumption to continue those beliefs. Quine resolved this ambiguity in favor of such a presumption, which effectively erased the sense of contingency attached to the fact that one conceptual scheme rather than another had been adopted in the first place. He did this largely by presupposing that a scientist was nothing but a member of her profession and, from that standpoint, has a vested interest in conserving her

---

36. On the epistemological significance of the boat metaphor, which Quine picked up as a visitor to the Vienna Circle meetings in the 1930s, see Cartwright et al. 1996, 89 ff.

"heritage." If Kuhn's revolutionary scientist was presaged in Lewis's portrayal of the inquirer as genuinely open to the future, potentially exchanging one conceptual scheme for another, Kuhn's normal scientist is captured in Quine's dogged portrayal of the scientist as carrying on with her paradigmatic pursuits until the weight of anomalous results forces a change upon her.

However much James may have endorsed the general idea that one should adapt one's preferences to match one's capabilities for action, it is unlikely that he would have licensed Quine's studied conservatism. The reason, I suspect, is that whereas Quine and Kuhn regard scientists as adapting to an environment mediated — or perhaps "insulated" — by other members of their community, James presented a more individualistic image of the scientist adapting to more direct encounters with the environment. This difference in setup serves to highlight complementary liabilities that can be committed in the name of "adaptationism." Put in terms borrowed from Jean Piaget's developmental psychology, Quine and Kuhn "adapt" by *assimilating* novel experience to the dominant belief system, while James "adapts" by *accommodating* one's beliefs to that experience: in the one case, too much rigidity; in the other, too much fluidity. In short, we see that the use of deductive logic in the understanding of scientific inquiry is not itself a problem. Problems arise only when logic is autonomized from the substantive processes of inquiry. C. I. Lewis is a useful witness because, unlike the logical positivists and especially their descendents in analytic philosophy, he refused to treat logic as a separate sphere, but at the same time, unlike Kuhn and his historical and sociological followers, Lewis did not dismiss the relevance of deductive logic in providing direction to scientific inquiry.

3. Underlaboring in a Kuhnian Key

So what exactly has been Kuhn's legacy to epistemology and philosophy of science? It has been to dull the critical edge of the concept of rationality that made philosophy's relationship to the natural sciences problematic, once they became the dominant form of knowledge in Western societies. Before 1800, philosophy and science were natural allies because not only were the same people often doing both but also they were commonly opposed to religious authority. However, we saw in chapter 1, section 6, as theology gradually yielded its cognitive authority to the natural sciences over the course of the nineteenth century, philosophy and science faced each other as potential foes, especially as academic gatekeepers like William Whewell appeared to be grooming the natural sciences to become the secular successor of a state church. The 1965 Kuhn-Popper debate, explored below, should be understood as a high-water mark in this tension between philosophy and science.

Here Popper defended "Science" as an ideal that potentially stood against the actual practices of scientists. He argued that only in its revolutionary phases—when scientists were inclined to question the philosophical foundations of their inquiries—has science lived up to its rational potential. Herd behavior ruled the rest of the time, when scientists were engaged in the business of tidying up the loose ends of the dominant paradigm. Moreover, Popper realized that the antiestablishmentarian character of his critique had a special relevance for the twentieth century's technologically scaled-up science that it probably would not have had before the First World War.[37] Of course, Kuhn promoted exactly the opposite viewpoint, and some of his followers have gone so far as to argue that the deep philosophical problems plaguing a paradigm may be excused once it has chalked up enough of its own empirical successes.[38]

Kuhn is a relative latecomer to the neutralization of philosophy's critical attitude toward science. Over 150 years ago, the man who coined both "positivism" and "sociology," Auguste Comte, presented the following chain of reasoning, which was meant as a rational reconstruction of the historical transition from "philosophical" to "scientific" modes of thought:

> 1. Philosophy leads us to prefer science as the highest form of inquiry by getting us to develop theories of truth and rationality that force us to justify our knowledge claims by the rigorous standards of logical reasoning, which, in turn, reveal the inadequacies in our taken-for-granted habits of thought.
>
> 2. But once we start to do science, we realize that these philosophical theories are themselves impediments to further inquiry, as they encourage us to rush to conclusions about the whole on the basis of knowing only a few of the parts. This tendency toward premature totalization is the result of trying to let reason do the work of empirical observation.
>
> 3. What this shows is that a mode of thought that persuades us of the value of science in the first place need not itself be of service to science, once we have been so persuaded. In fact, such a mode of thought may even be an obstacle and should be suspended, as it has outlived its usefulness.

For readers who do not immediately recognize *Comte's scenario*, consider the following story sometimes told about how the Scientific Revolution occurred:

---

37. Popper is most explicit on this point when directly responding to Kuhn. See Popper 1974, esp. 1146.
38. See esp. Laudan 1977, esp. chap. 2.

By relentlessly deploying his powers of ratiocination, Galileo revealed the artificiality and inconsistency of Aristotle's assumptions about physical motion. However, a better physics would not come simply by continuing to deploy those powers, as that only led toward legitimating new biases as "first principles," as in the case of Cartesian physics. What was needed was some empirically grounded first principles, "deductions from the phenomena," as Newton put it. Finally, science broke free from philosophy.

Setting aside the historical accuracy of this story, we can say that it clearly presumes that philosophy and science are compatible with each other only in some contexts but not others. There may be some temporary alliance between science and philosophy against a common foe such as Aristotle or Christianity. But once the foe has been vanquished, and science is ascendant, then a science of science would spend much of its time purging science of any philosophical residue.

According to Comte, the philosophical stage depicted in paragraph 1 occurred during the eighteenth-century Enlightenment, while the scientific stage coincided with the rise of positivism in his own nineteenth century. Since that time, both philosophers and sociologists have derived support from the Comtean scenario. Analytic philosophers have encountered this line of reasoning, whereby philosophy is presented as a form of cognitive therapy, throughout the work of Ludwig Wittgenstein. For example, in the *Tractatus*, Wittgenstein likened (his own) philosophy to the ladder that needs to be cast aside once it has been climbed. In his later work, he leans on the metaphor of letting the fly out of the bottle, after having drawn attention to the sound of its buzzing.[39] This convergence of Comte and Wittgenstein is most apparent in the work of David Bloor, the seminal thinker of the sociology of scientific knowledge (SSK). He does little to conceal his debt to Wittgenstein, and in fact most of his "antiphilosophical" sensibilities are more directly attributable to Wittgenstein than to his illustrious precursors in the sociology of knowledge, be it Mannheim, Durkheim, or even Comte himself.[40] Not surprisingly, philosophers have been especially riled by Bloor's proposal for eliminating their discipline.[41]

---

39. Wittgenstein 1922, prop. 6.54; Wittgenstein 1953, pt. I, 309.

40. Bloor 1983, chap. 9. For Bloor, all three classical sociologists sustain the philosophical impulse to make normative pronouncements that go beyond what is empirically warranted. Comte wanted sociology to anoint science the legitimate successor of the Roman Catholic Church, whereas Durkheim had sociology invest state education with a moral authority that could be used as a buffer against the anomic tendencies of capitalism, and Mannheim envisaged sociology as less a discipline than an order of free-floating intellectuals capable of mediating conflicts of value before they erupt into total war.

41. One of the more measured responses is Hacking 1984b. While it has become common for philosophers to represent their discipline as the *precursor* of epistemic practices that most epistemologists and philosophers of science would regard as positive cultural contributions,

If the reader doubts this characterization of Bloor's position, it might pay to recall an early formulation of the Strong Programme, which, a quarter century on, remains the signature statement of the SSK approach:

> The aim of the sociology of knowledge is to explain how people's beliefs are brought about by the influences at work on them. This programme can be broken down into four requirements. The first is that the sociology of knowledge must locate causes of belief, that is, general laws relating beliefs to conditions which are necessary and sufficient to determine them. The second requirement is that no exception must be made for those beliefs held by the investigator who pursues the programme. Special pleading must be avoided and causes located for those beliefs subscribed to, as well for those which are rejected . . . [Third, t]he sociology of knowledge must explain its own emergence and conclusions: it must be reflexive . . . [Fourth, n]ot only must true and false beliefs be explained, but the same sort of causes must generate both classes of belief.[42]

What we have here is a stereotypically positivist account of science that makes value-neutrality a desideratum of sociological explanations and stresses probative standards ("necessary and sufficient conditions") that greatly exceed what passes for normal natural-scientific practice — only now it is called *symmetry*.[43] Indeed, contrary to their reputation, SSKers are only *methodological* relativists. Their treatment of all cultures as having

---

Bloor held that philosophy is an *atavism* in an epistemic culture dominated by science. In the love-hate relationship that philosophers of science have had with scientists (e.g., Popper's wanting scientists to be more like philosophers, and the logical positivists' wanting philosophers to be more like scientists), philosophers have thought of themselves as ultimately on the same side as the scientists. Bloor denied this, suggesting that philosophy is on the side of the scientists *only in cultures where science is not the dominant epistemic practice*. Where Popper saw philosophers as keeping scientists on their toes by periodically awakening them from their dogmatic, normal-scientific slumbers, Bloor saw philosophers as indulging scientists in habits that make empirical inquiry so hard to do for any length of time. (It is worth noting that Popper and Bloor are meant to be giving conflicting interpretations of largely the *same* philosophical practices, such as the elaboration of methodological theories.) However, once the philosophers have been silenced, Bloor would not then presume to tell scientists what to do. The Wittgensteinian ladder will have been mounted and therefore ready to be cast aside. In that respect, Bloor is a *metarelativist*, someone who realizes that a methodological commitment to relativism has its own time and place and is not appropriate in all times and places.

Metarelativism relies on drawing a sharp distinction between the *content* and *function* of a form of knowledge, whereby the content of a knowledge claim — especially the claim that all knowledge is relative — is "progressive" or "reactionary," depending on the specific function that the claim performs in the society where it is uttered. But this judgment, in turn, depends on one's image of a "functioning society." Bloor supposes that the pursuit of normal science in discrete fields of inquiry is a mark of functionality, whereas I see it as a mark of dysfunctionality. Therefore, my sense of metarelativism implies that relativism may become obsolete, not when universalism has been defeated, but when relativism no longer adequately performs as dialectical foil to whatever happens to be the dominant mode of inquiry. I stress this ironic version of relativism in Fuller 1995b.

42. Bloor 1973, 173–74.
43. On SSK's "voodoo epistemology," see Roth 1987, 152 ff.

prima facie equal access to reality is a heuristic for understanding their knowledge claims. Whether or not each of these cultures then deserves to be promoted to the same extent is a normative matter that arises only once the cultures have been adequately characterized. In other words, it would not be unfair to say that SSKers adhere to a fact/value distinction with respect to their own methodology, even though the cultures they study seem to blur that distinction all that time. Indeed, some SSKers have admitted and justified this double standard as the mark of SSK's own scientific status.[44]

Consider the move in a larger epistemological perspective. In classical theories of knowledge, ones fixated on the individual as cognitive agent, a self-centered relativism is regarded as the initial liability that needs to be overcome. For example, Descartes knew what is in his own mind, but he suspected that there were things outside it that are different in nature. In that case, the relevant compensatory strategy is a kind of realism that looks for "primary" qualities that remain invariant under a variety of observations and cognitive transformations. In contrast, for SSKers, such realism is seen as the initial liability that needs to be overcome. Since SSKers concede at the outset that everyone inhabits the same world, it is all too easy to infer that everyone must think like oneself, at least if they are rational.[45] The relevant corrective here is a dose of methodological relativism: precisely because our reality is common, it cannot explain our palpable differences. We are thus better off regarding claims to a common reality much as Karl Mannheim did, as disguised partial perspectives, or "ideologies," that may gain certain local material advantage by capitalizing on our weakness for thinking in terms of totalizing forms of realism.[46] But once the ideological character of a society's knowledge claims is granted, does it follow that they are uncontestable? Courtesy of Kuhn, the answer has come to be yes. The next four sections are concerned with how this came to pass.

4. A Case of Mistaken Identity: How Reason Became Irrational

The first extensive critical review of *The Structure of Scientific Revolutions* is the one that emblazoned Kuhn in the minds of philosophers. Appearing in one of America's premier journals, it was written by a young Harvard-trained philosopher who had already published several articles showing that the logical analysis of language championed by the logical positivists did not do justice even to the science they cherished most, physics. Like all

44. See esp. Collins 1981; Pinch 1988.
45. This argument is explicitly made in Barnes and Bloor 1982, esp. 34.
46. I dub these two social epistemology strategies—the Cartesian "inside-out" and the Mannheimian "outside-in"—"A" and "B," respectively, in Fuller 1996d.

language games, Dudley Shapere (b. 1928) argued, physics too has an irreducibly pragmatic character that can be understood only by a historically sensitized version of ordinary language philosophy, that is, a study of what members of a particular physics community meant in the context of their utterance.[47] Although Shapere thought that Stephen Toulmin had made many of Kuhn's points—and in a philosophically more perspicuous manner—perhaps a decade earlier, the express generality of *Structure*'s account of science and its innocence of any formal logical apparatus made it an apt vehicle for enumerating the full range of omissions from the positivist account of science.[48] Moreover, by making Kuhn out to be proposing an "antipositivist" viewpoint, Shapere made himself appear less an opponent than a mediator of opposing camps. Thus, Shapere could demonstrate the value of philosophical sophistication in moderating the more wild-eyed claims of historians overly impressed with the remoteness of the past.[49]

The subsequent "historicist" turn in the philosophy of science—portrayed specifically as a "revolt against positivism"—proves Shapere to have been largely successful in his immediate intention.[50] Unfortunately, there

47. Shapere 1964, 1966. Shapere's Harvard mentor was Morton White, whose own work strongly anticipated the "pragmatist turn" associated nowadays with Richard Rorty. See esp. White 1956. White's attempt to encompass all English-speaking philosophy under a comprehensive pragmatism was overshadowed in his day by Quine's more focused and largely internal critique of positivism's conception of language. Shapere's early Harvard-influenced essays include Shapere 1960, 1963. Shapere was also the "special consultant" for the National Science Foundation's history and philosophy of science budget in the late 1960s, just before the field was given its own program. See Rossiter 1984, 100–101.

48. Shapere cites Toulmin's two books on the philosophy of science that predate Kuhn's: Toulmin 1951; 1961. Toulmin shares with Kuhn an early background in physics and with Shapere a philosophical training in linguistic analysis. While Toulmin seemed to attract a wide audience (e.g., both the positivist Ernest Nagel and the humanist Jacques Barzun endorsed his books), one reason these two relatively brief and easy-to-follow texts never garnered Kuhn-like attention might be that he wrote in the understated Wittgensteinian style of perceptively remarking on simple examples without issuing anything that might smack of a "theory" or a "scheme." The relationship between Kuhn and Toulmin remained a prickly one for the rest of their careers. Kuhn admitted having known of Toulmin 1961 while writing *Structure*, claiming that he could "understand why Toulmin may have been sore at me for stealing his ideas, but I don't think I did" (Kuhn et al. 1997, 176–77). Reflecting fifteen years after the "revolution" that he supposedly started alongside Kuhn, Feyerabend, and Hanson, Toulmin claimed that by the time historicism had started taking hold in the philosophy of science (and the larger society) in the early 1970s, his interests had already moved elsewhere. See Toulmin 1977, 161 n. 11.

49. Kuhn confirms Shapere's role in setting him up as the philosophical straw man of relativism and irrationalism in Kuhn et al. 1997, 185.

50. The canonical presentation of the story Shapere wanted his audience to believe is Suppe 1977, 3–232. In an afterword to the second edition, Suppe predicted that Kuhn's influence was waning and that the future of historicist philosophy of science belonged to Shapere and Toulmin. In retrospect, the most striking omission from Suppe's eight-hundred-page tome is any reference to the person who turned out to carry the torch for historicism, Larry Laudan, whose *Progress and Its Problems* was published in 1977, the very same year as the second edition of Suppe's book. Chapter seven of Laudan's book threw down the gauntlet at the emerging sociology of scientific knowledge, about which more in the next chapter.

is little reason to think that it was an intention shared by Kuhn's positivist patron, Rudolf Carnap, who allowed *Structure* to be published as part of the International Encyclopedia of Unified Science. As personal correspondence reveals, Carnap complimented Kuhn on having provided a historical grounding—in the contrast between "normal" and "revolutionary" science—for his own distinction between questions that are decidable within the terms of a given conceptual framework and those that require the introduction of extramural factors.[51]

51. The Carnap-Kuhn correspondence is analyzed in Reisch 1991. Thomas Kuhn informed me that he had not sought to write this particular volume of the positivists' encyclopedia, but was rather only the third person asked to do so. As Reisch informs me (personal communication, March 1994), the positivists' search for someone to write a volume on the history of science had extended back to the 1930s, when Federigo Enriques was first tipped for the job (but then died), followed several years later by I. B. Cohen, who had other commitments. In any case, Kuhn managed to fill the bill to Carnap's satisfaction, despite being largely innocent of the details of logical positivism while writing the volume. For his part, Carnap was quite explicit in declaring his belief that all scientific language change is either simply a matter of adjusting truth-values (cf. normal science) or a radical shift in the basic vocabulary and syntax of the language (cf. revolutionary science). Here it is worth remarking on the sense in which Kuhn upheld "the unity of science," which made his work attractive to Carnap. Despite his disavowal of any overarching teleology to scientific change, Kuhn nevertheless agreed with the logical positivists that science indeed has a "nature" that is common to a variety of disciplines and not seen elsewhere in society. Given Kuhn's inspirational role in the disunified vision of science, this point is easily overlooked. Here it pays to remember Conant's role in relocating the Unity of Science program at Harvard during World War II (see chapter 4, note 73, above). In contrast, the practices that Popper took to be essential to science were ones it shared with other instances of "the open society."

Like Kuhn, Carnap seemed to have difficulty imagining that *the very use* of a language may be a vehicle of its change. See Carnap 1963, esp. 921. Here both Kuhn and Carnap could have taken a page from the founder of modern structural linguistics, Ferdinand de Saussure, whose reflections on the conventional nature of language help explain how the linguistic sign can be both "mutable" and "immutable." The key point is that "conventional" does not mean "contractual": speakers do not freely assent to their languages, but rather tolerate its use. Consequently, while they are not especially motivated to change their language, neither are they motivated to arrest change when it happens. For an elaboration of this point, made expressly against the impulse toward static normative grammars, see Harris 1987, 84 ff. My own views of incommensurability diverge from Kuhn's in terms of my interest in following through the implications of Saussure's point. See introduction, note 65.

The 1990s have witnessed a veritable epidemic of attempts to show the similarity, or at least complementarity, of sensibilities between Kuhn and Carnap (as well as other historicists and positivists who have been traditionally portrayed as intellectual antagonists). Not surprisingly, this flurry of activity corresponds to the opening of archives at the Universities of Chicago and Pittsburgh that enable scholars to introduce evidence from personal correspondence that remove some of the sharp differences of perspective conveyed in the published works. Among the best of these works are Axtell 1993; Irzik and Gruenberg 1995; Uebel 1996. Axtell and Uebel represent two poles in this rapprochement: Axtell wishes to bring Kuhn closer to Carnap for purposes of criticizing Kuhn's epistemological legacy, whereas Uebel wishes to bring Carnap closer to Kuhn (and his recent philosophical followers) for purposes of politically rehabilitating positivism.

But from the standpoint of tomorrow's sociologists of knowledge, the attempt by a distinguished philosopher of physics to demonstrate that "Carnap plus Kuhn equals the philosophical agenda of the sociology of scientific knowledge" must count as one of the most curious

To be sure, over the years, Kuhn has wanted to leave the impression that he has been more influenced by positivist, or at least more broadly analytic-philosophical, considerations about the nature of language and knowledge than by the historicist ones that his work supposedly generated.[52] To understand the overarching significance of *Structure*, especially the sorts of projects it has helped and impeded, we need to start taking seriously that Kuhn's book constituted, pace Shapere, less a revolt against positivism than a continuation of positivism by other means. However, this is not to deny to Shapere the honor of having turned Kuhn into an object of philosophical fascination by parceling out *Structure* into analysis-sized problems of "meaning-variance," "theory-laden observation," and "absolute presuppositions." In this way, Shapere enabled Kuhn to become routinely discussed in the same breath as the ascendant philosophers of science of his generation, especially Toulmin, Russell Hanson, and, of course, Paul Feyerabend.[53]

In the Anglo-American philosophical community of the early 1960s, Shapere's constructed cohort of "historicists" were still perceived—to refashion a bit of Kuhn—as mere anomaly mongers, proffering elaborate counterexamples to positivist accounts of explanation and confirmation, anomalies that, once stripped of their historical superfluities, would be rendered tractable to logical analysis.[54] The idea that the historicists might

---

cases of this recent revisionism: Friedman 1998. I suppose that Friedman's interest should be taken to indicate that sociology has finally made a serious impression on philosophers, but to a sociologist without a vested interest in this achievement, the article reads like a fairly straightforward attempt at co-opting the opposition, now that it has acquired a certain academic respectability. Informed readers should judge for themselves.

52. For Kuhn's increasing tendency to embrace positivism, see Suppe 1977, 647. Also, contrast Kuhn's early disavowal of any connection to the sociology of scientific knowledge, about which more below (Kuhn 1977a, xxi–xxii), with Kuhn's eagerness to insert himself in recent philosophical debates in semantics and the theory of reference. See Kuhn 1989. Finally, consider Kuhn's own accommodating posture to the criticisms made of his work by such positivistically inclined philosophers as Carl Hempel and Wesley Salmon: Kuhn 1983. Eventually, Kuhn went so far as to regret publicly his failure to see the original affinity that Carnap asserted between their views of scientific change (see previous note): Kuhn 1993, 313, 331.

53. Kuhn, Toulmin, and Imre Lakatos (to be discussed below) were all born in 1922. Hanson and Feyerabend were born in 1924. Shapere himself aside, the clustering of these figures as an intellectual cohort is largely a product of the next generation of philosophers, such as Ian Hacking, Larry Laudan, Harold Brown, Fred Suppe, and Peter Machamer, all of whom were born in 1935–45. See, esp., Machamer 1975; Brown 1977. The latter, a textbook, received Kuhn's official endorsement.

54. Of all the historicists, this unflattering image stuck most readily to Feyerabend, who in 1964 was mostly known for having challenged Ernest Nagel's model of explanation by reduction: Feyerabend 1962. In retrospect, this challenge is normally regarded as the first time the authority of the historical record had successfully opposed a formal logical model. Although few philosophers relinquished the goal of reductionism at that time, rather than simply ignoring Feyerabend as raising irrelevant concerns, they endeavored to accommodate Feyerabend's challenge within the model. See Schaffner 1967. During most of the 1960s, Kuhn's

replace or *succeed* the positivists in a common lineage did not appear persuasive (or at least not an item discussed in the main philosophy of science journals) until Imre Lakatos published a volume of papers from a symposium held during the 1965 meeting of the International Congress of Logic, Methodology, and the Philosophy of Science, which was sponsored by Lakatos's home institution, the London School of Economics.[55] The symposium in question was officially devoted to the implications of Karl Popper's philosophy of science, but Lakatos was keen on using the forum to present himself as the heir apparent to Popper's chair.[56] However, as fate would have it, Kuhn stole the show, with his contribution to a Popper Festschrift being the one in terms of which all the other speakers—including Popper, Lakatos, Toulmin, and Feyerabend—rewrote their papers.[57] Thus, *Criticism and the Growth of Knowledge*, despite the Popperism packed into its title, is normally taught in seminars today as a repository of "famous responses to Kuhn."[58]

Despite Shapere's success at mounting a movement against the positivist regime in the philosophy of science, he failed to anticipate that his main vehicle, Kuhn, would turn out to be a Trojan horse. Because we still think of Feyerabend, Toulmin, Hanson, and Kuhn as of a piece, the views of the first three thinkers tend to be reduced to those of the fourth, the details

---

name typically appeared in philosophy journals as a corroborating footnote to a discussion of Feyerabend's challenges. It is important to remember that Feyerabend did not consolidate his own philosophy of science into a readily available text until the mid-1970s in Feyerabend 1975. Before then, Feyerabend was taken to be a radical philosopher of physics, whose views had interesting implications for the philosophy of science more generally. An apt comparison in our own day is with Arthur Fine. See Rouse 1991.

55. Lakatos and Musgrave 1970.

56. Thus, Lakatos's most famous and sustained postdissertation work appeared in this volume: Lakatos 1970. A relevant point here is that Popper was awarded his chair at the LSE in the Logic of the Social Sciences, based largely on such wartime works as *The Poverty of Historicism* and *The Open Society and Its Enemies*. Not surprisingly, Popper's main lobbyist for a permanent post at the LSE had been the liberal economist Friedrich von Hayek. The seminal work in methodology for which Popper is perhaps best known today, *The Logic of Scientific Discovery*, was a product of his younger days in Vienna, attracting little notice until it was translated into English in 1959. Thus, Popper's reputation as a general philosopher of science in the English-speaking world emerged as part of "the revolt against positivism," even though Carnap (again!) supported the original publication of Popper's book in the series associated with the positivist journal, *Erkenntnis*. Compounding the irony of the situation, in the debate that Lakatos staged between Popper and Kuhn, Popper was made to stand for positivism, despite his own lifelong insistence that it was he, not Kuhn, who "killed" logical positivism by "disproving" verification and induction as principles of scientific inference, two key planks of the positivist platform. Popper remained the stalking horse of positivism throughout the 1960s, as became even more evident in his debates with the Frankfurt School Marxists, which went by the name of the *Positivismusstreit*, about which more in note 60 below.

57. Thomas Kuhn, "Logic of Discovery or Psychology of Research?" in Lakatos and Musgrave 1970, 1–23.

58. It is also worth mentioning that the publication of *Criticism* coincided with the second edition of *Structure*, in which Kuhn answered some of his critics in a postscript.

of which are typically best known. However, this reduction obscures the criticisms that Kuhn's supposed comrades lodged against his views, which were more substantial than those that the positivists ever raised. A less pronounced, but similar, occlusion of differences has even occurred to the parallel band of antipositivists in Britain represented by Popper and his followers. They, in particular, detected an unsavory affinity between Kuhn and the positivists. Kuhn seemed to identify the essence of the scientific enterprise with puzzle solving, specifically filling in the gaps of a paradigm, a practice that suspiciously resembled the logical articulation and empirical confirmation of a theory. Where was the image of science as a *critical* enterprise, one—as Popper liked to say—continuous with Socratic questioning?

In Kuhn's scheme, the answer was to be found in the relatively rare instances of "revolutionary science," whose resolution depended entirely on which of the contesting sides eventually turns out to have the most followers. Thus, the most important moment in any inquiry—namely, determining how to proceed so as to best satisfy the ends of inquiry—was removed from the realm of rational deliberation to an apparently blind process akin to natural selection, the so-called *Planck effect*.[59] As we shall now see, the political imagery of revolution added an element of risk to this unpredictability, which subtly suggested that the questioning of ends should occur neither too often nor for too long.

5. *Structure*'s Great Metahistorical Scramble I: How the Enlightenment Never Happened

Let us start by dwelling on the significance that Kuhn's philosophical critics attached to the word "rational" and especially its counterpart, "irrational."

---

59. Max Planck, surveying his own career in his *Scientific Autobiography*, sadly remarked that "a new scientific truth does not triumph by convincing its opponents and making them see the light, but rather because its opponents eventually die, and a new generation grows up that is familiar with it": Kuhn 1970b, 151. The Planck effect has been subject to an empirical check, namely, the difference in age between British scientists who accepted and rejected Darwin's theory of evolution within ten years of the publication of *Origin of the Species*. The result was that only one aspect of the Planck effect held up, namely that scientists who continued to doubt Darwin after 1869 were on average significantly older than those who had been swayed. See Hull, Tessner, and Diamond 1978. However, as Hull, Tessner, and Diamond themselves admit, belief in the Planck effect seems to be a part of scientists' folk sociology of science. Still more importantly for our purposes, Kuhn regarded the Planck effect as a normatively acceptable feature of how science works. A secondary criticism of Hull, Tessner, and Diamond is that the ascendancy of Darwin's selectionist account of evolution was not meant to be covered by Kuhn's model of scientific change. On the curious absence of biology from Kuhn's examples, see chapter 3, note 79 above. Finally, for a comprehensive attempt to ground the Planck effect in the revolutionary tendencies of later-born members of a family, see Sulloway 1996, 20–54.

These words charged a great many academic debates in the 1960s, many of which involved Popper as the standard-bearer for "critical rationalism."[60] A common way of characterizing the significance of these debates is that they provided much of the early legitimation for the first field to declare itself the natural successor to Kuhn's project, namely the sociology of scientific knowledge (SSK), which is increasingly called by its philosophical label, *social constructivism*, now that it has acquired quite a broad interdisciplinary following across the humanities and social sciences. In the next chapter, we shall consider in more detail the way in which SSK and most of the STS community has interpreted its commitment to constructivism.

Of relevance here is that according to constructivism, rationalist philosophers had supposed that, without articulate knowledge of the norms of the scientific method, all order in science would dissolve and scientists would simply go about doing whatever they pleased. Nevertheless, so the sociologists argued, by closely attending to the laboratory practices of scientists, one can see that they get on quite well and regularly, even though they are typically unable to articulate the principles that govern their practices. What this seems to show, then, is that no obvious empirical place can be found in scientific practice for the methodological norms so cherished by philosophers. While philosophers should have met this line of argument easily, it is clear, in retrospect, that they did not.[61] As a result, a valuable opportunity was missed to explain the relevant sense in which the rational/ irrational distinction had to be drawn in order for the normative philosophical project to flourish. A quarter-century later, we find the following sentence in a presidential address to the Philosophy of Science Association: "Ironically, many philosophers of science have been attacking social constructivism in the sociology of science in the name of their own social construct: rationality."[62]

When one looks at the typical cases of "irrationality" that have worried Popperians and other philosophers so much, they are not associated with mass hysteria or other undisciplined forms of behavior. Rather, irrationality is attached to a certain complacency or inertia that arises from one's expectations being repeatedly confirmed or challenged only in ways that are easily accommodated or rebuffed. Rationality, then, is the turn of mind

---

60. The most famous was probably the German *Positivismusstreit* that pitted Popperian social theorists (including Ralf Dahrendorf and Hans Albert) against the Frankfurt School, especially Theodor Adorno and his understudy, Jürgen Habermas, who arose to prominence in the following decade for his skill in mediating and synthesizing alternative forms of rationality. See Adorno 1976.
61. A compilation of the generally acrimonious proceedings is Brown 1984.
62. Giere 1995, 12.

that resists habit, typically by criticism and actively confounding expectations.[63] Kuhn's apparent satisfaction with the preponderance of science being unreflective puzzle solving is the clearest indication to his philosophical critics that *Structure* presents an "irrationalist" picture of the scientific enterprise.[64]

One should not underestimate the extent to which this Enlightenment idea of "Reason" removing superstition and tradition remains a ready source of philosophical self-imagery even in our own day.[65] However, in the heat of philosophical debate, this eighteenth-century notion has been periodically confused with the conception of reason that came to the fore in the nineteenth century, especially with the rise of positivism. Here "reason" is a governing principle, one that regulates the growth of knowledge by directing and measuring the path of inquiry. Presupposed is that without such a governing principle, human energies would be dissipated into random motion. In that case, irrationality would indeed be akin to mass hysteria.[66]

At a deeper level, the differences between Enlightenment and positivist visions of reason reflect two nearly opposite notions of what the pursuit of knowledge is like. On the one hand, the Enlightenment vision is driven by

63. In fact, the distinguished Popperian anthropologist Ernest Gellner based his entire philosophy of history on the idea that because logical coherence is inversely related to social cohesion, any attempt to achieve a more logically comprehensive picture of reality has required challenging the normative integrity of one's own social order. See Gellner 1989. It should be clear that I concur with the spirit of Gellner's thesis, if not its exact letter.

64. See Popper's and Feyerabend's comments in *Criticism:* Popper 1970; Feyerabend 1970.

65. What has changed over the last two centuries is the metaphysical warrant for Enlightenment. There is less of an appeal to Reason as the essence of human nature (though it remains in Jürgen Habermas, today's leading Enlightenment thinker). For a nonessentialist version, see Fuller 1993b.

66. On the changing conceptions of reason, see Mandelbaum 1971. Kuhn is not alone in confusing the two senses of reason. Rorty 1979 refutes the notion of "foundationalism," which is tied directly to the nineteenth-century conception, but in part three (and his subsequent work) it becomes clear that he wants to bury the eighteenth-century conception as well. Rorty's role as an entry point for American understandings of recent French philosophical critiques of reason has compounded the confusion. Thus, while it is common for both friends and foes of "postmodernism" to lump Derrida 1976 and Lyotard 1983 as part of the same general antirationalist tendency, Derrida specifically aims to undermine the positivist conception of reason by engaging in an Enlightenment-like interrogation of that conception, whereas Lyotard indicts the Enlightenment conception of reason precisely for its totalizing sense of critique, which he sees as having destructive consequences for the diversified character of knowledge production in our day. I have elaborated the Enlightenment/positivist distinction in Fuller 1997d, chap. 3. When I invoke this distinction, the primary meaning of "positivism" is Comte's, but clearly logical positivism retained some elements of this original usage, as elaborated in Kolakowski 1972. Basically, the logical positivists retained the Comtean interest in providing foundations for knowledge, but by means clearing the rubbish that passed for knowledge in their day. I happen to believe that this Enlightenment impulse has been the better part of their legacy.

its ultimate end, an understanding of reality that is valid for all humanity. Criticism is thus like the forward momentum that enables a projectile to eliminate the air resistance along its path. On the other hand, the positivist vision is guided by an image of origins, specifically that of a building resting on secure foundations and climbing indefinitely into the heavens. Method functions here as a kind of logical scaffolding that gives structure to the epistemic edifice.[67] A good example of the conceptual havoc that is wreaked by a blurring of these two visions is the discussion that has surrounded Popper's notorious claim that the ability to "falsify" a hypothesis distinguishes scientific from pseudoscientific forms of inquiry: Does this claim imply that we come closer to the truth in the process of eliminating error (the Enlightenment spin), or simply that this process increases the credibility of the knowledge that remains after the false parts have been removed, thereby providing a firmer foundation for future growth (the positivist spin)? While I believe that Popper would rather firmly stick to the first interpretation because of his lifelong hostility to induction as a form of scientific inference, clarity on the matter has not been helped by his most famous encounter with Kuhn appearing in a book entitled "criticism and the growth of knowledge."

In terms of Enlightenment and positivist conceptions of reason sketched above, Kuhn's distinctly twentieth-century account can be seen as having worked in the following way to marginalize a strong sense of prescriptivism in the philosophy of science:

1. Through the Enlightenment spectacles with which the Popperians looked at Kuhn, Kuhn is an irrationalist because he valorizes the unreflective practices of normal science over the reflective ones of revolutionary science. Popperians see Kuhn as simply a traditionalist.

2. However, "normal science" is not quite the old idea of "tradition" in a new guise. Traditions aim to *preserve* the past in the present, but normal science exhibits *cumulative growth*. Traditional are the practices in which normal scientists engage; not traditional are the puzzle-solving domains to which those practices are applied. Here Kuhn trades on the positivist conception of reason, which predicates steady progress issuing from a firm epistemic base.[68]

3. In this way, reason is successfully contained, or relativized, to a paradigm, but inexpressible between paradigms. Thus, there is no place for a

---

67. The classic expression of the extended analogy between scientific inquiry and building upon a foundation appears in Carnap 1967, vii. A good historical account the architectural sensibilities that informed the logical positivist vision of philosophical foundations is Galison 1990.

68. The founders of the sociology of scientific knowledge have had the most acute understanding of Kuhn's complex debt to traditionalism and conservative ideology. See, e.g., Barnes 1994.

TABLE 6. *Scientific Rationality before the Great Kuhnian Scramble*

CONCEPTIONS OF RATIONALITY	IMAGE OF THE RATIONAL	IMAGE OF THE IRRATIONAL
Enlightenment	critique	tradition
positivist	method	disorder

critical form of rationality that is independent from, let alone oppositional to, the dominant epistemic practices of the time.[69]

This transition can be graphically represented as a "rotation of axes." Consider the two-dimensional representation of rationality's conceptual space in table 6. The difference between the logical positivists and the Popperians is that the former take the historical transition from the Enlightenment to the positivist conception of rationality as constituting "progress," whereas the latter regard it as reinstating a new tradition, this time in the name of science.

But after Kuhn, this disagreement has been sidelined by what amounts to a scrambling of historical allegiances. Specifically, Kuhn combined the tradition-method diagonal into the concept of normal science, and the critique-disorder diagonal into the concept of revolutionary science. The former is valorized and the latter demonized, as depicted in table 7.[70]

To illustrate the sort of rationality that drops out of Kuhn's picture of science, consider a set of analogies from the history of capitalist economics. Corresponding to the Enlightenment conception of reason is the entrepreneur, whose innovative spirit continually widens the sphere of production by breaking up cottage industries and guild mentalities. Corresponding to the positivist conception of reason is the manufacturer whose factory outproduces all competitors. Whereas the enemy of entrepreneurs — their vision of the irrational — is the suboptimally productive user of resources (say, one of Marx's *rentiers*), the enemy of manufacturers is the destroyer of resources (say, a Luddite attacking the machines in his factory).

The "Kuhnian" economist is essentially a friend to the manufacturer but an enemy to the entrepreneur. However, without a supply of entrepreneurs

---

69. Thus, while one can understand Sandra Harding's reliance on Kuhn to refute positivism, Kuhn's work itself provides little comfort to her own "standpoint epistemology," which implies a privileged critical perspective for groups marginalized from the dominant forms of power in society. See Harding 1986, 197–210.

70. In response to my presentation of these schemes at the 1996 American Philosophical Association, Eastern Division, meetings, Ofer Gal astutely observed that they may be applied to other developments in the history of modern philosophy. The work of Descartes can be understood as using "method" to oppose "tradition," and Kant as using "critique" to oppose the "disorder" of Humean skepticism.

TABLE 7. *Scientific Rationality after the Great Kuhnian Scramble*

PHASES OF SCIENCE	NORMAL SCIENCE (A.K.A. "RATIONAL")	REVOLUTIONARY SCIENCE (A.K.A. "IRRATIONAL")
research practice	method	critique
social backdrop	tradition	disorder

periodically upsetting existing trade patterns, engaging in "creative destruction" (in the inspired terms of Joseph Schumpeter, entrepreneurship's leading theorist), capitalism would itself soon become just another scheme for perpetuating inherited wealth, only this time wealth based on manufacture instead of agriculture. In that case, if one is not a member of a major industrial family, and emigration is not a possibility, then violent overthrow of the system would seem to be the only means available for the disenfranchised to acquire wealth.

This is where a Kuhnian economics would leave us — a place not so far from where Schumpeter thought capitalism was already heading.[71] However, Schumpeter had issued a proviso, one that may also be applicable to the fate of science in Kuhn's account. Schumpeter believed that violent overthrow was no longer a real possibility, given the risk to the relatively high standard of living that even the disenfranchised (supposedly) enjoyed in mature capitalist economies.[72] Rather, entrepreneurship would itself probably be routinized, as innovation becomes one more thing regulated by a state planning board. In chapter 5, we saw Derek de Solla Price tell a similar story about the diminished prospects for scientific revolutions in the future, given the heavy personnel and material investments that have already been made in current forms of Big Science.[73] In the next section, similar sentiments will be expressed by Paul Feyerabend. The question that needs to be asked of these gloomy forecasts is whether they are anything more than an artifact of an account of science that fails to recognize a place for rational criticism *outside* the existing epistemic power structure.[74]

---

71. The locus classicus of this view is Schumpeter 1950.
72. The difference between Marx and Schumpeter on this point can be traced to Marx's failure to anticipate the rise of national insurance and the welfare state as a buffer to the blows that capitalism's business cycles dealt to the average worker. Starting with Bismarck, the leaders of industrial countries realized that the taxes levied on the rich to provide universal social security were a small price to pay to keep the poorer citizens loyal to the nation even during financial crises.
73. For a sophisticated account justifying the slowdown of science, influenced by Kuhn and Price, see Rescher 1984.
74. One brave attempt to lay down an external rational critique of Big Science is Redner 1987, 252–53 (a severe critique of Kuhn). An emerging model for rationally criticizing Big Science from outside the epistemic power structure is the post–Cold War "downsizing" and "conversion" of the U.S. military establishment, which is currently (albeit painfully) being

6. *Structure*'s Great Metahistorical Scramble II: How We All Became Accountants

However, Popper was by no means the sole victim of reason's "double vision." Consider Lakatos's famous charge that Kuhn offered a "mob psychology" account of scientific revolutions. In response, calmer heads have remarked that Lakatos overstated his case, as Kuhn never really said that scientists take leave of their senses once shaken from the spell of a paradigm.[75] However, these critics presume that Lakatos's image of mob psychology was that of *mass hysteria*, when it was probably closer to that of the *herd mentality*.[76] At least, that would get at the nub of Lakatos's complaint, which can be understood only once we notice that his accusation was originally made in the context of an attack on the *positivists* (whom Lakatos calls "justificationists"). For a point that united Kuhn and the positivists was their unwillingness to associate radical criticism, or a choice between alternative normative systems, with anything they would call "rational," at least in the sense that they applied this term to science. For them, rationality is to be found not in *making (or breaking) a rule*, but merely in *following a rule*. Accordingly, the philosophical task of justification is limited to why a case does or does not fall under a rule rather than why the rule is the one under which the case either falls or not.

One charitable reason for the adherence to such a rigid distinction is the division of labor between, on the one hand, the philosophy of science and epistemology and, on the other, ethics and social and political theory. Such a distinction seems to have informed the logical positivist tendency

---

negotiated by career soldiers and civilian bureaucrats. Here, too, one finds the rhetoric of "great risk" that supposedly attends any attempt to radically alter the institution's scale and scope.

75. Lakatos 1970, 178. For some reassurance that Lakatos could not possibly be right about Kuhn, see Gutting, introduction to Gutting 1979, 7–8.

76. Moreover, on closer inspection, the "herd mentality" interpretation of mob psychology acquires support from the social psychology that informs Kuhn's most celebrated mythical progenitor. See Fleck 1979, 179–80, for an appreciative discussion of the father of mass psychology, Gustave Le Bon (1840–1930). Partly inspired by Le Bon, Fleck's concept of the "thought-collective" was a source for the disciplinary matrix sense of a Kuhnian "paradigm." Fleck criticized Le Bon for regarding the "collective hallucination" that underlies the mass mind as always a transient phenomenon, rather than one that can persevere under certain conditions (as would be needed for the continuity of a scientific community). Interestingly, here one finds a discussion of the gestalt switch as a model of paradigm change—specifically, how an uprooted tree looked like a disabled boat to a search team in the high seas. Given the Kuhn–Fleck–Le Bon genealogy, it is worth noting that Lakatos's sensitivity to the herd mentality may have come from his encounter with the ideas of "false" and "oppositional" consciousness, as an assistant to György Lukács. On the Lakatos-Lukács connection, see Dusek 1998. On Lukács's role in the development of critical social science, see chapter 5, section 2, above.

to separate questions "internal" and "external" to a framework. However, in practice, the distinction has served only to push questions outside a given framework—the very making or breaking of rules—from normative philosophical consideration altogether, perhaps to be exiled to a more purely descriptive discipline such as sociology or anthropology.[77] Lakatos understood well what was at stake here. The difference between making and following a rule corresponds to the *prescriptive* and the *evaluative* sides of the normative enterprise.[78] The distinction, which corresponds to the roles of legislator and judge in a legal system, was famously invested with substantial philosophical import by Kant, who in *Critique of Pure Reason* argued that our capacity to deal with reality is structured by a distinction that appears, in its simplest form, in the syllogism: namely, the difference between "All X is Y" and "This is an X." In logical terms, this is nothing more than the difference between the major and minor premise, or in Kantian terms, the *regulative* and *constitutive* uses of reason.[79]

77. For the internal/external questions distinction, see Carnap 1958, 205–6. The professionalization of the logical positivist movement in analytic philosophy can be traced in terms of the gradual withdrawal from prescriptive ("external") to evaluative ("internal") modes of normative discourse. This retreat in normative ambitions is marked in Carnap's own transition from the *Logical Structure of the World* (1929) to *Meaning and Necessity* (1950), Wittgenstein's transition from the *Tractatus Logico-Philosophicus* (1922) to the *Philosophical Investigations* (1951), the more general shift from artificial to natural languages as the basis of philosophical analysis, and even the shift in conceptions of formalism in the law from, say, Hans Kelsen to H. L. A. Hart.

78. See Lakatos 1981, 108, esp. n. 2. I rationally reconstruct Lakatos, who claims to be evaluating, not prescribing, but according to norms that, like Hegel's, only become apparent over the course of history, so our later judgments (about consequences) influence what should have been done—in that sense, against the idea that one can make "instant" judgments on the rationality of a historical episode. In this respect, Lakatos differed from the positivists and Popper, though the latter would allow for revision upon immediate feedback. But he also differed from Kuhn and SSK, who stuck exclusively to writing a natural history of science. Lakatos invoked Whewell as his historiographical model, but Whewell depended on the actual history of science to legitimate contemporary science, whereas Lakatos had the luxury of saying things could have happened more quickly (i.e., efficiently). See Hacking 1979a; Larvor 1998.

79. Kant draws the distinction most clearly in *Critique of Pure Reason* (1781), A644–47/B672–75. Kant associates the regulative use of reason with a commitment to an imaginary end that may be reached by several different means, and the constitutive use of reason with one of those means that is valued for its own sake. The modern locus classicus of the distinction is Rawls 1955. An interesting discussion of objections to separating these two sides of the normative enterprise (i.e., arguments against "mixed criteria" of moral judgment) made from a utilitarian standpoint is Johnson 1985, esp. 396 ff., which includes a critique of Toulmin. Among historicist philosophers of science, Toulmin has been the most astute in locating the origins of prescriptivist approaches to philosophy in legislation. He sees the task of setting up ideals and standards for comparing the claims of incommensurable theories as modeled on the lawmaking needed for a jurisdiction that encompasses communities with quite disparate local customs and entitlements: What Socrates did in theory, Solon had already done in practice. See Toulmin 1972, 86–87. For an extension of Toulmin's insight to include the social sciences as further theorizations of the legislative impulse, see Kelley 1990, 5–6. In our kinder,

Kant made much of the fact that the major and minor premises are justified in rather different ways. In today's terms, the major premise is freely chosen and justified pragmatically by the good that follows from its acceptance as a basis for thought and action. As a Kantian regulative principle, it defines the limits on our sense of reality until further notice, by specifying the fundamental terms of the world we presume ourselves to inhabit. If we are repeatedly disappointed while operating on the basis of this premise, then clearly it is time to adopt a new regulative principle.

However, packed into "repeatedly" is the assumption that the premise had been applied consistently to a significant number of cases, regardless of the consequences in each individual case. Such a consistent application of prior principles involves the constitutive use of reason. Constitutive principles give a rule-based system its fixed character, which has been described in a variety of ways that highlight the principles' rigidity and autonomy: "Gamelike," "a priori," and "innate" are three words that come to mind here. Whereas legislators typically design rules with an eye toward the future, judges are mainly concerned with applying rules in accordance with past practice. The intuition behind this sharp distinction between regulative and constitutive principles is clear enough, and we saw it in the previous section with C. I. Lewis's stress on the invariance of our conceptual scheme at any given moment as a precondition to its changeability in time. It is difficult to know whether a rule has efficacy (the legislator's goal) unless it has been given a trial run under relevantly similar conditions (the judge's goal).[80]

In contemporary analytic philosophy, the difference between the maker and the applier of laws — the legislator and the judge — tends to be seen in purely ethical terms: a steadfast judge places a systematic check on the legislator's utilitarian impulses, thereby enabling the emergence of a more edifying legal system: *rule utilitarianism*. However, if we return to Kant's original Enlightenment context, the first clear realization of the legislator/judge distinction turns out to be the U.S. Constitution of 1787, heralded in its day as the first instance of "philosophically designed" order. Since

---

gentler postmodern times, the term "interpreter" is typically used instead of "judge," though the role remains the same, namely, the application of a given framework to a case. For a similar transformation in sociology, see Bauman 1987. For a history of the judicially based sense of "interpretation," see Gadamer 1975.

80. Not surprisingly, the main objection to adjudicative consistency is its conflict with the directives of utility. See Johnson 1985, 397 nn. 7, 8. However, there are really two problems of consistency here: (a) the consistent application of the rule chosen; (b) the consistent application of the utilitarian principle. I believe that the latter poses the larger problem, since the tendency to respond immediately to the application of a rule ("act utilitarianism") would involve discounting the rule's more remote consequences. On the normative ambivalence surrounding the discounting of remote consequences, see Fuller 1997d, 95–101.

philosophy had not yet metastasized into the mutually incomprehending discourses of "ethics" and "epistemology," the Founding Fathers had no qualms about turning to Newton's popularization of the scientific method for a model of government.[81] If scientific hypotheses or civil codes were revised each time the consequences of their application went against their framers' intentions, there would never be a sufficient track record to determine the exact source of error. The corrections would always seem ad hoc and erratic, and their import endlessly ambiguous.

In this respect, a legal system that separates the powers of legislation from those of adjudication—in American terms, Congress from the Supreme Court—conceptualizes society as a laboratory in which legislators function as theorists who propose policies based on a hypothesized understanding of social action, whereas judges are the experimentalists who determine the environment in which those policies are given a trial run. In their ideologically opposed ways, the most distinctive schools of jurisprudence of the last hundred years—legal formalism and legal realism—are simply alternative articulations of the hidden scientific roots of modern constitutionalism, the former stressing the consistency of the judge needed for a fair test of the laws and the latter the foresight of the legislator needed to propose laws that are likely to increase society's well-being.[82] The differ-

---

81. See Cohen 1995.

82. On legal formalism and legal realism, see White 1957, esp. 59 ff. The difference between the logics of legislation and adjudication would be self-evident, were it not that the natural history of humanity suggests that normative regimes arise without any clear distinction between legislative and adjudicative functions. Rather, spontaneously generated patterns of behavior, typically resulting from a combination of contingent factors, are reinforced as norms once they have issued in sufficiently beneficial consequences to a sufficient proportion of the affected parties. The open question, then, is whether (1) the extension and continuation of those norms should depend exclusively on whether those affected by their implementation receive benefits comparable to those received regularly in the past, or (2) some other standard should apply, especially one that would allow for individual inconvenience in the short run, on the promise of long-term improvement of the collective situation. Let us consider each alternative in turn.

The first alternative is common to the literature on society arising by convention associated with the eighteenth-century Scottish Enlightenment (Adam Smith, David Hume, Adam Ferguson), which culminated in the twentieth century with the work of Friedrich von Hayek. The problem with this "bottom-up" approach as a complete account of norms is that it presupposes that just because norms emerge as by-products or unintended consequences of other activities, it follows that they can be maintained forever that way. This fallacy underwrites free-market capitalism, which says that markets emerge spontaneously (which may be true enough) and then concludes that the market must remain spontaneously organized forever, regardless of its consequences. The source of the fallacy is the failure to recognize that history can change the desirability of a norm, especially once people realize that they have been adhering to it all along and thereby start to strategically work around it in order to secure an advantage with respect to their fellows. In stylized form, it provides an explanation for how science could "evolve" into a "postepistemic" activity, once knowledge is seen as something

TABLE 8. *The Prescriptive and Evaluative Sides of the Normative Enterprise*

	NORMATIVE ORIENTATION	
	PRESCRIPTION	EVALUATION
defining virtue	efficacy	consistency
temporal orientation	forward-looking	backward-looking
type of activity	making (breaking) a rule	following a rule
type of rule (logical moment)	regulative (all X is Y)	constitutive (this is an X)
legal function (school of jurisprudence)	legislation (legal realism)	adjudication (legal formalism)
ethical orientation	utilitarianism	Kantianism
scientific orientation	revolutionary science	normal science
scientific function	explanation	confirmation

ence between the legislative and adjudicative perspectives is captured in table 8.

Generally speaking, as philosophy's command over the course of inquiry has receded in the face of academic specialization, the field's normative posture has shifted from prescription to evaluation, from legislator to judge—or still more modestly, to accountant, a keeper of someone else's books.[83] In that sense, the prescriptive side of the field has become trivial, since regardless of the philosophical faith one professes—be it realist or

---

that can be pursued as a means to political and private economic ends. This prospect was very vivid to the logical positivists. For more on this line of reasoning, see Fuller 1996e.

The second alternative, of course, presupposes that a boundary can be drawn around the affected parties to constitute a community. In modern economic theory, this line of thought culminates in the concept of "collective" or "public" good, an exemplar of which is supposed to be scientific knowledge. See Olson 1965. Not surprisingly, those who have favored this option such as the founders of the two leading normative metatheoretic frameworks of the modern era—Kant and Bentham—advanced their proposals in the context of strong nation-states where overriding concerns for social order (especially in light of the French Revolution) ensured that radical legislative impulses would be tempered by more cautious adjudicative ones. See Fuller 1998b.

83. The adeptness of philosophers of science in keeping the books of the special sciences has increasingly enabled them to function as the ultimate accountants, namely, journal referees. Sometimes they have even adjudicated significant disputes within a given specialty. The intellectual godfather of this approach is David Hull (b. 1935), a philosopher of biology who served in the 1970s as an editor of the journal *Systematic Zoology* and eventually became the president of the Society for Systematic Zoology. The philosophical lessons of that experience are drawn together in Hull 1988.

instrumentalist—the same cases are evaluated the same way. Thus, every science accountant balances the books in favor of Copernicus over Ptolemy by 1620, Newton over Aristotle by 1720, etc.[84] Those books ultimately belong to the special sciences, which now prescribe for underlaboring philosophers, not unlike Plato's topsy-turvy world of debased democratic regimes, in which elders defer to their juniors.

The lost philosophical image is that of a science accountant who is literally accountable, that is, one who takes responsibility for the norms by which specific scientific projects are evaluated. In other words, one would evaluate only if, in some sense, one participated in prescribing the norms in the first place.[85] In terms of instrumental rationality: you cannot say something is a more efficient means to an end unless you have had something to do with selecting the ends. Social contract theory is the model from politics of what it means to build prescription into evaluation. For example, in chapter 2, Max Planck and Ernst Mach were mainly arguing about the rules of the science game (prescriptive), which, at the same time, held implications for what were better and worse specific projects (evaluative).

Subject to a febrile intellect and an early death, Lakatos never managed to stem this tide. However, Lakatos's interest in resurrecting prescriptivism may ultimately explain his distinctive "rational reconstructionist" doctrines concerning the history of science, whereby the philosophical historian is enjoined to say how the history ought to have gone, had the historical agents abided by the philosopher's favorite rules. It would seem that if philosophers can no longer prescribe the course of knowledge production, the next best thing may be to fantasize how much better it would have been had they been allowed to do so! It is a mark of political correctness in our Kuhnified times that even sympathetic readers of Lakatos feel compelled to apologize for this feature of his thought.[86]

The idea that Kuhn helped *diminish* the normative dimension of the philosophy of science seems prima facie counterintuitive, since one usually says that Kuhn undermined the observation/theory, fact/value, empirical/normative, descriptive/prescriptive distinctions in sociohistorical accounts

---

84. One creative attempt to put a brave face on this evaluative uniformity is the concept of "preanalytic intuitions," which are the proper objects of inquiry for normative philosophy of science. See Laudan 1977, 160 ff. This strategy is more generally deployed to define the objects of analytic-philosophical inquiry in Cohen 1986.

85. I develop this more robust sense of science accounting in Fuller 1994g.

86. A striking case of apologetics is Hacking 1979a. Those interested in resurrecting a rational reconstructionist approach to scientific inquiry could do worse than to revisit the *Methodenstreit* fought between analytical and historical—in today's terms, "neoclassical" and "institutional"—economists in the German-speaking world, circa 1900, over the ontological status of *homo oeconomicus*. See Fuller 1993a, 180–86.

of science. However else one interprets the second term in each binary, Kuhn (on behalf of a litany of historicists) is said to have shown that it "loads" or "impregnates" the first term. Does not such imagery suggest the *ascent* of the normative and the theoretical?[87] Not necessarily. The metaphors of "loading" and "impregnating" imply a certain taken-for-granted character to the norm or theory in question. Yet, traditionally theorizing has been more than simply observing the world from within a given theoretical framework. Nevertheless, by shifting the task of theorizing so decisively toward accommodating phenomena within an existing framework, Kuhn may be seen as having ushered in a wave of *Theory Lite* scholarship. The special relevance of this point to science and technology studies will be critically explored in chapter 7, section 3C.

But we can also dig a bit deeper. Long before they spawned opponents, the positivists had themselves already discovered that observations are theory impregnated (in the sense that observational content is deducible from theoretical premises in a formal scientific language) and that facts are value laden (at least in the elliptical sense that the theoretical language one uses for inquiry involves a free value choice).[88] Moreover, the positivists and the antipositivists agreed on a generally instrumental view of scientific language, one that portrayed theories as rules for governing the conduct of inquiry. The key difference between the positivists and their opponents lay in the attitude they took toward these common elements. Whereas the positivists believed that theory- or value-ladenness threatened the objectivity of inquiry, unless explicitly circumscribed by an account of "testability," antipositivists tended to treat the uncertain objectivity of inquiry as a brute fact that could only be mitigated but never completely eliminated.

For his part, Kuhn masks this uncertainty by taking responsibility for the choice between scientific languages out of the hands of self-conscious deliberators and placing it in the "invisible hand" of the Planck effect. Thus, if you outlive — or, more precisely, outreproduce — your critics, you have earned the right to forget what may have been an epistemically sordid past, one full of conceptual indiscretions committed in the name of empirical expedience. Through the Planck effect, Kuhn effectively reintroduces a

---

87. This unduly charitable interpretation of Kuhn is given in Will 1988, an otherwise subtle treatment of the entire issue of normativity. Will treats normal and revolutionary science as phases enjoying equal normative status in Kuhn's scheme, when in fact Kuhn licenses revolutionary science only once normal science has failed on its own terms. In that sense, the "normative," as I use the term here, remains generally suppressed in Kuhn's scheme as being no different from the "natural" course of action.

88. What *are* "value free" and "theory neutral" are the possible observations that deductively follow from the chosen theoretical language, and their testability by a method whose validity does not depend on whether the tested theory is true or false.

sharp distinction between the "empirical" and "normative" questions of science in terms of a difference between the determinable present (normal science) and an indeterminate future (revolutionary science). Kuhn would seem to grant a presumption of validity to a research program that has already attracted followers. Such a move has made it that much harder for philosophers like Lakatos to motivate, let alone execute, a theory of "rationality" suitable for prescribing the course of inquiry.

In reducing the voice of reason to that of a scientific insider, *Structure* managed to satisfy an aspiration left unfulfilled by the original logical positivists. The key was that Kuhn built a certain limited kind of change into the very logic of science by reverting to a crypto-Aristotelian contrast between "natural" and "violent" motion, as captured by the normal/revolutionary science distinction. This point informs the necessarily Orwellian role that he assigned to the rewriting of history, one that erases any trace of a revolution from science's institutional memory.[89] Because Kuhn was convinced that science flourishes only in its sheltered normal phases, he wanted to dampen any long-term effects that these considerations might have, which explains his keenness to prevent scientists who work in the next paradigm from learning about their traumatic political origins.

As Lakatos perhaps realized most keenly, Kuhn feared not merely that scientists might start to blur the distinction between science and politics (say, by allowing larger social interests to direct their research agendas), but more importantly that they might come to realize that their field's survival

---

89. Kuhn's Harvard source for both his Aristotelian perspective on scientific change and its Orwellian consequences was Whitehead 1926, a popularization of process metaphysics that enabled Whitehead to portray the twentieth-century revolutions in physics as a neo-Aristotelian revival. Whitehead argued, "Nothing does more harm in unnerving men from their duties in the present than the attention devoted to the points of excellence in the past as compared with the average failure of the present day" (255). When Kuhn invoked Whitehead's famous quote—"a science that hesitates to forget its founders is lost"—he clearly interpreted it in the Orwellian sense to mean that science cannot progress unless its practitioners *take for granted* that it rests on secure epistemic foundations and hence feel no need to reflect on the achievements of the field's founders; "dogma" in that sense. See Kuhn 1963, 350. However, taking account of the original context of utterance, Whitehead may be given a more "revolutionary" reading: namely, that scientists should not invest so much significance in their predecessors that they end up inhibiting new lines of inquiry and ignoring the uniqueness of present-day concerns; in short, they should be willing to "forget" the past in the sense of *discarding* it when necessary. This difference between the conservative and radical readings of Whitehead is captured by the opposing senses of "burden" implied in the metaphors *burden of proof* and *burden of the past*. The former implies the ease and the latter the difficulty with which the past can be carried into the future. Whitehead's quote first appeared in the 1916 presidential address to section A of the British Association for the Advancement of Science. He was explaining how the Renaissance retarded the growth of science over the High Middle Ages by multiplying the number of classical authorities invoked and diminishing the critical role of logic in advancing beyond these authorities. See Whitehead 1949, 107. My thanks to Elihu Gerson and Val Dusek for helping me track down the original context.

has depended on periodic purges of intractable domains of inquiry that may be later picked up by another field or disappear from systematic treatment altogether. Contrary to the drift of Kuhn commentary, Kuhn did not object to the rational reconstruction of history — the story of how science *should* have happened — that philosophers like Lakatos have promoted. After all, this is the stuff of which Orwellian history is made. What is objectionable, however, is for scientists to be told (as philosophers are wont to do) that such reconstructions deviate from the actual historical record, as that could inject a measure of critical self-consciousness that could jeopardize normal science's puzzle-solving process. Thus, Kuhn would clearly give an affirmative answer to the provocative question posed by historian Stephen Brush: Should the history of science be rated "X" for scientists?

In retrospect, it is remarkable just how much of recent intellectual history has had to be rewritten to portray Kuhn as a radical theorist of science. Often great eloquence has been deployed:

> Popper was a cultural warrior in the grand manner, still fighting the battles of the Enlightenment and of the 1930s and 1940s against totalitarianism and religious obscurantism. Popper's annoyed reaction to Kuhn derived in part, I suspect, from his political engagements, from his authentically world-historical vision, and from his sense that reason is still a cultural force ranged against a series of historic enemies that the apparently naive Kuhn did not recognize. Popper spoke for a generation that saw the world almost lost to a regime that distinguished proletarian science from bourgeois science. If it is not too far fetched, one can imagine the worldly-wise Popper playing Cardinal Bellarmine to Kuhn's Galileo, conscious of how complex and fragile is a whole system of thought and practice built up through years of struggle, and trying to explain it to a youngish fellow who just wants to speak the "truth," oblivious to what in today's idiom would be called "the social responsibility of intellectuals."[90]

The eloquence of this passage notwithstanding, there is simply no evidence that Kuhn ever carried himself in Galilean defiance against the scientific orthodoxy of his day. If anything, Popper and his students have been known for the fiery wit that made Galileo a symbol of antiauthoritarianism, whereas Kuhn's intellectual demeanor is closer to Bellarmine's jesuitical circumspection. In any case, when Kuhn was first exposed to the Popperians, it was his fondness for the imagery of religious conversion that raised the most Popperian hackles, as it seemed to imply that scientific revolutions could not be resolved rationally *precisely because* they involved life decisions of the utmost importance. Ironically, the Popperians actually agreed with the positivists and Kuhn that decisions concerning the "absolute pre-

90. Hollinger 1995, 452.

suppositions" of one's practice are matters of free existential choice.[91] However, for that very reason, Popperians have held that the merits of such a choice must be evaluated in terms of its consequences (the extent of which is, unfortunately, rarely specified in practice). Rational beings act so as to be constrained only by the anticipation that they may want to reverse that decision at a later date, if the consequences of their actions do not turn out as planned.[92] An ethic of "piecemeal engineering" (Popper's phrase) naturally follows from this imperative. Therefore, an irreversible course of action is, in principle, irrational. From a policy standpoint, one would "rationalize" a process by making it more responsive to negative feedback.

Extending this standpoint to its most provocative application, Feyerabend would later declare that contemporary Big Science is irrational, as science's increasing dependency on large-scale military-industrial projects rendered any reversal of its current research trajectories politically and sociologically impossible: politically, because of the threat reversal posed to national security; sociologically, because of its threat to the skills and mindsets of the scientific community itself.[93]

7. The Result: Revolutions Rendered Both Invisible and Irrational

Although there is no evidence that Kuhn was *trying* to foist an irrationalist image upon radical criticism, *Structure* nevertheless does seem to have had this effect on its readers, most noticeably on those who have wanted to continue pursuing the critical project. A case in point is Paul Feyerabend, who openly converted to irrationalism in the 1970s, reveling in the "anarchy" that would allegedly be produced in a world without a uniform scientific method. In a sense, Feyerabend merely acquiesced to Kuhn's vision of what Popper's call for a "permanent revolution" in science would look like — a vision that was, to be sure, shared by the positivists, who always believed that Popper was courting cognitive chaos by eschewing definitions in philosophy and advising that scientists suspect all their knowledge claims

---

91. Neo-Kantianism, perhaps as filtered by Max Weber, was the source of this sentiment. See Proctor 1991, 151–54.

92. Popperian "conventionalism" about scientific facts (i.e., they are free choices joined to rigorously assessed consequences) can be seen as the philosophy of science equivalent of the Austrian economics account of the utility of rule of law in liberal societies. (See chapter 4, note 108.)

93. Feyerabend 1979. Ironically, in the same year that Kuhn's book was published, a work by a student of Popper's was published, initially to more fanfare, which argued against the irrationality implicit in the refusal of liberal Protestant theologians such as Paul Tillich to contest their most fundamental assumptions. See Bartley 1984.

as presumptions whose erroneous bases have yet to be discovered.[94] The difference between Popper and Feyerabend is, of course, Kuhn. Whereas Popper saw himself as subjecting the ends and means of inquiry to the same critical standards as the products of inquiry, Feyerabend takes Popper to be offering a politically naive strategy for wresting the course of inquiry from the Big Science power mongers. As Feyerabend might put it, the problem is not that the power mongers won't listen to reason, but that they listen all too well! Notice here that reason is presumed to be always "internal" to science. The only politically appropriate responses to Big Science, then, are forced entries, withdrawals of support, and displays of irreverence — all, in one way or other, violent departures from reason.[95]

But why must radical change occur *radically*? In other words, why is

---

94. Several philosophers have noticed that Feyerabend presupposes the positivists' rather limited conception of rationality so that he appears more outrageous than he really is. See Naess 1991. It is worth recalling that, of all the Popperians, Feyerabend was the only one whose early reputation was built as a technical critic of positivism, which suggests that he may have ingested more positivist dogma than he and his readers have realized. The positivists' own risk-averse attitude toward criticism comes out very clearly in recently unearthed correspondence between Carnap and Popper during the 1940s, shortly after Popper had finished *The Open Society and Its Enemies* and was seeking academic asylum in the United States (where Carnap had resettled during the Second World War; Popper was then in New Zealand). When Popper told Carnap that an essentialized distinction between capitalism and socialism obstructed the selective appropriation of what was good in both economic traditions, Carnap cautioned Popper not to be too critical of socialism, lest he offer comfort to its opponents. Carnap was specifically concerned that Popper was seeking refuge at the London School of Economics via Friedrich von Hayek, a well-known antisocialist. As it turns out, Carnap's letter of recommendation arrived too late to be of any use, whereas Hayek enabled Popper's passage to the LSE. My thanks to Mark Notturno for having uncovered this material, which he first presented at the Russian Academy of Sciences and was published in *Voprosy Filosofii* (1995) under the title (in English) of "Popper's Critique of Scientific Socialism, Or Carnap and His Coworkers," to accompany the Russian translation of *The Open Society and Its Enemies*.

From a sociology of knowledge standpoint, it would seem that Popper and Carnap occupied the positions of Galileo and Bellarmine, respectively. As the free-floating intellectual who was not beholden to any establishment, Popper enjoyed the luxury of denying what Carnap, as someone who epitomized professionalized philosophy in the United States, had to uphold, namely the image of inquiry as mutually oriented labor with collectively sanctioned products. It would be nice to think that one can sustain Popper's attitude even after achieving institutional clout comparable to Carnap's, but Popper's own example in Britain does not inspire optimism.

95. Dominique Lecourt, philosophical advocate of "proletarian science" during the French student uprising of 1968, has recently bemoaned the fact that inquiries into the ends of science have become monopolized by critics strongly influenced by the right-wing irrationalism of Oswald Spengler and Martin Heidegger. Even when the critics have had left-wing pretensions, as in the case of the Green movement, they suppose that science can be opposed only by engaging in practices *contrary* to those of science. Feyerabend certainly contributed to this sentiment in Europe, which goes against the self-critical rationalism of both Popperians and Marxists, despite their disagreement on so many other matters. However, as Lecourt rightly observes, these rationalists have taken today's "technoscience" as having greater facticity than "nature" itself. See Lecourt 1992.

reason specifically precluded from being a vehicle by which a radical departure can be made in the course of one's inquiries? When Popper remarked that the wonderful thing about humanity's place in evolution is that our theories could die in our stead, he was alluding precisely to this missing sensibility, that risking an idea is not the same as risking a life. To portray scientific revolutions in apocalyptic terms — as Feyerabend does for fun and defenders of science do in fear — is to overlook the potentially *invisible* character of such revolutions. My choice of terms here is quite deliberate, as "The Invisibility of Revolutions" is the title of chapter eleven of *Structure*. But Kuhn takes the relevant sense of "invisibility" to involve the following claims:

1. that the revolutionary character of paradigm change is obscured by subsequent textbook writers who make the transition appear continuous; and

2. that the outcome of a revolution is determined not by clashing parties coming to agreement, but by the research choices subsequently made by their students.

I considered claim 1 in the previous section, so let me now focus on claim 2, the Planck effect, which implies that argumentation over the scientific research agenda benefits the spectators more than the participants.[96] In one sense, Kuhn's point is undeniable, insofar as the champions of one research agenda rarely take themselves to be swayed by the arguments of their opponents. However, to leave the matter at that is to ignore the ways in which partisan positions shift, often unintentionally and imperceptibly, in the course of debate, as the stakes and implications of acceding to one argument over another appear in different contexts. A position that one would never have adopted at the start of a dispute may become easier to accept later, in large part because *the very practice of arguing* will have made one accustomed to the other's position. Moreover, the person may not believe that she has conceded anything "essential" to her position along the way. Only in retrospect can a historian detect that a subtle shift in the burden of proof took place that enabled the acceptance of a previously intolerable point of view.[97]

Thus, radical change can occur quite unradically, indeed, invisibly. However, conspicuously absent from Kuhn's account is any discussion of how *argumentation* may facilitate this transition. The absence is conspicuous because the other historicists in Shapere's original cohort — Hanson, Toulmin, and Feyerabend — explicitly made argumentation central to their

---

96. I explain the expression "Planck effect" in note 59 above.
97. I develop this point in Fuller 1988, 99–116, and Fuller 1993a, 101–6.

understanding of how scientific change occurs.[98] However much they differed over this process, they made a point of distancing what they meant by "argument" from formal logical deduction, in order to stress that the merits of alternative research agendas, or theories, are not already implicit in their ideal formulations, just waiting to be deduced; rather the theories' respective merits do not become evident until they are articulated publicly in terms of one another at a concrete decision point. Once again, like Lakatos's account of science, the prescriptive/evaluative distinction in normative functions figures crucially. Unlike the positivists, who portrayed scientific theories as generating predictions that are then tested against observations, both Toulmin and Hanson described scientists more like judges who look for the rule (theory) under which the test case at hand (observation) can be best subsumed. This reflected a shift in philosophical thinking from confirmation to explanation as the main business of science. If scientists are like judges, then philosophers of science, presumably, are like the legislators who construct (or discover?) the rules by which the contesting interpretations of the case can be resolved.[99]

98. Credit for first realizing this point about Kuhn goes to Scheffler 1967. Israel Scheffler was typical of philosophers who believed that only the sort of externally imposed methodological rules favored by the positivists would ensure that scientists' reasoning would conform to some common standard. He did not share Kuhn's faith that scientists would spontaneously reason according to a similar standard. In fact, it seemed (as Kuhn himself came to realize) that Kuhn inferred a rationality implicitly common to all scientists from the fact that science manages to reach closure after having passed its crisis and revolutionary phases. Scheffler correctly drew attention to this tension in Kuhn's thought: If incommensurability precludes scientists from agreeing on a standard of reasoning during a revolutionary phase, then whatever convergence or uniformity of reasoning is reached after a revolution has passed will bear no clear relation to the various forms of reasoning that the scientists brought to bear on the situation. In other words, Kuhn merely sidestepped the problem of translating incommensurable perspectives without biting the bullet and granting that the transition is, strictly speaking, irrational. Thus, Kuhn failed to show, in Hegelian terms, that Reason is ultimately cunning. Indeed, Reason looks rather impotent in Kuhn's hands. If nothing else, the ascendancy of Kuhn's perspective highlighted, by implicit contrast, a presupposition common to the historicists, the Popperians, and the positivists: namely, that normatively acceptable accounts of scientific rationality are *not* always already present in the scientific enterprise.

99. Hanson very explicitly declared his interest in reviving the distinctly philosophical art of "theory finding" over the day-to-day scientific task of "theory using." From the notes to *Patterns of Discovery*, it is clear that Hanson was inspired by Goethe's efforts to think his way beyond the Newtonian worldview and Charles Sanders Peirce's pursuit of "abduction," the elusive cognitive process that knits together abstract principles and concrete cases via claims to relevance, while he regarded the positivists as simply reproducing the scientist's natural attitude at a more abstract level. See Hanson 1958, 2–3. Both Hanson and Toulmin followed a strategy, familiar in constitutional law, of eliciting legislative principles from the conflict that arises when judges decide "hard cases," that is, cases that are arguably subsumable under any number of incompatible laws. In the case of the motion of elementary particles, Hanson argued casuistically that if the principles he elicited apply to this hard case, then a fortiori they apply to the more routine cases of scientific discovery (which, for Hanson, meant discovering

In contrast, Kuhn argued that scientists are not rationally justified in switching paradigms until *both* the unsolved problems of the old paradigm become unbearable and the basis for a new paradigm is in clear view. Instead of the image of arguing one's way to the next paradigm, we are presented with the collective inertia of tradition doing one's own thinking until it can no longer. The historian of political ideas John Pocock has observed that much of the excitement surrounding Kuhn's distinction between normal and revolutionary science came from a "romantic" view of history that relied on heroic revolutionary agents disrupting the otherwise inexorable reproduction of the past.[100] Here "tradition" captured this sense of the encumbered and inertial past. Its conceptual kin included "legacy," "inheritance," and, in a more logical vein, "presumption." However, the idea that the past weighs down the future, and hence that only revolution can overturn an oppressive *tradition* (as opposed to, say, an oppressive ruler), is barely two hundred years old.[101] More common has been what Pocock calls the "classical" view of history, which treats the past as raw material out of which the future is constructed. Accordingly, the openness of the future is defined by the variety of ways available to deploy the past. Despite its ancient pedigree, Pocock's classical view of history enables us to take a fresh look at the supposedly "essential tension" between tradition and innovation. The main differences between the classical and romantic approaches to history are depicted in table 9.

Seen through the classical vision, what counts as the forces of "tradition" and "innovation" is determined less by the content of the views as such, which in principle draw on the same cultural resources, than on the difference in access to those resources that traditionalists and innovators have as they compete to define a common future. For example, if we take volume of textual output as a rough measure of sustained interest, Isaac Newton would seem to have had a much greater interest in the Book of Revelation, the *Corpus Hermeticum*, and alchemy than in the astronomical and

---

a *theory* to explain some well-founded phenomenon). Toulmin subsequently recaptured the history of casuistry as a model of normative reasoning. See Jonsen and Toulmin 1988.

By contrast, the positivists focused on the logic of the more routine cases, which are essentially generated as observations by the dominant scientific theory. Kuhn stood with the positivists in seeing theory-driven research as mostly "mopping-up operations" that periodically turned up exceptional phenomena that only under extreme circumstances precipitate a "crisis" and "revolution." On the friendship between Hanson and Toulmin, see the foreword to Hanson's posthumous papers: Hanson 1971.

100. The following observations are drawn from Pocock 1973, 273–91. Pocock himself admits to having been initially enamored of Kuhn's revolutionary rhetoric when first introduced to it in the late 1960s.

101. The symbolic moment for this sense of revolution was, of course, the new calendar introduced after the French Revolution had overturned the ancien régime.

TABLE 9. *Classical and Romantic Historiography: A Neglected "Essential Tension"*

CLASSICISM	ROMANTICISM
The past consists of resources that are available in the present for constructing the future.	The past eventuates in the future unless disrupted in the present.
The past is raw material: it delivers an unformed potential to the present.	The past is an inheritance: it delivers the burden of tradition to the present.
History has no natural direction (one must be constructed in an agonistic field).	History has a natural direction (one that can be discovered by methodical inquiry).
Decay is natural unless order is actively maintained, be it to continue or change the past.	Development is natural unless impeded, to which radical disruption is an appropriate response.
Global rules need to be imposed in order to structure the pattern of local decisions.	Global rules simply articulate the order already emergent in local decisions.
A theory is tested in the expectation that it will be replaced as it fails to deliver on its promises.	A paradigm is developed in the expectation that it will thrive unless a stronger one overtakes it.

physical theories that have persuaded philosophers and scientists over the last three centuries that his was the finest mind that ever lived. Yet the fact that Newton continues to be treated as a paradigm of scientific reasoning and achievement means that those who would advance a radically alternative science must contest this selective appropriation of Newton's activities. One way would be to unearth suppressed contemporaries of Newton who would place his legacy in a different light. The point is that the future is literally constructed out of the past, as opposed to something one waits to happen when the moment is objectively right. Thus, on the side of the romantics stand those who perceive the difference between tradition and innovation as absolute, be it Lenin or Kuhn; whereas on the side of the classicists stand those who perceive the difference as relative, be it a constitutional lawyer or, as we shall now see, Toulmin.

8. The Road Not Taken: Toulmin's Route from Philosophy to Rhetoric

In Kuhn's hands, the "essential tension" between tradition and innovation in science meant that a scientist's colleagues are a purely reactive force that, through peer pressure, discourages the scientist from regarding her own contributions as decisive against their collectively held beliefs. Of course, there are all too many instances in the history of science when the relevant rivals have not directly confronted each other, and so their incommensurability remained until the Planck effect kicked in. However, a priori there is no reason to take this state of affairs as incorrigible. But was an alternative normative account of science dynamics available in the 1960s? The answer is that crucial elements were indeed available in the work of the polymath Stephen Toulmin. We can get an initial sense of Toulmin's alternative vision by culling the main theses of a pair of articles he wrote in response to Michael Polanyi's conception of a "republic of science" in the early 1960s:

> 1. Governments must make many decisions relating to the future of scientific research, but they are neither encompassed in a single "theory choice" nor even adequately met by a "ministry of science." Indeed, on matters concerning science, governments should be open to influences from the larger society, as decisions officially about science are really about the basis of civil authority in contemporary society, what Max Weber called the "legal-rational" mode of legitimation.[102]
>
> 2. An important source of the complexity in scientific decision making is the lack of any single legitimate source of scientific authority, contra Polanyi's policy of deferring to the judgment of elders (a conceit borrowed from Plato). This point is often lost because politicians typically make decisions only after consulting with senior scientists, ignoring the opinions of the rank and file altogether.
>
> 3. Given the uneven record of science's contribution to economic growth, one should distrust any argument that asserts a neat relationship between scientific and economic productivity. On the contrary, if scientific inquiry is regarded as valuable, then its pursuit should be treated as challenging the idea that endless growth in material wealth is required for the benefits offered by the pursuit of science.[103]

These three theses share a profound awareness that by the 1960s the ideal of science as an autonomous pursuit was both empirically unfounded and normatively disruptive of welfare state social policy. Of course, Kuhn himself never addressed science policy concerns as directly as Toulmin. Yet, as we saw in chapter 5, section 5, theorists as ideologically disparate as the finalizationists and Fukuyama inferred from the model presented in *Structure* that the trajectory of scientific change is, contra Toulmin, sufficiently self-determining to constitute — in the vivid medieval phrase for God — the "immutable mobile" of socioeconomic progress. Toulmin's response, as indicated in thesis 3, is that modern society's abiding interest in the autonomous pursuit of science is a standing anomaly to the capitalist paradigm of political economy, which funds only activities that explicitly enhance the

---

102. Significantly, an article that Toulmin published in the *New Scientist* in 1963 making these points inspired the establishment of a premier science policy research unit, SPRU, at Sussex University, United Kingdom. See Freeman 1992, 242–43.

103. Toulmin 1964, 1966. A convenient collection of science policy debate from the period that includes most of the pieces to which Toulmin responded and that responded to him is Shils 1968.

material well-being of society. In other words, if one had the will, one could mobilize the power of science *against* the logic of capitalism.

Toulmin and Kuhn were not only born in the same year (1922) but also trained as physicists in the most prestigious universities in their respective countries (Cambridge and Harvard). Both also fell under the spell of Wittgenstein at an early age. But before the early 1960s Toulmin had been known not only, and perhaps not primarily, as a philosopher of science but as a moral theorist and philosophical logician. His youthful work, *The Place of Reason in Ethics*, had successfully challenged the positivist dictum that value issues were emotively based attitudes—"mere ejaculations," in A. J. Ayer's words—not susceptible to rational argument.[104] In place of the sharp dualism between scientific rationality and moral rhetoric, Toulmin proposed that both discourses were "rational" and "rhetorical" in exactly the same way. In practice, this meant showing that ethics is an inquiry into what is *really* good, which signaled that reasons could be mustered and contested in the same way as in science. However, it also meant importing a conception of scientific reasoning that had a strongly rhetorical character, one in which "reality" defines a pragmatic limit to a particular inquiry, the moment when its participants no longer feel the need to pursue a matter any further. Of course, the inquiry can later be reopened, with "reality" then being the subsequent point of closure. Indeed, science has institutional mechanisms to ensure that that happens.[105] Ethics, Toulmin observed, is not so endowed, but that is only a matter of convention, one perhaps worth revising so that people are not led to think—as the positivist view might suggest—that important value issues are purely personal matters.

Toulmin made more explicit his endorsement of a rhetorically based, antipositivist conception of reason in *The Uses of Argument*, a book that started to become very influential among speech communication scholars in the late 1960s.[106] However, when it was first published in 1958, Toulmin's wholesale rejection of the concept of formal validity—the idea that there are forms of argument that are rationally acceptable regardless of their content—made the book very unpopular with its primary audience,

104. Toulmin 1986.
105. Toulmin would later call practices of this sort "compact." See Toulmin 1972, 378 ff. Toulmin's discussion of compactness has influenced the interest that sociologists of scientific knowledge have had with how scientific disputes are "closed" and then "reopened." See, e.g., Collins and Pinch 1982.
106. Toulmin 1958. The recent spate of interest in Toulmin among German philosophers is due to the incorporation of Toulmin's ideas about validity in the theory of communicative action put forth by Jürgen Habermas, who, in turn, learned of Toulmin's work from speech communication scholars. The canonical status of Toulmin's work in this field is symbolized by his prominence in van Eemeren, Grootendorst, and Kruiger 1987, 162–207.

philosophers, even among those who recognized that formal logic provided an inadequate model of human reasoning.[107]

Unlike Kuhn and the positivists, Toulmin never drew a sharp line between reasoning about empirical and normative matters. In fact, he was so keen on eliminating any such distinction that he challenged its ultimate philosophical basis, the Humean contrast between empirical content and logical form. He treated this difference as itself formal not substantive — in his words, a "field-dependent" distinction that can be drawn only in particular contexts of reasoning. In this way, Toulmin provided the philosophical bridge between ancient sophistic rhetoric and the constructivist philosophy that informs contemporary sociology of science.[108] In privileging the Sophists over Plato, Toulmin stepped over an invisible line of philosophical respectability, a transgression that has increasingly endeared him to readers outside analytic philosophy. But ultimately, Toulmin is a Mosaic figure in the quest to normalize the place of science in society. His forays into alternative models of scientific reasoning were united more by their opposition to scientific realism than by a consistent commitment to either relativism or constructivism.[109] In retrospect, it is clear that he has been groping toward a theory of rhetoric, with his original use of "paradigm" (one year before Kuhn's) recognizably falling under the rhetorical category of *topos*.[110] However, Toulmin never said how radical change could arise in the course of argument: How can people operating with incommensurable

---

107. See, esp., Manicas 1966.
108. The work that has most explicitly made this connection is Willard 1983. My own attempt at making this connection is clearest in Fuller 1995b.
109. A good example of Toulmin's ambivalence in his opposition to scientific realism is the field-dependent model of reasoning presented in *Uses of Argument*, which is plausibly seen as relativist, and the genealogical model of reasoning presented fifteen years later in *Human Understanding*, which is more strictly constructivist. The contrast marks a shift in Toulmin's thinking about the nature of rationality from spatial to temporal metaphors. The relevance of this point to science and technology studies will become clear in chapter 7, section 3C.
110. For Toulmin, a paradigm is an "ideal of natural order" in terms of which the adequacy of particular explanations are judged. As he himself has admitted on several occasions, here Toulmin was simply refiguring Collingwood's "absolute presuppositions" (see chapter 1, section 4). So defined, a paradigm anchored scientific argument by establishing a presumption, say in the case of Newtonian mechanics, in favor of bodies moving rectilinearly unless impeded. See Toulmin 1961, 56. Later Toulmin would credit Georg Cristoph Lichtenberg (1742–99), professor of natural philosophy at Goettingen, with having anticipated this use of "paradigm," interestingly in the context of a critique of the "chemical revolution" then being heralded by Antoine Lavoisier. Like Kuhn, Lichtenberg placed a positive value on "paradigm" and a negative one on "revolution." As might be expected from the cases of Whewell, Mach, and Duhem, Lichtenberg's "enlightened" methodological views were coupled with a "reactionary" orientation to what turned out to be the most important scientific development of his day (in this case, stemming from an anti-French bias). See Toulmin 1972, 106; I. B. Cohen 1985, 517–19. The *topoi* surrounding the breakdown of a paradigm have been explored more recently in Zagacki and Keith 1992; Keith and Zagacki 1992.

paradigms come to see each other as mutually relevant audiences worthy of sustained attention and perhaps even mutual incorporation?

To be sure, one reason for Toulmin's failure to answer this difficult question is his underestimation of the need for arguers (including himself, as aspiring revolutionary) to build ethos with their audiences. Unlike the classrooms where scientific instruction occurs, there are no captive audiences in the research sites of science. If the Planck effect is as prevalent as Kuhn maintained, then obviously scientists can never take for granted that colleagues holding contrary views will listen to them. Traditionally, the function of ethos in rhetoric has been to build community from audiences with little, if any, prior common cause.[111] The need for the different classes of a society to join forces in times of war would thus be a typical test of a national leader's ethos.[112] While this aspect of rhetoric would gain renewed prominence in the 1970s, with the reissuance of the works of Kenneth Burke (1897–1993), it was quite the opposite of what philosophers especially, but even some rhetoricians, had only a decade earlier taken the scope of rhetoric to be.[113]

The revival of Burke's community-building, action-getting conception of rhetoric rekindled the original spirit of rhetorical inquiry in the United States, which was to provide training in public speaking for democratic

---

111. Two types of answers may be given to this problem, both of which involve assumptions about the cognitive makeup of the audience. On the one hand, the leader's rhetoric may work because she taps into the audience's values and prejudices that already incline them in her direction. On the other hand, it may work because the audience enters the forum without any specific views on the matter and hence is open to persuasion. While the second answer would seem to be a prerequisite for rhetoric to work in a democracy, the defeat of Athens in the Peloponnesian Wars enabled Plato to demonize it as sophistic manipulation. The subsequent history of rhetoric has overcompensated by following the course of the first answer, which stresses that honest rhetoric must be built upon common value assumptions. Plato's success in this case has had regrettable long-term consequences (as argued in chapter 1), not the least of which has been to typecast the rhetorician as a conservative force — a teacher of eloquence and manners — in European society. An excellent recent history of rhetoric that ranges over these matters and includes incisive discussions of Toulmin, Burke, and Richards is Conley 1990. Fuller 1993b is devoted to reviving the original sophistic conception of rhetoric.

112. The question of leadership is especially acute in any philosophical attempt to revive a prescriptivist approach to science. One traditional rhetorical source of *ethos* is *kairos*, that is, a pressing occasion that calls for leadership. But what can be made so pressing that scientists listen to a philosopher — even if all the philosopher says is that they should listen to each other? Unfortunately, the prospect of philosophical legislation continues to carry the old Hobbesian baggage of a sovereign legislator who not only takes the initiative to propose laws but also has the will to enforce them. On *kairos*, see Kinneavy 1986.

113. Perhaps the most striking case in point is I. A. Richards (1893–1979), whose successive academic posts at Cambridge and Harvard made him the most prominent rhetorician in the English-speaking world from roughly 1930 to 1960. Richards was perfectly happy to see the academic study of rhetoric as a fellow traveler of logical positivism, in its ability to identify and correct misunderstandings. See Richards 1936.

citizens of all walks of life. This was a perfectly honorable setting for Toulmin to be regarded as a significant innovator. However, by reaffirming its democratic roots and public service orientation, rhetoric was effectively distanced from the issues that animated historians and philosophers of science as they increasingly came under Kuhn's spell. Generally speaking, the period since 1980 has been marked by a slow but significant devaluation of the role of language, especially argumentation, in the constitution of scientific authority among historians, philosophers, and even sociologists of science. Consequently, the democratic pulse that moves rhetorical inquiry is now almost completely absent from fashionable epistemologies of science.[114]

Indeed, we seem to inhabit a philosophical world where argumentation over scientific claims counts for little.[115] On the one hand, we have the "practice mongers" who devalue the role of language in science altogether and wax eloquent about the importance of bumping into things in the laboratory as implicit assertions of their reality. This is a great criterion for nostalgic ex-scientists. On the other hand, the natural sciences command

---

114. To appreciate the basis for my harsh judgment, start by considering some of the endearingly naive beliefs that the logical positivists held about the nature of science: namely, that one could contribute to it by speaking in terms of universally available observations and logically transparent claims that could be tested in terms of either other such observations or claims. Whereas the positivists agonized over an appropriate language for these things, the Popperians preferred "plain speaking" and ad hoc critiques of jargon and scientistic obfuscation. However one wishes to judge their specific efforts, these exiles from Vienna clearly believed that science was within the reach of more than just the people who happen to get advanced degrees in scientific subjects and regularly spend time in research sites. In our Kuhnified world, this is no longer the case.

But what about Feyerabend, whose study of Galileo's rhetoric is the case in point for the anarchistic epistemology advanced in *Against Method* (1975)? While Feyerabend grants rhetoric a more prominent place in exemplary episodes of scientific reasoning than the positivists were inclined to recognize, he nevertheless continues to portray rhetoric in the same negative light as the positivists saw it. After all, on Feyerabend's account, Galileo used rhetoric to cover up basic gaps in his arguments, which over the centuries philosophers have supplemented with idealized premises; however, these gaps did not escape the gaze of Galileo's Jesuit Inquisitors. The outrageousness of Feyerabend's account pertains to the short and long term of Galileo's fate. In the short term, the Inquisitors, normally the "bad guys" in positivist historiography, are presented as having offered cogent methodological criticisms of Galileo (much like the positivists themselves would). Yet in the long term, Galileo turns out to have gotten closer to our physics than the Inquisitors with their methodological niceties. The Feyerabendian moral, then, is that the positivists mistake rhetoric for method, but method doesn't win, either: *reality* does—and that's something that transcends both crafty rhetoric *and* rigorous method. If this conclusion seems to mix truth and long-term survival, it is because Feyerabend shares with Kuhn a certain unexplained fondness for Darwinian accounts of knowledge growth.

115. It is gratifying to see that even followers of Habermas, such as Rehg (1999), have begun to see this point. Taylor 1996 provides a good sense of what the classic philosophical problem of science—how to demarcate science from "pseudoscience"—looks like through the lens of argumentation theory.

TABLE 10. *The Epistemic Space of Acritical Science Studies*

	SOURCE OF KNOWLEDGE	
LEVEL OF ANALYSIS	DIRECT ACQUAINTANCE	DEFERENCE TO AUTHORITY
first order: what scientists do	scientific researchers (science in the lab)	scientific experts (science in public)
second order: what meta-scientists do	ethnographers of science	philosophical underlaborers of science

respect from people—even very educated ones—who have never been near a research site, but have been persuaded that airplanes are kept up in the air by "the laws of physics." On that basis, it is concluded that if one lacks direct acquaintance with how scientific knowledge is produced, then deference to the relevant scientific experts is the only "rational" course of action.[116] Missing here is a role for the old-fashioned idea of critical argument familiar from democratic politics, whereby restricted access to the production of knowledge claims—say, the closed-door setting of a legislative session—places a burden on the ensconced elites to submit their actions to public scrutiny and periodic ratification.

Instead, it would seem that argumentation is not so much delegated as privatized to the few experts in a highly circumscribed field of inquiry who must ultimately assume collective responsibility—in the form of professional "peer-reviewed" publication—for what any one of them claims to have found. I have dubbed this conceptual space, demarcated in table 10, as "acritical" to capture its failure to identify a standpoint outside of extant sciences from which to launch a critique of science.

There are at least two ways of characterizing the constitution of this space.[117] One is as the unholy alliance of those who see rank-and-file normal scientists as the hitherto exploited proletariat of science and those who see them as a medieval guild possessed of knowledge that deserves trust because it is transmitted only under restricted conditions. Here the scientific identity is constituted by a "conspiracy of silence," as the suppression of empirical traditions in science by a theory-privileging historical record is conjoined with the self-imposed tacit dimension of the traditions

---

116. I am embarrassed to say that much of what passes for "social epistemology" in analytic philosophy is devoted to legitimating this tendency. Thus, the need for investigating the social grounds of compulsive belief is short-circuited by arguing that we are endowed with a faculty of "common sense" that licenses deference to expert testimony as the epistemic path of least resistance. See Schmitt 1994, chap. 1, for a particularly bald expression of this view. This view's provenance is discussed in chapter 1, note 10, above; also see Fuller 1996c for a critique.

117. The key texts that mark the boundaries of this space are Kitcher 1993, Shapin 1994, Hacking 1983, and Pickering 1995.

CHAPTER SIX / 316

themselves.[118] Another way of characterizing this space is that the roles of metascientists—that is, sociologists and philosophers of science—reproduce at a higher level of abstraction the two signature roles of professional scientists: the laboratory researcher (cf. the laboratory ethnographer) and the public expert (cf. the underlaboring philosopher).

How did this state of affairs come about? Not least by Kuhn's notion of paradigm, which enables philosophers and sociologists to vacillate between two quite different senses in which science is a "practice."[119] On the one hand, practices are supposed to be like hidden premises or presuppositions, which suggests that they might be linguistically accessible and hence contestable. For example, with a certain measure of self-consciousness, should I not be able to reflect critically on the principles governing my paradigm (in the sense of weltanschauung)? Certainly, this question loomed large for Kuhn's philosophical critics, especially Karl Popper and Dudley Shapere. On the other hand, practices are also supposed to exist at a sort of permanently subconscious, maybe even visceral level, which *in principle* resists conscious expression. This is how both Kuhn and Michael Polanyi usually talked about scientists consumed by their paradigm. It often seems that today's practice mongers in science studies want to have their cake and eat it too: that is, a notion of practice that can justify (propositionally) what members of a society do, without their ever having to make that justification sufficiently explicit to invite criticism and hence possible revision of their practices.[120]

---

118. On the contradictory but no less constitutive role of silence in scientific practice, see chapter 4, section 6.

119. Turner 1994 provides a genealogy and critique of this mentality. See chapter 2, note 70, above for more details about Turner's thesis.

120. Why would anyone want to court such a potentially incoherent notion? Perhaps because it provides a mechanism for explaining why societies manage to survive and flourish even though they are not governed by our own principles. Practices work precisely because they do not need to be made explicit—and, in fact, when they are made explicit, a key part of their ability to bind action disappears. Thus, complementing nineteenth-century thought about practices and such related concepts as "traditions," "habits," and "cultures" were worries that modern critical rationality was eroding the basis for our own traditional forms of life. This helped explain, for example, the repeated failure of liberalism to take root in countries, particularly in Southern and Eastern Europe, that had no prior collective experience with democratic forms of government. (The argument would later apply to Africa and Asia with the rise of imperialism.) These difficulties were given concrete expression as the nation-states of Europe tried to subsume culturally diverse regions under a common body of law. Given that these regions thrived prior to the imposition of the "philosophically designed order" of a national constitution, the conclusion was widely drawn that, alongside a universal rationality consisting of explicitly formulated rules, there existed many tacit rationalities, access to which required living the life of someone from a particular region: *Gesellschaft* and *Gemeinschaft*, in Ferdinand Toennies's classic formulation. See Fuller 1999e.

A striking example of the lengths that some STSers will go to pursue the autonomy of practice from critique is Latour 1997, which argues that critique is morally objectionable

I wish to recover what lies outside these Kuhnified horizons of acritical science studies. This is the tricky proposition to which my project of social epistemology has been devoted for the past ten years. It involves defining the ends of inquiry in relative independence from whatever outcomes happen to benefit the research population at a given place and time. For example, it is self-serving to infer the democratization of science from its rootedness in a specific form of labor, if we do not also observe that both experts and nonexperts are ignorant and fated in sufficiently similar ways that we would *all* be better off pursuing an epistemology that avoided irreversible error than one purporting to determine what we can know with certainty.[121] However, before returning to this alternative future for science studies in chapter 8, we must consider how the field managed to succumb to Kuhnification in the first place.

---

because it presupposes a low opinion of scientific practitioners, whose alleged self-deception provides the only opportunity for the critic to practice her own trade. Critics treat practices as mere means to their own ends, while failing to recognize that the most perfect constructions are ones whose handiwork is hidden and hence without need of critical improvement. Latour's etiology of the critic's craft is based largely on Gaston Bachelard's defense of scientific labor from the philosophical exploitation, as discussed in chapter 7, section 4, below. However, Latour's equation of normal scientific practice with seamless construction harks to a version of the theological argument from design, namely the postulation of a *deus absconditus*—a God who builds the world so well that his services are no longer required. Thus, science works so well that its constructed character does not matter. On this line of thought in French philosophical histories of science—and their bearing on Kuhn—see chapter 8, section 2.

121. I have promoted this perspective as "shallow science" in Fuller 1993b, chap. 1, and the "outside-in" perspective on science in Fuller 1997d, chap. 1. A precursor to this stance is Ravetz 1987.

VII

Kuhnification as Ritualized Political Impotence
The Hidden History of Science Studies

1. The Contemporary Symptoms of Kuhnification

The science studies community currently suffers from self-inflicted Kuhnification. The main symptoms are a collective sense of historical amnesia and political inertia, which together define a syndrome, call it "paradigmitis." It is too easily forgotten that the original moral and political purpose for studying the social dimension of the natural sciences came not from social scientists looking for a new research specialty, but from natural scientists interested in repairing the perceived rift between professional scientific training and the skills in democratic citizenship associated with the classical liberal arts curriculum. In other words, like Conant's promotion of the history of science, the original scientific interest in science studies was pedagogically driven by a unified sense of inquiry. But while Conant sought to address the rift by making the public more receptive to science, the European sources of science studies largely engaged in the reverse strategy of rendering scientists more sensitive to the society supporting their work. In what follows, I shall focus primarily on the British precedent, but the chapter ends with a discussion of the Parisian origins of science studies. Its ascendancy in the field reflects the devolution of the welfare state and the associated weakening of academia's autonomy in the face of market forces.

Shortly after becoming prime minister of Great Britain in 1964, Harold Wilson called for the integration of science and technology into the mainstream of British society. The result was a series of courses instituted in the late 1960s to teach science and engineering majors about the social dimensions of their research, in the hope of tracking them into more socially beneficial directions. Among these new service teaching programs was the

Edinburgh Science Studies Unit, the original home of the sociology of scientific knowledge (SSK). Kuhn's book figured prominently in its courses. Indeed, the curriculum of what became known as the "Edinburgh School" was the first to integrate the teaching of history, philosophy, and sociology of science, the three fields that Kuhn was seen as having brought together in *Structure*, Kuhn's own discipline-based sensibilities to the contrary.[1]

However, the early success of the unit and the expansion of science enrollments at the peak of the Cold War justified the need for funding "research" in the sociology of science, which gradually autonomized the unit's interests from its original pedagogical mission. It will be no surprise to learn that this transformation happened as if *Structure* had provided the recipe. Thus, journals were established in which cross-citation to approved contemporary authors gradually replaced historical precedents laden with inconvenient political (i.e., Marxist) baggage. Not only has this strategy telescoped the field's sense of its own history (an Orwellian outcome that Kuhn deems necessary for motivating scientific activity), but more importantly, it has rendered problematic the field's relationship to contemporary social movements in which science figures prominently, as the aims, methods, and discourse of science and technology studies (STS) have come to be defined in ways that make it less permeable to larger political concerns.

A striking marker of this last tendency is the rhetorical currency in which the ongoing Science Wars is traded. I shall have more to say about this development over the course of this chapter. Here I would just note that when debates over the social dimensions of science and technology had a strong Marxist flavor in the late 1960s and early 1970s, both sociologists and scientists were expected to argue over the direction in which science should be heading.[2] However, now the debates focus almost entirely on who is academically authorized to pronounce on the nature of science. Alternative visions for socially situating inquiry have become overshadowed by demonstrations of technical competence — or lack thereof — about how

---

1. On the influence of Kuhn on the early Edinburgh School curriculum, see the course bibliography, Bloor 1975. According to Bloor, "Most of the course deals with a closely-knit set of issues connected with T. S. Kuhn's *The Structure of Scientific Revolutions*" (507). Kuhn's book, which had been used since the inception of the Science Studies Unit in 1967, was the only one students were required to purchase. Kuhn returns the compliment by voicing his objections to the sort of interdisciplinary metascientific work represented by the Edinburgh School in Kuhn 1977a, 4. Nevertheless, Barnes 1982 has drawn detailed connections between Kuhn's work and subsequent developments in the Edinburgh School.

2. An important book that exemplified this development was Rose and Rose 1970, authored by a sociologist and a biologist. The book was one of those originally assigned as a "set text" at the Open University, Britain's "university of the air" and Harold Wilson's most lasting educational legacy.

science is actually practiced.³ Instead of favoring one or the other side, public reaction to these new exchanges has been irritation at both, for they lack any sensitivity to the *future* of science. In that sense, Kuhn's famous "evolutionary" model of scientific change as a progress *from* that is not a progress *to* has become a self-fulfilling prophecy in contemporary science policy debate.

As we have observed in previous chapters, an unintended consequence of the abstract character of Kuhn's account is that it made it easy for all manner of inquiries to reinvent themselves as paradigms. Often the method of reinvention has been as simple as rewriting a discipline's history in Kuhnspeak and imposing seemingly strict but ultimately elusive criteria for membership and a regime of careful cross-citation of the journal literature.⁴ Together these actions easily fostered the impression — especially to

---

3. A good brief example from the STS side is Edge 1996. A more sophisticated version of this strategy is to argue that by denying the propriety of STS, scientists violate their own normative commitment to following the path of inquiry to wherever it may lead, even if it overturns established intuitions. Thus, before scientists condemn STS, they should see it as having arrived at the sort of insights that only specialized paradigm-driven studies can provide. See Fujimura 1998. However, Fujimura's choice of historical example — the original resistance of mathematicians to non-Euclidean geometries — is somewhat unfortunate because the objections came mainly from mathematicians like Edmund Husserl's teacher, Leopold Kronecker, whose views on the nature of mathematics partly anticipated those of the Strong Programme in the sociology of knowledge. See esp. Bloor 1973; Bloor 1976, chaps. 5–7; Barnes 1982.

Both Kronecker and the Strong Programme adhered to an austere ontological regime that cast doubt on entities that are not constructible in terms of a finite number of logical steps or empirical observations. In both cases, the aim was to keep mathematical inquiry grounded in ordinary physical intuitions, a constraint that non-Euclidean geometries were seen as having violated (at least until Einstein provided a physical basis for them in general relativity theory). Fujimura's neglect of this precedent for STS constructivism in the history of mathematics reveals a metalevel lack of constructivism, namely her easy acquiescence to Whig historiography, since the rhetorical force of her case study presupposes that critics of the non-Euclidean approach were simply wrong, as opposed to being engaged in "boundary work" over what to count as proper objects of mathematical inquiry. Generally speaking, those who adhered to the Euclidean orthodoxy wanted to preserve a holistic vision of mathematics linked to ongoing developments in philosophy, psychology, and physics. Thus, it was only with Einstein, nearly a century after the first geometries suspending Euclid's fifth postulate were proposed, that they came to be fully accepted within the field. While a constructivist might wish to endorse this delay in acceptance, Fujimura simply assumes the Whig view that Kronecker et al. were backward. See Fuller 1999a.

4. In a world where cross-referencing is used to trace the exchange of symbolic capital, the strategies for capital accumulation are easily mastered, once discovered. For example, one key metric is a journal's "impact factor," which is a function of how often articles published in the journal are cited in other articles. The most obvious strategy for boosting a journal's impact factor is for editors to ensure that articles published in their own journal cite work that has been previously published in that journal. For most authors, these citations, typically added as part of the refereeing process, are a small price to pay for access to a major international academic forum. Not surprisingly, the leading STS journal and house organ of the Edinburgh School, *Social Studies of Science*, has consistently had one of the highest impact

government policy makers and university administrators—that the discipline was purposefully engaged in the quest for knowledge. Kuhn actively distanced himself from these appropriations, arguing that *Structure* really only applied to the physical sciences. Nevertheless, the overall effect has contributed to the original mission of Conant's courses, since, as we saw in chapter 5, many humanists and social scientists who otherwise might have been critical of the natural sciences came to respect and even imitate those still-authoritative fields of inquiries by practicing some normal science of their own. In that case, the lesson taken from *Structure* was "If you can't beat 'em, join 'em."

All of the preceding observations about the social sciences apply with a vengeance to STS, where *Structure* is typically taken as a founding text, the content of which is sometimes read as a blueprint for organizing the field's research agenda—though Kuhn would have been the last to admit it. Indeed, Kuhn had the good fortune to arrive on the scene before STS began doing normal science because his text would have never made it through the peer review process. Just as no revolutionary scientist could stand to inhabit a world governed by the paradigm her work ostensibly started, so too a work as broad and speculative as Kuhn's *Structure* could not have survived peer review in contemporary STS, which has come to expect case studies cloaked in a strong empiricist rhetoric, with "theoretical reflection" amounting to token genuflection to the field's immediate founders and a sprinkling of their more aromatic verbal droppings. If political history happens twice, as Marx maintained, first as tragedy and then as farce, epistemic history follows a comparable pattern, first as genius (in the revolutionary moment) and then as error (in the normal scientific moment). In that spirit, consider these methodological failures of *Structure*:

- One model, adapted largely from physics, is presented as the model of all the sciences.
- The cyclical character of the model is based largely on about three hundred years of European history (roughly, 1620 to 1920), with examples chosen opportunistically to fit phases of the cycle.

---

factors of any journal in the social sciences, due in no small measure to the amount of internal cross-referencing.

The postmodern reader may urge that we interpret this result as a clever parody of the processes of scientific legitimation, a quarter-century preemption of Alan Sokal's recent parody of scholarship in cultural studies of science, this time performed by social scientists on natural scientists. However, even in Britain, irony knows limits: STS's scientometric success should be taken as a rather earnest application of the principles that STS has itself discovered for turning any field—including itself—into a science. See Sokal 1996. For an assessment of the Sokal hoax in the context of the larger issues considered in this book, see Fuller 1998f.

- The model acquires rhetorical force by Kuhn's claim to have demarcated the natural from the social sciences, though he fails to compare the two types of fields in any depth.
- There is little discussion of either the macro- or the microsocial character of science, aside from how novices are socialized into the practice of normal science.
- There is virtually no discussion of the economic and technological dimensions of science, even though Kuhn foregrounds the role of laboratory experiments.
- The narrative fails to adhere to SSK's famed "symmetry principle," in that Kuhn wants to account for science only when it is functioning as it should, which presumably implies that a dysfunctional scientific enterprise would be explained differently.

The most striking confirmation of STS's Kuhnification is the difference between the two handbooks that Sage has published in the name of STS over the last twenty years.[5] Sustained critical discussions of the history, philosophy, and politics of science in the first volume have been reduced, in the second, to the nostalgic prehistory of STS.[6] At a more routine level, it has become common for STSers to argue—against both scientists and sociologists—that theoretical considerations can never trump empirical case work, after which follows an lengthy enumeration of such cases, as if their sheer volume could meet the challenge posed by the theoretically inspired objector.[7] But exactly what was *Structure*'s role in this hardening of the intellectual arteries?

The transitory and suppressible character of the political in Kuhn's

---

5. The two handbooks are Spiegel-Roesing and Price 1977 and Jasanoff et al. 1995. As it turns out, I was part of the committee that supposedly "oversaw" the production of the latter work on behalf of the Society for Social Studies of Science (4S). After much hand-wringing and soul-searching, the committee decided to grant the professional society's official stamp of approval, more out of anticipated profits from sales than any genuine enthusiasm for the handbook's representation of the field. The decision seems to have had the desired fiscal effect, though it would be nice if, in the future, the intellectual integrity of the society's collective products is not compromised by financial considerations.

6. I especially have in mind Edge 1995.

7. A good case in point is Collins 1996. Given this response to criticism by so eminent an STS practitioner, I wonder whether I have missed a hidden ironic message to the conclusion to the leading social history of science done under the auspices of SSK: Shapin and Schaffer 1985, esp. 344: "Hobbes was right." The most natural reading of Shapin and Schaffer's account of the Hobbes-Boyle debates is that Hobbes—the defender of philosophical argumentation as a check on the excessive claims made on behalf of experimental observations—had a much better case than he has subsequently been given credit for. This lesson reflexively applied implies that SSKers should themselves be interested in recovering the role of theoretical interrogation in bringing into public view empirical findings that might otherwise be cloaked in the "tacit dimension" of the experimental craft. Yet a close examination of Collins's arguments suggests that his rhetorical lessons are taken primarily from Boyle's authoritative empiricism. Once again, if you can't beat 'em, join 'em. For more on the philosophical implications of this point, and especially the space for rhetoric that it opens up, see Fuller 1994c.

account has had a decisive impact on the development of SSK. In particular, three familiar sociological perspectives simply have no place in Kuhn's account of science:[8]

1. the view that science systematically reflects, or legitimates, specific class interests more generally represented in society, such that major changes in science are best explained by appealing to larger societal, even geopolitical, developments;[9]

2. the view that science lacks any clear integrity as a social practice, but is rather a loose collection of disciplines held together by a specifically "scientific" way of talking about them;

3. the view that science can be exhaustively reduced to a set of variables, drawn from the larger repertoire of social science, on the basis of which policy for science can be made as part of comprehensive social policy.[10]

This Kuhnian legacy has made the study of science a relatively self-contained project, innocent of any obvious interest in providing a comprehensive theory of society.[11] (An instructive exception to this claim, the Paris School's actor-network theory, will be considered in the final section of this chapter.) In recent years, some effort has been made to incorporate these more macrosociological factors, but typically this occurs at the level of "discursive resources," whereby the sociologist imputes, say, beliefs in class or status differences to the agents under study without committing herself one way or the other to the validity of those beliefs. The literature surrounding "boundary construction" and "boundary maintenance" captures this ten-

---

8. These three neglected features figure prominently in my own social epistemology, most explicitly perhaps in Fuller 1992c.

9. The textbook that has most successfully integrated STS with larger issues of critical science policy and a broader political economy perspective is Yearley 1988. Yet even here the philosophical debates surrounding STS are presented in separate chapters from those devoted to science criticism and policy, thereby creating an air of perfunctoriness to STS's relevance to the overall discussion. One line of criticism often found in Third World critiques of postmodernism that applies to Yearley is that STS can serve as a philosophical basis for science criticism only in societies where science is sufficiently established that positivism is already the dominant ideology (as in the First World, arguably). Otherwise, STS's relativism would inhibit science from ever gaining a foothold in the society. See Parayil 1992. This is an interesting criticism, for, as we shall see below, SSK's own claim to superiority over its philosophical rivals partly lies in its having become the best historical vehicle for the promotion of the scientific enterprise.

10. The most explicit attempt in this vein within the sociology of science has been Whitley 1984, which decomposes the differences between academic disciplines in terms of variables in organizational theory. An interesting integration of this perspective into a general sociology is Collins 1975, chap. 9. A worthy effort at reconciling this macrosociological approach to the microsociological one associated with SSK is Fuchs 1992.

11. Even Barry Barnes, the founder of SSK who has most successfully transformed himself into a social theorist, has been reluctant in drawing connections between sociology of science and social theory. A good case in point is his most sustained effort at social theorizing, from which one would never get the impression that he also studied science: Barnes 1990.

dency.[12] But in the main, the Kuhnian legacy is seen in STS's methodological commitment to providing an "internal sociology" of normal science, whereby the STS practitioner tries to understand the set of practices in a canonical research site—typically the laboratory—that constitute what the scientists themselves identify as their life world.[13]

2. Kuhnification Rooted in the Educational Mission: Britain versus America

In both America and Britain, the impulse to promote academic programs in the history, philosophy, and sociology of science was typically nurtured by senior scientists concerned with the sudden emergence of the natural sciences as a fixture of Cold War public policy. These sciences became the lightning rod for the public's most intense hopes and fears. The proposed remedy was to normalize the role of science in society, so that it did not seem irretrievably alien. Most significant for the present work has been Harvard's General Education in Science program. But to understand the process of Kuhnification, we must examine the British origins of STS that led to the field's first self-described paradigm, the Strong Programme in the Sociology of Scientific Knowledge.

Whereas the Harvard precedent stressed introducing science to nonscientists, the British one stressed introducing scientists to the larger nonscientific contexts in which their work was embedded. Indeed, curricular precedent for this way of seeing things had already been established at Oxford and Cambridge a decade or more before C. P. Snow's (1905–80) famous 1959 Rede Lecture, "The Two Cultures and the Scientific Revolution," drove home the need for scientists to enrich their training with an understanding of the broader cultural arenas in which they would increasingly figure.[14]

---

12. For a summary of mainstream STS literature on boundary maintenance in science, see Gieryn 1995. A more critical approach to the issue is provided in Fuller 1993b, 102–38; Taylor 1996.

13. The most philosophically interesting defense of this perspective is Lynch 1993. It is worth noting that Kuhn's influence in "internalizing" the sociology of science has even extended to social epistemology, most notably in the case of analytic philosophers who try to combine Kuhn and feminism. See Longino 1990, which draws an oversharp distinction between "constitutive" and "contextual" values in science, thereby effectively limiting the extent to which larger political considerations play a role in the constitution of scientific knowledge. Thus, Longino suggests that while all science is subject to contextual values, constitutive values may play a greater role in, say, biology than physics.

14. Snow 1959. A sensitive philosophical analysis of the responses to Snow's lecture may be found in Sorell 1992, 98–126. A comprehensive literary study of Snow's view of science (focusing on his novels) is Hultberg 1991. It would be a mistake to see Snow's lecture as merely the beginning of British attempts to integrate scientists into the civil service. The real pioneer

As Snow saw it, scientists needed to become more sensitive to public values before they can be persuasively entrusted with the fate of humanity. This was especially true in the case of Marx-inspired scientists who, following the lead of John Desmond Bernal, openly claimed that democracy may turn out to be an atavism in the scientifically planned society of tomorrow, since free elections, after all, produced Hitler. It did not help that Bernal and his cohort lagged behind the rest of the British public in acknowledging the atrocities to both science and society committed by the Soviet regime. All of this suggested that scientists needed to learn that values are not reducible to efficient outcomes but rather are invested in civic traditions that are designed to temper overbearing authorities, including science itself.

Although Snow himself supposed that the "two cultures" suffered from *mutual* incomprehension, his speech clearly left humanists with the impression that the burden was primarily theirs. Especially after a lecture given in 1962 by that redoubtable champion of amateurism, the critic F. R. Leavis, Snow has been usually read as arguing that the spiritual goals championed by humanists have been historically superseded by material

---

was Lord Richard Haldane (1856–1928), who held several cabinet posts in the liberal governments of David Lloyd George and Herbert Asquith in World War I, and is probably most remembered for his 1918 report "on the machinery of government." Haldane adapted the German model of science as the "spiritual infrastructure" of the nation, as discussed in chapter 2, section 3, resulting in the establishment of the United Kingdom's major research funding councils, the fortunes of which would wax and wane as military effort was mobilized and redeployed. Perhaps most striking in retrospect is that universities were treated mainly as training grounds for scientists who would then be employed by the state and industry. University-based research not explicitly tied to political-economic mandates did not receive much attention. This helps explain the context in which Bernal and Snow wrote. For more on the history of British science policy, see Rose and Rose 1970, esp. chap. 3.

On the establishment of courses in the history of science at Cambridge and Oxford, see, respectively, Hall 1984 and Crombie 1984. Hall observes that Joseph Needham (1901–95) had run the first courses in the history of science at Cambridge in 1936, over a decade before his epic study of the history of science in China had begun to appear. Needham wanted such courses to be a vehicle of internationalism in science, perhaps preventing future world wars while redressing the increasing disparity between rich and poor countries that has resulted from a period of "laissez-faire" science. See Elzinga 1996. Needham was UNESCO's first scientific director.

However, in the context of Cambridge politics, Needham, by training an embryologist, was just as marginalized as his closest counterpart, George Sarton, was in the United States. After the Second World War, an institutionally more powerful Herbert Butterfield justified the expansion of courses in the history of science as a stopgap against increasing specialization, to show that, even at its most technical, contemporary science was addressing questions of larger cultural significance that had been posed for centuries. An interesting feature of Crombie's account is the role that the search for hidden presuppositions (i.e., unexpressed questions) of past scientists played in justifying a role for research—not merely teaching—in the history of science. However, in this case, the concern was inspired by the work of Crombie's own Oxonian precursor, Robin Collingwood, who played a role in the United Kingdom comparable to that of Alexandre Koyré in the United States.

needs that only science can satisfy.[15] However, Snow was making a much more even-handed point, namely, that while scientific skills are singularly necessary for the survival of humanity, scientific training fails to instruct the moral imagination, especially the facility with alternative futures that is typically developed by the humanities. Snow's ideal civil servant would thus be equipped with a humanist's sense of ends and a scientist's sense of means.[16]

It is important here to see whom Snow had in mind as the "typical scientist" and the "typical humanist" whose worldviews had to be bridged. The contrast was *not* between a bloodless technocrat and an elitist *littérateur*. Rather, it was between someone like Bernal and someone like George Orwell (1903–50), the liberal journalist known nowadays as the author of *Animal Farm* and *1984*.[17] In the decade following the end of World War II, BBC Radio frequently held debates between scientists and humanists on the future of Western civilization (a.k.a. British society). With a few notable exceptions (e.g., Michael Polanyi), the scientists contended that the path charted by scientific materialism made it inevitable that political problems would be solved by technical means. This, in turn, would remove the volatility caused by protracted public debate, the supposed root of Fascism's mass appeal.

The humanists, again with some exceptions, were born-again liberals, former Communists who could not tolerate the excesses of the Stalinist regime, even if they were committed in the name of the proletarian revolution. Orwell, in particular, was scandalized by the degree to which the scientists bent over backward to excuse the Soviets. This included the ease with which the likes of Bernal would reverse the meanings of common words to make, say, "freedom" appear to be possible only in a collectivist regime, as people rationally forfeit their right to choose in order to enable

---

15. Leavis 1963.

16. It is worth noting that Snow's "Two Cultures" lecture finds a natural place in the genre of British literature devoted to "civilizing" scientists, which extends from Matthew Arnold's *Culture and Anarchy* (1869), in which "philistine" was first used to capture the utilitarian approach to legislation that tramples over tradition in the name of science, to Collins and Pinch 1993, which similarly captures science's cultural ineptitude in the image of the "Golem," a creature whose strength and clumsiness are two sides of the same coin. Unlike Germany and France, where science has been an integral part of state policy for at least a hundred years, in Britain science has been traditionally pursued in the private sector, for either leisure or profit, but in any case in arenas that were tangential to the arts of citizenship. This explains the special resonance of the expression "public understanding of science" in Britain, which is interpreted as the problem of integrating science into general education and the larger culture.

17. Here I follow Werskey 1988, 285 ff., written by a political historian of contemporary science who failed to congeal with the original Edinburgh Science Studies Unit, about which more below.

their intellectual superiors to decide policy on their behalf. Orwell was convinced that the unmitigated attraction of scientists to Marxism reflected deep totalitarian tendencies in the scientific mind, and indeed modeled the thinking of 1984's leading ideologue, O'Brien, on Bernal's on-air pronouncements.[18]

From Snow's standpoint, the scientists erred in trying to model political decisions on scientific problems, where it is normally assumed that nature dictates one optimal solution. Marx's philosophy of history obviously facilitated this conflation of science and politics, which only served to promote a public image of science as antidemocratic. If the natural sciences held the keys to human emancipation, then clearly their representatives were projecting the wrong image by leaning so heavily on the idea of nature speaking in one overbearing voice.[19] As it turns out, the Labour Party rode to power in 1964, inspired by Harold Wilson's famous 1963 party congress speech in Scarborough that popularized Snow's vision by linking Britain's future to a "scientific revolution" and "the white heat of technology." Soon thereafter, several interdisciplinary support teaching units were established in British universities to carry out Snow's proposals. The historically most important of these has been the Science Studies Unit at Edinburgh University, the birthplace of SSK, which by 1970 had evolved into a full-fledged graduate research unit.

Within Edinburgh, initial support for the unit came from Conrad Waddington (1905–75), professor of animal genetics, who early in his career had come to know Bernal by collaborating with Bernal's friend, Joseph Needham, on the genetic basis of embryology. Waddington broadly shared Bernal's commitment to scientists becoming more involved in public policy making, but he was closer to Needham in his appreciation of the diverse moral and religious contexts for this involvement.[20] Requiring science

---

18. Bernal was hardly alone in advocating the scientific suspension of democracy and other widely shared cultural values. None other than Kuhn's Ph.D. dissertation supervisor, John Van Vleck, went down the same path, especially in his trips abroad (though without the Marxist agenda, just as one would expect of a Harvard product of the period). See Van Vleck 1962.

19. However, it would be a mistake to conclude that Bernal's sense of the epistemic advance that science makes on politics completely disappeared from the British scientific community. Rather, it was made rhetorically more palatable, as in the following, taken from a modern classic of science popularization: "If politics is the art of the possible, research is surely the art of the soluble" (Medawar 1969, 97). Never has the distinction between preparadigmatic and paradigmatic inquiry been epitomized so succinctly.

20. I am alluding to Needham's life project, a multivolume history of "science and civilization" in China, which posed clearly the question of why a "scientific revolution" that marked seventeenth-century Europe had not occurred earlier in the technologically more advanced Chinese empire. For a sensitive assessment of his achievement, see Cohen 1994, 418–90.

students to take a course on "science in world affairs" would thus enable them to acquire Bernal's outward orientation to science without his normatively objectionable rhetoric.[21]

The teaching unit at Edinburgh was made up primarily of trained scientists who, through acquaintance with the works of Kuhn, Mary Douglas, and Bernal's archenemy, Michael Polanyi, came to realize that scientists had to be "socialized" and "acculturated" just as much as anyone else. These pioneers included a newly appointed director, David Edge (a radio astronomer who worked for the BBC and eventually founded the leading journal in the field, *Social Studies of Science*), Barry Barnes (a chemist who received postgraduate certification in social theory), and David Bloor (an experimental psychologist with philosophical proclivities). They constituted the core of the Strong Programme in the Sociology of Scientific Knowledge. However, before they became a recognizable research group, the Science Studies Unit managed to persuade science students to divert their career paths from basic to applied research, a mark of success in integrating scientific expertise into the larger culture.[22] But while the rhetoric of the Science Studies Unit was one of encouraging scientists to escape from a narrow discipline-based curriculum, the unit's justification for its own structure was anything but that:

---

21. Waddington had been a prominent member of the damage control team that tried to salvage a socialist sensibility to science from the ruins of Lysenkoism, once it became clear that the policies of Stalin's agriculture commissar were both false and oppressive. See C. H. Waddington, letter to *The New Statesman*, 25 December 1948, quoted in Werskey 1988, 297 n. Waddington may have been equally motivated by the pronouncements of his former colleague at the Institute of Animal Genetics, Herbert Muller (1890–1967), who, upon winning the 1946 Nobel Prize in physiology and medicine for discovering that X rays can induce genetic mutations, began calling for the provision of sperm banks for superior individuals to counteract the ambient nuclear radiation that was likely to deplete the genetic fitness of the human species. To his credit, though, Muller argued against updated appeals to Lamarck, which claimed that mass irradiation could *hasten* the course of evolution by increasing the rate of mutation in the population. See Graham 1981, chap. 8, which is a generally excellent source for the value implications of the life sciences that Waddington was trying to tackle.

The sociologist Zygmunt Bauman credits Waddington with having recognized that scientific specialization encouraged a diffusion of the ethical impulse that disabled even very educated people from placing their actions and attitudes in a comprehensive moral universe. Bauman has fashioned the term "adiaphorization" to describe this process, which would explain why, say, Bernal, despite his considerable intellectual sophistication, could remain so insensitive to traditional libertarian and democratic sensibilities. The Edinburgh science studies curriculum would presumably facilitate the needed normative reintegration. For a discussion of Waddington's *The Ethical Animal* (1960) in this context, see Bauman 1993, 68–69. My thanks to David Edge for providing Waddington's early intrauniversity memoranda, along with collateral material, concerning the establishment of the Science Studies Unit.

22. This point is reiterated in several of the documents that David Edge has kindly provided me. Perhaps the most prominently placed is Edge 1970, which characteristically hedges its conclusion so as to suggest that the students may have entered the course by "a prior career decision," so as to conceal the subversive potential of the program.

We do not look on our graduate training as a "broad education" or a preparation for careers in administration, science journalism, and the like: We believe that such educational aims are best attempted at the undergraduate level . . . At the graduate level, the disciplinary aspect of our interdisciplinarity is predominant: we want to *train scholars* in this area; this implies (we believe) personal supervision, careful selection and control of topics and students, and *small numbers*. A central, basically academic point is, in my view, involved here: the coherent frame which we have built up over the years is related to what coherence we have managed to achieve in our social structure; this, in turn, is related to the small size of the group, and the strength of its boundary. Much of the knowledge of the group is tacit: As I noted above, we have problems introducing group members to its subtleties, and *a fortiori*, we have considerable difficulty in communicating with outsiders (and a reluctance to attempt the task—which reinforces our "boundary"); we also know now, from experience, that those who cannot quickly become "insiders" by organic assimilation into the group can end up mystified and confused as to what's going on. This is at the heart of our strategy of "careful accretion," and small numbers. If the group became too large and differentiated, *the only thing distinctive which we had to offer would disappear.* (italics in the original)[23]

Three features of this quote are worthy of remark. The first concerns the subtly disparaging reference to "broad education," in contrast to the specialist "training" used to justify the Edinburgh program. The author of the piece, David Edge, had acquainted himself with the class-based critique of educational styles that had resonated in 1960s Britain through the academically domesticated Marxism of Raymond Williams and Basil Bernstein.[24] That Williams primarily studied the writings of literary figures and Bernstein the speech of school children did not prevent Edge from noting their common emphasis on the elitist character of the ability to embed one's utterances in a larger cultural narrative, speaking as if one were the rightful heirs of a publicly recognized tradition. In response, Edge portrayed the Edinburgh School in the manner of a minority whose subjugation is marked by restricted modes of expression that largely result from a lack of historical resources to legitimate its concerns. Here we reach the second feature of the quote via the allusions to the anthropologist Mary Douglas's preoccupation with group boundaries and Michael Polanyi's

23. Revised manuscript (1977, 17) of Edge 1975b. Thanks to Edge for providing this revealing document.
24. Edge 1975a. Also Williams 1961; Bernstein 1971. One popular text of the period that Edge cites, which brought Williams and Bernstein between the same book covers, is Young 1971. Interestingly, Young, who became famous in the late 1950s for his class-based critiques of the emerging technocratic "meritocracy," is singled out in a recent review of British contributions to sociology as an exemplar of the empiricist, atheoretical approach associated with a "normal science" mentality. See Albrow 1993, 85.

with tacit knowledge.[25] In Douglas's terms, the Edinburgh School was "high group, low grid." In other words, group inclusion was defined more in terms of what its members jointly rejected than any fixed relationship amongst the members themselves. Together they presented a "wall of virtue" against the polluted outsiders.[26]

Each of the founding members had a foot in a natural and a social science, with Edge's institutional seniority compensated by his staff's stronger academic orientations. Consequently, the group displayed little sense of hierarchy, and indeed, one could not remain a group member unless one

25. Douglas was the most famous student of Edward Evans-Pritchard, whose fieldwork influenced Polanyi's fiduciary sociology of science, as previously discussed in chapter 2, section 8. Her work appears on the early reading lists of the Edinburgh Science Studies Unit, alongside that of Kuhn, Polanyi, and Wittgenstein (Bloor 1975). In 1995, Douglas won the J. D. Bernal Award of the Society for Social Studies of Science for lifetime professional contribution to STS. Her original fieldwork was conducted in the immediate postcolonial period, specifically in the context of development aid for the former Belgian Congo. Douglas studied two neighboring tribes living under equally modest natural conditions, but one was flourishing while the other was starving. She located the crucial difference in the greater integration of young males in the former society, which inclined them to produce more for the tribe. Douglas's work cast doubt on the idea that tribal cultures were impoverished simply because of lack of natural resources—and that by implication would automatically improve with more resources, say, through development aid. Vast amounts of food and money cannot take the place of a tightly knit moral order, according to Douglas.

Moreover, Douglas had no qualms about applying this conservative perspective to contemporary Western societies. In ecologically conscious 1970, Douglas published a remarkable article in the *Times Literary Supplement*, in which cultural relativist rhetoric was used to justify a politics of environmental risk that superficially appeared radical, but upon closer inspection is profoundly conservative, if not reactionary:

> Our worst problem is the lack of moral consensus which gives credibility to warnings of danger. This partly explains why we fail so often to give proper heed to the ecologists. At the same time, for lack of a discriminating principle, we easily become overwhelmed by our pollution fears . . . Any tribal culture selects this and that danger to fear and sets up demarcation lines to control it. It allows people to live contentedly with a hundred other dangers which ought to terrify them out of their wits. The discriminating principles come from social structure. An unstructured society leaves us prey to every dread. (Douglas 1970a, 247–48)

This passage was included in a widely used "set text" in the sociology of science, Barnes and Edge 1982, 274. In this and other works, Douglas appears to argue that any order is better than no order, and that a particular order becomes better simply by virtue of its proven ability to provide order. The Hobbesian leviathan—"Might makes right"—has thus been sublimated. This sensibility, which Douglas shares with her mentor Evans-Pritchard, has been attributed to their shared Roman Catholic background. See Fox 1997, 337. In the previous paragraph of the excerpt, Douglas cites none other than Kuhn for the "comforting" idea that scientists are typically so integrated into their paradigms that change occurs only once the Planck effect kicks in and the older generation dies off. The commitment to relativism is thus revealed to be not an open invitation to participate in the constitution of society, but a gesture to fellow scholars simply to observe the spontaneous constitution of their object of inquiry, "society." For a trenchant critique of Douglas's anthropological work on the environment, see Marcus and Fischer 1986, 146–49.

26. The landmark presentation of the group-grid scheme is Douglas 1970b. The versatility of the scheme in analyzing social phenomena is best brought out in Thompson, Ellis, and Wildavsky 1990, a textbook.

could be treated as a peer. Finally, the third feature comes into view, namely, the remarkable cohesiveness of the Science Studies Unit's staff—except for the member whose interests probably best represented those of Waddington and the unit's other early scientific backers. Although Don K. Price, by then the doyen of American science policy thinkers, gave the Science Studies Unit's inaugural address in 1966, the unit failed to retain a specialist in contemporary politics of science. Eventually, the unit stopped trying to fill the gap, settling instead on the geneticist-turned-historian, Steven Shapin. It is perhaps no small coincidence that Shapin has devoted much of his professional career to studying the processes by which one achieves the level of trust needed to become a credible producer of knowledge in relatively closed communities, such as the early Royal Society.[27]

3. Long-Term Effects of Kuhnification

The three effects of Kuhnification that I analyze below are all instances of the depoliticization of inquiry. The first case concerns the historiography of science, especially the way post-Kuhnians draw the distinction between "internal" and "external" histories of science. I then examine the way in which David Bloor draws the distinction between philosophy and sociology of science, which largely coincides with preparadigm versus paradigm, respectively. Finally, I consider the subtle but insidious way in which Kuhnification has infected the writing style, and hence thought pattern, of STSers.

A. Internal versus External History of Science

The distinction between internal and external accounts of science has been substantially transformed since it was first adumbrated in the 1930s. In a nutshell, the change has been from *externalizing the internal* to *internalizing the external* of science. As a historical reality check, the reader should recall that at the start of this intellectual odyssey, 1931, the capitalist world was in the throes of the Great Depression that for many marked the long awaited preconditions of World Communism. Of course, the next fifty years witnessed the rejuvenation of capitalism in the West, which enabled it to outlast, largely by outspending, the perceived Communist menace.

---

27. Shapin, who studied history and sociology of science at the University of Pennsylvania while Erving Goffman was at the peak of his popularity, began at the Science Studies Unit in 1972 and stayed fifteen years. His predecessors, the political scientist Leonard Schwartz and the political historian of science Gary Werskey, were each in post for only a few months. A good sense of the continued insularity of SSK from larger developments in STS (let alone larger developments in the intellectual world) is their recent collective opus: Barnes, Bloor, and Henry 1996. This point has not been lost on the book's reviewers. See Sardar 1997.

In 1931, two prominent Soviet officials, Nikolai Bukharin (editor of *Izvestia*) and Boris Hessen (director of the Moscow Institute of Physics), created a big stir in London by purporting to have demonstrated the ideological character of the history of science when recounted as the logical unfolding of timeless ideas. They argued that without acknowledging the operative political and economic conditions at each point in the history, the temporal succession of these ideas would appear undermotivated. Thus, in the case of the notorious "Hessen thesis," the ascendancy of the Royal Society and the acceptance of Newtonian mechanics were explained as key elements in the building of the British nation.[28]

As the excesses of Stalinist political practice counteracted the Soviet Union's original liberatory promise, this robust externalism came to be attenuated, mainly through Robert Merton's treatment of the internal and external as complementary "factors" that influence the course of science to varying degrees at different moments in history.[29] Merton granted that the internal trajectory of the history of science was much more than mere ideology, but external factors still had the power to either hasten or delay that trajectory. Indeed, twenty-five years after Merton's original thesis, Rupert Hall was able to claim that no social history of science had been done since 1953. Some would place the date at the start of World War II.[30] During the Cold War, external factors became increasingly associated with sources of ideological distortion deemed capable of subverting science at any moment.[31] At that point, the philosophical project of "demarcating" science from nonscience started to influence the historiography of science.

Also around this time, the late 1960s, a "professional historian's" sensibility emerged that portrayed the entire internalist/externalist distinction as itself ideological and, in any case, diversionary from the more technical concerns of historians coming to grips with the past. This view, which still predominates among historians of science, was launched with a couple of well-placed surveys on the historiography of science written by Kuhn.[32]

---

28. Bukharin 1971. See also Graham 1985. Graham argues that Hessen's Marxist credentials were in doubt at home, given his support of Einstein's theory of relativity. Consequently, Hessen had to construct his argument to show that Newton's achievements remained relevant, despite their bourgeois origins. Since Soviet scientists did not question the validity of Newtonian mechanics, it became especially important to stress as much as possible its politically unsavory origins.

29. Merton 1970.

30. Hall 1963. For the connection between World War II and the end of social history of science, see Crowther 1968, 288–91.

31. Perhaps the most important work to read Merton in this light was Barber 1952.

32. Kuhn 1968, 1971. Both are reprinted in *The Essential Tension*, 105–61. For a latter-day version of this sensibility, though written from an avowedly "externalist" perspective, see Shapin 1992a.

Here the historian is not a demystifier of ideology or an agent of social transformation, but rather someone devoted to understanding the past on its own terms.

This may involve demythologizing certain philosophical conceptions of history ("Whiggish" ones) that portray the present as the inevitable outcome of the past—but nothing more. Kuhn portrays the historians traditionally assembled under the banner of "externalism" (including the Marxist ones) as striving for a culturally embedded understanding of science that enriches the internalist picture, typically by drawing on a wider range of scientists (not just geniuses) and sources (not just published books). Subject to this soft-focus treatment, externalists merely complete the project originated by internalism, but they do not challenge its fundamental premises. In Kuhn's hands, "internalism" itself becomes a very broad church whose members include Auguste Comte, George Sarton, Pierre Duhem, and Alexandre Koyré—in other words, individuals who seem to have little in common other than a general belief in the overall positive cultural significance of science. At that point, the internal/external distinction had come full circle, with the external becoming fully assimilated to an internalist historiography of science.[33]

B. Philosophy versus Sociology of Science

Throughout *Knowledge and Social Imagery*, Bloor characterizes "philosophical" conceptions of truth and rationality in such terms as "ideological," "divisive," and "coercive"—the last in connection with Emile Durkheim's view that a truth is a belief on the basis of which a community obliges an individual to act in a certain way.[34] The presumption in each case is that the philosophical notions are impediments to scientific inquiry. But, in fact, Bloor tended to make a stronger claim, which comes out most clearly in his account of the Kuhn-Popper debate, namely, that *philosophy politicizes science unnecessarily*. Consider Bloor's diagnosis of why this debate never made much headway, thereby telling the sociologist of knowledge more about ambient cultural concerns in the 1960s than anything specific about the nature of science:

> The claim I want to put forward is that unless we adopt a scientific approach to the nature of knowledge, then our grasp of that nature will be no more than a projection of our ideological concerns. Our theories of knowledge

---

33. This strategy is reflected in Kuhn's efforts to reconcile the self-determination of normal science (cf. internalist history) and the role of political intervention in revolutionary science (cf. externalist history) by ensuring that scientists dictate when and how politics can enter their sphere. By this means, the external is internalized. See Kuhn 1970b, 168.

34. Bloor 1976, 174–75.

will rise and fall as their corresponding ideology rises and declines; they will lack any autonomy or basis for development in their own right. Epistemology will be merely implicit propaganda.[35]

It is common nowadays to fault Kuhn for exaggerating the discontinuity of conceptual change with the term "revolutions." Interestingly, however, Bloor originally faulted Kuhn's appeal to revolutions at a deeper level, namely, that it fed into the tradition of philosophy as politics by other means. According to Bloor, Kuhn and Popper opened the door for philosophers to use the history of science as yet another source of examples for articulating their preferred regime. While *The Structure of Scientific Revolutions* had the makings of a testable empirical model of scientific change, its galvanizing political imagery caused Kuhn, Popper, and the rest of their interlocutors to slip out of naturalism and back into the world of "ideology and utopia" that Karl Mannheim had identified some four decades earlier. The not-so-hidden agenda, then, of the Strong Programme is that it would depoliticize our understanding of science, in part by avoiding the political rhetoric that undermined Kuhn's own aims along these lines.

Bloor was not so naive as to believe that science can purge itself of *all* politics. However, to acknowledge the persistence of social interests in science is not necessarily to *endorse* the presence of such interests, especially when they potentially vitiate the sociologist's understanding of scientific practice. When Bloor attacked the influence of philosophy on science, he presumed that the practices of scientists have a historically based integrity of their own. This was the result of his reading, through late-Wittgensteinian lenses, that science has a distinct "form of life." Of course, scientific practice may be subject to legitimate change by its practitioners. The larger social- and self-interested concerns that scientists bring to their work typically contribute to such change, but only when they actually *improve* scientific practice. But who decides whether practice is helped or hurt? That, according to Bloor and his SSK followers, is an empirical matter to be settled by the scientists in a particular community, the result of which the sociologist is then in a position to discover empirically.

For Bloor, philosophy can muddle the perspective of either the sociologist or the scientific agent, should one be led to believe that the significance of the interests at play for inquiry may go unrecognized—as in Hegel's cunning of reason—by most practicing scientists. In that case, scientists have lost their power to represent science, and politics in a sense objectionable even to Bloor has triumphed. In the final analysis, despite their divergent starting points, Bloor concurred with Kuhn and most contemporary philos-

35. Bloor 1976, 80.

ophers of science that the political should be disentangled from the scientific and purged from at least one's own (i.e., the sociologist of science's) scientific practice.[36]

Missing from this consensus, however, is the idea that a recognition of the role of politics in science might invite a reconceptualization of science as intrinsically political, which would then be reflected back on the sociologist of science's own practice. This is the prospect favored by Marxists and feminists—as well as myself. But how might one do the initial spadework of sowing the seeds of dissent in the judgment that Bloor shares with Kuhn and other philosophers of science? Perhaps the best line of attack would be to aim at Bloor and Kuhn's shared Wittgensteinian commitment to the *game metaphor* for understanding the nature of scientific practice.[37]

One putatively attractive feature of the Wittgensteinian strategy of treating social practices as language games is its implication that such practices exist only insofar as their practitioners act as if they believe in them. However, such a view suffers from two unrealistic assumptions. One is that social practices are so clearly demarcated from each other that they can be individuated on the basis of the beliefs of their practitioners. But is it so obvious where "science" stops and the rest of society begins? The players often seem to be the same in the two cases. The second unrealistic assumption is that the actions licensed in a given language game do not have consequences for other language games. But of course they do, in which case the practitioners of the recipient language games may wish to have a say in what passes through their borders. Here I have in mind the effects of science that spill over into the larger society, be it industrial pollution or standards of educational achievement.[38] I shall resume with this point in chap-

---

36. The desire for apolitical purity in the sociology of science was present even in the earliest formulations of the Strong Programme: e.g., Barnes 1975. An excellent critique is provided in Proctor 1991, chap. 16. However, it would be unfair to imply that all of the outreach work inspired by STS has tended to reinforce an uncritically positive image of science. For example, a series of booklets published by Deakin University Press are designed to enable Australian citizens to interrogate the role of science and technology in their lives, often by drawing on the critical questions asked by philosophers and sociologists of science in their inquiries. Especially noteworthy is Albury 1983.

37. For an influential endorsement of the science-as-game metaphor that explicitly draws on Kuhn for support, see Lyotard 1983, 26. Bloor 1983 is devoted to his Wittgensteinian influences. Bloor's more recent reflection on Wittgenstein can be found in Bloor 1992.

38. An interesting slant on this discussion is provided by Alasdair MacIntyre, who argues that the ritualistic and noncognitive character of religion in the modern world—as depicted not only in the such positivist tracts as Ayer 1936 but also in the works of many anthropologists—is a mark of *secularization*. Because the contexts in which the outcomes of theological disputes would make a practical difference have been culturally eliminated, belief in God has effectively retreated from the public to the private sphere, thereby giving the public expression of such disputes the appearance of little more than a game that may elicit enthusiasm from its players but little claim on a reality beyond the context of play. It may be argued that the

ter 8, when I propose the concept of movement as an antidote to the gamelike implications of paradigms.

C. Kuhnification as a Literary Style: Its Metaphysical Costs

Post-Kuhnian depoliticization has probably made its most insidious inroads in the writing styles of STSers, be they recently minted Ph.D.s or seasoned professionals. This style deserves a name appropriate to its status: *contextualist boilerplate*.[39] Consider the following, which is drawn from the concluding paragraph of a recent article, culled from a doctoral dissertation at an Ivy League university and published in one of the leading journals in the history of science. I have very lightly edited the piece to enable its author to escape with anonymity:

> Three characteristics of the process of creating and adapting scientific knowledge stand out. First, as we have seen, the process was multivalent, strongly shaped not only by those who sought to produce scientific knowledge, but also by the audience for which it was produced. Second, the granting of authority to a particular unit of knowledge — be it a concept, an instrument, or a technique — occurred only after complex evaluations took place assessing validity, relevance, and costs vis-a-vis the needs of its targeted public, and here the outcome was determined by particular and local distributions of social, economic, and epistemic power. And third, knowledge production involved transformation, not only of the producers of the knowledge and of the public that was being persuaded, but also of the knowledge itself, as it was continually redirected and modified in order to accomplish its persuasive tasks.

I do not wish to suggest that this summation is untrue to the historical episode recounted earlier in the article. On the contrary, it is true to the point of triviality; hence, its boilerplate character. This chunk of text could have come at the end of virtually any recent article in the social history of science — and probably even before any research had been done. You probably cannot tell which discipline or century the author is writing about — and that is *not* because these have been deliberately masked.

Here we see the ultimate irony of the contextualist character of STS: It is all too easily universalized and rendered into an a priori historiographical

---

Wittgensteinian turn in Bloor's work, while not true to science as it exists in today's world, constitutes an implicit normative appeal to secularize science and hence restrict its relevance to players of its language games. See MacIntyre 1970. On the secularization of science, see Fuller 1996a. The theme of secularizing science is resumed in chapter 8, section 4.

39. My thanks to Deirdre (formerly Donald) McCloskey for introducing me to this use of the term "boilerplate," which usually refers to the form of words that lawyers mechanically use when drafting contracts. The term itself comes from the late nineteenth-century process by which syndicated articles were transmitted to newspapers for immediate publication without editing.

scheme.[40] The telltale sign is that our young historian's language veers between (what is excerpted here) rather vague, general, and fairly commonsensical explanatory principles and (what preceded it in the article) detailed descriptions of particular episodes that are laced together by these principles as narrative convenience dictated. Sometimes the general principles are associated with a fashionable contemporary theorist, but even then, the theorist's words are considerably watered down from their original meaning. No effort is spent trying to split hairs with the theorist's exegetes to focus whatever light the theorist might throw on the topic. Thus, Michel Foucault may be invoked in accounts that appeal to "power" and "discipline," but the distinctive theoretical spin that Foucault gave to these terms—which brought them *out* of the realm of the commonsensical—is rarely retained in these accounts. Instead, the reader is delivered the generic message that knowledge is power, and that power turns out to be much more complicated than first thought. Unfortunately, this conclusion does not get us much beyond the first moment of inquiry, when we marvel at the complexity of empirical reality and then seek theoretical guidance to distinguish between more and less salient aspects of that complexity. STS seems incapable of moving to that second stage, and so its overtures to theory rarely escape banality.

The main casualties of this vacuous appeal to theory are subtle conceptual distinctions that could help STS develop a genuine critique of science and technology. In effect, the homogenizing effects of Kuhnification—in a paradigm that routinely espouses "heterogeneity"!—has made the average STSer theoretically tone-deaf. Consider the first and third "characteristics" in the excerpt from our young historian above. The author appears to be describing the same thing in slightly different words. Nevertheless, this rewording would flag significantly different phenomena to the discerning theoretical ear. In point one, the author speaks of knowledge construction as a "multivalent" process, whereas in point three it is described as undergoing "transformation." There is a big difference between the idea that knowledge has multiple aspects at any given time and that it changes over time. Of course, the two ideas are logically compatible, but they are not the same. Their difference matters because one might wish to distinguish the *intended* from the *unintended* consequences of a particular knowledge construction. Talk of "transformation" implies that people are

40. Biagioli 1996 is alone among pieces by mainstream science studies practitioners to recognize this point—and see it as an epistemological problem for the field. Anticipating my efforts to distinguish contingentism from relativism, Biagioli weds an evolutionary epistemology to a Bourdieuan sense of rivalry that I originally discussed here in chapter 5, note 28, and to which I return in chapter 8, section 1. A sense of the mainstream resistance to this line of thinking may be gleaned from Schaffer 1996.

shaping knowledge to their interests, and those whose interests are stronger will get the knowledge shaped to their wishes. However, talk of "multivalence" implies that the shape taken by a given body of knowledge may ultimately elude the interests of the dominant shapers because they cannot anticipate all the ways its various aspects might be taken up by others. By clarifying this distinction, the path is open to alternative conceptions of the assignment of responsibility and the proposal of future courses of action.

In fairness to junior colleagues, I must also point out that contextualist boilerplate is just as common among distinguished senior STS practitioners. Indeed, students may pick it up as a stylistic tic from their teachers. Here is a good example that appears in a review essay of a recent major work in the field:

> Generalities like these detract from the book's more complex and provocative message: that "objectivity" in science and decision making is a *contingent* social product, sought after perhaps in every democratic society, but realized only in ways that reflect particular historical circumstances. Democracy does not simply promote objectivity; it both constitutes and is constituted, in specific guises, by the strategies that advance impersonal rationality.

I have highlighted "contingent" to alert the reader to another incapacitating conceptual confusion, one that neglects the following point of logic: *the opposite of contingency is necessity, not universality.* Contingency is the favorite modality of political revolutionaries of a scholarly bent. By declaring that things could have been otherwise in the past, the contingency monger hints that things can be otherwise in the future. The STS methodological commitment to *constructivism* amounts to just this: one reveals the seams in an apparently seamless web of belief in order to suggest that the web somehow can be rewoven.[41] But how does one empirically demonstrate that things, indeed, could have been otherwise? Showing, say, that the relationship between objectivity and democracy has varied from place to place, or time to time, refutes the idea that the relationship is always and everywhere the same.[42] In short, the *universality* of the relationship is disconfirmed. However, this result is logically compatible with the improbability, in each specific case, that objectivity and democracy could have been interrelated other than they were. In other words, even if many other significant features of the historical situation had been different, the rela-

---

41. In this respect, I do not exempt myself from the ranks of the "contingency mongers." See Fuller 1994c.

42. Of course, in one sense, this variation can easily be explained in terms of the multiple meanings attached to "objectivity" and "democracy." But insofar as people draw inspiration from one set of meanings to promote another that travels under the same name, the deeper question remains.

tionship between the two concepts in each case might have remained largely the same. In that case, *particularity* (i.e., nonuniversality) would be compatible with necessity. Conversely, one might empirically demonstrate a contingently universal relationship between objectivity and democracy. This would mean that the same relationship between the two concepts can be found in virtually all times and places, but in each case, that relationship is forged and maintained by a different social process.

The conceptual space I defined in the previous paragraph is mapped out more explicitly in table 11. To disambiguate the two dimensions of universality/particularity and necessity/contingency, as I have just done, is to acknowledge that *variety* and *change* are two distinct ways of expressing difference. The one implies difference in *space*, the other in *time*. This allows for a more nuanced understanding of social reality beyond the block universe of "necessary universality" and the endless flux of "contingent particularity," neither of which provides an adequate premiss for purposeful action, as the very first metaphysical skirmish between Parmenides and Heraclitus a century before Socrates should have already made clear. Missing from this crudely Manichaean division are "contingently universal" and "necessarily particular" forms of knowledge: on the one hand, beliefs that are very widespread but do not run very deep in any given society and hence may be reversed under the right circumstances; on the other, beliefs so specific to a particular society that they constitute its very identity, such that reversing them would be tantamount to erasing the society's culture. Among the contingent universals may be the globally pervasive character of scientific modes of legitimation (dubbed threateningly as "hegemonic" or "imperialist"), especially if we accept a discourse-analytic view of science as equivalent to a set of linguistic strategies for legitimating a whole

TABLE 11. *Four Epistemic Modalities for Expressing The Compatibility of Democracy and Objectivity*

	NECESSARY	CONTINGENT
universal	Knowledge claims are valid regardless of who, if anyone, makes them: e.g., democracy and objectivity are mutually implied, and so should always appear together, if either appears at all.	Knowledge claims are widely adopted in the manner of a lingua franca — out of mutual convenience: e.g., democracy and objectivity are positively correlated, but for cross-culturally different reasons.
particular	Knowledge claims are shared by group members to maintain the group's unique identity: e.g., democracy and objectivity are jointly essential to the identity of our society but not all societies.	Knowledge claims are indexed to specific places and times: e.g., in a given culture, "democracy" and "objectivity" coexist in rather unique senses of those terms.

host of contradictory and potentially reversible activities.[43] Among the necessary particulars, I would include the belief that the rise of science and the West are inextricably bound up with one another, which (perhaps understandably) exerts an especially strong hold on Western societies.[44]

Currently, STSers collapse the variability and changeability of knowledge claims into the semantic black hole of "contingency," leaving the impression that any belief anywhere is always up for grabs. In this way, strong grounds are provided for producing indefinitely many case studies of "science in action," in order to display the full range of contingency, but no grounds are provided for thinking that an overarching socioepistemic strategy could ever significantly alter science and technology's place in the world. Moreover, by referring to this depoliticized form of contextualism in terms of "contingency," STSers create some authoritative distance between their "analyst's" perspective and the perspectives of the agents under study in particular epistemic contexts, which is not unreasonably seen as the epistemic colonization of the agents. After all, contingency is a feature of knowledge claims that is normally not directly experienced by the knower. It is only when the knower performs a metalevel inquiry relating her beliefs to the conditions sustaining them that a sense of their contingency becomes apparent.[45] The difference between the agent's and the analyst's perspectives in STS effectively formalizes a division of labor between object level (or "first order") and metalevel (or "second order") inquiry. This is unlike the sense of the "particularity" of one's beliefs, which arises as part of everyday cognition, whenever one runs across a significant number of people who hold different beliefs on similar matters.

A reader impatient with abstractions may wonder what is to be gained by the conceptual sophistication I have been urging here. My main point is clearly a philosophical one. To a large extent, STS's metaphysical commitments have remained implicit, defined by the field's opposition to the view that science provides a universally valid account of a reality that exists independently of human beings—a position sometimes mischaracterized as "positivism" but better called simply *realism*. In the next section, I shall argue that this opposition constitutes STS's originary myth, the driving force of its internal history. In that sense, the conceptual sloppiness that continues to mar writing in this field reflects STS's captivity to its founding

43. The canonical presentation of discourse analysis as a methodology in the sociology of science is Gilbert and Mulkay 1984. It is applied to the "imperialism" that supposedly informed India's scientific achievement in Raj 1988.
44. I develop this point in Fuller 1997d, 137–44.
45. This fact provides the pretext for Karl Mannheim's view that the sociology of knowledge is an "oppositional science" in the sense promoted by György Lukács. See chapter 5, section 2, for a more detailed discussion.

myth. In earlier chapters, I alluded to two distinct metaphysically inspired strategies that STSers have used to oppose realism: *relativism* (first raised in section 4 of the introduction and then elaborated in chapter 6, section 2) and *constructivism* (discussed as a dissenting metaphysics within the scientific community in chapter 2, section 2). While both relativism and constructivism are rightly associated with STS in its opposition to realism, they do not oppose realism in the same way; to wit:

> *Scientific realism* involves two distinct claims, each of which can be denied separately:
>
> 1. A scientific account is universally valid. Therefore, if a scientific theory T is true, it is true everywhere and always. The denial of this claim is *relativism*. It implies that reality may vary across space at any given time.
>
> 2. A scientific account is valid independently of what people think and do. Therefore if T is true, it is true even if nobody believes it. The denial of this claim is *constructivism*. It implies that, for a given place, reality may change over time.

This, in turn, suggests two general strategies for designing an antirealist theory of science, as outlined in table 12. At the moment, the two antirealisms presented in the table are collapsed into one critique of realism because STSers tend to presume that the crucial opposition for understanding the role of science and technology in society is between "necessarily universal" and "contingently particular" knowledge claims, with the realists among philosophers and scientists standing for the former and an undifferentiated relativist-constructivist coalition in STS standing for the latter. Admittedly, Western philosophical discourse since Plato has been often conducted in such stark terms. Thus, knowledge claims incapable of meeting the standards of deductive validity required of a geometric proof—in which the conclusion necessarily follows from a set of explicit and universally accepted premises—have been routinely regarded as ephemeral and arbitrary.[46]

However, from the standpoint of contemporary cultural politics, the more relevant opposition is between the "contingently universal" and the "necessarily particular."[47] Already the language of social theory has made some significant semantic shifts in this direction. Consider two examples.

46. For a witty critique of much recent philosophy of science—including Popper, Lakatos, Feyerabend, and Kuhn—that has been led to skepticism by imposing impossibly high standards of knowledge, see Stove 1982, which ultimately puts the blame on David Hume.

47. My version of social epistemology can be seen as pursuing a contingent universalist stance, whereby I deny the independence of reality without denying its potential universality: in short, a *nonrelativist constructivism*. A cognate position that draws on the role of biotechnology as the crucible for constructing new worlds may be found in Haraway 1997, esp. 99, which formally contrasts relativism and constructivism.

TABLE 12. *Two Difference-Based Strategies for Designing an Antirealist Theory of Science*

DIMENSION OF DIFFERENCE	VARIETY	CHANGE
ideal basis for difference	one moment across space (synchronic)	one place over time (diachronic)
modalities of difference	universality vs. particularity	necessity vs. contingency
corresponding antirealism	relativism	constructivism

First, the Holy Grail of "cultural universals" long sought by anthropologists has been replaced by an interest in processes of "globalization," which imply a concern for how disparate factors conspire to produce a synchronized whole, such as a "world system." Second, the traditionally ephemeral character of cultural particulars has been hardened into what often seems to be a politically nonnegotiable concept of "identity." These two examples have been crystallized in public discourse in terms of what the American political theorist Benjamin Barber has called "McDonaldization vs. the Jihad," while academic prophets of the impending "information age" have conjured up a struggle between global computer networks and local identities forged in resistance to the messages that such networks convey. But the Theory Lite character of STS manages to avoid engaging with these issues, despite their palpably practical import.[48]

In this section, I have argued that contextualist boilerplate in STS writing is the primary means by which the field's Kuhnification is signaled on a regular basis. The style licenses an atheoretical empiricism that aims to represent the situated character of knowledge production while denying the knowers under study any real capacity for change. Ironically, as practiced in this mindlessly generalized fashion, as a "style," contextualism exemplifies what logicians call a "pragmatic paradox": The content of what one says ("context is all") is contradicted by the fact that one is saying it (in this case, because one says it always *without* concern for context).[49] If we regard Kuhnification as an epistemological syndrome, then this logical aberration may be seen as symptomatic of a more specific disorder: *paradig-*

---

48. For the first example, see Wallerstein 1996. For the second, see Barber 1995; Castells 1996–98.

49. Indeed, I am surprised that philosophers, usually easy marks for the facile paradox, have not attempted to invalidate STS simply on grounds of self-contradiction for arguing that knowledge is "necessarily contingent" or "universally local." After all, how could STSers know such a thing, given that so few cases have actually been examined? Perhaps we need to rejig the Platonic faculty of "intellectual intuition" so that "some A is B" licenses us to infer not that "all A is B," but rather that "most A are not B." Here STSers may make a valuable contribution to cognitive science—as themselves phenomena meriting explanation.

mitis, the tendency to reason that if a methodological strategy works in one case, it must work for all.

Thus, while most STS scholars officially oppose the very idea of a universal theory of science, they are nevertheless inclined to promote a universal *method* for studying science, be it Bloor's four tenets of the Strong Programme, Harry Collins's eleven propositions on experiment, or Latour's seven rules of method. The supremacy of this method is then justified by a self-serving historical myth, the basic structure of which I shall make explicit in the next section. Moreover, this way of thinking feeds into the standard demonstration of worth in a grant-driven research culture: for example, "Latourian actor-network theory has been applied to many different cases that received grants and produced results; ergo, it will probably work for the case for which I want a grant." Indeed, here may lie the proper sociology of knowledge explanation for the spread of paradigmitis throughout academia in our time.

4. STS's Own Internalist Myth and the Problem of Nature

An elementary lesson of the sociology of knowledge is that seemingly similar ideas can be generated from radically different social contexts, and unless care is taken, much of that original context may be unwittingly transferred to new settings where the ideas are planted. As a result, the ideas may develop in unexpected and unwanted directions. The book before you may be read as a demonstration of this lesson, as *The Structure of Scientific Revolutions* has proven to be a Trojan horse carrying much more than simply a general theory of scientific change.[50] However, the extent to which STSers have failed to reflect on the disparate sociological conditions that have called forth a broadly STS-style response is remarkable. Even a cursory understanding of the different post–World War II science policies pursued in the founding nations of STS—the United Kingdom, the United States, and France—should be enough to cast doubts on any straightforward history of the field. Nevertheless, again in classic Kuhnian fashion, STS is prone to conceptualize its own history as an evolving response to theoretical and methodological issues surrounding the nature of science that transcend national boundaries. Indeed, contemporary STS rhetoric regularly bears witness to this founding myth.

The myth goes like this. We may think of STSers and philosophers of science as the two sides of a dialectic. Both are fixated with the two main public symbols of science—the special laboratory site where knowledge is

50. Hence the title of my Kuhn obituary, Fuller 1997a.

produced and the special talk of truth, objectivity, and reason that goes on around that site. Both suppose that the secret of science lies in revealing the relationship between the two symbols. Whereas the philosophers invest powers of "referentiality" in the talk that enable science to transcend the site, the sociologists impute "indeterminacy" to such talk to explain how it obscures access to that site. To be sure, the loose and often reconstructed (perhaps even fabricated) character of what we say in relation to what we do is a commonplace feature of everyday life. Consequently, to draw attention to it under the rubric of "indeterminacy" serves a critical and polemical function only for those who would otherwise think that the language of science is somehow exempt by virtue of its unique transparency and rigor.

While many philosophers—and the scientists who take them seriously—have held such views in the past, with the decline of philosophical commitment to "positivism" or "realism" (two quite different positions that agree on the point at issue), it is no longer clear how much more mileage STS can gain by highlighting the "underdetermined" character of scientific discourse, aside from the sheer accumulation of case studies that confirm this virtually conceded point of referential indeterminacy. In that respect, even accepting STS's founding myth on its own terms, the field remains stuck in the second moment of its dialectic with philosophy of science.

There has been a self-declared radical strategy from within the STS community to dissolve this dialectical predicament. Its immediate source is Bruno Latour, but its ultimate provenance is that godfather of French structuralism and poststructuralism, Gaston Bachelard (1884–1962).[51] Bachelard had a peculiar way of envisaging the relationship between scientists and philosophers that was compelling to a French intellectual culture that has never gotten over its initial love affair with Karl Marx. According

---

51. The best introduction to Bachelard's philosophy of science, especially stressing its reception in Paris of the late 1960s, remains Lecourt 1975. For its relevance to STS, see Bowker and Latour 1987. Bachelard's work bears a passing resemblance to Kuhn's because of a common emphasis on scientific change through radical disjunction. However, Kuhn does not seem to have gotten much from his textual and personal encounters with Bachelard. See Kuhn et al. 1997, 166–67. Kuhn attributed this to Bachelard's explicitly structuralist methodology, but there may be another reason. All the great French historical philosophers of science, from Pierre Duhem and Emile Meyerson (about whom more in chapter 8, section 2) to Koyré, Bachelard, Bachelard's student Georges Canguilhem, and Canguilhem's students Louis Althusser and Michel Foucault, have been interested mainly in the ruptures in worldview that have enabled scientific perception to break free of ordinary experience, thereby creating an artificially self-contained world. For Kuhn, this really captures only the most radical transition, namely from preparadigmatic to paradigmatic inquiry, but not transitions from one paradigm to the next within a scientific discipline. To be sure, this distinction in ruptures is neater analytically than empirically, but it does help explain why the French philosophers seem to find far fewer revolutions in science than Kuhn and his followers have.

to Bachelard, scientists most closely approximate the ideal of *homo faber*, humanity in a state of world making, whose work is organically integrated into the world: we are all constructivists before we are anything else. From the academic Marxist perspective that gave Bachelard his widest reception in France of the late 1960s and early 1970s, he seemed to portray scientists as a proletariat exploited by bourgeois philosophers who held them accountable to standards not of their own making. At the institutional level, Bachelard's defense of scientists from philosophers should be read as a moment in the resistance of the polytechniques from the domination of the universities. He was trying, in the French context, to reverse the tendencies that were represented in nineteenth-century Britain by William Whewell (as discussed in chapter 1, sections 6–8), who sought to portray technical innovations as "always already" theorized. In short, every legitimate discovery would be seen as born justifiable.

For Bachelard, philosophers derive the conceptual equivalent of surplus value from performance standards that transcend the scientists' own horizons. These standards typically adopt a frame of reference that refers either to the beginning or the end of inquiry, that is, when science is still an idea in the scientist's mind or in fully finished form as an artifact or experimentally reproducible effect: roughly speaking, "idealism" (subjectivism) versus "realism" (objectivism). From either temporal end of the process, actual scientific work appears to fall short of the presumed philosophical norm. This provides the raison d'être for philosophical intervention: scientists are berated whenever they fail to meet the norm, while philosophers claim credit whenever scientists conform to it. Thus, Mach's failure to accept the existence of atoms and discover relativity theory is traceable to his opposition to scientific realism, whereas Einstein's success on both counts is attributable to his commitment to said philosophy. In this respect, the philosopher is part-ventriloquist, part-alchemist. Metaphysical ideas are projected into scientific work, only to be retrieved as the "essence" of that work, with the residual aspects then treated as in need of philosophical mediation.[52] In this way, any scientific achievement can be reduced to its rational essence and various arational deviations.

That the ultimate meaning of science comes from philosophy is persuasive just as long as philosophy speaks in one voice, and hence science is shown to be single-minded in its aims. However, Bachelard's point was that philosophy pulls science in many contradictory directions, more in the manner of ideology than an account of reality supposedly superior to sci-

---

52. Of course, this philosophical practice is not limited to France. Plenty of it can be found on display in the anglophone world, say, in Leplin 1984. In this context, Ian Hacking and Arthur Fine correspond to Bachelard and Latour in offering antidotes.

ence. In his characteristically analytic "philosophical topology," Bachelard—and Latour after him—depicts this phenomenon in terms of gradients of "displacement" or "dispersion" from any number of presumed philosophical norms.[53] What centers these displacements and dispersions is the actual work of science. By so privileging science as the "immutable mobile" of philosophical inquiry, Bachelard had overturned the entente cordiale between philosophers and sociologists, which involved treating natural science as an object teleologically drawn toward the philosophical domain of reason, absent the interference of sociologically defined forces. In effect, this Aristotelian image was rendered Newtonian, as Bachelard showed that natural science was driven by its own inertial impulse, subject to the competing pulls of various philosophies acting at a distance.[54]

In the wake of Bachelard, STS has not only privileged science over philosophy, but has even uncannily reproduced the value orientation of contemporary Big Science in the sites it has chosen for study.[55] Moreover, the field has continued the misleading impression that one is a scientist only in research, but not in teaching or administration. (Could this book have been written had I adhered to these scruples?) Of course, STS typically tells a rather different story from the one scientists or philosophers are inclined to tell about these privileged research sites. Usually, more people and things are incorporated into the STS narratives, which complicates the picture of how science manages to do as well as it does. But at the same time, the added complication diffuses responsibility for any of the actions taken in the name of science. On the one hand, this helps redistribute the credit for scientific work from the few "geniuses" who normally receive all the glory; on the other hand, it makes it difficult to hold anyone accountable for

53. On the use of this imagery by Bachelard and Latour, compare Lecourt 1975, 40 ff.; Latour 1993, chap. 3.
54. The entente cordiale between philosophy and sociology speaks to the lingering influence of Vilfredo Pareto (see chapter 3, section 5), which is probably stronger in France than in any of the other major national sociological traditions. He is typically read as having established a division of labor between sociology and whichever discipline is given responsibility for studying rationality, be it philosophy, economics, or evolutionary biology. The only reason French sociologists have not regarded this division of labor as demeaning is that, in true Paretian fashion, *most* social actions turn out to be irrational, so sociologists are left with much to explain. A good case in point is Boudon 1981. On this view, sociological analysis consists mainly of the analysis of variance from a normative model. An interesting cultural benchmark is that when, in the anglophone context, Larry Laudan proposed the very same division of labor between sociology and philosophy of science under the rubric of the "arationality assumption," he was widely seen as *demeaning* sociological inquiry. See Laudan 1977, 196–222.
55. Simply consider the sites of the classic case studies: Latour and Woolgar 1979, Knorr-Cetina 1981, Collins 1985. Karin Knorr-Cetina has done the most to continue this tradition with impunity. See Knorr-Cetina 1999, a comparative ethnography of the European particle accelerator (CERN) and a major molecular biology laboratory.

anything.⁵⁶ Consequently, the interpenetration of science and society so vaunted by our field rhetorically functions to discourage inquirers from looking far beyond those objects of fascination—the laboratories—to see how science reflects larger societal forces. Instead, science is portrayed as "always already social," which implies that whatever larger forces need to be taken into account will be "inscribed" in the people and things located in the laboratory. Not surprisingly, STS practitioners have endured an uneasy relationship with Marxist and feminist science critics. They have been united more in terms of a common foe—the scientific establishment—than a common methodological and axiological orientation.⁵⁷

More generally, STS conveys a surface radicalism, in that the analyst is supposed to suspend any technical knowledge that she might have of the practices under study. The subversive epistemological consequence, already noted, is that the STS practitioner often notices things, especially incongruities between word and deed, that escape the practicing scientist's attention. But the subversion only goes so far, as STS's own critical potential is truncated by the symmetrical tendency to suspend any technical knowledge of the *social sciences* that STS practitioners might bring to bear on the case.⁵⁸ While such methodological asceticism is not without precedent in the sociological literature, it has had the effect of precluding political factors that do not explicitly enter the scientists' own deliberations.⁵⁹

56. This problem is made central to understanding the role of "agency" in social theory in Fuller 1994b. Take, for instance, the leading STS model of science in society, actor-network theory. What, on a sympathetic reading, may appear to be an amorphous network of highly contingent nodes (an instance of Granovetter's "strength of weak ties") may be portrayed, less sympathetically, as an all-pervasive system whose general structure cannot be purposefully altered by some strategic intervention, let alone a social movement. In this way, STS practitioners may be able to continue their steady stream of detailed case studies for both collegial and cliental consumption without offering counsel to those interested in a fundamental renegotiation of science's social contract. It may be, then, that the joke is on us when we fail to recognize that actor-network guru Bruno Latour played it straight when he told an interviewer that STS does not pose any serious threat to the scientific establishment. See Crawford 1993. I shall return to the strategic amorality of actor-network theory in the final section of this chapter.

57. Among those involved in the uphill struggle to integrate STS into more comprehensive traditions of science criticism are Restivo and Loughlin (1987); Aronowitz (1988); Haraway (1991); Harding (1991). See also the contributors to Ross 1996.

58. Kuhn is explicitly charged with *retarding* the sociologization of science in Restivo 1983. Also, see section 1 above for the aspects of social-scientific knowledge absent from STS, at least in its SSK form.

59. One sustaining methodological precedent for SSK's noncritical practice from American sociology is "grounded theory," as put forward in Glaser and Strauss 1967. Here sociologists are enjoined not to introduce any more theory than the subjects themselves do when accounting for their own actions. An interesting—though probably dubious—epistemological assumption of grounded theory is that an inductivist methodology that reveals the multiplicity of perspectives of a social situation is *in general* best suited for challenging the status

For example, STS practitioners employ discourse-analytic techniques to reveal the various voices in a language game played by a community of scientists, but they do not use the techniques to engage in an ideology critique of science that appeals to factors that sustain the game but transcend the scientists' control or awareness. It is as if a postmodernist aversion to proffering master narratives has forced the STSer into a role of simply stripping away narratives that "others" have imposed on the subject under study, in the hope that something meaningful will remain to be said.[60] In the event that this does not happen, one can simply revel in the "chaotic" or "rhizomatic" character of the subject's behavior.

In short, STS seems to have inherited from Kuhn and SSK a sense of normative confusion, an incapacity to pronounce on whether it likes or dislikes what it so perspicuously sees. If "relativist" is the name that a philosopher or scientist gives to someone who raises an inconvenient fact against an incontrovertible truth, then "universalist" is the name that a historian or sociologist gives to someone who insists on deciding whether she would want to live in the world that she describes. At the risk of being accused of universalism, then, I would like to query the locus of normativity in the scientific enterprise. Many recent social histories of science stress the role played by technicians and other on-site laboratory personnel in the production and maintenance of apparatus needed to persuade onlookers that an experiment has worked properly. These people generally went unrecognized in their day. What is the normative conclusion that should be drawn from this? Is it something akin to a labor theory of value—those who do the work deserve the credit? Yet, as suggested above, the more populated the world of technoscience, the more diffuse the assignment of credit and blame—that is, unless some explicit attempt is made to remove credit or blame from others who have held it before.

Moreover, even this incipient labor theory of value omits the role that those away from the site of original knowledge production—the colleagues, policy makers, teachers, students—play in conferring a "scientific" status on the fruits of the site's labors. Given the constructivist methodological maxim of defining scientific practices by their consequences rather than

---

quo. For a recent incarnation of the same spirit, see Star 1995, which I critiqued in Fuller 1996b. While it may be a good heuristic when the perspectives have been traditionally suppressed, it is not at all clear that grounded theory has the same salutary effects on traditionally hegemonic perspectives. Like the Marxists and feminists, I worry that "going native" among elite laboratory scientists merely serves to render them "kinder and gentler" than their structural capacity for power would imply. For the methodology of an alternative "critical ethnography," see Harvey 1990.

60. For a explicit defense of this point, see Latour 1993, esp. 5–8, 122–27. See also Latour 1988b.

their causes, it would seem that regardless of the effort spent in the laboratory — be it by big-name scientists or no-name technicians — the final determinant of scientific status is the community of recipients, which may well be more democratically distributed than the community of producers.[61] In that case, should not STS practitioners ultimately prefer a theory of scientific value based on *utility* rather than labor? Unfortunately, this question remains both unanswered and, more tellingly, *unasked* in the STS literature.[62]

Of course, things need not be that way, since the possible ways forward are fairly obvious. Just focusing on the relationship between scientists' words and deeds, at least three gambits are available to the STS community:

> 1. Openly admit that science is no more or less truthful, rational, objective, etc. than other social practices, and conclude that either the status of science needs to be lowered or the status of other social practices needs to be raised.
>
> 2. Argue that the special talk surrounding science may have no binding force on the actions performed at the research site but it does constrain the possibilities for action in administrative and educational settings, where the appeal to science serves a more explicitly legitimatory function. In that case, those interested in witnessing the distinctive power of science would do better focusing their attention on these distribution points rather than the original "hands-on" sites of production.

61. Marxists have the most to learn from their ideological antagonists when it comes to specifying the exact locus of value. A good starting point is Sowell 1980, which claims to be elaborating Hayek 1945. Of relevance here is what Sowell calls the "physical fallacy": the idea that the value of a good is determined by the people who handle it just before it goes to market, as if their sheer contact is all that matters. This is the line of thought that underwrites the labor theory of value common to Thomas Aquinas, John Locke, and Karl Marx. The labor theory of value becomes a basis for critique when one assumes a disproportionate control of the market by those who own or manage the means of production. In that case, those who work for these "captains of industry" appear to be singularly exploited. But given that most businesses fail, the captains very often go down with their ships. Until this inconvenient fact is taken on board, all the politically correct talk about the unsung heroes of the scientific workplace will seem rather hollow. I am sufficiently impressed by the physical fallacy to believe that those who control the *consumption* of science — from teachers to advertisers — should be the main targets of STS critique. Of course, many science producers also fall under this category, but it is not their role *as producers* that would be of interest. See Fuller 1993b, 24 ff.

62. Perhaps the most popular advertisement for the consequentialist character of science is Latour 1987. Yet this book fails to explore the normative implications of the consequentialist thesis. Perhaps if it did, it would not enjoy its current popularity. Specifically, while Latour clearly dissociates himself from Marxist indignation that was born of the labor theory of value, he stops short of endorsing the obvious alternative, namely a neoclassical or Austrian economic determination of science's value by its users. Some may call this a strategic strength, others a moral weakness, of his text. I was initially captivated by *Science in Action*, but given its massive yet largely uncritical reception outside the scientific community, I must conclude that the book's evasiveness on this crucial point outweighs its strengths in challenging positivist and realist visions of science. I return to this point at the end of this chapter.

3. Somehow try to regiment scientific discourse to live up to its own normative ideals by subjecting scientific claims to greater scrutiny than one would ordinary claims. This would entail a level of suspicion and discipline that would effectively undermine the so-called "tacit dimension" that has traditionally conferred on scientific knowledge its status as expertise.[63]

It turns out that these three gambits define the parameters of *social epistemology*, an orientation to inquiry designed to check the spread of paradigmitis by making STS "reflexive," in the sense of subjecting its ends to *routine* revaluation, as opposed to the Kuhnian "if it ain't broke, don't fix it" mentality.[64] For example, the mode of revaluation stressed the most in this book would have STS researchers rethink their past as a guide to future action. Thus, as we saw in chapter 2, Ernst Mach excavated historically suppressed research traditions in order to chart a course that diverged sharply from the streamlined future promised by Max Planck.

As it stands, however, STS treats its past unproblematically, always tracing its proper origins to the Edinburgh School after having paid tribute to Kuhn as the mythical progenitor.[65] The argument from there simply concerns the step that logically follows this collectively assumed history. Thus, the most wide-ranging and seemingly radical professional debate within STS in recent years, the so-called Epistemological Chicken controversy, is ultimately about nothing more than alternative extrapolations from a common past.[66] The rhetoric of the debate is most naturally explained as niche differentiation in what has become a crowded field for STS research, much as one would expect of a paradigm that has become, in Pierre Bourdieu's words, "a world apart."[67]

On the one hand, we have Harry Collins and Steven Yearley, who

---

63. I have put forward this view most explicitly in Fuller 1992c. For a like-minded historical study, see Porter 1995.

64. See Fuller 1996c, where I introduce the thesis of the "Normative Underdetermination of Social Regularities" (NUSR), which is designed to undermine the presumption that a widely tolerated social regularity — even within the scientific community — is ipso facto normatively desirable. Applied to the present context, this means that the paradigmatization of STS is not necessarily a mark of epistemic virtue, if what is excluded is of greater significance than what is included. (Putting the point this way clearly implies that I regard "significance" as a contestable matter.) Rather, the ends served by those regularities first need to be evaluated. I provide an account of how social regularities in science acquire a "positive" rather than a "negative" spin in the educational system in Fuller 1997d, 63–67. Mulkay 1979b is normally credited with first promoting the normative underdetermination of research practice.

65. This remains striking even in such well-informed "second-generation" texts as Golinski 1998, 13–27. For a more balanced view of Kuhn's significance on STS, see Hess 1997, esp. 22–27, 48–51.

66. The original papers were Collins and Yearley 1992; Latour and Callon 1992. See also Fuller 1996f.

67. Bourdieu 1988. For a social-psychological account of academic niche differentiation, see Fuller 1994e.

propose to stick to the letter of the field's Edinburgh origins by extending the signature SSK methods of ethnography, discourse analysis, and critical historical scholarship to more domains of inquiry—beyond the usual academic research sites to environmental movements and knowledge-engineering firms—without deepening or challenging SSK's fundamental assumptions. In their hands, STS would remain autonomous not only from the folk theories that scientists and technologists use to explain their activities but also from the more sophisticated theories that social scientists have developed to explain other aspects of social life.

On the other hand, we have the self-avowed radicals of the debate, the Paris School of STS represented by Michel Callon and Bruno Latour. They believe that adhering to the original Edinburgh spirit requires breaking with its letter; hence, they call for a generalized application of SSK's symmetry principle. The complexity of technoscientific networks revealed in their studies cannot be accounted for simply by invoking social factors, however symmetrically (i.e., evenhandedly) they are applied to successful and failed courses of action. *Natural* factors need to be invoked as well— and just as symmetrically. Not surprisingly, scientists who follow the STS literature have welcomed the Parisian turn, since it clearly reopens the door to traditional, even commonsensical, explanations of science that incorporate both social and natural factors "interacting" to produce, say, an experimental outcome.[68] It would seem, then, that we have reached one of those all-too-familiar Molièrean moments in academic life when a move that appears radical within the terms of a paradigm is equivalent to the prose that everyone else outside the paradigm has been always speaking (albeit now with a French accent).[69]

---

68. See, e.g., Labinger 1995. For a critique of commonsense talk of social-natural interactionism modeled on that of mind-body interactionism, see Fuller 1996c.

69. The locus classicus of discussions of the relationship between symmetrical and asymmetrical sociocognitive practices is Lévi-Strauss 1966, esp. 30–33. Lévi-Strauss distinguishes between practices that aim to create a hierarchy of primary and secondary qualities out of an otherwise ontologically flat world (e.g., reductionist approaches to science) from practices that aim to remove the advantage that a spiritual force has gained at the expense of a community (e.g., elaborate native death rites). The former, which Lévi-Strauss categorizes as "games," convert a symmetry into an asymmetry, whereas the latter, "rituals," convert an asymmetry into a symmetry. (Drawing on the language of number theory that lay behind much of the discourse of the "structuralist" turn in the human sciences, Lévi-Strauss himself spoke in terms of the conversion of symmetries.) Thus, football played as a game aims to distinguish winners from losers, whereas performed as a ritual (in, say, New Guinea) it aims to create an equal number of winners and losers, thereby dissolving the inequity represented by the winner/loser distinction.

The difference between Lévi-Strauss and Latour on this score is that the former was an asymmetrist who drew distinctions between the "civilized" and "savage" minds (albeit much less judgmentally than past anthropologists), while the latter is a "symmetrist" who aims to dissolve such distinctions once they have been recognized. In other words, from a Lévi-

Here the social epistemologist would take a step back from this incestuous dispute and consider the broader dialectic of history. From the standpoint of considerations raised in chapter 2, the Parisian turn to Nature is a historical step backward, if not an outright leap into nostalgia. The increasing reliance of physicists and chemists on the laboratory manufacture of experimental effects had already led several reflective practitioners at the dawn of the twentieth century to question the cognitive significance of the category of Nature for scientific inquiry.[70] However, these practitioners—including Ernst Mach, Pierre Duhem, and Wilhelm Ostwald—were generally outcasts from the physics establishment. They were acutely aware that the continued appeal to Nature allowed physics to enhance its social status, despite the increasingly artificial settings and arcane mathematical formulations that characterized physical research. Nevertheless, from a rhetorical standpoint, these appeals suggested an ultimate end to scientific inquiry and that a physicist will be the one to reveal it, whether it be an insurmountable barrier (e.g., the unsplittable subatomic particle) or an intended destination (i.e., the stuff out of which everything else is unreplaceably constructed). While this enabled the physics community to preserve its autonomy in the face of co-optation by the state and industry, it also sidelined, on the one hand, theological accounts of reality that purport to transcend physical inquiry (Duhem's concern) and, on the other, instrumental accounts that subserve physics to human ends (Mach's and Ostwald's concern).

Of course, the category of Nature need not be suspect when studying science and technology. First, one can imagine a "naturalization" of science that would involve studying the practices of natural scientists as one would other activities and processes in the natural world, namely, by a full range of empirical methods: direct observation to see how things are, experimentation to see how they might be, and history to see how they have been. Indeed, a combination of the last two methods may lead to some

---

Straussian standpoint, Latour 1993, chap. 4, tells us how a savage would write a book called *The Civilized Mind.* The question, of course, is exactly *who* would want to read such a book, and why. From what I have said in the text, one obvious audience is scientists wishing to be relieved of the burden of legitimating the social order (by generating asymmetries of what is true/false, rational/irrational, etc.) without having to change their actual research practices.

70. This issue has been raised anew in recent years, given the increasing cost of conducting Big Science research. Based on his interviews with distinguished scientists in several fields, one journalist has concluded that science is currently in the process of taking one more conceptual step away from Nature by transferring its activities from the artificial world of the experimental laboratory to the virtual world of computer simulation. In that case, hypothesis testing will become more like literary criticism, in which aesthetic criteria are used to assess what will have become technologically enhanced constructions of the imagination. See Horgan 1996.

surprisingly critical conclusions, such as that the distinction between the natural and social sciences is an institutionally entrenched historical accident.[71]

But a craftier approach to Nature as an analytic category involves driving a semiotic wedge between the term "Nature" and the function it serves in our conceptual scheme. The French psychoanalytic theorist Jacques Lacan (1901–81) used the expression "floating chain of signifiers" to describe the phenomenon whereby words acquire meaning by referring, not to an extralinguistic reality, but to still more words. For Lacan, a Freudian nominalist, this was the structure of the unconscious.[72] Examples of the Lacanian unconscious at work include slips of the tongue, malapropism, euphemism, and most generally, metonymy—the rhetorical trope that allows the speaker to refer to something by mentioning something else conventionally associated with it. Lacan regarded this continual deferment of meaning as a closed system. The trick then is to identify a series of terms that eventually return to the one originally displaced. Without making any particular commitment to the peculiar etiology that accompanies Lacanian psychoanalysis, the circuit of displacements that ground contemporary controversies surrounding STS are captured in table 13.

Note that Nature plays an important role in this chain of signification, but its exact significance is determined by where one starts to recount any of several overlapping accounts of our times that is enabled by the chain. Here is one such story. The devolution of the nation-state has transformed the public from a taken-for-granted object of political discourse to a specialized form of knowledge that requires the work of academics, marketing researchers, and spin doctors for its elicitation. Perhaps the most concrete demonstration of the public's newfound elusiveness is the gradual decline in electoral participation in the world's oldest democracies. Meanwhile, the role previously played by "public opinion" in centering political discourse has been replaced by the natural environment, resulting in a politics that is manifested not in the voting booth, but in localized protests and alternative lifestyles. This is due to the environment's impact on the background conditions that enable a dispersed polity to pursue its increasingly customized goals. Even if the state can no longer bend the populace to its will, the threat of mass contamination from a nuclear disaster is usually sufficient to focus a society's collective attention toward action.[73] Finally, advances in computer technology, combined with a general demystifica-

71. This is the main thesis of Fuller 1993a.
72. A relatively accessible introduction to this notoriously difficult writer is Lacan 1972.
73. This theme has been popularized in recent years in Beck 1992. See also Eder 1996. (My thanks to Roy Boyne for alerting me to the significance of this literature.)

TABLE 13. *The Unconscious of Science Studies*

METAPHYSICAL FUNCTION	OLD TERM	MEANING	NEW TERM	CONVERSION PRINCIPLE
form	"expertise"	knowledge that is elusive yet essential for survival	"public"	political devolution
goal	"public"	concerns that are common to members of a population	"nature"	environmental degradation
matter	"nature"	scarce resource that is susceptible to alienation and commodification	"expertise"	intellectual automation

tion of knowledge-based forms of authority, have shifted the terms in which expertise is discussed from ones of *cultivation* to those of *preservation*, both of which borrow from traditional discourses of Nature. In this context, it is common to bank on the hope that the value added by having a human rather than a machine execute an intelligent task offsets the considerations of efficiency that have traditionally marked the reception of labor-saving technologies.[74]

However, none of these metaphysically clever ways of accommodating Nature into contemporary studies of science and technology quite captures the modus operandi of the Parisian turn in STS. The rest of this chapter is devoted to the cultural and political positioning of the STS researcher in the ongoing "Science Wars," culminating with an extended examination of the Paris School itself. I shall argue that the Parisian appeal to Nature can be regarded as a preemptive attempt at co-optation: that is, an attempt to lure scientists and technologists to the STS cause before STS suffers a similar fate, even if it means that STS researchers must yield some of the ground gained by their social-scientific colleagues.

5. STS on Cruise Control: Diagnosing the Science Wars

A good way of characterizing a mode of inquiry in the throes of Kuhnification is *epistemic cruise control*. Specifically, the Kuhnified field loses any sense of historical reflexivity, a precondition to effective political engagement: that is, a sense of where one has come from and where one should be going. Instead, it simply continues along the path that has been laid down by presumed past successes. Any suspicion that STS might be on

---

74. See Fuller 1998e.

cruise control in this sense has been confirmed with a vengeance by the ongoing Science Wars, in which the disparate images of science promoted by scientists and STSers have been contested in the public sphere.[75] An impartial survey of STS responses to date would have to include the words "surprise" and "confusion," even though it should have been perfectly evident that the diminished status STSers assign to institutionalized science as a societal arbiter of rationality and objectivity would eventually meet with resistance by the scientific community. In what follows, I distinguish between what may be called the *amnesic* and *inert* sides of cruise control: respectively, a forgetfulness of the past and an obliviousness to the future.

A. Amnesia

In 1994, a book appeared by two scientists who had steeped themselves in STS and affiliated literatures and surfaced with the conclusion that, the field's pretensions to the contrary, STS is poorly placed to contribute to a progressive politics, given its refusal to countenance a knowledge base independent of its social origins.[76] If STSers believe that knowledge is no more than what those in authority claim, then how can it serve as a basis for liberating oppressed minorities? The implicit answer — that those minorities constitute themselves as communities bound by traditions of local knowledge — is unrealistic in a world whose local affairs are irretrievably entangled with global ones. The authors of *Higher Superstition*, Paul Gross and Norman Levitt, the one a marine biologist and the other a mathematician, leaned heavily on examples from the medical and environmental sciences, where the failure to adopt a "scientific" perspective was supposedly responsible for untold disasters. The critique is ironic in two senses, both of which speak to the spell that Kuhn has unwittingly cast over STS.

The first irony was betrayed in the knee-jerk response of many leading STSers who accepted the Kuhnian premises of Gross and Levitt's critique, namely that an inquiry that admitted the influence of its sociopolitical setting was inherently suspect. Thus, STSers simply denied that their claims to knowledge were inextricable from the interests that informed the context of their production by pointing to well-established traditions of research, as evidenced in the grants, results, and awards secured by the field's practitioners, none of which had anything to do with the original context of knowledge production. Moreover, all of this was carefully distinguished from the pseudopractitioners of STS in such para-academic endeavors as "cultural

---

75. The expression "Science Wars" was coined by the cultural studies scholar, Andrew Ross, originally as editor of the special issue of *Social Text* that included the notorious Sokal hoax (Sokal 1996).

76. Gross and Levitt 1994.

studies" and "social activism," which (of course!) fully deserved Gross and Levitt's scorn. Indeed, the tenor of this argument has been that the more carefully STS delved into the social character of knowledge, the more closely its own mode of knowledge production would approximate that of the "normal science" practices of the disciplines they studied. In other words, elements potentially disruptive to the conduct of inquiry, because of their socially specific character, would eventually come to be laundered or internalized as part of the disciplinary regime of science — and thereby no longer provide a source of disturbance. As Bruno Latour himself might put it, STS reveals the epistemic purity of its own inquiries, as it locates the social taint in the inquiries of others.[77]

    77. This response was accentuated with the Sokal hoax, since Sokal's highly documented piece threw together works by core STS researchers with those by cultural studies and feminist scholars, as well as various recent French intellectuals, for whom science is at least as much a metaphorical resource for their own thinking as an object of critical scrutiny. For an extended version of Sokal's argument, see Sokal and Bricmont 1998. Fuller 1998f provides a direct response.

    Was this lumping together of "academic leftists" justified? Notwithstanding the instinctive distancing moves of more orthodox STSers, the answer is far from straightforward. Whatever obscurity of expression is evidenced by French intellectuals today is less due to ignorance, let alone antipathy, to the natural sciences than to the long-term effects of arrested intellectual exchange between France and Germany after the rise of Nazism. (My argument presupposes that Germany has been the principal source of philosophical and scientific inspiration, since the rise of the modern university in the early nineteenth century.) This has led to an involution of French philosophical prose that intensifies as the philosopher finds her native market niche. For example, it is now hard to believe that, while still in his thirties, the psychoanalytic theorist Jacques Lacan was in regular contact with the Bourbaki mathematicians or that he drew lasting lessons from Alexandre Koyré's seminar on Galileo (see chapter 1, note 65). Even Jacques Derrida developed his celebrated thesis on the primacy of writing in Western thought in an extended introduction to the French translation of Edmund Husserl's *The Origin of Geometry*. Derrida 1973 is a good transitional work in his corpus that reveals the traces of his attempt to deal with the metaphysical foundations of formal scientific thought. A more general insider's account of the intercourse between the human and nonhuman sciences in postwar French thought is Piaget 1970.

    Whereas Henri Bergson and Emile Meyerson, two prominent and antithetical figures in French thought before World War II, wrote demanding but still relatively accessible prose that enabled them to debate with colleagues across national boundaries, such debates now occur only with great difficulty, even though arguably the writings of French intellectuals have never been more easily available in translation. Instead, the French and German "masters" are mediated by exegetes and epigones who engage in virtual exchanges in the language of translation. My thanks to George Gale for initiating this discussion of the sources of obscurity in contemporary French thought. Fuller 1988 defends the general thesis that communication breakdown of the sort witnessed in Europe during World War II provides the main source of radical conceptual change.

    Lest I be accused of an anti-Continental bias, I must observe that the cessation of contact with Germany probably also explains the preciousness of ordinary language philosophy that flourished in the United Kingdom in the immediate postwar period before the discipline was colonized by U.S. developments in semantics and the theory of reference, starting in the late 1950s and continuing throughout the Cold War period. It is worth noting that this anglophone philosophical involution was savaged by such cosmopolitan émigrés of the old Austro-Hungarian empire, including Karl Popper, Friedrich von Hayek, Imre Lakatos, and Ernest

The second irony is that from their brief but concentrated study of the STS literature in relation to other recent movements in the American academy, Gross and Levitt managed to acquire a better understanding of the historical backdrop to the peculiar trajectory of STS than many of the field's practitioners seem to have. Again, given the historical amnesia that normally characterizes scientists according to Kuhn, it is perhaps unsurprising that STSers themselves were caught off guard when Gross and Levitt reinvented the modernist response to postmodernism in the "culture wars" that had erupted in humanities departments nearly twenty years earlier. On that occasion, English translations of such French "poststructuralist" theorists as Jacques Lacan, Michel Foucault, and Jacques Derrida were becoming widely available for the first time. Back then, critics influenced by these theorists deconstructed attributions of "value" in literature and "validity" in literary criticism, using arguments similar to those now being used by STSers against science. It would seem, then, that just as senior natural scientists have caught up with the original wave of postmodernism, the latest—and most professionally oriented—generation of STSers have lost contact with this body of work. Thus, whereas it was common in the 1970s for my teachers to worry that I would treat Foucault and Derrida as the new Parmenides and Heraclitus, today's STS recruits routinely become familiar with, say, Bruno Latour before having assimilated Foucault or Derrida, let alone Latour's philosophical mentor, Michel Serres. Little wonder that they fail to anticipate, meet, and sometimes even comprehend the sorts of objections that people like Gross and Levitt are trying to make.[78]

B. Inertia

Consider the special issue of the field's founding journal, *Social Studies of Science*, devoted to "The Politics of SSK: Neutrality, Commitment, and

---

Gellner—most notably in Gellner 1959. For them "Britain" stood for the Scottish Enlightenment, John Stuart Mill's liberalism, and Bertrand Russell's cultured skepticism—not the Oxbridge academic establishment where ordinary language philosophy flourished. A start at articulating the details of these tendencies behind today's split between anglophone and Continental European philosophy may be found in Collins 1998, chaps. 13–14.

78. A good example of this "déjà vu all over again" is that a supposed mark of STS radicalism is the tenet that Nature is the product, not the cause, of what scientists decide. (STS students know this as Latour's "third rule of method.") This inversion of the normal relationship between cause and effect, ultimately taken from Nietzsche, is no more than a standard trope in postmodernist deconstructions. For a textbook demonstration, along with a refutation that was widely noted in its day, see Culler 1982, 86–88; Searle 1983. My own early attempts to apply deconstruction to science after having read the relevant French theorists but not much STS are captured in Fuller 1983. The most obvious point of contact between then and now is the stress that both poststructuralists and STSers place on the "indeterminacy" or "uncertainty" of interpretation, what is sometimes called the "semiotic" dimension of social life. Semiotics is explicitly used to structure STS theorizing in Bijker and Law 1992, esp. pt. 3.

Beyond."[79] Common to the array of positions presented is a sense that the inquirer's politics is largely a personal matter that lies outside the proper modes of STS inquiry. STS practitioners are forced to reflect *methodologically* on their status as political agents only once they are thrown into situations that require them to defend the integrity of their research. This typically happens when someone who has been made a subject of STS research uses that research to promote her own ends. At that point, leading STS practitioners gather in print to reach consensus on *the* stance that one is "professionally" licensed to take. Discussion of normative issues is effectively transferred to the "meta" level, which shields STS practitioners from having to confront their own internal political differences on particular substantive issues. Thus, the sense of "politics" discussed in this context seems abstracted from what normally passes for "cultural criticism" in society at large. Whatever else STSers might be when they engage with political issues, they are *not* public intellectuals.[80] This point will acquire special significance in the conclusion to this book, where I contrast the dynamics of inquiry in a paradigm and a *movement*. In the latter case, the terms of criticism are not confined to what can be expressed in the paradigm's discourse but rather resonates with ambient political currents. In other words, STS would reproduce within itself—homeopathically, as it were—the conflicts embedded in the larger society. But before we explore the sense in which STS might be a movement, we must first understand the sense in which political differences are now paradigmatically contained in STS.

The first point to note is that, despite the field's polemical reputation, its empirical studies are rarely subject to professional scrutiny or criticism. Researchers tend not to reanalyze the results of earlier research; instead, they colonize different, if sometimes overlapping, domains.[81] In that way, potential theoretical and normative disagreements are sublimated as differences in one's acquaintance with the case at hand, with the advantage invariably going to the person who saw the natives (or their documents) "firsthand," especially when this required a considerable expenditure of mental and perhaps even physical energy. Thus, the main discursive moves in a STS conference consist of comparing and contrasting cases, as one

79. Ashmore and Richards 1996. The special issue was edited by Malcolm Ashmore and Evelleen Richards and includes contributions by the editors, Harry Collins, Brian Martin, Dick Pels, Brian Wynne, and Sheila Jasanoff.
80. A valiant attempt to reverse this trend is the editorial introduction to Ross 1996.
81. Thus, it would seem that STS has taken only half the advice for critically creative inquiry offered by the great social science methodologist Donald Campbell: it has accepted the "fishscale model of omniscience" but not the "competitive reanalysis of data." On these two matters, see, respectively, Campbell 1969 and Campbell 1984.

presumes that differences in the conclusions reached are due to differences in the cases themselves and not to differences in the competences or the background normative commitments of the researchers. In this respect, STS conducts its business in a thoroughly "objectivist," normal scientific manner.[82]

What STS practitioners typically fail to consider (at least publicly) is that politics may have entered at the very moment they decided to pair an STS standpoint with a particular research topic. As a result, the motives for doing STS remain conspicuously obscure. On the one hand, if pure scholarship were the primary aim, then why not apply STS to less politicized domains? On the other hand, if political activism were the aim, then why not simply do advocacy-based research? An uncharitable interpretation would suggest that STS researchers simply want to have their cake and eat it: they want to occupy a buffer zone within which they can occasionally influence policy decisions without ever having to expose their partisanship, as that might render them vulnerable to more powerful forces in society. Even resisting this appeal to such anxious motives, we can still regard STS as enabling its practitioners to walk the following two tightropes:

- attacking the Kuhnian mythology of science as a self-determining community of inquirers at a point where the image is most vulnerable (e.g., cutting-edge research, science-based public controversies), without themselves having to relinquish that image in their own normal scientific activities;
- aligning themselves with what, from a vaguely leftish standpoint, would be regarded as the underdogs in the societal struggle under study, without having to commit to either a general theory of politics or a specific party platform.

82. Moreover, this attitude has spread, most markedly in recent historical writing on science. The bellwether episode is the historiography of what, since the end of World War II, has been called the Scientific Revolution of the seventeenth century. The first half of the twentieth century witnessed several attempts by historically inclined philosophers and philosophically inclined scientists to chart and contest the chronologically vague but suggestive terrain surrounding that expression. The late medieval and early modern periods effectively became Rorschach tests for projecting one's views of what enabled Western culture to develop the unique social practice called "science," which could then be used as a standard against which to evaluate later and even contemporary practices. The names of Duhem, Mach, Koyré, Crombie, and Needham continue to dominate this discussion, since, with the exception of Shapin and Schaffer (1985), no new competitors have entered the fray in the last generation to define the Scientific Revolution. This point becomes very clear upon examining Cohen 1994, a compendium of perspectives. One implicit lesson of Cohen's book is that, in recent years, non-Western scholars have made most of the original contributions to understanding the West's distinctiveness. For their part, Western historians have shifted from framing their narratives around a normatively charged expression like "Scientific Revolution" to specific nations and figures of seventeenth-century Europe. Indeed, Shapin has denied the existence of the Scientific Revolution in a recent popular work (1996) designed to introduce readers to the historically distinctive features of science.

In light of all this nimble-footedness, a possibly more useful exercise than speculating on the political implications of STS research would be a comparison of science-based controversies that STS has covered with those it has not. In that case, the "emergent" political agenda of the field can be discerned and then made a topic for reflection. For example, no officially sanctioned STS-style analyses have yet appeared that improve the epistemic standing of Nazi science or Creation science.[83] In public discourse about science, both heterodox forms of inquiry are routinely subject to "asymmetrical" treatments that presume their ill-foundedness, which is in turn explained by the nonrational beliefs of its practitioners that are not shared by practitioners of more esteemed cases of science. Yet, these are typically the sorts of cases that STS would rehabilitate.

Even in the realm of environmental politics, which has increasingly absorbed STS practitioners over the past few years, it is striking just how much more critical attention is focused on the dogmatism of government and industrial scientists who play down the degree of uncertainty in their judgments than on the dogmatism of environmental activists (and their scientific spokespeople) who play *up* the degree of uncertainty — and sometimes even assert the virtual certainty of imminent disaster. Of course, there are many good reasons for wanting to tilt one's account against prevailing scientific opinion in environmental politics, but they involve going beyond the self-imposed normative limits of STS.[84] The exact point of STS research would be made more transparent if such reasons were explicitly stated. Among such reasons may be:

- The inquirer may see herself as an agent of democratization who levels the epistemic playing field by privileging the perspective that starts from the weakest power base (or greatest burden of proof).
- The inquirer may harbor her own theory of risk that says it is always

---

83. To be sure, Creationism and Nazism provide interesting cases in which scholars in neighboring fields have appealed to STS concepts to question the legitimacy of our accepted views about science. For example, American rhetoricians have appealed to STS concepts to rehabilitate the public status of Creationism as an implicit criticism of the scientific community for its ham-fisted treatment of opponents (Taylor 1996, 135–74). Nevertheless, this work has been largely ignored by mainstream STS practitioners who share the same religion-bashing tendencies of philosophers of science and practicing scientists. Indeed, Gross (1997) has remarked on this peculiar convergence of opinion, where he would have expected STS to back the Creationists' "struggle for identity." In the case of Nazism, the historian Robert Proctor has used STS as a springboard for pushing the "bad politics can lead to good science" thesis to its logical limits. Proctor (1999) suggests that the Nazis pioneered what we now take to be enlightened views concerning the promotion of organic foods, vegetarianism, smoke-free environments, and related forms of "healthy living," because they believed that a superior race should practice a superior lifestyle.

84. Radder 1992. This thesis is expanded in Radder 1996, 1998.

TABLE 14. *Officially Sanctioned STS Attitudes to the Politics of Its Own Research*

	FOCUS ON WHAT IS INSIDE STS	FOCUS ON WHAT IS OUTSIDE STS
assert own epistemic authority	1. expert mediator	4. arms merchant
deny own epistemic authority	2. Zen master	3. parasite

better to err on the side of caution and hence to privilege the perspective that estimates the most uncertainty in a situation.
• The inquirer may be a devotee of a localist epistemology that is, in principle, antagonistic to universalist models that fail to incorporate the nuances of the particular case.
• The inquirer may presuppose a "discourse ethics" that refuses to accept that the need to maintain social order is sufficient reason for compromising the truth, say, by downplaying levels of risk.

Nevertheless, as is to be expected of a field in the grip of a Kuhnian paradigm, STS researchers gingerly avoid broaching these difficult political, ethical, and epistemological issues as grounds for legitimate disagreement over the different purposes to which the field's resources may be put. This is symptomatic of the field's suffering from paradigmitis. In the special issue of *Social Studies of Science* mentioned above, the disease appears in two forms. One form stresses the distinct expertise that STS can impart to science-based controversies; the other the uncertainty (about, say, what the future may bring, how policy should proceed) that lies beyond the sphere of STS's empirical research base.

In other words, one talks about what is either contained in or excluded from the presumed paradigm, but not the two simultaneously, as that would draw undue attention to the artificial boundary between them and hence threaten the paradigm's integrity. Moreover, these alternatives appear in conjunction with judgments as to whether STS's historical rejection of the epistemic authority that philosophers have sought in science should apply to STS itself. The result is the matrix of ideal types depicted in table 14.

Below I characterize each ideal type in terms of its attitude toward the politics of its own research. The contributors to the special issue of *Social Studies of Science* can be seen as combining these attitudes to varying degrees, according to the context of their research:

1. *Expert mediator:* Because STSers study how closure is reached in scientific knowledge claims, they possess an expertise that none of the

contesting parties to a publicly relevant scientific controversy is likely to have, which then enables STSers to serve as mediators in such disputes. This is the most straightforward claim to STS as a discipline that deserves a place alongside the other disciplines in the university curriculum—a claim it gladly makes without problematizing the idea of disciplinized knowledge. It is a version of the social-scientific appeal to paradigms that we saw in chapter 5.

2. *Zen master:* The only politics that STSers officially recognize is a principled opposition to all master narratives, regardless of the "real-world" political consequences of such opposition in each case. Of course, any given STSer may have her own views about real-world politics, but these are regarded as outside the proper sphere of STS. Two points are worth noting. First, as has already been suggested, the "principled" nature of this opposition is in practice compromised by the selective character of the cases taken up by STSers. Second, the Zen master assumes the impossibility of a unifying vision of science and society that at the same time enables the flourishing of multiple perspectives. Clearly, then, a Hegelian synthesis is not on the cards.

3. *Parasite:* The expertise of STSers consists in the identification of hidden ambiguities and uncertainties in ongoing scientific controversies. Unlike the STSer as expert mediator, the parasitic STSer does not claim to possess any expertise of her own, merely the ability to reveal the lack of expertise in scientists in matters of public concern. The intended result is to make scientists more modest and nonscientists bolder in their participation in policy matters. A charitable interpretation of this situation is that policy initiatives become more experimental—bolder both in the possibilities they explore and the checks they impose. A less charitable interpretation is that STSers simply perpetuate the open character of scientific controversies so that decisive action is never taken, while their own services are rendered essential.

4. *Arms merchant:* Depending on the social position of a given group in relation to a particular STSer, some social agents will be better placed than others to exploit STS knowledge. While STSers knowingly provide resources for others to empower themselves, addressing existing inequalities in the distribution of access to those resources lies beyond STS's jurisdiction. Moreover, this inability to control the flow of one's own knowledge products has been heightened as STS research is increasingly funded on a contract basis. Although STS research has always had a strong funding basis outside academia (in government, the nonprofit sector, and even some industry), ideologically soothing appeals to STS "professionalism" tend to obscure the ease with which specifically contract-based research lends itself to political captivity by the client. In that sense, the STSer unwittingly offers legitimatory resources to the highest bidder.

In the next section, I consider the development of the group of STSers who have most effectively combined these four attitudes into a distinctive research practice commanding the respect of policy makers and academics alike in many countries. But before discussing the Paris School, let me make fully explicit the precedent I have used to launch this critique of STS's Kuhnification. It comes from American sociology's soul-searching in the 1960s: Alvin Gouldner's stinging indictment of "the theory and practice of cool" that he accused Howard Becker of promoting under the banner of symbolic interactionism and labeling theory in his 1966 presidential address to the Society for the Study of Social Problems.[85] Tellingly, Becker's address has been presented as a mainstream sociological precursor of the STS perspective.[86]

To his own question, "Whose side are we on?" Becker seemed to answer society's underdogs, those psychosocial deviants whose very existence constituted "social problems" for those who funded much sociological research in the era of the welfare state: the insane, the poor, the addicted, the criminally inclined, and the variously strange. On the surface, Becker's argument called for the articulation of these deviant perspectives, giving them a clarity and coherence that "the establishment" would otherwise deny them. But Gouldner questioned Becker's sincerity. The accounts promoted by Becker's "cool" approach to social problems cast the deviants as living in self-contained "worlds" where deprivation effectively becomes a mark of positive identity. But exactly who benefits from such a portrayal? Here is Gouldner's bracing answer:

> The new underdog sociology propounded by Becker is, then, a standpoint that possesses a remarkably convenient combination of properties: it enables the sociologist to befriend the very small underdogs in local settings, to reject the standpoint of the "middle dog" respectables and notables who manage local caretaking establishments, while, at the same time, to make and remain friends with the really top dogs in Washington agencies or New York foundations. While Becker adopts a posture as the intrepid preacher of a new underdog sociology, he has really given birth to something rather different: to the first version of new Establishment sociology . . . It is a sociology that succeeds in solving the oldest problem in personal politics: how to maintain one's integrity without sacrificing one's career, or how to remain liberal although well-heeled.[87]

Gouldner regards Becker much as I regarded (in chapter 2, section 8) Edward Evans-Pritchard's attempts to understand the Azande and Nuer "in their own terms." An epistemologically liberating relativism, upon closer scrutiny, made it easier for a distant power—the inquirer's client in both

---

85. Gouldner 1968, which was a response to Becker 1967.
86. Star 1995, 1. For a critique, see Fuller 1996b.
87. Gouldner 1968, 111.

cases—to contain the natives by redressing the balance of power enjoyed by a more immediate (and competing) oppressor of the natives. At the very least, the lesson to be taken from these examples is that the people who are given voice in a sociological narrative are not necessarily the ones best positioned to exploit their newly articulated identities. They can be easily turned into pawns in a power struggle. Indeed, whatever strength the natives derived from their invisibility and marginality is generally lost. If the natives are not then connected with a social movement capable of extending their sphere of action, their interests will have been effectively rendered a target for ventriloquism by a higher authority.

This perverse consequence would appear even more troubling were the sociologist not its direct beneficiary, much in the manner of what Georg Simmel called the *tertius gaudens*, the third party who benefits from the miseries of others. For by stopping short of taking any direct responsibility for what her client does with the information gathered on the natives, the sociologist articulates the self-imposed limits of her paradigm, which in turn reinforces (for the benefit of current and future clients) her "professionalism." Moreover, as Gouldner astutely observed, when Becker acknowledged the inevitability of bias and value commitment in his research, he made it seem—as STSers continue to do—that these are matters of personal predilection that would vary across inquirers. But in fact, the structural position of the sociologist in relation to both clients and subjects virtually ensured the reproduction of specific power relations, regardless of the content of the sociologist's findings.[88]

To be sure, the sociologist's *tertius gaudens* role is based at least as much on survival as self-aggrandizement. An apt comparison is with the moral situation of the informant in wartime, except that here we are talking about a marketplace environment where increasing amounts of academic research is done by contract workers whose employment prospects are determined on a grant-to-grant basis. Under such a regime, if researchers do not provide quality information about their subjects to clients, they will be quickly replaced by someone more willing and able to do so. Thus, the researcher's credibility as a witness is always at issue. Gouldner detected

88. Gouldner 1968, 112. Among the science studies disciplines, historians of science have been probably most reflective about the structural character of the power relations in which their work is embedded. That they are often commissioned to write histories *for* the client—a museum, a laboratory, an industrial firm, a professional association—enables the tension between their interests as critics and celebrants to be felt with unusual vividness. In contrast, sociologists are rarely hired as sociologists but rather as high-grade information gatherers and analysts. Thus, it is easier for them to compartmentalize their activities—writing one kind of report for the client and another for a professional journal. This difference reflects the extent to which history still has a public face that sociology lacks. See Soederqvist 1997, esp. the chapters by Jeff Hughes, Skuli Sigurdsson, and Joseph Tatarewicz.

this tendency in Becker's concern that sociologists might succumb to sentimentalizing their subjects and thereby "lose their cool." For Gouldner, the object of such sentimentalization, people's suffering, is morally and politically significant but, for Becker, they prove practically and theoretically inconvenient. The idea that suffering might be integral to people's experience implies that the oppressed might see themselves not in terms of some positive identity (or distinct "social world" they inhabit) but exactly as their oppressors see them—driven primarily by their subjugated status—and that their suffering was either avoidable in the past or eliminable in the future.[89]

For the client-centered sociologist to introduce such a measure of indeterminacy and changeability about the natives into her report would be tantamount to admitting that she had not mastered the means of controlling them. It would also draw undue attention to the client's role in maintaining the power relations revealed in the sociologist's report. Of course, there is a metaphysical remedy for such uncomfortable revelations. It involves flattening the ontology of the social world so that structures are replaced by networks, and all parties are presented as exerting their own kind of power over each other, according to the alliances they can form in a given circumstance. Claims that the natives are subjugated or suffering are thus converted into ones about their hidden competences and agency. As a result, the contingency of the natives' condition may not be reduced, but the client's responsibility for it is. With that in mind, we turn to the Paris School of STS.

6. STS Captured by Its Own Cleverness: A Social Epistemology of the Paris School

For the last ten years, the major source of intellectual excitement in science and technology studies has come from France, particularly the Center for the Sociology of Innovation (CSI) at L'Ecole Nationale Superieure des Mines in Paris, which is one of *les grandes ecoles*, the elite institutions of higher education. The two principal theorists, Michel Callon (b. 1945) and Bruno Latour (b. 1947)—the one trained in engineering and economics and the other in philosophy and anthropology—have developed a version of STS known as "actor-network theory," which promises nothing less than

89. In this respect, *theodicy*—the theological explanation for the distribution of goods and ills in the world irrespective of human intention—has yet to be fully accommodated within sociological inquiry, even though the first "naturalistic" accounts of the social order (the "invisible hand") were basically secularized versions of theodicy. See Fuller 1998a. The classic sociological discussion of theodicy is Weber 1964, 138 ff. On the importance of theodicy to understanding social attributions of meaningfulness, see Berger 1967.

a complete makeover of the social sciences, specifically by defining "networks" as the stuff out of which both individual identity and social organization are constructed.[90]

From a sociology of knowledge standpoint, the most striking feature of this development is that it has been more popular as a research agenda in the English-speaking world than in France itself, even though the French context for studying the social character of science is rather unique. It has included from the outset a very strong state-led "strategic research" initiative designed to promote what the late socialist leader François Mitterand in his first successful bid for the French presidency in 1981 called "technoculture." Here Mitterand had taken a page from Harold Wilson's 1964 rhetorical playbook in using science and technology as rallying points to unite the country around a vision of economic progress.[91] As part of this campaign, STS would be instrumental in tracking the application and reception of technoscientific innovation.

In fleshing out his conception of technoculture, Mitterand contrasted his own *politique de filières* with the *politique de créneaux* of his predecessor, the neo-Gaullist Valery Giscard d'Estaing. This switch in metaphors would become second nature in STS by the end of the 1980s, as the mark of "technoscience" becomes the strength of its networks rather than the achievement of specific goals. As it turns out, Giscard had also promoted science and technology, but in the context of triaging French productive capacities by targeting research funds to industries—especially aerospace, arms, nuclear energy, and transport—where France enjoyed a comparative advantage on the world scene.

However, France's world share in these areas did not grow during the Giscard presidency (1974–81), and overall the economy slumped. Mitterand's remedy was to make information technology integral to *all* sectors of economy, including the old industrial warhorses of steel, mining, and shipbuilding. In the end, France continued to lag on the world economic stage and its industrial sector imploded. Yet, as unprecedented numbers of workers lost their jobs under Mitterand, they could at least take comfort in their high-tech skills that, in principle, enabled them to move into the booming

---

90. The role of networks in social life has long been recognized in the social sciences, but usually as an intermediate level of social organization between, say, a face-to-face group and an institution. However, the Parisian proposal would redefine these multiple levels as networks of varying lengths, resiliency, and rates of growth. The advent of global electronic communications has reinvigorated the appeal to "network" in social theory. The most ambitious attempt to see the entire contemporary scene in these terms is Castells 1996–98. Interestingly, Castells makes no reference whatsoever to actor-network theory, though he appears to be familiar with recent work in the social history of technology. For a social-epistemological critique of Castells's project, see Fuller 1999d.

91. On Mitterand's refashioning of Wilson, see Turney 1984, 221–22.

service sector of the economy. Indeed, among major industrialized nations, the distinctiveness of France's economic profile over the last half century has been its rapid conversion from agriculture to service as its leading source of wealth and employment.[92]

Although the CSI had been founded toward the end of the de Gaulle era in 1967, 1980 had marked the beginning of the collaboration of Callon and Latour in what they first called "the sociology of translation."[93] "Translation" was meant broadly to cover the process whereby one thing represents another so well that the voice of the represented is effectively silenced. Central to this process is the capacity of something to satisfy—and thereby erase—a desire. Callon and Latour exploited the Latin root of "interest" as *interesse* ("to be between") to capture this capacity, which reverses the ordinary meaning of interest by implying that it is the presence of an object that creates (or perhaps reorients) a desire that the object then uniquely satisfies. That object is the mediator.

Actor-network theory was built on case studies of the success—and especially the failure—of translation in this sense. Significant in the French science policy context were three failures: that of the electric car to be made publicly available; the Minitel to become integrated into global computer networks; and a computer-driven customized rail system to appeal to Parisian commuters.[94] In each case, the failure was traceable to an exaggerated confidence in what top-down management could accomplish without attending to the "interests" of those whose cooperation would be required for the policy's implementation. It gradually became clear that these mediators held the balance of power. This point was underscored by the increased unemployment of all workers in the 1980s, with the exception of those who repaired and maintained the information technology that helped enable France's rapid transformation from an industrial to a postindustrial economy.[95]

These facts, combined with the powerful role played by minor parties in French elections, undermined the myth that had been regularly deployed by both Bonapartists and Republicans for nearly two centuries to justify any number of French policy swings. According to this myth, France is a unitary nation run from the top by scientifically informed civil servants,

---

92. Gildea 1996, 86 ff.

93. The sociology of translation was first presented as a general methodology firmly rooted in scientometric techniques. See Callon, Law, and Rip 1986. However, it started to receive general notice in the anglophone world with the publication of Latour 1987, which is now probably the most popular book in science and technology studies.

94. On these three science policy failures, see, respectively: Callon and Latour 1981; Castells 1996–98, 1: 342–45; Latour 1996.

95. Gildea 1996, 102–3.

quite literally "civil engineers."[96] Political parties provide alternative holistic visions of the best strategy for realizing France's national destiny, which is typically pursued by a combination of world-historic diplomatic initiatives and protectionist economic policies. However, the combination of de Gaulle's departure from public life in 1968 and the decline of Stalinist and Maoist factions soon thereafter placed these holistic visions in disrepute. It quickly became clear that opposing parties of the right and left could legitimate what turned out to be in practice largely the same policies. While public awareness of such ideological transparency has rendered the French electorate cynical of all professional politicians, at the same time it has instilled a spirit of compromise in those same politicians that had been previously lacking, which in turn has enabled coalition governments to stay in power longer than ever in the postwar period.[97]

This encapsulation of French politics displays a form of systemic rationality that would have met with Pareto's approval: a circulation of elites who are sufficiently self-conscious that neither they nor their constituencies take their totalizing visions too seriously, given the political bottom line of "power" (in the case of the rulers) and "order" (in the case of the ruled).[98] Under the circumstances, it is no surprise that the man most closely associated with "the end of ideology" thesis, Daniel Bell, has been the most influential American social theorist in contemporary France, the ultimate source of French theories of postmodernism, postindustrialism, and technocracy. In this postideological world, there are a new set of "angels" and "demons."

The angels are intermediaries who, by adding or subtracting their support, keep the elites in constant circulation, thereby reinforcing the appearance of justice in the system: i.e., that every dog eventually has its day. The angels maintain equilibrium by preventing any party from fully dominating the system: in short, a power that only the weak can exert in a dynamic field of play. The demons are those who declare the entire system corrupt and propose in its place a radically new order that will end the need for all this political restlessness. In the French context, these demons have been clearly marked: the right posed Jean-Marie Le Pen's National Front Party, the left posed the Greens. Actor-network theory was conceived from the standpoint of the angels, which is to say, it provides a metaphysical

---

96. Perhaps the most philosophically interesting precursor to actor-network theory's demystification of the French civil engineer is Hayek 1952, the classic genealogy of methodological holism in the social sciences, which traces the myth back to Saint-Simon and Comte.

97. Gildea 1996, 189–90.

98. An early attempt in English to establish the connection between actor-network theory and Pareto's Machiavellian tradition was Clegg 1989.

justification for the political system remaining Paretian. Its critical edge lies solely in reminding policy makers not to get carried away by their own rhetoric. Thus, the stress on networks that extend unexpectedly across national boundaries can be seen as an antidote to hyperbolic claims for central planning that hark back to France's founding unitarian myth.

In that respect, the power that Callon and Latour themselves exert as social analysts is a by-product of their having tempered the claims of seemingly more powerful actors. Their work, in turn, helps establish a rhythm to the circulation of elites that is tolerable to the society as a whole. This message has had a special resonance in France, given the status degradation of knowledge workers ever since de Gaulle damned academics and professionals by giving them exactly what they had asked for, namely, democratized entry into their ranks. So implemented, this legacy of the "Spirit of '68" simply confirmed the Machiavellian maxim that the multiplication of allies is the best way to divide loyalties and dissipate power. Of special relevance to L'Ecole des Mines was that once their ranks swelled, engineers found it more difficult than ever to act as a united front in political negotiations. The divergent educational backgrounds and career trajectories of engineers seemed to render the idea of professionalism obsolete. Some have turned to unionlike behavior, while others, fearing their chances of promotion, have sought forms of organization that do not threaten management's ultimate control of the means of production.[99]

De Gaulle's original policy of democratization—some would say proletarianization—has intensified with each successive French government, regardless of party. Moreover, it has taken several forms. At the most basic level, university enrollment grew threefold from 1960 to 1970, and then doubled again by the end of the 1980s. The number of universities and research clusters, though still concentrated in metropolitan Paris, has increased, thereby reducing the public image of academic leaders from trusted mandarins to feuding warlords. (In the field of STS alone, there are at least six research units in the Paris area.) But most significantly, *les grandes ecoles* have been forced to change their mode of domination in French society. While the admissions policies of these institutions are no less elite than in the past, the power their graduates exert in French society has generally declined, and the locus of power has shifted from the humanities and public administration schools to those like L'Ecole des Mines, devoted to applied science, technology and business.[100]

99. A strongly recommended work on the strategic uses to which Western governments have put "democratization" in order to dissipate the collective power of the learned professions is Krause 1996. The case of post-1968 engineers in France is discussed at 158 ff.
100. The details of this transiton are elaborated in Bourdieu 1996.

To be sure, Mitterand had failed in the 1980s to democratize entry into *les grandes ecoles*. Nevertheless, the value of professional degrees in engineering declined with the expansion of less prestigious engineering schools. This has had profound implications for STS, and sociology more generally, in France. First of all, it drove home a point of logic: it is possible for most top managers to be drawn from elite schools, while the likelihood of an elite graduate becoming a top manager diminishes. This is possible in a society where the elite can control *only* entry into its own immediate ranks but not the constitution of the field of play, which may include additional competitors who bring other qualities that alter the criteria of, say, leadership potential.[101] Perhaps the most important intellectual legacy of this shift has been Pierre Bourdieu's concept of "symbolic capital" as a form of knowledge-based power that is only partially determined by "cultural capital" (e.g., the quality of one's upbringing and training) yet potentially convergent with "economic capital" (e.g., where competition for top posts approximates a free market).[102]

The locally effective but globally indeterminate sociology of translation offers a realistic vision for engineers whose "civil" status has been challenged by this self-inflicted neoliberal state of nature that forces each person to look after her own interests. But even if engineers no longer draft the blueprints for the future governance of society, they nevertheless remain instrumental in whether or not particular emerging tendencies acquire forward momentum.[103]

"At the level of theory," to recall a phrase of the last great French Marxist, Louis Althusser, many recent strands in philosophy and sociology have

---

101. The pattern described here, which was characteristic of the growth of academic life across the Western world at the height of the Cold War, led the political economist Fred Hirsch to coin the phrase *positional good*, the paradigm case of which is the university degree. The possession of such a credential increases one's chances in the labor market only if access to it is somehow restricted. See Hirsch 1976. Most attempts to "socialize" the concept of capital stem from this insight, which has heavily influenced my own thinking about the power of knowledge in general. See Fuller 1994a, 1994b, 1994i, 1996c, 1996f.

102. Considering Bourdieu's largely antagonistic stance to the actor-network approach, it is ironic that his conceptual innovation was introduced to the anglophone world by Latour and Woolgar 1979, esp. chap. 5. This point is elaborated above in chapter 5, note 28. Bourdieu and Latour can be seen as trying to capture the same transformation from opposing perspectives: Bourdieu, the director of the leading state-supported research institute in the social sciences, critiquing the ways the state has buckled under external economic pressures; Latour, the resident sociologist at a leading beneficiary of the emerging neoliberal order, denying that the state ever had much control in the first place.

103. This conception of the role of engineering in the processes of social reproduction is most closely connected with the American historian Thomas Hughes, who provided a valuable contact between actor-network theorists and more mainstream historians of technology in the 1980s. For an epitome of his corpus, situated in one of the seminal STS anthologies, see Hughes 1987.

drawn attention to the role of mediators who, though weak in themselves, are able to block or enable major alignments of knowledge and power.[104] The distinctive feature of Callon and Latour, which has undoubtedly contributed to their appeal among STSers, is their explicit claim that the analyst herself accrues power in this mediating function. In contrast to Max Weber's image of the sociologist who proposes feasible means to ends that have been selected by the policy maker, Callon and Latour propose to show just how little real power the policy maker exerts and thereby demonstrate their own indispensability to the policy process.[105]

The idea that an initially weaker party may directly benefit by displaying the weakness of others is the mark of what Latour's philosophical mentor, Michel Serres (b. 1930), has called the *parasite*, a concept simultaneously modeled on the presence of noise in a communication channel and the generosity of hosts who indulge an unwanted guest.[106] The STSer thus becomes the ultimate noisy guest. The general strategy behind this move is fairly straightforward: Weber has been turned on his head. If the "modernity" of the state is marked by its reliance on scientifically authorized modes of legitimation, then instead of indulging their masters in the belief that policy regimes can be rendered efficient, duly authorized social scientists can both prove their usefulness and run interference on state policy by highlighting unforeseen obstacles on the way to policy implementation. The sociologist is thus able to manufacture a sense of integrity and even value-neutrality—along with a hint of radicalism—in a client-driven world: she can stare down her master while reinforcing the master's need for her services. A not-inappropriate comparison may be the psychotherapist who strings along the patient for the material benefit of the former and the spiritual benefit of the latter.

Beyond this rather strategic approach to research promoted by the Paris School, there has been, of course, a more generalized cynicism toward the university. Most striking in this regard is the provenance of the "postmodern condition" in which many intellectuals think we live. The famous report by that name that Jean-François Lyotard wrote for the higher education council of Quebec in 1979 is dedicated to the department where he

---

104. See Rouse 1996, chap. 7. Of particular interest is Urry 1995, 33–45 (originally published in 1981, as a critique of Anthony Giddens). Urry, a distinguished British sociologist, appeals to the image of the parasite but in a less self-aggrandizing fashion than his French colleagues. Wisely disavowing Kuhn as providing a model for sociology (42), Urry maintains that the field's strength comes from making apparent the assumptions of other discourses in a situation that more nearly resembles Popper's open society (34).

105. Callon and Latour 1981, 300–301.

106. Serres 1980. Serres and Latour 1995 presents Latour's conversations with Serres.

held a chair in one of the new universities of Paris, wishing that it may flourish while the university itself withered away.[107]

From Lyotard's Parisian perch, de Gaulle's attempt to placate the academic radicals who demanded more open admissions to elite institutions served only to co-opt them and compromise the independent standing of academia in French society. It was the state's last desperate attempt at maintaining social order in a world that was quickly exceeding its control. In this context, the appeal to specifically academic standards of discourse — including its more philosophically embroidered forms, such as Habermas's "ideal speech situation" — appeared as a disguised reactionary ideology for arresting the cross-fertilization of ideas and the novel developments they breed, which had increasingly come from outside academy.[108] Thus, in Lyotard's hands, the university was reduced from a transcendental condition for the possibility of critical inquiry to a cluster of buildings where representatives of these discourses have chance encounters and set up temporary alliances, subject to the terms set down by buildings' custodians, a.k.a. academic administrators. Notwithstanding Latour's frequent protests to the contrary, actor-network theory should be seen as extending the postmodern condition from the humanities to the science and engineering faculties.

My excavation of the social epistemology informing the Paris School of STS should remind the reader of an elementary lesson in the sociology of knowledge, namely, that *a seemingly radical innovation that quickly acquires widespread currency probably serves some well-established interests that remain hidden in the context of reception.* This much Marxism understood, which is why the mobilization of class consciousness — consolidating the powerless into a source of power — would have to precede any genuinely progressive revolution. The failure of Marxists to follow through on this strategy is now commonly seen to imply that the strategy itself was in error. This results in a political complacency that explains the curious fate common to paradigms and actor-networks, whereby surface ruptures of the status quo are accompanied by the containment of political possibilities. A good case of this "containment in action" is a well-regarded piece of STS normal science, Donald MacKenzie's *Knowing Machines*, which applies a version of social constructivism influenced by actor-network theory to the careers of various forms of high technology, ranging from jet airplanes to mainframe computers.[109]

MacKenzie, probably the most academically prominent student of the

107. Lyotard 1983, xxv; for a critique, see Fuller 1999c.
108. Lyotard 1983, 65 ff.
109. MacKenzie 1995. For a detailed review of this work, see Fuller 1997c.

Edinburgh School, is a self-styled socialist who is not afraid to debunk determinist theories of technology associated with orthodox Marxism. This is a fair task for social constructivism, and given what we saw earlier in this chapter, it should eventuate in an enlargement of the sphere of public action. In the spirit of constructivism, MacKenzie argues that when the exact identity of a technological innovation has yet to be consolidated, competing perceptions of the technological horizon delimit the field of possibilities for what counts as technological change. The alternative lines of development can be shown to empower different constituencies. However, MacKenzie's own case studies leave the impression that once the identity of a technology becomes relatively stable, the only way to induce further change is by capitalizing on what he calls "insider uncertainty," namely, internal disagreement among the technology's recognized experts. In actor-network jargon, the only people qualified to open a "black box" are those who have been in a position to close it. This is a remarkably elitist image of the prospects for technological change, one that owes more to Pareto or Schumpeter than Marx.[110] (Yet it is precisely the image that we have found endemic to the Harvard environment in which Kuhn spent his intellectually formative years.)[111] Pareto and Schumpeter were quite clear that the coalition of elites needed to maintain power is always unstable and hence subject to cycles of subversion, or "circulation." But they were equally clear that the populace at most provides the official pretext for change (e.g., an election, a policy initiative taken "on behalf of the

---

110. MacKenzie 1995, 16–17. This line of thought taken to its logical conclusion produces the following critique of feminist deconstructions of technology, taken from Grint and Woolgar 1995: "If Foucault is right that truth and power are intimately intertwined, then those seeking to change the world might try strategies to recruit powerful allies rather than assuming that the quest for revealing the truth will, in and of itself, lead to dramatic changes in levels and forms of inequality" (306). While there is certainly nothing wrong with exhorting constructivists and feminists to do more than simply publish their critiques in academic journals, it is telling that the only avenue recommended for getting action on those critiques involves courting the actually powerful few, as opposed to organizing the potentially powerful many. Thus we encounter the limits of the elitist political imagination, which takes *what has been* to be the totality of *what can be*. Another indicator is the tendency to repackage opportunism as audacity. In this vein, Grint and Woolgar fault feminists and constructivists for "timidity" when they refuse to adapt their critiques to changing circumstances and audience (305). No doubt feminists and constructivists would regard Grint and Woolgar as advising a sellout. At the very least, Grint and Woolgar's flexibility must be seen in the context of the ends justifying the means, perhaps the main Machiavellian motif. However, one virtue of their dogged pursuit of the Machiavellian argument is that it forces would-be social critics, reformers, and radicals to respond by focusing on the *ultimate ends* of their activities. As we shall see in chapter 8, section 3, Grint and Woolgar may be charitably read as promoting an "American" style of social movement thinking, whereas the feminists and constructivists they criticize — Judy Wajcman and Langdon Winner are discussed explicitly — are promoting its "European" counterpart.

111. See chapter 3, sections 3–5.

people") and more often functions simply as pawns, or "deployable resources," in Latourspeak.

This last point is worth stressing because actor-network theory is full of emancipatory-sounding talk that claims to reveal the "missing masses" needed for any large-scale sociotechnical achievement. However, the masses are presented as if they were literally physical masses whose movement is necessary to give an elite forward momentum. The agency of these masses is thus limited to the extension or withdrawal of collaboration, not the initiation of action. The current fashion for distributing agency across both people and things merely underscores the value of the masses as means to the ends of other parties, since in many cases nonhumans turn out to be at least as helpful as humans in achieving those ends.[112] Although actor-network enthusiasts often make much of the innovative political vision implied in this extension of agency from persons to things, some disturbingly obvious precedents for this practice seem to have been suppressed from STS's collective memory, the first from capitalism and the second from totalitarianism.

The first precedent was touched upon in section 4, namely, the affinity with the metaphysics of capitalism, which, through the process of commodification, enables the exchange of human and machine labor on the basis of such systemic values as productivity and efficiency. This is the sense in which technology is normally regarded as a "factor of production": i.e., a potentially efficient replacement of people. Indeed, the metaphysically distinctive tenet of *socialism* in modern political economy has been its revival of the medieval doctrine that human beings are the ultimate source of value in the world. But like capitalist cost accounting, actor-network theory knows no ontological difference between humans and machines. Consequently, the subtext of the title of Latour's *We Have Never Been Modern* might have read "We Have Never Been Socialist," to capture the increasingly neoliberal climate of French science policy that makes ontological leveling seem so attractive. This point is lightly veiled in Latour's refashioning of the word "delegation" to capture the process whereby humans and nonhumans exchange properties, which legitimates the treatment of humans as cogs in the wheels of a machine, and machines as natural producers of value.

Here we might compare the Parisian treatment with the most developed set of arguments for extending agency to nonhumans. These fall under the rubric of "animal liberation," as popularized by the Australian moral

112. The classic case study of the distribution of agency to humans and nonhumans is Callon 1986. This view was subsequently popularized in Latour 1988a. For subsequent applications, see esp. Ashmore and Harding 1994.

philosopher Peter Singer (b. 1946).[113] In this guise, the politics of agency veers toward restraint and caution rather than mobilization and facilitation. An important difference between Singer and Latour is that the animal liberation movement has gravitated toward a conception of "animal rights" modeled on the civil rights accorded to humans. Significantly, a sentient creature, usually a mammal, is the paradigm case of a "nonhuman." In contrast, the various Parisian exemplars of a "nonhuman" have typically resided much lower on the evolutionary scale: scallops, microbes, and even mechanical door closers all serving as examples at various points.[114] The overall effect is that in its proliferation of agency, actor-network theory dehumanizes humans, while animal liberation humanizes animals.[115]

Animal liberation's excesses are regularly documented in the forced entries into university laboratories to "free" animals that have been caged for experimental purposes. Yet there is an even less savory precedent for the

113. Singer 1975.
114. See note 104. On door closers, see Bruno Latour (a.k.a. Jim Johnson), "Mixing Humans and Non-humans Together: The Sociology of the Door-Closer," in Star 1995, 257–89. Interestingly, when Latour turns his attention to primates, he suggests that they may actually confer value on their activities much as socialists think humans do, because apes do not inhabit a world mediated by technology. See Latour 1994.
115. The contrast implied here can be captured more formally by considering the process by which the agency of the humans ("A") may be subtly transferred to the nonhumans ("M"):

(a) A can achieve its ends by a variety of means, such that if one means is absent, another is available and sufficient to the task.

(b) The variety of means at A's disposal shrinks, such that A is now limited to utilizing a means, M, that cannot be easily replaced if unavailable.

(c) Given (b), in order for A to achieve its ends, it must first ensure that M's needs are satisfied.

(d) Satisfying M's needs absorbs enough of A's resources to make it difficult for A to achieve its own ends in the manner it would like.

(e) A has the option of either altering its ends so as not to have to rely on M (or M exclusively) or adopting the satisfaction of M's needs as its own ends.

When Hegel, following Spinoza, said that freedom fully realized is the recognition of necessity, he had in mind an idea that can easily be lost in the liberatory rhetoric associated with the extension of agency to nonhumans, namely, that to increase the number of agents is *not* to increase the amount of agency in the world. On the contrary, it is to limit or redefine the agency of the already existing agents. In stage (e) of the translation process outlined above, A's full recognition of M's agency requires that A either make room for M as a separate agent or merge with M into a new corporate agent. In both cases, A is forced to alter its own identity. In the former case, the change may be rationalized as A's coming to lead a simpler life, whereas in the latter, it may be rationalized as A's now having access to more power than before. The former corresponds to Animal Liberation, the latter to actor-network theory. The former retains the human as unique agent (at least at the species level) but at the cost of diminished wants and power, whereas the latter magnifies the wants and power of the human but at the cost of rendering each individual a (potentially replaceable) part of the larger corporate machinery. In my earlier treatment of the actor-network conception of agency (Fuller 1996f), I had not distinguished these two possibilities and, if anything, assumed that the actor-network conception would converge with the Animal Liberationist's stance. However, the cumulative weight of case studies done under the influence of actor-network theory has convinced me otherwise.

extremes to which an actor-network perspective can be taken, namely the twentieth century's unique contribution to political theory and practice: *totalitarianism*. Contrary to Latour's oft-repeated claim that politics has never taken technology seriously, totalitarian regimes stand out from traditional forms of authoritarianism precisely by the role assigned to technology as the medium through which citizens are turned into docile subjects, specifically parts of a corporate whole. While attention has usually focused on totalitarian investments in military technology, of more lasting import have been totalitarian initiatives in the more day-to-day technologies associated with communication, transportation, and building construction. As argued in chapter 2, section 3, the early stages of these developments already informed science policy debate in Continental Europe at the dawn of the twentieth century. Ultimately these technologies enabled unprecedented levels of mass surveillance and mobilization, all in the name of configuring the national superagent. In the course of this configuration, any sharp division between humans and nonhumans was removed. An important consequence was that a subset of the human population— say, the Jewish race or Communist ideologues—could be excluded from the corporate whole as such great security risks that the rest of the human population would agree to submit themselves to sophisticated invasive technologies in order to become "incorporated."[116]

Actor-network theory can be understood as the account of society that results once there is no longer a hegemonic state apparatus in charge of this technostructure: a devolved totalitarian regime; in a phrase, *flexible fascism*. Instead of a unitary state that renders everyone a means to its specific ends, now everyone tries to render everyone else a means to her own ends. The former members of the corporatist state may have lost their sense

---

116. This point was made by Carl Schmitt (1888–1985), the Weimar jurist who provided the original legal justification for the one-party state that became Nazi Germany. Schmitt held that technology was the latest and most durable corporate glue because its apparently neutral character seemed to impact on everyone equally, thereby enabling conflict to metamorphose from the elite cross-border confrontations of the past to "total war" involving a nation's entire population. Schmitt envisaged that the threat of an external foe more powerful than any internal foe would lead citizens to submit to the application of mass technologies for purposes of defeating that foe, however much their own personal freedoms may be constrained. For Schmitt, this situation marked the perfection of politics, which presupposes a common good very much as the Athenians envisaged it (see chapter 1, section 2 above). Schmitt carried this idea to its logical conclusion, namely, that something must be defined as mutually hostile to those who constitute the commonwealth in order for a society to achieve political self-consciousness. See Schmitt 1996, 103. Of course, from an actor-network standpoint, technology's neutrality is merely apparent because anyone capable of, in Latourian terms, "opening the black box" of technology can use it to her advantage against others, even within a given society. To be sure, Schmitt had in mind a state monopoly on mass technologies.

of common purpose, but they retain the personal ethic that attended that purpose.[117] The difference in actual outcomes is much less predictable than under a totalitarian regime, but ultimately explainable in terms of the agents' differential access to the resources needed to attain their ends. Thus, the necessitarian myths that originally propped up Mussolini, Hitler, and Stalin have now yielded to contingent narratives centered on Pasteur, Edison, and Seymour Cray (the inventor of the mainframe computer).[118]

One of the most remarked-upon features of fascist ideology is its easy combination of an animistic view of nature, a hyperbolic vision of the power of technology, and a diminished sense of individual human agency. The same could be said of the "delegations" and "translations" that characterize the accounts of sociotechnical systems provided by actor-network theory.[119] However, actor-network theory's resonance with totalitarianism loses its shock value once we recall its origins in the training of an engi-

---

117. A good way of thinking about this devolution is that the fixed functions performed by parts of a corporate whole are replaced by contingently defined complements of a distributed labor process. The continued anglophone domination of the Internet—even after the U.S. Defense Department has relinquished formal control—may be seen as the ultimate expression of this devolved state. The English Channel makes a crucial difference in the explanation of the emerging neoliberal environment of which actor-network theory is an important Continental expression. The difference turns on which "chicken-and-egg" story one tells about the relationship between Society and the Individual. Regarded from the Continental side of the channel, liberalism appears as the result of devolved corporatism. Austrian economics exemplifies this tendency, as observed in chapter 4, note 108. But seen from the British side, inherently autonomous individuals need to be coerced into a corporate whole. Thus, where the Continental European political theorist regards anomie as the biggest threat to social order (i.e., the loss of duty), her Britannic counterpart focuses on oppression (i.e., the loss of liberty).

118. These three technoscientists have been the subject of actor-network accounts by, respectively, Bruno Latour, Thomas Hughes, and Donald MacKenzie. One of the eerier similarities between the predilections of totalitarian and actor-network theorists is the glorification of the heroic practitioner—be it the power-politician or the heterogeneous engineer—whose force of will overcomes the self-imposed limitations of superstitious citizens and academics in the grip of a theory. Thus, comparable to Pareto's disdain for the planning pretensions of social democrats is Michel Callon's contempt for the sociologists Pierre Bourdieu and Alain Touraine, who define in words the contemporary state of French society, something engineers supposedly do much more effectively in their daily practice. See Callon 1987, esp. 98 ff. This sentiment undoubtedly goes down well at L'Ecole des Mines and may help explain the Theory Lite character of much STS research, critiqued in section 3C above.

119. From Latour's brief discussion of totalitarianism, it is clear that he regards it as one of the "surplus value" philosophies discussed in section 4 above, except that value here is added to political, rather than scientific, practice. See Latour 1993, 125–27. Here he comes closest to endorsing the Pirandellist "It is so, if you think so" form of relativism of which his critics have often accused him. Specifically, he explains the formidability of totalitarian regimes in terms of a widespread belief in their underlying philosophies, rather than, say, the collective impact of the actions taken under their name. Latour officially wants to ensure that people are not inhibited by philosophies that stray too far from the scene of action, but his argument also implies that one ought not be inhibited from forming alliances with people to whom such philosophical labels as "totalitarians," "capitalists," and "imperialists" are conventionally attached. Thus, in Latour's hands, nominalism all too easily slides into opportunism.

neering profession that over the past quarter-century has lost its custodianship of a unitary French state. This lost world is the technocratic side of French thought that began with Napoleon's establishment of the *polytechniques*, was apotheosized by Comtean positivism and has now reached its decadent phase with actor-network theory. Its spirit moves imperceptibly through an anglophone STS community that still harbors the stereotype of "French intellectuals" as the consciences of their times — individuals with a spiritual paternity traceable to Emile Zola and Jean-Paul Sartre — even though fifteen years have passed since Michel Foucault's death.

VIII
Conclusions

1. The Canonization of Saint Thomas Kuhn

Thomas Kuhn's *The Structure of Scientific Revolutions* was one of the most influential academic books of the second half of the twentieth century, and arguably the one that has done the most to shape both academic and public perceptions of science. However, *Structure* was the product of a particular context and its influence has been of a particular kind. The context may be roughly divided into *personal* and *situational* factors. The key personal factor was Kuhn's membership in a generation trained in physics who came of age at the dawn of World War II. During that time, the discipline that had attracted Kuhn and others as the continuation of natural philosophy by experimental means was rapidly transformed into the paradigm case of the sociotechnical behemoth, "Big Science." Like others of that generation, Kuhn found this transformation profoundly disillusioning.

The key situational factor that enabled Kuhn to channel his disillusionment productively was the General Education in Science curriculum, designed by Harvard president and U.S. atomic bomb administrator James Bryant Conant. As one of the masterminds behind the transition to Big Science, Conant was concerned that the public might become suspicious of science if they understood its subsequent developments exclusively through the horrific effects of the bomb. Kuhn developed the argument of *Structure* while an instructor in Conant's curriculum, the aim of which was to enable students to abstract a distinctive scientific mind-set that has remained constant throughout the vast social and technical changes that science has undergone in its history.

Kuhn and Conant were clearly using the General Education curriculum for somewhat different, yet overlapping, ends. Kuhn was given the opportunity to articulate the ideal of scientific inquiry that had originally motivated him to pursue a career in physics, while Conant had found a reliable medium for normalizing science's role in contemporary society. What they shared was an interest in promoting a normatively desirable understanding of science that was grounded, in some sense, in its history.

However, this story is complicated by the exact way Kuhn attempted to ground his normative ideal, namely in three hundred years of the history of European physical sciences, while at the same time refusing to comment on the failure of those very sciences (not to mention the biological and social sciences) to conform to his ideal for most of the twentieth century. Kuhn's response to the chord that his book struck—his silence and increasing withdrawal from the communities that embraced him—is at least partly explained by his awareness of a double-truth doctrine in the writing of history of science, which he himself called "Orwellian": on the one hand, a heroic history to motivate scientists in their daily activities; on the other, a messy, dispiriting, yet more down-to-earth history that the professional historian uncovers mainly for consumption by other historians. Ironically, what Kuhn presented as the "real" history of science in *Structure* itself turned out to be a myth, not only because its own empirical basis was suspect, but more importantly its narrative was used uncritically by social scientists and other inquirers to legitimate their activities as paradigms on the same footing as those of the physical sciences.

The overall effect has been that *Structure* diverted emerging tendencies in the 1960s to question the role of Big Science in the academy and society at large, while reinforcing the ongoing fragmentation and professionalization of academic disciplines. Both developments marked a decisive turn away from the ideal of a unified science that probably motivated Kuhn's original interest in physics as the continuation of natural philosophy by more exact means. The net social conservatism of *Structure*'s impact could well have pleased Conant, but not its support for an intensified division of academic labor. However, the latter explains the book's appropriation by a broad church ranging from "normal scientists" to self-avowed "postmodernists." The point of my book has been to explore the background social, philosophical, and historical conditions that have allowed this strange turn of events, in the hope that we may still be in a position to remedy whatever damage has been caused by an unreflective acceptance of the account of science given in *Structure*.

Although I have been chiefly concerned with the career of a book, *The Structure of Scientific Revolutions*, the contexts in which I have embedded the book's origins and impacts, and especially the normatively charged language I have occasionally used to explain these developments, suggest that I wish to pass judgment on its author, Thomas Kuhn. To be sure, it is difficult *not* to do so. Historical figures so close to our own time invite consideration of their suitability as role models, and the more I learned the less I approved. I admit to favoring figures who display an awareness of the sociohistorical setting in which they stake their claims to knowledge. In this respect, James Bryant Conant and Alexandre Koyré, in rather different but equally reflexive ways, appear more exemplary figures than their protégé, Thomas Kuhn. Indeed, reading *Structure* in light of his two mentors easily leaves the impression that their value choices constitute taken-for-granted premises for Kuhn's account of scientific change.

Nevertheless, I have neither the interest nor the evidence to deliver a verdict on Kuhn's life, let alone indict the man of crimes of the intellect. As far as intellectual personalities are concerned, my main interest is in evaluating "Kuhn" as an ideal type of how academics respond to their social environment—indeed, the sense in which Kuhn was "there," as raised in the article that originally motivated my inquiries, as recounted in the preface to this book. It will become clear in what follows that Kuhn's mode of response to his environment marks a profound transition in the nature of academic life. In short, what did it mean to be *someone in Kuhn's position?*

In the anglophone outposts of French social theory, it is nowadays fashionable to speak of *habitus*, the set of attitudes and expectations one acquires through the successive forms of discipline that constitute one's upbringing, which are subsequently reinforced by others over the course of a lifetime.[1] In this book, there have been occasional glimpses into Kuhn's habitus, especially his lengthy incubation period at Harvard, which encompassed undergraduate and graduate training, as well as his induction into the newly created Society of Fellows and ultimately his first regular teaching post. In his last major interview, Kuhn left little doubt that his years at Harvard were the most formative in his life. His father and uncles had attended Harvard. Harvard was where Kuhn first found a circle of friends and felt he fitted in. Failure to achieve tenure at Harvard also nearly caused Kuhn to have a nervous breakdown. Last but not least, Kuhn met Conant, whom Kuhn regarded as the brightest person he had ever known—a judgment that forced him to shift his father, a clever and energetic engineer-turned-businessman, down to the second position.[2]

---

1. The concept of habitus is developed in Bourdieu 1977, 72–95.
2. On these Harvard-related details, see Kuhn et al. 1997, 146–48, 163, 170. Conant's displacement of Kuhn's father invites further psychoanalytic exploration. In the interview, Kuhn contrasts his father, a quick-witted man of action, and his mother, a socially inept intellectual. Kuhn identifies with the latter but admires the former. However, Kuhn also observes that his father failed to fulfill his potential, in part because his efforts came to be dispersed after World War I, where his talents had been concentrated in the U.S. Army Corps of Engineers.

Conant's style of recruitment politics reflected an aristocratic orientation to the social order, one to which Kuhn acquiesced, albeit without ever having engaged in its active promotion.[3] Kuhn's passive acceptance was probably facilitated by the different values that Conant and Kuhn assigned to Conant's actions toward Kuhn. For example, Kuhn reports being very impressed that Conant wanted him to do a case history on mechanics for Natural Sciences 4, given the significance of mechanics for the history of science.[4] No doubt, for his part, Conant appreciated the efficiency of having the case history done by someone with the relevant knowledge at his fingertips.

Moreover, Harvard's willingness to deny Kuhn tenure shortly after Conant's departure from the presidency testifies to a general impression that Kuhn was beholden, however passively, to Conant's patronage and the vision of the world with which it was associated. This vision assigned to elite American universities the unique role of consolidating and protecting the heritage of Western civilization, especially as it underwent the twin twentieth-century threats of Nazism and Communism. More specifically, like C. P. Snow's depiction of the "two cultures" divide in Britain, Conant's vision presupposed that the humanities had handed over the task of cultural preservation to the sciences, or at least to those trained in the natural sciences who could make that heritage come alive to nonscientists destined for positions of power. Conant made this vision most explicit in his foreword to Kuhn's first book, *The Copernican Revolution*, but its most concrete legacy was in the General Education in Science curriculum where Kuhn honed the core theses of *The Structure of Scientific Revolutions*.[5]

Generally speaking, the need for achievement figures prominently in Kuhn's interview. In fact, he associates his discovery of the power of normal science puzzle solving with a watershed moment in his education, namely when he realized that the unstructured setting of his "progressive" primary school days had prepared him poorly for solving physics problems in high school, thereby undermining his "straight A" average. See Kuhn et al. 1997, 148.

Those interested in pursuing the psychoanalytic dimension of Kuhn's thought should find two features of the interview of note. First, as a young man, Kuhn underwent psychoanalysis for his difficulty in relating to women, though his description of the relationship to his father and Conant may well strike the psychoanalytically inclined as "feminized." My thanks to Stephanie Lawler for talking me through this psychoanalytic possibility. Second, he claims that his interest in "climbing into people's heads" was triggered by this experience in psychoanalysis. See Kuhn et al. 1997, 163. The latter point bears an intriguing relationship to Jacques Lacan's admission that his own distinctive approach to psychoanalysis was triggered by Koyré's interpretation of Galileo. See chapter 1, note 65.

3. On "recruitment politics," see the underrated Cook 1991, 65–66.

4. Kuhn et al. 1997, 159.

5. Although Kuhn studiously refrained from acknowledging any specifically intellectual debts to Conant, he nevertheless admitted that he found it hard to cope when Conant's various administrative duties forced him to turn over the lecturing of Natural Sciences 4 to Kuhn, which suggests how much he had relied on Conant's intellectual (and other) leadership in the

Conant's sense of science's world-historic mission did not especially endear him to Harvard's doyens, most of whom still operated with a liberal arts college model of the university in which the humanities reigned supreme and even the natural sciences were viewed more as teaching than research subjects. Indeed, resentment periodically surfaced in the minutes of the General Education meetings at how the liberal arts were becoming subordinated to the needs of Big Science research, be it the special deals that researchers negotiated on teaching loads or the way in which Conant himself generally saw teaching as a conduit for promoting the aims and products of research. The formative experiences in Kuhn's professional life occurred in the midst of this particular culture war. This point comes out very clearly in the protracted debate over his tenure, which centered on Kuhn's drift from the sciences to the humanities without having made a clear mark in any field.

In Whiggish hindsight, we may be tempted to conclude that the doyens were unduly harsh or downright obtuse in their judgment of *The Copernican Revolution* as a good teaching text but not much more.[6] However, without a clearly established history of science profession in the United States, let alone at Harvard, it was genuinely difficult to determine the disciplinary criteria that applied in Kuhn's case. The only frame of reference that would have been common to both the allies and enemies of Conant was the nature of Harvard itself. Kuhn simply failed to deliver on Harvard's fifteen-year cultural investment, at least as Harvard's humanist doyens measured it. The most natural interpretation of the tenure committee's assessment, then, is that Kuhn was not sufficiently attentive to the obligations that were incurred by the privileges he enjoyed as a recruit to the American intellectual aristocracy.[7] What the committee failed to anticipate was that Kuhn

---

course. Kuhn wrote out all his lectures, a compensatory reaction that, by his own account, inhibited his subsequent efforts at writing. See Kuhn et al. 1997, 166.

6. Remarks by Edwin Kemble and Leonard Nash, in *Minutes*, 8 November 1955.

7. In the lengthy deliberations over Kuhn's tenure, this point was explicitly raised by Harry Levin, who eventually became the Irving Babbitt Professor of Comparative Literature at Harvard. Levin said he sat on many university committees over the years in which Kuhn's lack of performance was repeatedly justified by extenuating circumstances. See *Minutes*, 8 November 1955. Later Levin would reflect that during this period his own field was undergoing an identity crisis, as a largely American attempt to construct an abstract literary science, the "New Criticism," clashed with the synthetic historical scholarship of such émigré Europeans as Erich Auerbach and Leo Spitzer, who were aligned with the iconographic tradition discussed in chapter 1, section 3. Kuhn's work did not fall comfortably into either category. See Levin 1969, esp. 479.

In these self-professed democratic times it is awkward to invoke an aristocratic ethic that revolves around the exchange of privilege and obligation. Indeed, according to the *Oxford English Dictionary*, when the expression *noblesse oblige* was coined in 1837, it was already meant ironically. Nevertheless, the injection of an elitist dimension into normative discourse suggests that some people, in virtue of their social position, deserve to be held accountable to

would later publish a book, *Structure*, that accomplished much of what the committee had wanted — indeed, with the help of the cultural investment that constituted Kuhn's habitus (specifically his stint as a General Education instructor), but without Kuhn's deliberate involvement.

We face a subtle interpretive problem here, one that reflects the historical transition from an *aristocratic* to a *capitalistic* field of play in academia. Perhaps most indicative of this problem is what Robert Merton (b. 1910) has called "the principle of cumulative advantage," which we first encountered in chapter 1, note 1, under its nickname the "Matthew effect." According to this principle, the more benefits one receives, the more one will continue to receive. In what sense does this characterize Kuhn's rise to prominence? Merton, himself a Harvard man somewhat older and less privileged than Kuhn, recognized Conant's style of recruitment politics for what it was.[8] However, Merton tends to give the principle a much stronger capitalist spin than would seem appropriate in Kuhn's case, which leaves the principle's general normative implications radically unclear. Kuhn, the product of an aristocratic culture, provides Merton's most elaborate illustration of the Matthew effect at work, yet Merton's principle is usually associated with a capitalized scientific environment, where one's academic credentials clearly prove to be a good predictor of the quantity and quality of both one's own and one's students' long-term research productivity.[9]

In this context, the principle is normally read as marking an invisible-

---

a somewhat different standard from that of the run of humanity. Modern moralists are uncomfortable with this idea because it presupposes that we cannot all be judged by a common standard, be it deontological or utilitarian, that marks us as children of the same God or ape. In other words, to accept "privilege" and "obligation" as reciprocal terms of moral appraisal is to acknowledge the failure, or at least the shortfall, of the democratic project whereby people are judged entirely on the basis of their own intentions and actions, without factoring the cultural burden they have inherited and acquired.

However much we may wish academia to be constituted as a democracy in the sense presupposed by modern ethical theory, it is not now and it certainly was not in Kuhn's lifetime. By pretending otherwise, we may salvage the honor of those who have occupied Kuhn's position, but we do our nonelite colleagues a severe injustice in the process. If someone groomed to rule fails to provide the expected form of leadership, then that is prima facie grounds for believing that such a person has morally failed. The most articulate and systematic challenge to modern ethical individualism is still Bradley 1927. Bradley's conception of one's "station and its duties" should be read as the philosophical counterpart to Bourdieu's sociology of habitus.

8. Perhaps the reader will not be surprised to learn that Merton's own professional progress did *not* exhibit Kuhn's streamlined trajectory. Though himself a Harvard graduate, upon completing his Ph.D. in 1936, a tight job market forced Merton to the backwater of Tulane University, New Orleans, out of which a torrent of publications catapulted him into a distinguished position in an Ivy League university, Columbia. A good intellectual biography is Crothers 1987.

9. The contrast in Merton's presentation of the principle of cumulative advantage may be seen in the Kuhn-oriented Merton 1977, 71–108 and Merton 1973, esp. 439–59, which focuses on the natural sciences.

hand process, the uncanny ability of the scientific community to pick its winners without explicit criteria or external supervision. However, those generally skeptical of invisible-hand arguments immediately pounce on the word "uncanny." They question whether the predictions are sufficiently independent of the outcomes at successive stages of the process—who gets into graduate school, who gets a job, who gets an article published, who gets tenured and promoted, etc.—to constitute a series of fair tests. In an ideal capitalist environment of perfect competition, they would. The people making the predictions (admissions officers, personnel committees, editorial boards) would literally place bets (in the form of scholarships, grants, salaries, and journal space) on particular players. But these bettors do not control the outcomes of the game. The outcomes emerge from interaction of the players themselves, depending on who turns out to make the most brilliant advances, as seen through the eyes of their peers.

What sort of evidence would one seek to demonstrate that the principle of cumulative advantage is, indeed, the result of the capitalist process just described? Two facts stand out as very relevant to the natural sciences (and increasingly relevant to the other academic disciplines). The first is that, in the game outlined above, the vast majority of players—including those with prestigious pedigrees—lose in the long run. That the winners are most likely to come from prestigious backgrounds is certainly compatible with the fact that most of those with prestigious backgrounds drop out of the game after a certain point: they fail to complete their degrees, they fail to get and keep good jobs, they fail to publish, or finally, if they publish, they fail to be recognized for having published. In short, the amount of waste in individual human talent tolerated by the science game speaks to a process that is not subject to the designs of any human agency.

The second relevant fact is that the players must demonstrate their skill as soon as they enter the field of play, endlessly showing that they can provide return on investment, so as to continue to enjoy the affections of future investors. The combination of these two facts can leave the impression that the science game is not rigged. Promise must be quickly backed by product. It is tough to win simply because of the scarcity of prizes relative to the pool of contestants. Such an understanding of the environment invites the ascription of credit and blame to the skills and efforts of individuals, both on the field and in the betting parlors, so to speak. Scientists can only blame themselves for not making the most of their cultural capital, and similarly university officials can only blame themselves for having invested their institution's cultural capital on the wrong scientists. Such is the individualized moral universe of the capitalist field.

In contrast, someone in Kuhn's position is most naturally understood as having operated in an aristocratic field of play, one in which cumulative advantage incurs cumulative obligations. The normative presuppositions here are markedly different from the capitalist ones just enumerated. (However, determining that Harvard and other elite American universities from the early 1940s to the late 1950s operated in an aristocratic, rather than a capitalist, field of play would require demonstrating that they could have reliably placed their recruits in influential academic — not to mention nonacademic — posts. Certainly, Conant and Harvard's humanist doyens acted as if they could.) The basic scheme is that each generation of academic leaders is actively recruited by those in a realistic position to select them. Thus, while the aristocrat and the capitalist concur on the highly stratified nature of academic success, they explain it in radically different terms: the former by design, the latter by effect.

Aristocratic recruitment typically involves a period of incubation during which recruits are not expected to produce any independent work, but rather are to become imbued with the doctrine that they will spend the rest of their lives extending and defending. But there come moments of truth, when recruits must do something that reveals their induced capacity to lead. For example, they may spontaneously rally to a defense of the realm when it comes under attack, even when the realm is the notional one of "academia" or "science."[10] For better or worse, Kuhn never actively engaged in this strategy. Lest we forget, Kuhn was ultimately judged a *failure* by those who managed the aristocratic recruitment process, and it is only once more strictly capitalistic criteria are introduced that the overall impact of *The Structure of Scientific Revolutions* can be explained, namely, in terms of what I described in section 7 of the introduction as the book's "servant narrative" status, which attracted a wide range of intellectual consumers.

However, the question remains why such a success of the marketplace, which depends entirely on the use that others make of a work, should then serve to confer attributions of profundity on its author. Here Kuhn's habitus

10. The ideal type for this sense of aristocratism is provided by the Japanese samurai, who successfully translated the unconditional loyalty and discipline demanded of their warrior ethic from feudal administration to research management in the late nineteenth and early twentieth centuries. See Fuller 1997d, 123–29. However, similar precedents can be found among some European aristocrats, especially during the seventeenth and eighteenth centuries. See Bauman 1987, 25–34. Closer to Kuhn is the career of his rival, Gerald Holton (b. 1922), who has mitigated the tensions between science and its social environment for the better part of a half-century. Holton's guardianship of scientific virtue extends at least to 1958, when he was asked to turn the Proceedings of the American Academy of Arts and Sciences into the quarterly journal, *Daedalus*. The early articles from his editorship were collected together as Holton 1967, a set of mostly Harvard-based responses to C. P. Snow's "two cultures" thesis. Most recently, Holton's samurai impulses have been expressed in a set of essays directed at the ongoing Science Wars (Holton 1993).

plays a crucial role: it was much easier to take Kuhn out of Harvard than Harvard out of Kuhn. I mean this in two senses. The first relies on a very perceptive attempt to understand the aristocratic mentality as an ideal type of political action. In the case of fallen aristocrats, *saintliness* is often the interpretation that nonaristocrats have projected onto the chosen one's alternative lifestyle. The second sense in which Harvard could not be taken out of Kuhn, which will be discussed in the next section, concerns Kuhn's propensity to find himself at the right place at the right time to be led in directions that turn out to be fruitful, or at least suggestive, for his research.

Terrence Cook has analyzed not only those who adhered to their elite calling but also those who, in one way or another, strayed from the appointed path.[11] Among the many ways in which saints reveal their aristocratic bearing is their ability to endure, evade, and exit from disagreeable social situations without their own status becoming diminished in the process. Saints typically ignore criticism and stoically suffer injustice because they believe that a more active response would compound damage that has been already done. Only those with considerable control over their own fate who also believe in the larger significance of their actions are entitled to think in such terms. Lesser mortals have no choice but to respond, regardless of the consequences for either themselves or others. This point can be illustrated with Kuhn's response to the perversion of normal science that accompanied the atomic age. As we saw in chapter 4, section 6, the uncritical pursuit of highly technical work that enables paradigmatic puzzle solving to proceed apace has also enabled scientists to be easily co-opted into projects where their prowess is subserved to often dubious military-industrial ends. Yet, instead of reflecting critically on the ends of their inquiries, Kuhn would seem to have scientists either stick to their work or, as Kuhn himself did, withdraw from it entirely.

Saints are perceived as leaders in direct proportion to their rejection of the obligations imposed by their aristocratic habitus. Usually this rejection is deliberate but it can also be unselfconscious, which then leads saints to spurn their followers. Often this serves only to encourage the followers to apply and develop the saint's ideas, as if to prove their own worthiness.[12]

11. Cook 1991, esp. chaps. 4–5.
12. At the risk of courting charges of cynicism, I have observed that attributions of saintliness are most easily made by those who have never experienced the aristocratic lifestyle and hence have only witnessed the freedom—but not the constraints—that such a lifestyle entails. In other words, the sense that a saint's followers have of their own imagined inability to resist the temptations of aristocracy contributes significantly to the aura of holiness that surrounds the fallen aristocrat. Because this sense is based more on ignorance than knowledge of the aristocrat's actual situation, very personal forms of resistance that, were they made by someone in a less exalted setting, would be regarded as simple expressions of irritation, inconvenience, and avoidance can be easily interpreted as bold political gestures, if the person experiencing

Whether the followers are justified in this course of action depends on the hold that the aristocratic imperative has on the rest of society. One indicator is a widespread belief that good can arise only from pure motives. Thus, if change is unlikely to occur without an initiative from a disaffected aristocrat, then it is not surprising that saintliness is attributed to that person. With the decline of hereditary monarchies in politics, academia may be the only globally pervasive institution that still has pockets of aristocracy in this sense. In that case, the aristocrat's purity of motives applies to inquiry. Given Kuhn's personal disdain for ideological extensions of his views, it is not far-fetched to interpret Kuhn's actions as those of a saint, albeit a rather unselfconscious and secular one. At least, I shall assume this diagnosis in what follows.

2. A Career of Lucky Accidents and Studied Avoidance

It is clear from interviews conducted with him over the years that Kuhn turned away from a career in theoretical physics after becoming profoundly disaffected by the routine and destructive uses to which science was put in World War II. Conant's curriculum seemed to provide a way of reinstating his original interest in science. But as the years passed, Kuhn distanced his concerns from those of the historians and sociologists of science who had derived inspiration from his work, many of whom were overtly concerned with understanding the changing contemporary scene, even when the past was their nominal topic. Thus, when asked explicitly whether the story presented in *Structure* would need to be altered in light of the changing character of science in the twentieth century, Kuhn had this to say:

> I see no reason to suppose that the things I think I have learned about the nature of *knowledge* are going to be disturbed by the need to change the theory of *science*. I could be all wrong with respect both to science and to the nature of knowledge, but I would make this separation to explain why I'm less concerned about the question, "Is science changing?" than I might be if studying the nature of science weren't in the first instance simply a way of looking at the picture of knowledge.[13]

Interestingly, Kuhn did not treat the question as a provocation either to modify his model (for failing to match the contemporary scene) or to condemn the contemporary scene (for failing to live up to his model). Instead, he respecified his project at a level of abstraction that escaped having to decide between the two. Moreover, Kuhn's retreat to the "nature of

---

these feelings has an aristocratic background. In recent public consciousness, the career of Lady Diana Spencer, the late princess of Wales, perhaps best illustrates this phenomenon.
13. Sigurdsson 1990, 24 (italics in the original).

knowledge" invited scrutiny in just those features of his work that philosophers have found most objectionable: questions of meaning and reference, especially in relation to how scientists come to acquire a specific orientation to the world. But from the standpoint of the anti-Whig historiographies discussed in the introduction to this book, Prig and Tory, this strategy made perfect sense, as it effectively shifted the salient epistemological difference from the forward-looking "true vs. false" (i.e., how scientific claims are ultimately received) to the backward-looking "understood vs. misunderstood" (i.e., how those claims were originally intended). Thus, whereas philosophers of science structured their arguments around alternative sets of criteria (e.g., "realist vs. instrumentalist") that can justify more or less the same set of more or less progressively correct theory choices in the history of science, Kuhn took exactly the reverse tack of showing how the same set of criteria can justify quite different theory choices depending on how the criteria are interpreted and applied at specific moments. On that basis, incommensurability between paradigms appeared inevitable.[14]

No matter how much Kuhn recanted his more radical rhetoric about scientists in different paradigms inhabiting different worlds, his own research agenda always kept this possibility open—certainly more so than the possibility that science may exhibit some normatively desirable sense of "progress." A typically Kuhnian line of reasoning that attempted to put some distance between himself and his radical admirers was to grant the plausibility of the underdetermination of theory by data or the theory-ladenness of observation, and then wonder why self-styled "Kuhnians" would want to conclude that the validity of scientific claims is relative to the social conditions of their production or that nature plays a negligible role in scientific theory choice.[15] Kuhn was correct to observe that these conclusions do not deductively follow from their premises.[16] Yet Kuhn's own failure to address exactly how nature makes itself felt in a socially conditioned science hardly set a good example for his would-be disciples.[17]

14. Credit for making this point explicit goes to Doppelt 1978.
15. Sigurdsson 1990, 22–23; Kuhn 1992, 8–9.
16. After all, the validity of scientific claims may be relative to the social conditions of their *distribution*, which would require looking at the political-economic relations in which they figure, such as the spread of capitalism, imperialism, democracy, etc. This is the view I happen to hold. Alternatively, there may be limits to the human condition—be they Kantian or Darwinian in nature—such that, presented with the same evidence and background information, humans will respond within a relatively narrow range of possibilities.
17. The most philosophically sophisticated defense of bracketing considerations of an external reality from sociological accounts of knowledge remains Barnes and Bloor 1982, which argues that reality plays a negligible role in sociological explanations precisely *because* it is presupposed by all such explanations, and hence it offers no way of explaining the differences that arise in people's beliefs. The type of argument represents strategy B social epistemology, as I called it in chapter 6, note 46.

Of course, Kuhn could not have foreseen all the ways his readers would interpret what he did and did not say in *Structure*. Nevertheless, on several occasions after the book's publication, he was invited to reflect on these matters. But more than that, as I observed in chapter 1, section 5, Kuhn was actually afforded a clear opportunity to anticipate the consequences of his book, namely in response to Paul Feyerabend's prescient remarks on the 1960–61 draft of *Structure*, which led him to deem the manuscript "ideology covered up as history."[18] Kuhn characteristically failed to understand how Feyerabend's concerns bore on his own project. As Kuhn saw it, "The quasi-sociological elements of my approach were overwhelmed by [Feyerabend's] desires for society in the ideal."[19] Undeterred by Kuhn's obtuseness, Feyerabend once again raised this objection at Imre Lakatos's famous 1965 conference where Feyerabend's mentor, Karl Popper, formally confronted Kuhn in debate.[20]

That Kuhn had so rapidly risen to the rank of Popper's debating opponent on Popper's home turf surprised British observers at the time.[21]

---

18. For the context and correspondence relating to Feyerabend's remarks, see Hoyningen-Huene 1995.
19. Kuhn et al. 1997, 187.
20. Feyerabend 1970, esp. 202–3.
21. In personal communication, several professional philosophers—who, as students, traveled from Oxford to attend Lakatos's celebrated conference—recall having wondered why such a fuss was being made about Kuhn. Without Lakatos's conference, they believe, *Structure* might have left little impression on British philosophy. Since virtually all the core logical positivists had moved to the United States with the rise of Nazism, Britain lacked much of a positivist establishment for somebody like Kuhn to be portrayed as revolting against. Moreover, British higher education provided an especially poor soil for cultivating a cadre of professional philosophers of science. Philosophy was done rarely alone and usually with either classics or mathematics. The standard vehicle of academic professionalization, doctoral-level training, was virtually nonexistent. The spirit of G. E. Moore's antinaturalism had been inherited by the dominant "ordinary language" philosophy, whose learned ignorance of science was eventually savaged in Gellner 1959. Moreover, the influential Viennese émigrés turned out to be outliers from mainstream positivist thought. While Ludwig Wittgenstein and Karl Popper were not especially enamored of the humanistic bent of British philosophy, they were downright hostile to attempts to specialize the discipline. In this respect, the career of A. J. Ayer (1910–89) is instructive. After attending some Vienna Circle meetings, Ayer became the positivists' first anglophone proselytizer with the publication of *Language, Truth, and Logic*. But despite the book's enormous notoriety and sales, Ayer himself soon withdrew from the philosophy of science to deal with more classical questions in epistemology and the philosophy of mind, and eventually succeeded Bertrand Russell as Britain's "public philosopher."

The substance of Kuhn's own views was seen from Britain as a hodgepodge of such familiar Britain-based thinkers as Wittgenstein, Polanyi, and Toulmin. Moreover, another philosopher had already begun to triangulate a position around their philosophical concerns: the ex-engineer Rom Harre (b. 1927), whose work was grounded in the ordinary language philosophy of 1950s Oxford. However, Harre crossed an invisible line of philosophical respectability when he admitted Aristotelian realism back into the conceptual arsenal of contemporary science. Harre's metaphysical revanchism appeared to overreact to a logical positivism that probably never had the support in Britain that he supposed. Miller 1972 is a trenchant review of his systematic treatise, Harre 1970. Since that time, Harre has become a philosophical guru

Nevertheless, it reflected the Popperian perception that Kuhn had been anointed by Conant to provide a philosophical defense of the Big Science initiatives that increasingly characterized American research in the Cold War era.[22] Yet given Kuhn's ostracism from Harvard in 1956 and the dismantling of Conant's General Education curriculum shortly after the launch of Sputnik (see chapter 4, section 7), the view from London seemed to be nearly ten years out of date. In 1965, Kuhn probably did not warrant such exalted treatment. Nevertheless, the star billing helped convert the Popperian conjecture about Kuhn's status into a self-fulfilling prophecy—a most ironic fate, considering Popper's own heightened awareness of the havoc that publicly promoted predictions can wreak on the reliability of our knowledge of human beings.[23] The irony is only compounded by Kuhn's failure to realize, until after *Structure* was already in press, that logical positivism had moved on from its extreme Vienna Circle formulation to a position so close to his own at the time that he openly wondered whether he would have written *Structure* had he known of that shift.[24]

A curious feature of Kuhn's self-understanding was the ease with which he acknowledged the accidental character of what turned out to be decisive influences in his intellectual development. In his last extended interview,

---

among social psychologists who eschew experimental methods in the study of human beings in favor of discourse analysis. (Aristotle would have approved.) Yet here too Harre's revolutionary designs have been seriously challenged. See Tibbetts 1975, a review of Harre and Secord 1972. As to be expected, Harre's objectionable metaphysical ideas have been largely reinvented (with hardly any mention of Aristotle—or Harre, for that matter) by Nancy Cartwright (b. 1944) and received much more warmly in that guise. Her ongoing attempts to rehabilitate the logical positivists into post-Kuhnian philosophy of science have no doubt aided in that cause.

Among Kuhn's earliest British defenders was Mary Hesse (b. 1924), who eventually held the first chair in history and philosophy of science at Cambridge. A student of the philosophy of nineteenth-century physics, and especially its continuities with natural theology, Hesse had found Kuhn's interest in the religious character of the scientific community a relief from the secular humanist image of science popularized by Russell and Ayer. After David Bloor did postgraduate work with her in the late 1960s, she went on to champion—again almost alone in the British scene—the Strong Programme in the Sociology of Scientific Knowledge. A representative selection of her work is Hesse 1980. My M.Phil. dissertation (on "reductionism" in phenomenology and logical positivism) was supervised by Hesse in 1980–81.

22. Interview with Jagdish Hattiangadi in Toronto, October 1994. Hattiangadi had been a graduate student at the LSE in 1965 and was originally tipped to be Kuhn's respondent, until Kuhn objected. A good sense of the Popperian view of Kuhn's social significance, which takes his Harvard habitus very seriously albeit critically, is Jarvie 1988, esp. 314 ff.

23. The locus classicus for this argument is Popper 1957.

24. From his final interview, it is clear that, while writing *Structure*, Kuhn was working with the conception of logical positivism he received as a Harvard undergraduate in the early 1940s. See Kuhn et al. 1997, 183–84. Curiously, although Kuhn never hid Quine's influence on his own thought, he did not seem to recognize the role that Quine's long-term engagement with Rudolf Carnap played in modifying the logical positivist position to one that would enable Carnap, by the early 1960s, to see Kuhn as a kindred spirit, as noted in chapter 6, section 4. See Creath 1990.

Kuhn gave the impression that he was in endless need of guidance to focus his thoughts. Yet, he never seemed to appreciate the tension between having a continuous epistemological project and its particular expression being determined by chance events. More comprehensively reflexive thinkers would have incorporated this tension, however abstractly or symbolically, in the account of knowledge they produced. For example, the account would probably *not* portray the crises that occasion major epistemic change as internally generated. However, Kuhn tended to present these fortuitous episodes more as signs that he was already on the right track, what in more religious times would have been associated with signs from "above," especially given that these "accidents" have been largely responsible for defining Kuhn's project in his interpreters' minds. Thus, in Kuhn's last interview, we learn the following:

1. A footnote in Hans Reichenbach's *Experience and Prediction* led Kuhn to Ludwik Fleck's *The Genesis and Development of a Scientific Fact*;
2. A footnote in Robert Merton's Harvard Ph.D. thesis, "Science, Technology, and Society in Seventeenth Century England," led Kuhn to Jean Piaget's *The Child's Conception of Movement and Speed*;
3. James Bryant Conant led Kuhn to Britain, where he learned about history and philosophy of science as a field of inquiry and met Mary Hesse, who would turn out to be his strongest British champion;
4. I. B. Cohen led Kuhn to Alexandre Koyré's *Galileo Studies* and thereafter the man himself;
5. Alexandre Koyré led Kuhn to Gaston Bachelard;
6. Karl Popper led Kuhn to Emile Meyerson's *Identity and Reality*.[25]

Of all these chance encounters, I believe that the last left the most indelible impression in Kuhn's intellectual orientation. This claim calls for substantial comment, since it implies that the person normally seen as Kuhn's most formidable antagonist—Karl Popper—actually provided him with the principal resource to bolster his position. To be sure, fifteen years separated the time that Kuhn first met Popper as William James lecturer at Harvard (1950), when Kuhn was told of Meyerson's work, and then confronted Popper on his own turf in London (1965). But even more remarkably, that source turned out to be one of this century's strongest champions of continuity as a theme in the history of science, a theme ostensibly alien to Kuhn's disjunctive account of scientific change. So, who exactly was Emile Meyerson?

Meyerson (1859–1933) was no less than France's most influential nonacademic scientific intellectual of the first half of the twentieth century. He

---

25. All of these fortuitous contacts are mentioned in Kuhn et al. 1997, 162–68.

inaugurated a role that was subsequently usurped by Gaston Bachelard, discussed in chapter 7, section 4. An industrial chemist by training and a man of letters by disposition, Meyerson was the darling of antipositivists across the European Continent.[26] Popper's recommendation of Meyerson to Kuhn was prescient in at least two respects. First, it drew attention to the hidden French roots of modern anglophone philosophy of science. As I observed in chapters 1 (note 136) and 6 (note 32), Popper's conception of science as the open society and of reality as an open-ended process were indebted to Henri Bergson, whom Meyerson took to be his main academic rival. Here we need to recall when Kuhn first met Popper, namely the latter's invitation to deliver a set of lectures in honor of a philosopher—William James—whose compatibilist attitudes toward religion and science and popular touch made him America's answer to Bergson. As Bergson and James had regarded thought in general, Popper located the essence of scientific inquiry in an endless quest for self-transcendence. In terms of this process, established facts and theories are little more than way stations that potentially obscure the course of inquiry if they are taken as final products in their own right. Popper told Kuhn to read Meyerson precisely because he had already detected in the young Kuhn disagreements on this point.

Meyerson never hid his debts to Leibniz and Kant, both of whom were inclined to treat established facts and theories as direct evidence for the processes by which they were—indeed, had to be—produced. Thus, according to Meyerson, the best way to understand the nature of science is not to observe the actual conduct of science, as social constructivists subsequently would, because that could lead to so many dead ends and unscientific directions. Rather, one should start with unproblematic scientific achievements, because they provide the threshold for what it is that competing scientific theories have historically tried to achieve.[27] This point is

26. Meyerson was plugged into the major scientific networks of his day, regularly corresponding with Einstein and de Broglie. Nevertheless, he remains a neglected figure, even in France, but especially in anglophone philosophy. The most thorough examination of his corpus is still La Lumia 1966. My thanks to George Gale for sharing his own interest and knowledge of Meyerson with me.

27. A good example of Kuhn's attachment to this Meyersonian doctrine is his response to Shapin and Schaffer 1985, the most influential social constructivist history of science. Kuhn accuses them of not knowing, or ignoring, the technical details of hydrostatics that "everybody now learns in high school" in their explanation of why Boyle's account of the air-pump was preferred to Hobbes's (Kuhn et al. 1997, 192). Here Kuhn takes what is now a long-standing scientific finding as the goal toward which both Boyle and Hobbes were aiming in their seventeenth-century dispute, clearly abstracting what Kuhn presumes to be the common goal of their scientific inquiry from whatever other personal and political goals distinguished them in their day. In contrast, a social constructivist would refuse to grant any clear distinction between scientific, political, and personal goals, until a canonical account of the episode is constructed, whereby the various goals would be disentangled for purposes of vindicating the dominant research trajectory. Thus, from Shapin and Schaffer's standpoint, it would not have

easily overlooked because, like today's social constructivists, Meyerson himself invoked the distinction between what scientists say and do—but he meant *what scientists do once it had been done, not as they were doing it.* Historically, this mentality is closely associated with the argument from design in theology. It capitalizes on the psychological fact that after an event has occurred it is harder to imagine equally probable alternatives than before it occurs. This post facto perspective, in turn, suggests that the event was caused for a reason; hence, the need for a rational agent behind the scenes.[28]

The argument from design is most persuasive when the world is seen as both rational and complex, since taken together, these two factors diminish the probability that sheer chance could explain why things are as they are. Thus, whereas Bergson read history back to front, so that at any point in the past the future appeared indefinitely open, Meyerson read it front to back, so that the present appears to be the logical culmination of the past. Methodologically, Bergson shadowed the stream of consciousness, while Meyerson diagnosed the textual trace. That Kuhn stood with Meyerson on this point is demonstrated by his own method of research (noted in chapter 4, section 2), which privileged finished works in the public domain over draft manuscripts in the archives.[29]

Given Meyerson's exclusive interest in proven scientific achievements, it would be easy to conclude that he regarded his theory of scientific change as normative rather than descriptive. However, this would clearly mistake the spirit of his enterprise, as Meyerson's most heated polemics were directed against the logical positivists and behaviorists, both of whom

---

made sense to mention matters currently settled in hydrostatics, since it was the resolution of the Hobbes-Boyle dispute that helped to settle them.

In these terms, Popper may be seen as having agreed with social constructivists about the actual nature of science; hence, his insistence on interpreting the "basic observation statements" of the logical positivists as revisable "conventions" for the conduct of inquiry, not indubitable foundations of knowledge. However, Popper differed from the social constructivists in his insistence on an explicit normative standard, against which such ongoing developments in science may be judged. My own social epistemology agrees with Popper on both scores, though I am more explicit than Popper about the political character of the means by which this normative standard is determined. See Fuller 1999b, but also chapter 6, section 1, above.

28. Far from being anomalous, Meyerson's design-oriented approach to the history of scientific achievement finds good company in the history of scientific reasoning, especially the Reverend Thomas Bayes (1702–61), whose eponymous theorem aimed to formalize inductive (or in Charles Sanders Peirce's more precise terms, "retroductive" or "abductive") inference in order to show that the probability of a divine intelligence increased as science revealed just how well-ordered nature is. See Hacking 1975, esp. chap. 18. Whewell's interest in putting the "theos" back in scientific theorizing also fits in this tradition (see chapter 1, section 6, above).

29. See Lecourt 1975, 53, who wisely places Kuhn closer to Meyerson than to Bachelard, despite superficial resemblances.

he accused of normative heavy-handedness. Meyerson understood the expression "the nature of science" very literally to mean that science is an activity having intrinsic ends, no mere means for predicting, controlling, or even representing something outside itself, called "Nature." Indeed, in response to Moritz Schlick of the Vienna Circle, Meyerson argued that to claim that science aims at the comprehensive representation of Nature is to introduce a transcendent normative standard that sidelines the empirical search for the ends that science has set for itself as the embodiment of the human intellect.[30] We begin to see why Kuhn might have been confused about Feyerabend's charge that *Structure* engaged in an ideological whitewash of science. Kuhn insulated his account from anything — be it called "Nature" *or* "Society" — that might externally direct the development of science as a form of knowledge. It left Kuhn with what struck Feyerabend as an artificially, perhaps even strategically, suspended conception of science.

Meyerson — and Kuhn after him — perceived a much harder boundary between science and Nature than his interlocutors. For example, whereas a positivist, behaviorist, or Popperian would equate the "self-correcting" character of scientific inquiry with how it responds to phenomena in Nature, Meyerson saw science's self-correction more strictly as a purging of its own past, the ongoing conversion of empirical findings into logical deductions, which placed under continual erasure any evidence for science's existence in a world outside its own rationalizing tendencies. Philosophers of science may see here an attempt to turn the distinction between the contexts of discovery and justification into a two-stage developmental process. Speaking more fashionably, we might say that Meyerson had an "autopoietic" conception of scientific inquiry, as did Kuhn. Meyerson himself invoked the term "conservation" — as of number, matter, and energy — to identify the principles that have historically functioned as the transcendental basis for knowledge of the physical world: to wit, that whatever happens in this world is the result of something else that happens in the same world. For Meyerson, the one great revolutionary moment in intellectual history came when the pre-Socratic philosophers abandoned the appeal to supernatural agency and grounded their inquiries in a generalized conservation principle.[31]

Alexandre Koyré also shared this suspended view of science. Koyré was one of several Jewish émigrés from the Russian borderlands of Eastern Europe whose cultural background was sufficiently similar to Meyerson's to

---

30. This debate is outlined in La Lumia 1966, 11–12.
31. The point comes out most clearly in Meyerson's magnum opus, *Identity and Reality* (1908).

appreciate the Platonic hermeticist precedent of the perspective that Meyerson brought to bear in the salons he conducted from his home, which were the talk of Paris before the rise of Nazism.[32] However, Koyré's affinities with Meyerson are easily obscured if we focus too much on the fact that Koyré was Kuhn's proximate source for the idea that seventeenth-century Europe witnessed a scientific revolution that culminated in the Newtonian paradigm. In fact, Koyré's historiographical stress on rupture was ultimately consonant with Meyerson's on continuity.

Take Koyré's portrayal of Galileo, previously raised in chapter 4, section 4. The principal rupture occurred between the underlying structures of reality that were available only to the intellectually adept and the realm of empirical phenomena that made the art of experiment a suitable foil to the commonsense forms of observation underwriting Aristotelian science. In Galileo's day, Scholastic scientists, with one eye on spiritual governance, favored empirically based forms of knowledge that smoothed over the epistemic differences between the governors and the governed. In fact, those harboring a more strictly Platonic concern for ensuring the integrity of knowledge over time sought protection from such contamination through practices that admit of esoteric interpretations. The great breakthrough that constituted Galileo's approach to experimentation was that it met the Platonic need, while at the same time enabling the conversion of those who are moved only by their senses. The former is illustrated by the potential access that experimental intervention allows to the mechanisms that underwrite empirical regularities; the latter by the import attached to an experimental observation that confirms a prediction. Together these two aspects systematically purge scientific thought from extrascientific contaminants.

In one sense, Kuhn helped update the psychology that informed this perspective by introducing Piaget's "genetic structuralist" account of child development in his contribution to Koyré's Festschrift.[33] Piaget recognized the tension between science's overarching interest in what Piaget, following Meyerson, called the "conservation" of knowledge over time with the periodic reconfiguration of the terms under which that conservation occurred.[34] According to Kuhn and Piaget, the resistance that experience

32. See Collins 1998, 1024 n. 20.
33. Kuhn 1977a, 240–65 (originally 1964). For more on Kuhn's debt to Piaget and its bearing on Koyré's influence, see introduction, note 37, above.
34. Piaget routinely motivated his account of cognitive development as a reaction to Meyerson. See Piaget 1952, 13; Piaget 1970, 21, 39, 122. Koyré 1978, 2, cites Meyerson for the classic Piagetian example of the modern notion of inertia appearing self-contradictory to ancient and medieval physicists who had failed to abstract the principle from its empirical realizations. For his part, Meyerson was attracted to Hermann von Helmholtz's project of naturalizing the normative dimension of cognition by translating Kantianism into experimental psychology.

poses to our conceptual scheme is not simply an instance of the irrational, as Meyerson thought, but rather marks of a reality that exists beyond our concepts, to which we must somehow "accommodate," in Piagetian terms. However, as the word "accommodate" suggests, this recognition of an external reality is not entirely welcomed. Indeed, given his own experience in psychoanalysis and its admitted influence on other aspects of his work, Kuhn may have been moved by the Freudian concept of "trauma" when trying to capture these unwanted contacts with a world beyond one's reach, which culminate in a paradigmatic crisis.[35]

Had Koyré not died the year his Festschrift appeared, he probably would have responded to Kuhn's piece by drawing attention to the "straightness" of his interpretation of Piaget, which focused exclusively on how his experiments induced in children the equivalent of paradigmatic crises, without commenting on how the experiments occluded their own manipulative character. Indeed, by failing to account for the role of the Piagetian experimenter in the constitution of directed epistemic change, Kuhn lost the political side of Meyerson's project, which Koyré had uncovered through his deep knowledge of the history of Platonism. This may be summed up as the ongoing construction of epistemic continuity and progress, a process whose significance Kuhn downplayed by discussing only its products, the histories recounted in postrevolutionary scientific textbooks. Indeed, Kuhn so minimized the significance of effort needed to maintain the distinction between the scientist's history of science and the historian's that he made the two appear to subsist in parallel universes. The irony here is that shortly before the publication of *Structure*, Kuhn had shown that the energy conservation principle, far from being a transcendentally knowable feature of physical reality, began life as a set of incommensurable interpretations of ongoing research and only gradually acquired its current canonical form.[36]

So far my account has attributed Kuhn's significance largely to things that other people thought and did. As we have just seen, this extends beyond the reception of *Structure* to its actual composition. This pronounced state of uninvolvement suggests that he did indeed suffer from a state of diminished cultural responsibility that makes the sense of "being there" I raised in the preface to this book more than just a nasty dig. In any case, we need a term for the incapacity to do what is expected of someone in a given

---

However, according to La Lumia 1966, chap. 9, Meyerson never satisfactorily reconciled the transcendental and empirical elements of his historical epistemology, a fate perhaps also suffered by Kuhn. On Helmholtz's project, see Hatfield 1990b.

35. See note 2.
36. Kuhn 1977a.

social position, a failure to acknowledge from where one had come to where and was supposed to go. Let us call this condition *culturopathy*. Culturopaths lack reflexive engagement with what they say and do. They go through life as if in a vacuum or a bubble. Academic training unwittingly renders its subjects susceptible to this disorder while preparing a "universal class" of pure inquirers: the so-called Ivory Tower mentality. The symptoms may range from the crudely comic to the more subtly pathetic: on the one hand, the proverbial absent-minded professor; on the other, the scholar who supposes that publication ipso facto secures readership. The disorder takes a more specific form in terms of the relationship between historians and philosophers of history, or "metahistorians" in Hayden White's terms.

White uses the expression "cognitive responsibility" to distinguish the two groups of inquirers.[37] Metahistorians display cognitive responsibility for their narratives in ways that ordinary historians do not. They introduce epistemological complications that historians normally avoid because historians generally do not refer to the contexts in which their own texts are written and read.[38] By this criterion, both Conant and Koyré were metahistorians. Conant's reflexive engagement appeared in disparate presentations of the nature of science to disparate audiences; Koyré's in scholarly works clearly aimed for highly specialized audiences mentally prepared to receive uncomfortable truths. These acts of cognitive responsibility incurred costs: Conant's message became diffuse and widely attacked, Koyré's esoteric and nearly ignored. Kuhn lacked this sense of responsibility because he took their two visions as the background conditions for his own seamless narrative for an audience largely unfamiliar with both the quotidian science policy struggles that concerned Conant and the transhistorical worries about truth preservation that concerned Koyré.

---

37. White 1973, 14 n. 7, 23 n. 12. White's source for "cognitive responsibility" is Stephen Pepper, *World Hypotheses* (1948). Pepper, who trained at Harvard under the critical realist Ralph Barton Perry, was chair of the Berkeley philosophy department when Kuhn arrived as a junior faculty member in 1957. Indeed, Kuhn credits him with having wanted a historian of science to be hired within the philosophy department. See Kuhn et al. 1997, 174. Pepper distinguished between those who defended their knowledge claims in ways that were compatible with their own presuppositions and those who did not. The latter were deemed cognitively irresponsible. In this way, all the bugaboos of rationalism—animists, mystics, and skeptics—could be dismissed in one fell swoop as soon as they engaged in reasoned argument. Although many suggestive comparisons have been made between Kuhnian paradigms and Pepperian world hypotheses, at most Pepper emboldened Kuhn to continue along a path he had already discerned. Indeed, given Pepper's intent, Kuhn would have to be seen as having ironically appropriated Pepper. After all, Kuhn licensed concerns about cognitive responsibility only after a world hypothesis was threatened with anomalies.

38. Here I follow White's discussion for purposes of elucidating the history/metahistory distinction without necessarily endorsing his blanket characterization of historians and philosophers. Clearly, "historians" and "metahistorians" in his sense can be found on both sides of the disciplinary divide between history and philosophy.

Of course, Kuhn's specific style of culturopathy bore the traces of his habitus. He disengaged aristocratically. Upon the publication of *The Structure of Scientific Revolutions*, Kuhn did not call for the immediate overthrow of the intellectual and political regime associated with Big Science. Instead, he withdrew his services from *both* the regime's supporters its and opponents. Only someone already in a privileged position could have afforded the luxury of disengaging in this fashion without being consigned to oblivion. For despite having failed at Harvard, Kuhn managed to move to Berkeley and then Princeton, where he finally received tenure. He retired as the Laurence Rockefeller Professor of Philosophy and Linguistics at MIT.

If Kuhn was indeed a culturopath, then the challenge for his admirers will be to disentangle whatever insight they have derived from *Structure* from any attributions of intentionality to its author. The difficulties that lie ahead in neatly drawing such a distinction speak to the peculiar way in which academic knowledge claims are legitimated in our times. In what sense does a text remain a useful legitimatory tool once it becomes generally known that its author meant few, if any, of his readers' interpretations and that the social contexts of the text's production and distribution are regarded as suspect by the text's most discerning readers? If readers leave my book troubled by this question, then I shall have succeeded in conveying my main critical point. You will have begun to appreciate *Structure* as an ironically self-exemplifying text: a work that grounded science's success in paradigmatic pursuits that was itself the creation of a mind-set that could not see beyond the paradigms under which it labored.

3. Getting over Kuhn: The Secularization of Paradigms as Movements

Consider the first lesson in the average Western epistemology course: knowledge consists of a truth that is believed for good, if not the best possible, reasons; in philosophical shorthand, knowledge is "justified true belief."[39] This definition, which is usually presented with a gesture to Plato or Descartes, fossilizes the opposition that has characterized the "essential tension" of Western culture since the sixteenth-century Protestant Reformation, specifically the process by which civil authority became autonomous from religious control, or *secularization*.[40] That knowledge claims command one's belief harks back to tests of religious commitment, whereas the demand that such claims be justified recalls the legal procedures of trying cases in secular courts. In this sense, the philosophical definition of

39. See, e.g., Chisholm 1974.
40. Fuller 1997e.

knowledge is a negotiated settlement between secular and sacred authorities.[41] The two poles of the tension, which stress the "justified" and "belief" side of the definition, respectively, are epitomized as follows:

> (A) *Because knowledge is ultimately a justified truth claim, it does not require a personal commitment of belief, simply conformity to the procedural rules of evidence and inference.* Example: legalism, or the public acceptance of secular authority.
>
> (B) *Because knowledge is ultimately a matter of belief, it can never be fully justified, except by the strength of the commitment and its consequences for action.* Example: voluntarism, or the private acceptance of sacred authority.

It may seem that (B) has virtually disappeared from scientific discussions of knowledge. However, as we saw in chapter 6, section 2, such a verdict would be too hasty. In excavating Kuhn's pragmatist roots, I mentioned William James's "will to believe" version of pragmatism as a precursor, which is the exception that proves the rule. Current debates between realists and instrumentalists also turn on whether one truly needs to "believe" in the entities referenced in one's theories or simply act as if one believed in them. One plausible way of encapsulating recent philosophical debates over scientific rationality is in terms of when one should make and forsake commitments to particular research programs, especially in the face of less than adequately justified knowledge claims.

Kuhn, following Polanyi, located the "genius" of science in the personal commitment that each scientist presumes of her colleagues. This mutual presumption then creates a climate of tolerance for somewhat divergent paths of research and even temporary disagreements over matters of fact and interpretation. In that sense, (A) and (B) remain bound together because (A) is taken to govern the microlevel of day-to-day research and (B) the macrolevel of the paradigm's overall direction. Kuhn's account of paradigmatic change in *The Structure of Scientific Revolutions* manifests the latent instability of the classical definition of knowledge. As puzzle solving proceeds apace in a paradigm, scientists who profess a commitment to a certain vision of the truth and have played by a set of rules for justifying claims to the truth will inevitably encounter anomalous phenomena that eventually cause them to diverge on the appropriate direction for inquiry. This, in turn, precipitates the "crisis" that eventuates in a "revolution" and a new paradigmatic regime.

For Kuhn this tension — the source of collective disenchantment associated with secularization — is potentially divisive; hence its presence should

---

41. On the idea of belief as commitment or faith (from the Latin *fides*, as in "fidelity"), see Smith 1977. On the rise of secular law in the emerging nation-states of Europe after the fall of the Roman Empire, see Kelley 1970.

be minimized at all costs, most notably in the "progressive" histories of science that students are taught in their introductory science textbooks. However, an alternative social epistemology of science would forgo this Orwellian solution and embrace the tension as productive, along the lines of Popper's model of "conjectures and refutations" as the model of rational inquiry, whereby one would be both the best proposer of her own knowledge claims and the best refuter of such claims made by someone else. But here I must quickly add that the desired metatheory would justify the participation of the *entire* society in the process of mutual criticism rather than just a self-selected community of experts.

Instead of first tampering with people's biases as they are trained to be "objective" in their personal assessment of each other's knowledge claims, I believe that "objectivity" should be a continuously emergent property of the interaction of proponents and opponents of knowledge claims. Biases, such as they are, would then be negotiated, canceled out, or otherwise overcome in open discourse, not prior restraint. The model social entity of this collective dialectical process is the *movement*, which gains strength not by resolving its internal differences but by involving ever larger segments of society in the articulation of those differences. A good image here is that of a whirlpool that draws more attention to itself as discussion acquires more intensity.[42] The closest that academia currently gets to this arrangement is the constitution of the social sciences, which (in sharp contrast with the natural sciences) do not launder out ideological disagreements in professional training, but rather enable those disagreements to align with, and often alter, conflicts in the society at large.[43]

The American sociologist Robert Wuthnow has shown that the three most socially significant intellectual movements in the West's modern era—the Protestant Reformation, the Enlightenment, and nineteenth-century socialism—were successful to the extent that a fairly esoteric group of inquirers extended their arguments to the wider society, so that others found their categories relevant to describing their own lives and situations.[44] These movements lost their creative transformative energy once they became sectarian and paradigmlike. The difference between a movement and a paradigm may be seen as a shift in the relationship between

---

42. For an impressive synthetic treatment of movements as the core social formation in terms that resonate with the key role I assign them in knowledge production, see Melucci 1996. My thanks to Gerard Delanty and Sujatha Raman for alerting me to the significance of this work.

43. On the movementlike character of the social sciences, see (in addition to chapter 5, section 1, above) Fuller 1997d, 20–23. The idea of movements as the natural opposite of paradigms was first suggested to me by Sujatha Raman.

44. Wuthnow 1989.

presumption and burden of proof. Whereas a movement shoulders the burden of trying to persuade people who are not yet true believers, a paradigm's members presume the strength of their common commitments and then wonder how a substantial change in direction would be possible. The question, then, becomes how to get those who do not spontaneously share the movement's core beliefs and experiences to act in ways that promote the movement. *In any case, I urge that we turn Kuhn on his head and demonstrate that a paradigm is nothing more than an arrested social movement.*

This inversion entails that we regard inquiry as an especially focused form of political action. Whereas a paradigm-based approach to knowledge would declare politics to be vulgar metaphysics, a movement-based approach treats metaphysics as an inchoate politics. Thus, a stable body of knowledge is simply what political action becomes once the public space for contestation has been restricted. (In a similar vein, a functioning artifact—a technology—is simply what political action becomes once patterns of access and usage have become regimented.) Movements wither and die when "true believers" of various persuasions break off the debate and form sects that invite discourse only from the like-minded. In that case, the knowledge becomes esoteric and the artifacts fetishes. Sometimes sectarianism is a legitimate response to a debate that has devolved into violence. But with a little luck, by then the movement will have left its long-term mark, as the major groupings in society—most of whose members are casual observers to the movement's activities—reconfigure themselves in terms defined by the movement's discourse.

Thus, it is crucial for understanding what follows that readers forget any monolithic connotations that the Reformation, the Enlightenment, and socialism have acquired since their heyday as movements. Following Wuthnow, I am exclusively concerned with the multifarious activities of those who identified themselves in terms of these three movements, and *not* the activities of those upon whom these terms have been foisted once the movements had been reduced to "grand narratives" that capture little more than a rough periodization of modern history.

Traditionally, social movements have been conceptualized as purely reactive entities composed of disgruntled—if not downright irrational—individuals who lack the sustained purposefulness enforced by proper institutions, such as scientific paradigms. Perhaps because professional sociologists have often worked on behalf of public administration or industrial management, even they have tended to treat movements as transient or degenerate social formations. Wuthnow reverses this negative image by charting the trajectory by which discourse fields manage to acquire the political and economic resources that enable them to become vehicles for large-scale social change. Accordingly, a movement gestates during a period of economic expansion, which allows many people to enter discourse-intensive occupations, such as the clergy, academia, and the state bureaucracy. The proliferation of these occupations implies, at the very least, that people feel they need to know what others are doing before they themselves can act, but they cannot fathom for themselves how those others think. In the final sections of chapters 2 and 7, this feeling was captured in terms of a sense of increased "social complexity" and the attendant need for these salaried scribes to engage in social "intermediation," but Alvin Gouldner and other critical theorists have tried to give a more radical spin to their labors.[45]

This emergent communicative complexity is followed by a period of economic contraction that causes considerable status dislocation as different sectors of society adapt differently to their new situation. Professions that had been prestigious or rich lose status, and vice versa, thereby providing the condition of "relative deprivation" that is often seen as a precondition for social revolution. People in the discourse-intensive fields, who have themselves become dislocated, compete with one another in offering new criteria of legitimation. They convert their collectively threatened position into an opportunity for expansion and very often risk taking (hence the frequency with which political revolutions are associated with alienated intellectuals). Whether any major social change actually occurs depends on the ability of dislocated groups to identify a common foe, such as a neighboring country or a vulnerable minority, even if, upon reflection, this supposed foe is clearly little more than a pretext for change: a scapegoat.

Wuthnow's account represents a recent trend toward treating movements as "flexibly organized cognitive praxes" that produce knowledge for enabling and disabling certain transformations of social life.[46] What differentiates movements from paradigms is their sense of organization—not necessarily their goals, their longevity, or even their commitment to inquiry. Successful movements manage to retain their dynamism, their distinctive form of consciousness, as they gain credibility in the course of achieving concrete goals. They do not simply "evolve" into paradigms. However, because credibility is popularly measured by the degree of stability that one contributes to the social order, the *dynamic credibility* required of successful movements would seem to strain the imagination. Two styles of recent theorizing about movements define the "essential tension"

45. For an early realization of this phenomenon, see Gouldner 1979. See also chapter 5, section 2, above.
46. For a theoretically sophisticated elaboration of this point, see Eyerman and Jamison 1991.

needed for this dynamic credibility to be maintained. Between them they display the tension between (A) and (B) in the classical definition of knowledge, corresponding, respectively, to what I call the *North American* and *European* styles, so-called for where the relevant researchers tend to come from, but undoubtedly also a willful exploitation of cultural stereotypes for analytical purposes.[47]

The North American style stresses the *justification* side of the classical definition of knowledge, while the European style stresses the *belief* side. The European style centers on the consciousness-raising function of movements. It is primarily studied by social-psychological methods. The North American style focuses on the goal achievement function of movements. Its studies have been grounded most recently in rational choice economics. Each style of movement thinking is necessary, but not sufficient, for maintaining a movement's dynamic credibility, as can be seen in table 15.

The European style emphasizes the role of movements in forming a collective identity among people who may be disparately located (in both space and status), but who nevertheless share experiences that heretofore have been ignored or trivialized — even by the individuals themselves. The original example that Marx used to discuss this feature is particularly instructive. Martin Luther campaigned to get the German peasants to stop discounting the cognitive significance of their own sensory and spiritual experience. This campaign was at once directed against Catholic theology and heliocentric astronomy, which, in their quite different ways, were bastions of cognitive authoritarianism. However, as Marx himself had already realized in *The German Ideology*, a movement that thrives entirely on consciousness raising is likely to be confined, ever more dogmatically, to just those people who have had the relevant sensitizing experiences. In short, it becomes cultish to the point of losing all hope of establishing society-wide credibility.

In contrast, the North American style focuses on the instrumental side of movements, their ability to achieve the goals on their agendas. Here we find the efforts to distill utopian aspirations to planks on a party platform, which enable the movement to make a series of short-term alliances with more mainstream interest groups. Not surprisingly, the sheer increase in the movement's dimensions is taken by its members as a sign of progress, even if it involves diluting the movement's identity and exaggerating the significance of getting a compromise bill passed as part of an omnibus legislative package. The advantage of seeing movements as agenda-pushing

47. A good sourcebook for research into the two styles of movement thinking is Morris and Mueller 1992. The distinction between what I have called "European" and "North American" styles of movement thinking is normally credited to J. Cohen 1985.

TABLE 15. *The Essential Tension That Defines Social Movements*

	MOVEMENT STYLE	
	EUROPEAN	NORTH AMERICAN
epistemology	belief-oriented	justification-oriented
sociology	ideology	technology
practice	consciousness raising	agenda pushing
status	end in itself	means to an end
norm	intensity of commitment	breadth of support
economy	resource generating	resource mobilizing
rationality	communicative	instrumental
corrupt version	cultishness	co-optation

vehicles is that it provides concrete reference points for the movement's activities, continually reminding the movement's members—especially those who have *not* had the relevant sensitizing experiences—that it is heading the entire society in the right direction. However, a movement that is exclusively focused in this way easily falls victim to its own success, as the movement's ability to adapt to the mainstream gets mistaken for its ability to bend the mainstream to its will. In short, the movement becomes captive to its immediate context.

Thus, the dynamic credibility of movements depends on creatively resolving the tension between cultishness and co-optation. Of course, this is easier said than done. Most contemporary movements display both tendencies at once. The sharp divide in the strategy and tactics of so-called "radical" and "liberal" feminism within the women's movement may be the clearest contemporary exemplar of the difference between European and North American styles of movement thinking. Many radical feminists ground the distinctive consciousness of women in their biological differences from men, whereas liberal feminists regard gender as a one of several sociohistorical markers of inequities in a system that aims to eliminate all such inequities. The history of black activism in the United States has reproduced the difference between the European and North American styles in successive generations, but each time in a new register: consider W. E. B. Du Bois versus Booker T. Washington, Malcolm X versus Martin Luther King, and Molefi Asante versus Cornel West. To be sure, the specific terms of each disagreement were strongly colored by the major political issues of its day, but not so as to obscure the fundamental difference between the European and North American styles of movement thinking in each case.

When it comes to sustaining a movement's dynamic credibility, the tension in need of resolution is rather different from Kuhn's "essential tension" of tradition and innovation, referred to in the title of his collected essays, which defines a paradigm's form of knowledge. According to Kuhn, for the latest generation of scientists to remain motivated, they must be led to believe that even a revolutionary theory that came from "left field," such as Darwin's theory of evolution by natural selection or Einstein's theory of special relativity, could have just as easily come from establishment science. This leads to the double truth that distinguishes the historical consciousness of historians and scientists. In contrast, the essential tension defining movement knowledge involves showing that the disparate historical origins of various interest groups in fact converge in common cause. On the one hand, the goal is to keep an already existing community intact by homogenizing the more disparate features of its history; on the other, it is also to enlarge the community's constituency by integrating its disparate strands into one trajectory. The way to meet both goals at once is to recruit the larger society, so that differences within the movement become the terms in which those outside the movement define themselves. This is the ultimate trick for the public intellectual to turn.

Movements are especially effective in this regard during periods of socioeconomic dislocation, when the old social categories fail to capture emerging political realignments. Regardless of what people think of a movement on its own terms, the movement's discourse may nevertheless provide the only publicly accessible framework for understanding the full range of ongoing changes. A good case in point is the legacy of socialism, no small part of which was that factory owners came to think of themselves as a "class" in systemic opposition to the class represented by their employees.

Of course, the factory owners did not become card-carrying socialists once they started to think in terms of class. However, by accepting this movement-inspired designation as their own, they unwittingly opened themselves to certain ways of describing and explaining existing divisions in society that eventually made it easier to justify the intervention of the state in economic affairs. Wealth that would have appeared, early in the nineteenth century, to be the result of the factory owner's individual initiative was more commonly seen, by the end of the century, as the product of some sort of exploitation. This transition enabled the taxation of factory owners and the protection of workers to be regarded as reciprocal policy measures in the emerging welfare state. Thus, owners have increasingly had to shoulder the burden of showing that they are entitled to keep all the wealth produced under their name. In short, the discourse community created by a social movement can be politically effective simply by altering the "spin" that different social groups give to one another's activities, which in turn opens new spaces for action, especially by third-party regulatory agencies.

Considered in light of Kuhn's overriding concern for consensus formation in science, a striking feature of the trajectory common to Wuthnow's three movements is that the peak of their influence corresponded to a high level of internal division. In each case, opinion was divided over an abstract philosophical question in ways that had clear implications for the parameters of legitimate collective action. The Protestant Reformers disputed interpretations of the Bible and the writings of the Church Fathers. The Enlightenment wits argued about humanity's capacity for self-governance. Socialists vied over whether industrial capitalism and parliamentary democracy were preconditions or impediments to the ideal society. Rather unlike the professional posture of a Kuhnian paradigm, the parties to these three movements did not presume that concerted practical action had to await resolution of these fundamental questions. On the contrary, the more the movements increased their transformative capacity, the wider the circle of people who felt that their interests were somehow implicated in the swirl of opposing discourses.

Here it is worth recalling Wuthnow's own roots in the sociology of religion, which, following Max Weber, has regarded institutionalization — the formation of doctrinal consensus and its ritualized reinforcement — as sapping the spirit that marked a religion's charismatic origins.[48] Under the Weberian gaze, established churches appear as the domestication of more ecstatic forms of religious experience. Similarly, the kind of divisiveness that eventually diminished the impact of the movements that Wuthnow studied was that of sectarian withdrawal, often under the guise of "purity": that is, either a refusal to argue with doctrinal opponents or a refusal to acknowledge the legitimacy of *any* existing authority.[49] Indeed, it would not be far-fetched to think of consensus-based normal science as a strategic retreat from the spirit of inquiry in just this sense, especially if "inquiry" is conceived in the Popperian sense of a sustained willingness to challenge the status quo and to entertain opposing arguments: "permanent revolution," as he put it in a bit of anti-Kuhnian pique.[50] In that case, what Kuhn regarded as a mark of collective self-discipline on the part of the founders

---

48. An excellent recent attempt to put Weber's perspective in the context of fin de siècle fears of degeneration is Herman 1997, esp. 128.
49. For contemporary corroboration of this point, see Frey, Dietz, and Kalof 1992.
50. See esp. Popper 1975.

of the Royal Society to exclude politics, religion, and morals from their purview would come to be seen as an institutionalized failure of nerve. And this is precisely the image that I wish to promote.

The Reformation, the Enlightenment, and socialism each left the state stronger, not because the intellectuals supported the status quo (often they did not) but because their disputes reinforced the idea that there was a single, albeit elusive, source of authority, control over which could be determined by publicly contestable means.[51] In theory the ultimate source may have been Truth, but in practice the state turned out to be the unintended beneficiary of each movement's relentlessly critical inquiry. Despite the ambiguous lessons contained in this conclusion, the fact that the beneficiary was the state—and not a private sector of society—offers a ray of hope of movements contributing to the revival of the public sphere.

The unique sociological success of science in the twentieth century has been its ability to dictate to the state the terms of its preservation. In effect, the scientific community has required that the state adopt its leading theories as a civil religion in return for providing authoritative means for organizing and mobilizing the populace.[52] An unusual feature of this process is that whereas religion is typically integrated into the daily lives of people who can provide religiously sanctioned justifications for their practices (e.g., food intake in terms of dietary laws), science maintains its hold on society largely through public accounting procedures—such as examinations of mental and physical competence—that still have relatively little connection to people's lives, as reflected in their persistent ignorance of what informs those procedures.

This helps explain the current crisis in the "public understanding of science" in the English-speaking world, the analogue of which would be difficult to imagine in the case of religion—not because people are more secure in their religious beliefs but because those who reject religion have a better grasp of what they are rejecting than those who reject science.[53] Since science was embraced by state agencies before it gained much grassroots support in the general public, it has continued to appear an artificial feature of contemporary societies, with the Enlightenment ideal of the "citizen scientist" who can think for herself proving increasingly elusive.

---

51. See Wuthnow 1989, 577. The idea of the state as the repository of Truth is, of course, a feature of Hegel's philosophy of history.

52. We discussed the origins of this process in chapter 2, section 7, when examining the long-term implications of Planck's victory over Mach.

53. See Fuller 1997d, esp. chaps. 1 and 4.

4. High and Low Church Secularizations of Science

The process by which Christendom came to be secularized may prove a useful guide to what lies ahead for science. As the European states secularized, they refused to grant any religion a monopoly over political and economic resources, while protecting the rights of any religion to profess its creed within state borders. The immediate cause of secularization was the destabilizing effect of religious wars on the emerging nation-states of Europe in the sixteenth and seventeenth centuries. The decoupling of political legitimation from religious affiliation was just as much the product of Machiavellian survival instincts as of any interest in ensuring maximum freedom of expression. And while the institutional ascendancy of the natural sciences in the second half of the nineteenth century is often regarded as a major vehicle of secularization, we may have reached a point at the end of the twentieth century—given the concentration of state resources on scientific research—that calls for the secularization of science itself.[54] To paraphrase the Enlightenment critic Gotthold Ephraim Lessing, the true test of science as a form of knowledge may be whether it can command believers even after state support has been removed.

If the secularization model is an apt one, we may speak of two "waves" in the critique of the social dimensions of science and technology, akin to the waves of secularization in the history of modern Christianity. I have called the two waves, *Low Church*, which resembles the Protestant Reformation of the sixteenth and seventeenth centuries, and *High Church*, which is akin to the radical hermeneutics of the "Higher Criticism" of the Bible in the eighteenth and nineteenth centuries, which I discussed in chapters 1 and 2 under the rubric of "critical-historical theology." In terms of the "essential tension" in Western epistemology raised in the previous section, the Low Church is belief oriented, while the High Church is justification oriented.[55]

---

54. Among major mainstream economists, the idea that, by the second half of the twentieth century, science had become "the secular religion of materialistic society," has been most clearly observed in Johnson 1965, esp. 141.

55. I was inspired to draw the Low/High Church distinction in STS in response to Juan Ilerbaig, a Spaniard studying in the United States who was concerned that the field was becoming a victim of its own academic success by increasing the incestuousness of its theoretical debates. Consequently, the original Low Church concern with a reform of the functions of science and technology in society at large was at risk of being completely lost. See Fuller 1992d; also Fuller 1993b, xiii. As of this writing, the only STS textbook that clearly grounds the origins of the field in Low Church movement-oriented concerns is in Spanish: González Garcia, López Cerezo, and Luján 1996. To avoid confusion, I must observe that González Garcia, López Cerezo, and Luján draw a distinction between "European" and "North American" STS that reverses the most natural reading of "European" and "North American" approaches to social movements I discussed in the previous section. This reversal makes sense,

In the first wave, just as Luther, Calvin, and their associates called for the Church to recover its spiritual roots from corrupt material involvements, the 1960s witnessed the rise of scientists who "conscientiously objected" to their colleagues' complicity with the state in escalating the Cold War. A secularized science would never have given us the nuclear arms race, just as a Protestantized Christianity in the Middle Ages would not have been able to mobilize the material and spiritual resources needed to field a series of Crusades against Islam. Indeed, these insider critics of science were more likely to speak in terms of programs in "Science, Technology, and Society," in which courses in the history, philosophy, and sociology of science were part of the core of the science curriculum, not merely enrichment courses taught outside science departments — let alone in autonomous science and technology studies graduate programs that award doctorates for research that shadow the activities of scientists without ever coming to terms with their normative implications.[56] In this context, the late epistemological anarchist Paul Feyerabend appears as the purest of Protestants in calling for the complete divestiture of state support for science as the best way of retrieving the spirit of critical inquiry from Big Science's inhibiting financial and institutional arrangements. Extending the Protestant analogy would involve including the recent charges of scientific misconduct, which have precedent in the personal corruption of Church officials that made the calls for reform most vivid for the average devout Christian.

The second wave of secularization occurred once the Enlightenment transformed the intellectual orientation of academic theology from the professional training of clerics to a form of critical inquiry conducted independently of religious authorities. The last and most accomplished generation

---

once we keep in mind that González Garcia, López Cerezo, and Luján are talking about the location of STSers (the Europeans tending to be High Church, and the North Americans Low Church), whereas I was referring to the location of social movement researchers, i.e., the people who would study STS as a social movement. Needless to say, the reason for this reversal in location between the studiers and studied would be worth pursuing.

In the United States, Low Church members tend to affiliate with the National Association for Science and Technology in Society (NASTS), whereas the High Church is associated with the Society for Social Studies of Science (4S). I first publicly presented this schism in STS at a conference in Copenhagen in October 1992, to which Bruno Latour responded that he had not realized that a branch of STS was rooted in the social activism of the 1960s (and then went on to speak disdainfully about the 1960s, for reasons that may reflect the French experience detailed in chapter 7, section 6).

56. For this phase of STS history, see Cutcliffe 1989. Among the science critics were, in the United Kingdom, the husband-and-wife team of brain scientist Steven and sociologist Hilary Rose, and, in the United States, materials scientist Rustum Roy, marine biologist Rachel Carson, and botanist Barry Commoner. These latter-day descendants of the Protestant Reformers had little to do with the establishment of the science and technology studies departments.

of these theologians constituted the Young Hegelians under whose spell Karl Marx fell during his student years. David Friedrich Strauss's *Life of Jesus* and Ludwig Feuerbach's *The Essence of Christianity* were the texts from this period (the 1830s) that have had the longest impact. Much in the spirit of post–Strong Programme sociologists of science who have subjected the laboratory to ethnographic scrutiny, these theologians applied the latest techniques of literary archaeology and naturalistic social theory to demystify the Scriptures. Far from blaspheming God, they believed their demystified readings of early Church history liberated genuine spirituality from the superstition and idolatry that remained the primary means by which the pastoral clergy kept believers in line. However, the ironic style of these authors put them seriously at odds with both political and religious authorities, causing many of them to lose their professorships and preventing others—such as Marx himself—from ever pursuing academic careers.

The young Karl Marx wrote *The German Ideology* as a series of didactic reflections on how it was possible for the Young Hegelians, despite the attention they paid to the material conditions of Christianity, to be so oblivious to the material conditions of their own times, and hence be caught off guard by those who accused them of sacrilege.[57] Perhaps a similar book is now in order, given the surprise that STS practitioners have expressed about the reception that the scientific community has given their work, which has culminated in the recent Science Wars, as recounted in chapter 7, section 5. It would seem that today's science secularizers have underestimated the extent to which threatening the transcendental rhetoric of science threatens science itself.

So far, in terms of political effectiveness, the High Church would seem to suffer in comparison with the Low Church. However, in retrospect, some Low Church attacks on the scientific establishment may also have been misdirected, at least insofar as they imputed much more power to the sheer possession of scientific knowledge than to the social conditions that enable science to have its momentous consequences. The most extreme version of this sentiment is a scientific version of Luddism that argues that it would have been better never to have developed atomic physics because of the key role it later played in the development of nuclear weapons.[58] In the face of these extreme sentiments, the High Church tendency to reduce science to a language game has been helpful. It draws attention to the fact that when it comes to the sciences' social disposition, what matters is not

57. Marx 1970. On *The German Ideology* as an attempt to propel the hermeneutical differences among the Young Hegelians into a political context, see Meister 1991, esp. 86 ff.
58. This viewpoint is by no means limited to the Low Church. Consider the stand taken by the Oxford metaphysician, Michael Dummett, as recounted above in chapter 2, note 106.

the set of statements and equations most closely associated with a science's cognitive dimension, but whether these tokens in the scientific language game are made of "glass beads" (as in the Hermann Hesse novel by that name) or money and people. In short, the High and Low Church may ultimately compensate for each other's deficiencies.

5. Reinventing the University by Rediscovering the Contexts of Discovery and Justification

Apropos the Low Church tendency to conflate the propositional content of scientific knowledge with the social conditions that enable it to possess worldly power, a great advantage of characterizing the history of science under the rubric of movement rather than paradigm is that it draws attention to the alignment of words and deeds—"ideology" and "technology," as Alvin Gouldner would have put it—that determines the precise social form that science takes: Is it a Platonic academic cult insulated from the material world or the technoscientific infrastructure through which the material world transpires? The historic site for resolving that tension has been the *university*.

Until quite recently, the folk ontology of the university has been that of a relatively staid and stable institution, reflecting its dual role in extending the frontiers of knowledge in research, while reproducing the existing social order in education.[59] In chapter 5, section 3, we saw that already by the mid-1960s, *Structure* bolstered this dual process in the face of student revolts by legitimating academic disciplines, as institutionalized in the department structure of universities. But as was equally shown in chapter 2, section 7, the duality of research and teaching, especially as pursued by paradigm-driven sciences, has been increasingly in tension over the course of the twentieth century. Indeed, truth be told, stability has rarely been the hallmark of university life. At various points in history, especially in late thirteenth-century Paris and early nineteenth-century Germany, the university was quite explicitly the crucible in which social change was forged—more a home for movements than paradigms. These were periods of considerable socioeconomic dislocation and political unrest, which forced people to seek new categories for understanding their new life situations. As guardians of society's reproduction process, educators were in an especially good position to influence the character of the changes. Back then the universities controlled "the means of analysis." Today they are controlled by think tanks.

59. Fuller 1999b, pt. 2, elaborates this thesis.

Alasdair MacIntyre has argued that academics were at the peak of their collective public influence in thirteenth-century Paris.[60] He tells of how the Parisian doctors were worried that the early universities were devolving into mere training centers for legal, medical, and theological practitioners: the graduates would be masters in certain strategic forms of reasoning but unskilled in the general use of reason to criticize their own practices and to pursue non-self-interested inquiry. The solution was to subsume all the faculties under a single "universe of discourse"—a Christianized Aristotelian idiom—within which criticism ("dialectic" in the classical curriculum) would be explicitly encouraged. A doctor's initial submission to this discourse would be sufficient to certify his faith. Everything he argued after that—however radical or skeptical his conclusions—would be treated as devout inquiry. According to MacIntyre, the Enlightenment seriously erred in targeting the Church-dominated universities as opponents to free inquiry. The result (again according to MacIntyre) has been the fragmented and socially inert institution we have today: many autonomous faculties that politely ignore each other and collectively have little effect on society as a whole.

MacIntyre is clearly indulging in revisionist history, especially when he claims that the Enlightenment destroyed, rather than promoted, the public sphere by being a little too hostile to Christianity and a little too friendly to autonomy. However, his account serves to remind us of the historically important role that the university has played as a vortex for general social change.[61] The one Enlightenment philosopher who remained an academician all his life, Immanuel Kant, recaptured MacIntyre's lost image in his last book, *The Conflict of the Faculties* (1798), which helped cause the resurgence of academic inquiry as a political force that characterized the German idealists and Hegel and his followers, most notably Marx.

However, the history of the university over the last 150 years has been, for the most part, a transformation of "academic freedom" into a jealously guarded guild right. Instead of providing the models for public debate, academicians have made a point of discouraging public-minded attitudes and actions among themselves. The benefit has been to shield the university from direct political interference; the cost has been to disable the university from functioning in *any* political capacity, *especially one of its own initia-*

60. MacIntyre 1990.
61. One aspect of this situation that we cannot explore here is the role of the thirteenth-century university in laying the groundwork for the seventeenth-century Scientific Revolution, a favorite thesis of Pierre Duhem. Even those who refuse to trace Galileo's discoveries back to Scholastic speculations in Paris and Oxford generally admit the university's uniqueness in incubating revolutionary ideas. One recent learned defense of this thesis is Huff 1993. See also Fuller 1997d, chap. 4; Fuller 1999b, chap. 3.

tive. Consequently, most of the drive to change the structure of the university today has come from the outside, often from corporate sponsors who want to blur hallowed distinctions between basic and applied research, not to mention between academic and vocational training.

Rather than seizing these external socioeconomic changes as opportunities for the university to take control of the forces of social change, both liberal and conservative academics have too often recoiled from the challenge, treating it entirely as a threat to the university's integrity—again, presupposing a stability to the institution that it has lacked when at its best in the past. A notable exception to this tendency is *feminism*, which arguably has done more to dynamize the structure of the Western university— to put administrators and faculty more in the mind-set of a social movement—than any school of thought since the original Humboldtian call to Enlightenment as the mission of the university in early nineteenth-century Prussia.

In particular, feminism has drawn systematic attention to the ways in which disciplinary divisions obscure both complex problems and alternative forms of life. Over the course of this century, Marxism has attempted a similar emancipatory mission in the universities. However, feminism is unique in that the movement's theorists truly embody their own theories. Instead of upper-middle-class white males just talking about the revolution in their classrooms, the female professoriat actually live it. This has opened the door to more diverse elements of society entering university life, a development that generally falls under the rubric of *multiculturalism*, much of which challenges residual elitist elements in even Marxist and even Western feminist thinking.[62]

The maxim that the Gospel should not be spread before the ink on the last page is dried is a familiar trope from the history of disciplinization, and indeed has been a useful way of demarcating the pure academic fields from the liberal professions.[63] Pure academics believe that practical action— be it policy advice or political activism—is only as good as the quality of

---

62. On the development of feminism and multiculturalism in the contemporary academy, see Fuller 1999b, chap. 4. Sandra Harding has been probably the most prominent feminist to have developed this point. See Harding 1991, 1993. However, there are also signs that even feminism is suffering from paradigmitis—admittedly of a most peculiar kind. A good example is Nicholson 1994. This book is a collection of exchanges among four of contemporary feminism's leading theorists: Seyla Benhabib, Judith Butler, Drucilla Cornell, and Nancy Fraser. As Fraser herself points out (158), they unwittingly reproduce the positions already staked out by, respectively, Jürgen Habermas, Michel Foucault, Jacques Derrida, and Richard Rorty.

63. A full-blown theory of disciplines based on this perspective is Abbott 1988. The perennially vexed status of "psychology" as both a therapeutic and a research practice is perhaps the paradigm case of this kind of boundary work.

the knowledge on which it is based. If there is room for doubt, then that may be enough for the action to do more harm than good. Yet these High Church tenets are little more than superstitions that presuppose the very sort of epistemological foundationalism that STS researchers have been keen to reject on the basis of constructivist scruples. If the practical lessons of STS research were reduced to two maxims, it would be these: *that you do not need to be an expert to understand expertise*, and, moreover, *that the experts themselves may not live up to their own standards*. Thus, the possibility that we may be wrong or change our minds is taken as not merely given but unavoidable. What we need, then, is an ethic of accountability for the knowledge claims we decide to make, in light of such a fallibilist epistemology. This may mean making it institutionally easier to admit error and change one's mind in public, as well as to compensate those who have been wronged by the actions we have taken.[64] The overall result would be to make a general reluctance to participate in politics unjustifiable on a purely epistemological basis.

Taken together, the radical lessons contained in High and Low Church STS provide the core ideas for reversing the embushelment of Western intellectual life, which, as we saw in chapter 1, section 6, reached the university's doors in the 1830s with William Whewell's coinage of "scientist" as the name of a profession.[65] Closely associated with this coinage is the distinction in the contexts of discovery and justification in scientific research. In chapter 1, section 7, we witnessed the initial deconstruction of that distinction by Kuhn and other historicist philosophers of science. Sociologists of science then dealt the final blows in the 1980s. Common to the

---

64. For a defense of the need to preserve what I have called "the right to be wrong," see Fuller 1999b, esp. chaps. 1, 8. It was also instrumental in the original modern debates concerning the "public sphere" in Germany. See Broman 1998.

65. The rest of this section retraces ground covered in Fuller 1999b, chap. 6, sec. 3. The key difference between what is said there and here turns on the role assigned to *republicanism* in that book versus that of *social movements* in this one. (On the provenance of republicanism, see chapter 1, section 2, above.) To be sure, they are not the same but they share a concern for the material conditions under which knowledge can have empowering consequences for those who would lay claim to it. In republican societies, those who engage in public debate need not worry about the consequences of their knowledge claims on their own well-being: staking an idea is kept separate from staking a life. In that sense, one can err with impunity. For this reason, property ownership was often made a requirement for political participation.

Social movements adopt a rather opposed strategy but to similar effect: instead of assigned well-defined "inalienable" (property) rights to individuals, movements "deindividuate" their members, such that each individual can count on the rest of the collective to compensate for whatever problems she runs into as they all strive toward realizing the movement's ideals. Were scientific inquiry conceptualized as a movement in this sense, then the fact that a discovery originally emerged from a particular research program would be taken not as an achievement warranting the reinforcement of that program, but rather a problem that is redressed only by making that discovery available to as many other research programs as possible.

historicists and sociologists has been the view that a research tradition justifies its continuation by the number of robust discoveries that are made under its auspices. In effect, a research tradition enjoys intellectual property rights over the knowledge claims it originates. Thus, if a scientist working in, say, the Newtonian or Darwinian research tradition happens to make an important finding, then the finding counts as a reason for promoting the tradition, and soon the impression is given—especially in textbooks—that the finding could have been be made *only* by someone working in that tradition. In other words, priority quickly becomes grounds for necessity.

This view presupposes a highly competitive model of scientific inquiry that gravitates toward the dominance of a single paradigm in any given field. It does not entertain cases in which knowledge claims originating in one research tradition have been adapted to the needs and aims of others. One important reason is that ultimately the historicists and sociologists believe that alternative research traditions are little more than ways of dividing up the labor in pursuit of some common goals of inquiry, such as explanatory truth or predictive reliability. Thus, they presume that there is some automatic sense in which a discovery made by one tradition is "always already" the property of all—though access to this supposedly common terrain requires that one exchange allegiances first.

Consider the treatment of Darwinian evolution and Creation science as mutually exclusive options in the U.S. public school curriculum. Although two-thirds of Americans who believe in evolution *also* believe that it reflects divine intelligence, such compatibility has yet to be seen as a philosophically respectable option, and consequently has no legal import.[66] But what exactly would be wrong with teachers trying to render biological findings compatible with the Creationist commitments of most of their students? One common answer is that the presupposition of a divine intelligence or teleology has retarded biological inquiry in the past and has not contributed to evolutionary theory since the time of Darwin's original formulation. Yet the contrary presuppositions of mechanistic reduction and random genetic variation are also likely to have led to error.[67]

66. Carter 1993, 156–82.
67. For more on the broad church of maverick biologists who advance this line of argument, see the interviews with Brian Goodwin, Lynn Margulis, Stuart Kauffman, and Stephen Jay Gould in Horgan 1996, 114–42. These biologists are attracted, to differing degrees, to holistic, preformationist, and even quasi-teleological accounts of evolution. Typically, they justify their revision of the Darwinian canon by widening the scope of evidence and reasoning taken to be relevant to a comprehensive explanation for the development of life on Earth. Here it is worth observing that one of the founders of the neo-Darwinian synthesis was himself a Russian Orthodox Christian who made a valiant attempt to reconcile his own research in genetics with contemporary existential theology: Dobzhansky 1967.

Here arises the social responsibility of the science educator, either a professional scientist or a philosopher or sociologist of science: Should students be forced to accept the current scientific canon in the spirit of a Kuhnian paradigm—that is, as a total ideology that would deny the legitimacy of whatever larger belief systems they bring to the classroom? Or should students learn how to integrate science into their belief systems, recognizing points of compatibility, contradiction, and possible directions for personal and collective intellectual growth? If we favor the latter movement-based perspective over the former paradigm-based one, then we need to reinvent to discovery/justification distinction. According to the old distinction, an ideally justified discovery would show how anyone with the same background knowledge and evidence would have made the same discovery. The role of justification was thus to focus and even homogenize the scientific enterprise through a common "logic of scientific inference." In practice, however, "the same background knowledge and evidence" was an understatement of what was actually needed for people to draw the same conclusions, namely, involvement in a particular research tradition.

In contrast, the new distinction I propose conceptualizes scientific justification as removing the idiosyncratic character of scientific discovery in a deeper sense than the old distinction pursued—not simply the fact that a discovery was first reached by a given individual in a given lab, but the fact that it was reached by a particular research tradition in a given culture. In other words, the goal of scientific justification would be to eliminate whatever advantage a particular research tradition or culture has gained by having made the discovery first. Its overall import would be to remove the objectionable, exclusionary features associated with "scientific progress" without erasing the undeniable insights that have been achieved under that rubric.

This project would safeguard scientific inquiry from devolving into a form of expertise whereby, say, one would have to be a card-carrying Darwinian before having anything credible to say about biology. It would also have the opposite effect of the old distinction, in that it would aim to render a discovery compatible with as many different background assumptions as possible, so as to empower as many different sorts of people. Models for this activity can be found in both the natural and the social sciences. In the natural sciences, there are "closed theories" (e.g., Newtonian mechanics) and "dead sciences" (e.g., chemistry), which can be learned as self-contained technologies without the learner first having to commit to a particular metaphysical, axiological, or perhaps even disciplinary orientation. Perhaps the historically most interesting models of this renovated sense of discovery/justification at work are the hybrid forms of inquiry that emerged originally as a defensive response to Western colonial expansion in the nineteenth century, which have reappeared in late twentieth-century non-Western resistance to "postmodern" approaches of science studies.[68]

In the social sciences, conceptual and technical innovations originating in one tradition are typically picked up and refashioned by other traditions, so as to convey an overall sense of history as multiple partially intersecting trajectories. Indeed, just this cross-fertilization has historically given the social sciences the appearance of a field fraught with unresolvable ideological differences. However, from the social movement approach to knowledge production I advocate, this is a *good* thing.[69] It means that the universal value represented by "science" starts to resemble that of "democracy," in that both may flourish in a variety of social settings but, at the same time, must be actively maintained and renewed because of the ease with which the ideal can turn corrupt, especially as particular sciences or particular democracies become victims of their own success: e.g., governments whose mass popularity renders them authoritarian, sciences whose consensuality renders them dogmatic.

In table 16, I contrast my vision of movement-driven "citizen science" with that of a paradigm-driven "professional science" in terms of alternative ways of articulating the discovery/justification distinction. The relationship of discoveries to their scientific justification has been traditionally compared with tributaries leading to a major river. Sticking with the fluvial metaphor, I counter with the image of a major river opening up into a delta in which multiple traditions can make use of a body of knowledge that

---

68. In Fuller 1997d, chap. 6, I considered the cases of modern Islam and Japan, where the instrumental power of the natural sciences has been neither denied nor anathematized, but rather systematically reinterpreted so that these sciences become a medium for realizing the normative potentials of their respective cultures. Along the way, some telling critiques of the historicist perspective on science are made. Basically, Islam criticized the West for not anticipating the destructive and despiritualizing consequences of its "science for its own sake" mentality, whereas Japan lodged the reverse charge that the West superstitiously clings to the stages undergone by its own history as a global blueprint for the advancement of science. In Fuller 1999e, I bring the story up to date, as I report on the first global cyberconference on "public understanding of science," which I organized in February 1998. Here the non-Western participants questioned the postmodern tendency to lump science and technology together as "technoscience," while cultures are seen as autonomous entities. On the contrary, these interlocutors argued that the balance between the West and non-West will be redressed only once technology is divested from an ideological sense of "science" that presupposes the adoption of a certain cultural or theoretical perspective before making full use of the technology in question.

69. In this respect, I do not share the misgivings occasionally expressed in Deutsch, Markovits, and Platt 1986, which documents the tradition-jumping tendency of innovation in the social sciences.

TABLE 16. *Redrawing the Contexts of Discovery/Justification Distinction*

KNOWLEDGE-PRODUCING UNIT	PARADIGM (CLOSED SOCIETY)	MOVEMENT (OPEN SOCIETY)
metaphor guiding the distinction	convergence: tributaries flowing into a major river	divergence: a major river flowing into a delta
prima facie status of discovery	disadvantage (because of unexpected origins)	advantage (because of expected origins)
ultimate role of justification	concentrate knowledge through logical assimilation	distribute knowledge through local accommodation
background assumption	discoveries challenge the dominant paradigm unless they are assimilated to it	discoveries reinforce the dominant paradigm unless they are accommodated to local contexts
point of the distinction	turn knowledge into power (magnify cumulative advantage)	divest knowledge of power (diminish cumulative advantage)
definition of "contemporary science"	present is continuous with the future—the past is dead and best left to historians	present is continuous with the past—the future is open to the retrieval of lost options
scientist exemplar	Max Planck	Ernst Mach

originated in only one of them. At stake in both cases is the ease with which knowledge can be used as an instrument of power.[70]

6. Final Strategic Remarks

In chapter 7, section 4, I briefly discussed the "Epistemological Chicken" debate on the future of STS. There I observed that the two sides extrapolated from a common past to alternative futures, both of which inhibited the field's political and intellectual dynamism. In light of the movement model of inquiry developed in section 4 of this chapter, some significant nuances can be added to this critique. On the one hand, Harry Collins's appeal for STS to plow ahead in a business-as-usual fashion, extending the breadth of its empirical work without plumbing its theoretical depth, is the sort of parochialism one would expect from the corrupt version of the European style of movement thought. On the other hand, Bruno Latour's appeal for the field to incorporate the scientist's folk ontology into its own

---

70. The table refers to "assimilation" and "accommodation." The terms, also found in Kuhn, are due to Jean Piaget and are used here to distinguish the relative openness (low and high, respectively) of a knowledge-producing community to fundamental change.

narrative framework smacks of co-optation, an instance of the corrupt version of the North American style of movement thought.

The impasse between Collins and Latour is symbolized by the Janus-faced character of STS's much-vaunted case study methodology. On the one hand, case studies create intellectual entitlements for the STS practitioner that effectively restrict the "community of inquirers" simply to those with similar training and experience; on the other, because case studies are typically evaluated merely in terms of their descriptive adequacy ("Does it tell a good story?"), and not some larger normative context, they can be of potential use to a wide range of users, most notably those who do *not* share the STSer's personal or professional commitments. Either way the dynamic spirit of inquiry loses.[71]

Nevertheless, beyond my gloomy diagnosis of the Epistemological Chicken debate lies a rosier prognosis for how a debate of this sort may be turned into a movement catalyst. On the surface, Collins and Latour appear to be arguing about the future of a specialized field of inquiry called "science and technology studies," but in fact their attitudes reflect a fundamental disagreement about the prospects of their own knowledge production site, the *university*. Collins has steered clear of collaborating with the state and industry, whereas Latour has been housed in an institution that has had to develop such networks in order to sustain its research programs. There is nothing especially mysterious about this difference. Their respective national academic contexts largely explain it. But the difference also reflects an emerging schism within academia more generally, which STS is in an especially good position to explore, given its professional interest in the social conditions of knowledge.

The schism is between what fashionable science policy theorists call "mode 1" and "mode 2" conceptions of knowledge production.[72] Collins represents the mode 1 conception of university-protected, paradigm-driven research, whereas Latour represents the mode 2 conception, which welcomes the university's permeability to extramural concerns. There is much at stake, once the dispute is amplified in these terms. What does it mean to be "political" or, for that matter "public," in this volatile context of inquiry?

---

71. The success of case studies in driving out "grand narratives" from science studies—namely, the philosophical histories mourned in the introduction to this book—is much like the success enjoyed by a country that remains neutral during a time of war. While historians, philosophers, and sociologists of science can nowadays congratulate themselves for not reifying Capitalism, Progress, and even Science, the forces behind these hypostases continue their work in the world uninterrupted. All that has changed is that a group of academics have voluntarily removed themselves from the fray.

72. This jargon comes from Gibbons et al 1994. For a critique of the historical sensibility that informs this work, see chapter 2, note 20.

How are the universal aspirations of inquiry to be reconciled with the pressure to be both professional and client-driven? What fate awaits the flickering spirit of criticism in an intellectual world increasingly beholden to the idea that there is safety in numbers? The influences that have flowed through and from *The Structure of Scientific Revolutions* provide one set of answers to these questions, which I have presented and opposed in these pages.[73]

Perhaps the most disturbing feature of the debate surrounding mode 1 vs. mode 2 knowledge production is the negative stereotyping of the university that it suggests, especially the association of academic inquiry with disciplinary specialization (mode 1) and of nonacademic inquiry with interdisciplinary exploration (mode 2). As we saw in chapter 5, section 3, this stereotype has been one of the Kuhnian legacies to higher education—that the natural development of science consists of a potentially endless division of disciplines into mutually exclusive domains of inquiry. Any diversion from that plan must come from the larger society, which is the ultimate source of the excitement and relevance that characterizes interdisciplinary research. Yet the "problem-centered" nature of such research means that once the work of an interdisciplinary team is completed, its members return to pursue normal science in their home disciplines. There is little expectation, and virtually no institutional incentive, to convert that research into a form of knowledge that challenges existing specialties. Indeed, like the biological hybrids to which it is often compared, interdisciplinary research is not supposed to spawn offspring. I have spent much of my career contesting this view of interdisciplinarity.[74]

Notwithstanding the proven persuasiveness of this Kuhnian mythology, disciplinary specialization was not endemic to either scientific inquiry or academic life before the onset of the Cold War. Indeed, with the end of the Cold War, and a general contraction of funds for higher education, this point is being learned the hard way. But it is just as much a lesson in history as economics. The task before us as academics is to reverse the image of the university that the mode theorists project, so that instead of being an institution that must be forced from the outside to change, the university becomes the site where major social change is initiated.[75]

---

73. Another recent attempt to address these questions is Brown 1998.
74. See Fuller 1993b, esp. chap. 2.
75. Although the argument cannot be pursued here, I would go so far as to claim that, given the current surplus of academically overqualified people (i.e., people with Ph.D.s working outside of universities), the potential exists for reversing the effects of mode 2 knowledge production, as academic norms are injected into state and business cultures that have become *too* adaptive through an uncritical pursuit of short-term goals. An instance of this injection is the emergence of the "chief knowledge officer" (CKO) in corporations, often sporting the

Of course, this is a very tall order, but the history of STS provides one clear way of seeing it through. It requires that our field overcome its amnesiac sense of novelty and inertial sense of autonomy (see chapter 7, section 3) by rekindling its institutional roots in service teaching for scientists and engineers. A perhaps more attractive way of putting the point is that STS is better positioned than any other field to reinvent liberal arts education, which originally aimed to equip the elite with the ability to think independently about a broad range of topics where their opinions would matter.[76] Today the charge has been democratized and the students are not so much ignorant as encumbered with technical knowledge whose strengths are much better known than its weaknesses. In our mode 2 world, it is increasingly obvious that private firms can provide specialized training much more effectively than a university bureaucracy that would have students take courses that have little bearing on their anticipated future careers. Yet the precarious employment market of high capitalism and democracy's need of a fully functioning citizenry points to the unique value of a specifically university-based education.

Ironically, STS talks a lot about the importance of "science in the making" but says little about "*scientists* in the making." Consequently, STSers typically confront scientists — be it as subjects in the laboratory or adversaries in the Science Wars — as fully finished and all-too-resilient products. In contrast, the insertion of STS into science education at a mass level would erase the field's alien character. Indeed, eventually STS thinking would reappear in the self-understandings of the scientific traditions, as respected past scientists are assigned credit for anticipating important STS findings.[77]

---

credential "executive Ph.D.," who treats the scientific literature as part of the economy's extractive sector, i.e., a natural resource that can be cultivated and prospected in ways that the scientific community itself has failed to do because of its habitualized normal scientific patterns of behavior. (In this context, the scientific community functions much as indigenous peoples sitting on a rich mineral deposit.) The interesting feature of this development is the extent to which businesses will allow their training, practices, and expenditures to be structured by academic considerations, which are seen as helping to stabilize their dynamic environments. Of special note is the "Fenix" initiative between the Stockholm Business School and Chalmers Institute of Technology, which is now supported by most of the Swedish multinational corporations. My thanks to Thomas Hellstrom and Merle Jacob for introducing me to this post–mode 2 setting.

76. I must thank Bill Keith for this insight into the implications of my interest in distinguishing paradigm-based from more broadly university-based knowledge production.

77. This strategy was first made vivid to me by Graeme Gooday, who spoke at a conference I organized on "Science's Social Standing" at Durham University in December 1994. This was the first formal meeting of representatives of the various sides in the simmering Science Wars. Gooday impressed the scientists in the audience by recalling that an icon of experimental science no less than Thomas Henry Huxley objected to the routinization of laboratory training on the grounds that it would stifle the creativity of fledgling inquirers. See also chapter 2, note 34, above, where these remarks are placed in the context of Mach's

CONCLUSIONS / 423

Earlier in this book, I have had occasion to remark that the Orwellian sense of history that Kuhn ascribed to scientists was partly inspired by Alfred North Whitehead's famous quote, "A science that hesitates to forget its founders is lost."[78] In closing, let me suggest a relevant counter-quote from another Harvard sage of the early twentieth century, George Santayana (1863–1952), who declared, "Those who cannot remember the past are condemned to repeat it."[79] After Kuhn, it has become common to think that history needs to be left behind if science is to make progress. But it may be that what Kuhn and especially his followers have taken to be the mark of intellectual resolve on the part of the scientific community is really, as Santayana thought, indicative of either infancy or, more likely, senility. In either case, Santayana's saying points to a conception of the past as the living repository from which today's heresies and tomorrow's orthodoxies may be forged.[80] Raymond Aron made the point with characteristic elegance: "The past is never definitively fixed except when it has no future."[81] This book may be seen as having pursued the complementary thesis: *The future is never definitively fixed except when it has no past.* Ironic though it be, a forthright attempt to make the past contemporaneous with the present may be the best strategy for progressive thinkers in any field of inquiry to keep the future forever open. If there is a subversive message hidden in The Structure of Scientific Revolutions, this is it.

---

approach to science education. The theoretical basis of my approach to STS education is given mostly in Fuller 1993b and is sympathetically applied and extended in Collier 1997.

78. See chapter 6, note 89.

79. Santayana 1905, 284. I first came across this quote in my earliest exploration into the nature of historical consciousness, a meditation on the relationship between Gibbon's ambivalent account of the burning of the Library at Alexandria in 641 and David Hume's contemporary call to commit to the flames every book that does not contain logic or evidence. See Fuller and Gorman 1987.

80. Compare Santayana 1936, 94. This point marks somewhat of a shift from an earlier statement in which I claimed that history (and the humanistic disciplines more generally) could play only a critical role in reconstituting the conditions of knowledge production. See Fuller 1993a, xiii–xiv.

81. Aron 1957, 150.

REFERENCES

Abbott, Andrew. 1988. *The System of Professions*. Chicago: University of Chicago.
Adams, Hazard, ed. 1971. *Critical Theory since Plato*. New York: Harcourt Brace Jovanovich.
Adorno, Theodor, ed. 1976. *The Positivist Dispute in German Sociology*. London: Heinemann.
Agassi, Joseph. 1988. *The Gentle Art of Philosophical Polemics*. La Salle, Ill.: Open Court Press.
Ahmad, Aijaz. 1992. *In Theory: Classes, Nations, Literatures*. London: Verso.
Ainslie, George. 1992. *Picoeconomics*. Cambridge: Cambridge University Press.
Albisetti, James. 1983. *Secondary School Reform in Imperial Germany*. Princeton: Princeton University Press.
———. 1987. "The Debate on Secondary School Reform in France and Germany." In Mueller, Ringer, and Simon 1987, 171–96.
Albrow, Martin. 1993. "The Changing British Role in European Sociology." In *Sociology in Europe*, ed. B. Nedelmann and P. Sztompka. Berlin: Walter de Gruyter.
Albury, W. Randall. 1983. *The Politics of Objectivity*. Victoria, Australia: Deakin University Press.
Anderson, Perry. 1992. *A Zone of Engagement*. London: Verso.
———. 1998. *The Origins of Postmodernity*. London: Verso.
Apel, Karl-Otto. 1984. *Understanding and Explanation*. Cambridge: MIT Press.
Appleby, Joyce, Lynn Hunt, and Margaret Jacob. 1995. *Telling the Truth about History*. New York: Norton.
Arditi, Jorge. 1994. "Geertz, Kuhn, and the Idea of a Cultural Paradigm." *Sociological Review* 45:597–617.
Areen, Judith, Patricia King, Steven Goldberg, and Alexander Capron. 1990. *Law, Science, and Medicine*. 3d ed. Mineola, N.Y.: Foundation Press. 1st ed. 1984.
Arnold, Thurman. 1966. "A Philosophy for Politicians." In *New Deal Thought*, ed. H. Zinn, 35–43. Indianapolis: Bobbs-Merrill. First published in 1935.
Aron, Raymond. 1957. *The Opium of the Intellectuals*. Garden City, N.Y.: Doubleday.

———. 1966. *Peace and War: A Theory of International Relations*. Garden City, N.Y.: Doubleday. First published in 1961.
———. 1970. *Main Currents in Sociological Thought*. Garden City, N.Y.: Doubleday. First published in 1967.
Aronowitz, Stanley. 1988. *Science as Power*. Minneapolis: University of Minnesota Press.
Ash, Mitchell. 1991. "Gestalt Psychology in Weimar Culture." *History of the Human Sciences* 4:395–416.
Ashmore, Malcolm. 1989. *The Reflexive Thesis: Wrighting Sociology of Scientific Knowledge*. Chicago: University of Chicago Press.
Ashmore, Malcolm, and Stella Harding, eds. 1994. "Humans and Others: The Concept of Agency and Its Attribution." Special number of *American Behavioral Scientist* 37:731–856.
Ashmore, Malcolm, and Evelleen Richards, eds. 1996. "The Politics of SSK: Neutrality, Commitment, and Beyond." Special number of *Social Studies of Science* 26:219–418.
Axtell, Guy. 1993. "In the Tracks of the Historical Movement: A Reassessment of the Kuhn-Carnap Connection." *Studies in the History and Philosophy of Science* 24:119–46.
Ayer, A. J. 1936. *Language, Truth, and Logic*. London: Victor Gollancz.
Baars, Bernard. 1986. *The Cogntive Revolution in Psychology*. New York: Guilford Press.
Baigrie, Brian. 1995. "Fuller's Civic Republicanism and the Question of Scientific Expertise." *Philosophy of the Social Sciences* 25:502–11.
Baker, John. 1961. "The Controversy on Freedom in Science in the 19th Century." In *The Logic of Personal Knowledge*, ed. J. Baker, 89–96. London: Routledge & Kegan Paul.
Banner, Michael. 1990. *The Justification of Science and the Rationality of Religious Belief*. Oxford: Clarendon Press.
Barber, Benjamin. 1995. *Jihad vs. McWorld*. New York: Ballantine Books.
Barber, Bernard. 1952. *Science and the Social Order*. New York: Macmillan.
———, ed. 1970. *L. J. Henderson on the Social System*. Chicago: University of Chicago Press.
Barnes, Barry. 1969. "Paradigms, Scientific and Social." *Man* (March): 94–102.
———. 1975. *Scientific Knowledge and Sociological Theory*. London: Routledge & Kegan Paul.
———. 1982. *T. S. Kuhn and Social Science*. London: Macmillan.
———. 1990. *The Nature of Power*. Cambridge: Polity.
———. 1994. "Cultural Change—The Thought-Styles of Mannheim and Kuhn." *Common Knowledge* 3, no. 2: 65–78.
Barnes, Barry, and David Bloor. 1982. "Relativism, Rationalism, and the Sociology of Knowledge." In Hollis and Lukes 1982, 21–47.
Barnes, Barry, David Bloor, and John Henry. 1996. *Scientific Knowledge: A Sociological Analysis*. Chicago: University of Chicago Press.
Barnes, Barry, and David Edge, eds. 1982. *Science in Context*. Milton Keynes, U.K.: Open University Press.

Bartley, W. W., III. 1974. "Theory of Language and Philosophy of Science as Instruments of Educational Reform: Wittgenstein and Popper as Austrian Schoolteachers." In *Boston Studies in the Philosophy of Science*, vol. 14, ed. R. Cohen and M. Wartofsky, 307–37. Dordrecht: Reidel.
———. 1984. *The Retreat to Commitment*. 2d ed. La Salle, Ill.: Open Court. 1st ed. 1962.
Baudrillard, Jean. 1983. *Simulations*. New York: Semiotexte.
Bauman, Zygmunt. 1987. *Legislators and Interpreters*. Cambridge: Polity Press.
———. 1993. *Postmodern Ethics*. Oxford: Blackwell.
Bazerman, Charles. 1988. *Shaping Written Knowledge*. Madison: University of Wisconsin Press.
Beck, Ulrich. 1992. *The Risk Society*. London: Sage. First published in 1986.
Becker, Howard. 1967. "Whose Side Are We On?" *Social Problems* 14:239–47.
Beiser, Frederick. 1987. *The Fate of Reason*. Cambridge: Harvard University Press.
———. 1992. *Enlightenment, Revolution, and Romanticism*. Cambridge: Harvard University Press.
Bell, Daniel. 1960. *The End of Ideology*. New York: Free Press.
———. 1966. *The Reforming of General Education: The Columbia College Experience in Its National Experience*. Garden City, N.Y.: Doubleday.
———. 1973. *The Coming of Post-industrial Society*. New York: Basic Books.
Bell, Daniel, and Irving Kristol, eds. 1969. *Confrontation*. New York: Basic Books.
Ben-David, Joseph. 1972. "Scientific Growth: Reflections on Ben-David's 'Scientist's Role.'" *Minerva* 10:166–78.
———. 1984. *The Scientist's Role in Society*. 2d ed. Chicago: University of Chicago Press. 1st ed. 1971.
———. 1991. *Scientific Growth: Essays on the Social Organization and Ethos of Science*. Berkeley and Los Angeles: University of California Press.
Bender, John, and David Wellbery, eds. 1991. *Chronotypes: The Construction of Time*. Palo Alto: Stanford University Press.
Bentley, Eric. 1946. *The Playwright as Thinker*. New York: Reynal & Hitchcock.
Berger, Peter. 1967. *The Sacred Canopy*. New York: Doubleday.
Berger, Peter, Brigitte Berger, and Hansfried Kellner. 1973. *The Homeless Mind*. New York: Random House.
Berger, Peter, and Thomas Luckmann. 1967. *The Social Construction of Reality*. Garden City, N.Y.: Doubleday.
Bergson, Henri. 1935. *The Two Sources of Morality and Religion*. London: Macmillan. First published in 1932.
Berkson, William, and John Wetterstein. 1984. *Learning from Error: Karl Popper's Psychology of Error*. La Salle, Ill.: Open Court Press.
Berle, Adolf, and Gardner Means. 1932. *The Modern Corporation and Private Property*. New York: Harcourt Brace and World.
Berlin, Isaiah. 1969. "Historical Inevitability." In *Four Essays on Liberty*, 41–117. Oxford: Oxford University Press. First published in 1954.
Bernal, John Desmond. 1935. "If Industry Gave Science a Chance: The Boundless Possibilities Ahead of Us." *Harper's*, February, 258–59.
———. 1939. *The Social Function of Science*. London: Macmillan.

———. 1971. *Science in History.* 4 vols. Cambridge: MIT Press.
Bernal, Martin. 1987. *Black Athena: The Afroasiatic Roots of Greek Thought.* New Brunswick: Rutgers University Press.
Bernstein, Barton. 1993. "Seizing the Contested Terrain of Early Nuclear History: Stimson, Conant, and Their Allies Explain the Decision to Use the Atomic Bomb." *Diplomatic History* 17:35–72.
Bernstein, Basil. 1971. *Class, Codes, and Control.* Vol. 1. London: Routledge & Kegan Paul.
Bernstein, Howard. 1981. "Marxist Historiography and the Methodology of Research Programs." *History and Theory* 20:424–49.
Bernstein, Richard. 1976. *The Restructuring of Social and Political Theory.* Oxford: Blackwell.
———. 1983. *Beyond Objectivism and Relativism.* Philadelphia: University of Pennsylvania Press.
Biagioli, Mario. 1990. "The Anthropology of Incommensurability." *Studies in History and Philosophy of Science* 21:183–209.
———. 1993. *Galileo, Courtier.* Chicago: University of Chicago Press.
———. 1996. "From Relativism to Contingentism." In Galison and Stump 1996, 189–206.
Bijker, Wiebe, Thomas Hughes, and Trevor Pinch, eds. 1987. *The Social Construction of Technological Systems.* Cambridge: MIT Press.
Bijker, Wiebe, and John Law, eds. 1992. *Shaping Technology/Building Society: Studies in Socio-technical Change.* Cambridge: MIT Press.
Blackmore, John. 1973. *Ernst Mach.* Berkeley and Los Angeles: University of California Press.
———, ed. 1992. *Ernst Mach: A Deeper Look.* Dordrecht: Kluwer.
Bloom, Allan. 1987. *The Closing of the American Mind.* New York: Simon and Schuster.
———. 1990. *Giants and Dwarfs.* New York: Simon and Schuster.
Bloor, David. 1973. "Wittgenstein and Mannheim on the Sociology of Mathematics." *Studies in History and Philosophy of Science* 4:173–91.
———. 1975. "A Philosophical Approach to Science." *Social Studies of Science* 5: 507–17.
———. 1976. *Knowledge and Social Imagery.* London: Routledge & Kegan Paul.
———. 1983. *Wittgenstein and the Social Theory of Knowledge.* Oxford: Blackwell.
———. 1992. "Left and Right Wittgensteinians." In Pickering 1992, 266–82.
Bok, Derek. 1982. *Beyond the Ivory Tower: The Social Responsibility of the Modern University.* Cambridge: Harvard University Press.
Boring, Edwin. 1950. *A History of Experimental Psychology.* 2d ed. New York: Appleton Century Crofts. 1st ed. 1929.
Boudon, Raymond. 1981. *The Logic of Social Action.* London: Routledge & Kegan Paul. First published in 1979.
Bourdieu, Pierre. 1975. "The Specificity of the Scientific Field and the Social Conditions of the Progress of Reason." *Social Science Information* 14, no. 6: 19–47.
———. 1977. *Outline of a Theory of Practice.* Cambridge: Cambridge University Press. First published in 1972.

———. 1988. *Homo Academicus*. Cambridge: Polity Press. First published in 1984.
———. 1996. *The State Nobility*. Cambridge: Polity Press. First published in 1989.
Bowker, Geof, and Bruno Latour. 1987. "A Booming Discipline Short of Discipline: Social Studies of Science in France." *Social Studies of Science* 17:715–47.
Bradley, F. H. 1927. *Ethical Studies*. 2d ed. Oxford: Oxford University Press. 1st ed. 1876.
———. 1968. *The Presuppositions of Critical History*. New York: Quadrangle. First published in 1874.
Brannigan, Augustine. 1981. *The Social Basis of Scientific Discoveries*. Cambridge: Cambridge University Press.
Braverman, Harry. 1974. *Labor and Monopoly Capitalism: The Degradation of Work in the Twentieth Century*. New York: Monthly Review.
Brennan, Teresa. 1993. *History after Lacan*. London: Routledge.
Bricmont, Jean. 1997. "Science of Chaos or Chaos in Science." In Gross, Levitt, and Lewis 1997, 131–76.
Brinton, Crane. 1950. *The Shaping of the Modern Mind*. New York: New American Library.
———. 1952. *The Anatomy of Revolution*. New York: Random House.
Broman, Thomas. 1998. "The Habermasian Public Sphere and 'Science in the Enlightenment.'" *History of Science* 36:124–49.
Brooke, John Hedley. 1991. *Science and Religion*. Cambridge: Cambridge University Press.
Brown, Harold. 1977. *Perception, Theory, and Commitment: The New Philosophy of Science*. Chicago: University of Chicago Press.
Brown, James Robert, ed. 1984. *The Rationality Debates: The Sociological Turn*. Dordrecht: Reidel.
Brown, Richard Harvey. 1998. "Modern Science and Its Critics: Toward a Postpositivist Legitimization of Science." *New Literary History* 29:521–50.
Bruner, Jerome. 1983. *In Search of Mind*. New York: Harper & Row.
Brush, Stephen. 1975. "Should History of Science Be Rated X?" *Science* 183:1164–83.
———. 1995. "Scientists as Historians." *Osiris* 10:215–32.
Buchdahl, Gerd. 1969. *Metaphysics and the Philosophy of Science: The Classical Origins from Descartes to Kant*. Cambridge: MIT Press.
Buchwald, Jed. 1993. "Design for Experimenting." In Horwich 1993, 169–206.
———. 1995. "Conclusion." In *Scientific Practice: Theories and Stories of Doing Physics*, 345–51. Chicago: University of Chicago Press.
———. 1996. "Memories of Tom Kuhn." *History of Science Society Newsletter* 25, no. 4: 4.
Buchwald, Jed, and George Smith. 1997. "Thomas Kuhn. 1922–1996." *Philosophy of Science* 64:361–76.
Buck, Peter, and Barbara Rosenkrantz. 1981. "The Worm in the Core: Science and General Education." In *Transformation and Tradition: Essays in Honor of I. Bernard Cohen*, ed. E. Mendelsohn, 371–95. Cambridge: Cambridge University Press.
Bukharin, Nikolai, ed. 1971. *Science at the Crossroads: Papers Presented to the Inter-*

national Congress of the History of Science and Technology held in London from June 29th to July 3rd 1931 by the Delegates of the Soviet Union. Reprinted with new foreword by Joseph Needham and new introduction by Gary Werskey. London: Cass. First published in 1931.

Burke, Kenneth. 1969. *A Grammar of Motives*. Berkeley and Los Angeles: University of California Press. First published in 1945.

Burnham, James. 1941. *The Managerial Revolution*. New York: Day.

Burnham, John. 1987. *How Superstition Won and Science Lost: Popularizing Science and Health in the United States*. New Brunswick: Rutgers University Press.

Burtt, Edward. 1954. *The Metaphysical Foundations of Modern Physical Science*. New York: Doubleday. First published in 1924.

Butterfield, Herbert. 1931. *On the Whig Interpretation of History*. Cambridge: Cambridge University Press.

———. 1955. *Man on His Past: The Study of the History of Historical Scholarship*. Cambridge: Cambridge University Press.

Buxton, William, and Stephen Turner. 1992. "From Education to Expertise: Sociology as a Profession." In *Sociology and Its Publics*, ed. Terence Halliday and Morris Janowitz, 373–408. Chicago: University of Chicago Press.

Callebaut, Werner. 1993. *Taking the Naturalistic Turn, or How Real Philosophy of Science Is Done*. Chicago: University of Chicago Press.

———. 1996. "Thomas Kuhn as an Evolutionary Naturalist." *Evolution and Cognition* 2:127–38.

Callon, Michel. 1986. "Some Elements of a Sociology of Translation: Domestication of the Scallops and the Fishermen." In *Power, Action and Belief: A New Sociology of Knowledge?* ed. John Law. 196–229. London: Routledge & Kegan Paul.

———. 1987. "Society in the Making: The Study of Technology as a Tool for Sociological Analysis." In Bijker, Hughes, and Pinch 1987, 83–106.

Callon, Michel, and Bruno Latour. 1981. "Unscrewing the Big Leviathan: How Actors Macro-structure Reality and How Sociologists Help Them to Do So." In *Advances in Social Theory and Methodology*, ed. K. Knorr-Cetina and A. Cicourel, 277–303. London: Routledge & Kegan Paul.

Callon, Michel, John Law, and Arie Rip. 1986. *Mapping the Dynamics of Science and Technology*. London: Macmillan.

Campbell, Donald. 1969. "Ethnocentrism of Disciplines and the Fishscale Model of Omniscience." In *Interdisciplinary Relationships in the Social Sciences*, ed. M. Sherif and C. W. Sherif, 328–48. Chicago: Aldine.

———. 1984. "Can We Be Scientific in Applied Social Science?" In *Evaluation Studies Review Annual*, vol. 9, ed. R. Connor, D. Attman, and C. Jackson, 26–48. London: Sage.

———. 1988. *Methodology and Epistemology for Social Science*. Chicago: University of Chicago Press.

Cantor, Norman. 1991. *The Inventing of the Middle Ages*. New York: William Morrow.

Cao, Tian Yu. 1993. "The Kuhnian Revolution and the Postmodern Turn in the History of Science." *Physis* 30:476–504.

Cardwell, D. S. L. 1972. *The Organisation of Science in England.* 2d ed. London: Heinemann. 1st ed. 1957.
Carnap, Rudolf. 1942. *Introduction to Semantics.* Cambridge: Harvard University Press.
———. 1958. *Meaning and Necessity.* 2d ed. Chicago: University of Chicago Press. 1st ed. 1950.
———. 1963. "Replies and Systematic Expositions." In *The Philosophy of Rudolf Carnap,* ed. P. Schilpp, 859–1016. La Salle, Ill.: Open Court Press.
———. 1966. *An Introduction to the Philosophy of Science.* New York: Harper & Row.
———. 1967. *The Logical Structure of the World.* Trans. R. George. Berkeley and Los Angeles: University of California Press. First published in 1928.
Carter, Stephen. 1993. *The Culture of Disbelief.* New York: Doubleday.
Cartwright, Nancy, Jordi Cat, Lola Fleck, and Thomas Uebel. 1996. *Otto Neurath: Philosophy between Science and Politics.* Cambridge: Cambridge University Press.
Cassirer, Ernst. 1923. *Substance and Function.* La Salle, Ill.: Open Court Press. First published in 1910.
———. 1960. *The Logic of the Humanities.* New Haven: Yale University Press.
———. 1963. *Individual and Cosmos in Renaissance Philosophy.* Oxford: Basil Blackwell. First published in 1927.
Castells, Manuel. 1996–98. *The Information Age: Economy, Society, and Culture.* 3 vols. Oxford: Blackwell.
Cavell, Stanley. 1976. *Must We Mean What We Say?* Cambridge: Cambridge University Press.
———. 1979. *The Claim of Reason.* Oxford: Oxford University Press.
Ceccarelli, Leah. 1995. "A Rhetoric of Interdisciplinary Scientific Discourse: Textual Criticism of Dobzhansky's *Genetics and the Origins of Species.*" *Social Epistemology* 9:91–112.
Chisholm, Roderick. 1974. *Theory of Knowledge.* Englewood Cliffs, N.J.: Prentice-Hall.
Chomsky, Noam, Ira Katznelson, Richard Lewontin, David Montgomery, Laura Nader, Richard Ohmann, Ray Siever, Immanuel Wallerstein, and Howard Zinn. 1997. *The Cold War and the University.* New York: New Press.
Chubin, Daryl, and Edward Hackett. 1990. *Peerless Science.* Albany: SUNY Press.
Clegg, Stewart. 1989. *Frameworks of Power.* London: Sage.
Cohen, H. Floris. 1994. *The Scientific Revolution: An Historiographical Inquiry.* Chicago: University of Chicago Press.
Cohen, I. Bernard. 1984. "A Harvard Education." *Isis* 75:13–21.
———. 1985. *Revolution in Science.* Cambridge: Harvard University Press.
———. 1995. *Science and the Founding Fathers.* New York: W. W. Norton.
Cohen, I. Bernard, and F. G. Watson, eds. 1952. *General Education in Science.* Cambridge: Harvard University Press.
Cohen, Jean. 1985. "Strategy or Identity: New Theoretical Paradigms and Contemporary Social Movements." *Social Research* 52:663–716.
Cohen, L. Jonathan. 1986. *The Dialogue of Reason.* Oxford: Clarendon Press.

Cohen, Morris, and Ernest Nagel. 1934. *An Introduction to Logic and the Scientific Method*. London: Routledge & Kegan Paul.

Coleman, James. 1978. "Sociological Analysis and Social Policy." In *A History of Sociological Analysis*, ed. T. Bottomore and R. Nisbet, 677–703. New York: Basic Books.

———. 1990. *The Foundations of Social Theory*. Cambridge: Harvard University Press.

Collier, James, ed. 1997. *Scientific and Technical Communication: Theory, Practice and Policy*. Thousand Oaks, Calif.: Sage.

Collingwood, Robin. 1972. *An Essay on Metaphysics*. Chicago: Henry Regnery Company. First published in 1940.

Collini, Stefan, Donald Winch, and J. W. Burrow. 1983. *That Noble Science of Politics*. Cambridge: Cambridge University Press.

Collins, Harry. 1981. "What Is TRASP? The Radical Programme as a Methodological Imperative." *Philosophy of the Social Sciences* 11:215–24.

———. 1985. *Changing Order: Replication and Induction in Scientific Practice*. London: Sage.

———. 1987. "Certainty and the Public Understanding of Science: Science on Television." *Social Studies of Science* 17:689–713.

———. 1996. "Theory Dopes: A Critique of Murphy." *Sociology* 30:367–74.

Collins, Harry, and Trevor Pinch. 1982. *Frames of Meaning*. London: Routledge & Kegan Paul.

———. 1993. *The Golem: What Everyone Needs to Know about Science*. Cambridge: Cambridge University Press.

Collins, Harry, and Steven Yearley. 1992. "Epistemological Chicken." In Pickering 1992, 301–27.

Collins, Randall. 1975. *Conflict Sociology*. New York: Academic Press.

———. 1998. *The Sociology of Philosophies: A Global Theory of Intellectual Change*. Cambridge: Harvard University Press.

Commoner, Barry. 1971. *The Closing Circle: Nature, Man, and Technology*. New York: Alfred Knopf.

Conant, James Bryant. 1947. *Understanding Science*. New Haven: Yale University Press.

———. 1950. *The Overthrow of the Phlogiston Theory: The Chemical Revolution of 1775–1789*. Harvard Case Histories in Experimental Science, case 2. Cambridge: Harvard University Press.

———. 1952a. Foreword to Cohen and Watson 1952.

———. 1952b. *Modern Science and Modern Man*. New York: Columbia University Press.

———. 1959. *The American High School Today: A First Report to Interested Citizens*. New York: McGraw Hill.

———. 1961. *Science and Common Sense*. New Haven: Yale University Press.

———. 1970. *My Several Lives: Memoirs of a Social Inventor*. New York: Harper & Row.

Conant Presidential Papers. 1948–50. Harvard University Archives.

Conley, Thomas. 1990. *Rhetoric in the European Tradition*. Chicago: University of Chicago Press.
Constant, Edward. 1973. "A Model of Technological Change Applied to the Turbojet Revolution." *Technology and Culture* 14:553–72.
Cook, Terrence. 1991. *The Great Alternatives of Social Thought*. Lanham, Md.: Rowman & Littlefield.
Cooley, Mike. 1980. *Architect or Bee?: The Human/Technology Relationship*. Slough, U.K.: Langley Technical Services.
Crawford, T. Hugh. 1993. "An Interview with Bruno Latour." *Configurations* 2: 247–69.
Creath, Richard. 1990. *Dear Carnap, Dear Van*. Berkeley and Los Angeles: University of California Press.
Crombie, Alaistair, ed. 1963. *Scientific Change*. Oxford: Oxford University Press.
———. 1984. "Beginnings at Oxford." *Isis* 75:25–28.
Crothers, Charles. 1987. *Robert K. Merton*. Chichester, U.K.: Tavistock.
Crowther, J. G. 1968. *Science in Modern Society*. New York: Schocken Books.
Culler, Jonathan. 1982. *On Deconstruction*. Ithaca: Cornell University Press.
Curtius, Ernst Robert. 1989. "Max Weber on Science as a Vocation." In *Max Weber's "Science as a Vocation,"* ed. P. Lassman and I. Velody, 70–75. London: Unwin Hyman.
Cutcliffe, Stephen. 1989. "The Emergence of STS as an Academic Field." In *Research in Philosophy of Technology*, ed. P. Durbin. Greenwich, Conn.: JAI Press.
Dahl, Robert. 1963. *Modern Political Analysis*. Englewood Cliffs, N.J.: Prentice-Hall.
———. 1989. *Democracy and Its Critics*. New Haven: Yale University Press.
Dallmayr, Fred, and Thomas McCarthy, eds. 1977. *Understanding and Social Inquiry*. South Bend: University of Notre Dame Press.
D'Amico, Robert. 1989. *Historicism and Knowledge*. London: Routledge.
Daniel, Hans-Dietrich. 1994. *Guardians of Science: Fairness and Reliability of Peer Review*. Weinheim, Germany: VCH.
Danziger, Kurt. 1990. *Constructing the Subject: Historical Origins of Psychological Research*. Cambridge: Cambridge University Press.
Daston, Lorraine. 1987. *Classical Probability in the Enlightenment*. Princeton: Princeton University Press.
Davidson, Donald. 1982. "On the Very Idea of a Conceptual Scheme." In *Relativism: Cognitive and Moral*, ed. M. Krausz and J. Meiland, 60–80. South Bend: University of Notre Dame Press. First published in 1973.
De Mey, Marc. 1982. *The Cognitive Paradigm*. Dordrecht: Kluwer.
Dennett, Daniel. 1995. *Darwin's Dangerous Idea*. London: Faber & Faber.
Dennis, Michael. 1987. "Accounting for Research: New Histories of Corporate Laboratories and the Social History of American Science." *Social Studies of Science* 17:479–518.
Derrida, Jacques. 1973. *Speech and Phenomena*. Trans. D. Allison. Evanston: Northwestern University Press.
———. 1976. *On Grammatology*. Trans. G. Spivak. Baltimore: Johns Hopkins University Press.

Descombes, Vincent. 1980. *Modern French Philosophy.* Cambridge: Cambridge University Press.
Deutsch, Karl, Andrei Markovits, and John Platt, eds. 1986. *Advances in the Social Sciences. 1900–1980.* Lanham, Md.: University Press of America.
Deutsch, Karl, John Platt, and D. Senghaas. 1971. "Conditions Favoring Major Advances in Social Science." *Science* 171:450–59.
Dewey, John. 1920. *Reconstruction in Philosophy.* Boston: Henry Holt and Company.
———. 1938. *Logic: A Theory of Inquiry.* New York: Henry Holt.
Dinneen, Francis. 1967. *An Introduction to General Linguistics.* Boston: Holt, Rinehart & Winston.
Dobzhansky, Theodosius. 1967. *The Biology of Ultimate Concern.* New York: Doubleday.
Doel, Ronald. 1997. "Scientists as Policymakers, Advisors, and Intelligence Agents: Linking Contemporary Diplomatic History and with the History of Contemporary Science." In Soederqvist 1997, 215–44.
Doppelt, Gerald. 1978. "Kuhn's Epistemological Relativism: An Interpretation and a Defense." *Inquiry* 21:33–86.
Douglas, Mary. 1970a. *Implicit Meanings.* London: Routledge & Kegan Paul.
———. 1970b. *Natural Symbols: Explorations in Cosmology.* London: Barrie and Rockliff.
———. 1980. *Edward Evans-Pritchard.* Harmondsworth, U.K.: Penguin.
———. 1986. *How Institutions Think.* Syracuse: Syracuse University Press.
Dretske, Fred. 1981. *Knowledge and the Flow of Information.* Cambridge: MIT Press.
Drucker, Peter. 1993. *Post-capitalist Society.* New York: Harper Business.
Drury, Shadia. 1988. *The Political Ideas of Leo Strauss.* New York: St. Martin's Press.
———. 1994. *Alexandre Kojeve: The Roots of Postmodern Politics* London: Macmillan.
Dublin, Max. 1989. *Futurehype.* New York: Dell.
Ducrot, Oswald, and Tzvetan Todorov. 1979. *Encyclopedic Dictionary of the Sciences of Language.* Baltimore: Johns Hopkins University Press. First published in 1972.
Duhem, Pierre. 1954. *The Aim and Structure of Physical Theory.* Princeton: Princeton University Press. First published in 1914.
———. 1969. *To Save the Appearances: An Essay on the Idea of Physical Theory from Plato to Galileo.* Chicago: University of Chicago Press. First published in 1908.
———. 1991. *German Science.* La Salle, Ill.: Open Court Press. First published in 1915.
Dummett, Michael. 1981. "Ought Research to Be Unrestricted?" *Grazer philosophische Studien* 12/13:281–98.
Dupre, John. 1993. *The Disorder of Things: The Metaphysical Foundations of the Disunification of Science.* Cambridge: Harvard University Press.
Dusek, Val. 1998. "Brecht and Lukács as Teachers of Feyerabend and Lakatos." *History of the Human Sciences* 11:25–44.
Dvorak, Johann. 1991. "Otto Neurath and Adult Education: Unity of Science, Ma-

terialism, and Comprehensive Enlightenment." In *Rediscovering the Forgotten Vienna Circle*, ed. T. Uebel, 265–74. Dordrecht: Kluwer.

Easton, David. 1991. "Political Science in the United States: Past and Present." In *Divided Knowledge: Across Disciplines, Across Cultures*, ed. D. Easton and C. Schelling, 37–58. London: Sage.

Eder, Klaus. 1996. *The Social Construction of Nature*. London: Sage. First published in 1988.

Edge, David. 1970. "Career Choices by Science Studies Students." *Nature* 225: 506–7.

———. 1975a. "On the Purity of Science." In *The Sciences, the Humanities, and the Technological Threat*, ed. W. R. Niblett, 42–64. London: University of London Press.

———. 1975b. "The Science Studies Unit, Edinburgh University." In Group for Research and Innovation in Higher Education, *Case-Studies in Interdisciplinarity*. Vol. 2, *Science, Technology, and Society*. London: Nuffield Foundation.

———. 1995. "Reinventing the Wheel." In Jasanoff et al. 1995, 3–24.

———. 1996. "Stop Knocking Social Sciences." *Nature* 384 (14 November): 106.

Edgerton, David. 1996. *Science, Technology, and British Industrial "Decline," 1870–1930*. Cambridge: Cambridge University Press.

Edwards, Paul N. 1996. *The Closed World: Computers and the Politics of Discourse in Cold War America*. Cambridge: MIT Press.

Elkana, Yehuda. 1980. "Of Cunning Reason." *Transactions of the New York Academy of Sciences: Science and Social Structure*, ed. T. Gieryn, 32–42. New York: New York Academy of Sciences.

———. 1987. "Alexandre Koyre: Between the History of Ideas and Sociology of Knowledge." *History and Technology* 4:111–44.

Elster, Jon. 1979. *Ulysses and the Sirens*. Cambridge: Cambridge University Press.

———. 1983. *Sour Grapes: Studies in the Subversion of Rationality*. Cambridge: Cambridge University Press.

Elzinga, Aant. 1988. "Bernalism, Comintern, and the Science of Science: Critical Science Movements Then and Now." In *From Research Policy to Social Intelligence*, ed. J. Annerstedt and A. Jamison, 92–113. London: Macmillan.

———. 1996. "UNESCO and the Politics of Scientific Internationalism." In *Internationalism and Science*, ed. A. Elzinga and C. Landstroem, 89–131. London: Taylor Graham.

Ericsson, K. Anders, and Herbert Simon. 1984. *Protocol Analysis: Verbal Reports as Data*. Cambridge: MIT Press.

Etzkowitz, Henry, Andrew Webster, and Peter Healey, eds. 1998. *Capitalizing Knowledge*. Albany: SUNY Press.

Evans-Pritchard, Edward. 1964. *Social Anthropology and Other Essays*. New York: Free Press.

Eyerman, Ron, and Andrew Jamison. 1991. *Social Movements: A Cognitive Approach*. Cambridge: Polity Press.

Ezrahi, Yaron. 1990. *The Descent of Icarus*. Cambridge: Harvard University Press.

Fallows, James. 1993. "Farewell to Laissez-Faire! Clinton Pulls a Reagan on Free-Market Republicans." *Washington Post*, 28 February, p. C1.

Ferretti, Silvia. 1989. *Cassirer, Panofsky, and Warburg: Symbol, Art, and History.* New Haven: Yale University Press.
Festinger, Leon. 1957. *A Theory of Cognitive Dissonance.* Palo Alto: Stanford University Press.
Feuer, Lewis. 1969. *The Conflict of Generations.* London: Heinemann.
Feyerabend, Paul. 1962. "Explanation, Reduction, and Empiricism." In *Minnesota Studies in the Philosophy of Science*, vol. 3, ed. H. Feigl and G. Maxwell, 28–97. Minneapolis: University of Minnesota Press.
———. 1970. "Consolations for the Specialist." In Lakatos and Musgrave 1970, 197–229.
———. 1975. *Against Method.* London: New Left Books.
———. 1979. *Science in a Free Society.* London: Verso.
———. 1991. *Three Dialogues on Knowledge.* Oxford: Blackwell.
Fine, Arthur. 1984. "The Natural Ontological Attitude." In Leplin 1984, 83–106.
Fisch, Menachem. 1991. *William Whewell, Philosopher of Science.* Oxford: Oxford University Press.
Fisch, Menachem, and Simon Schaffer, eds. 1991. *William Whewell: A Composite Portrait.* Oxford: Oxford University Press.
Fischer, Frank. 1992. "Participatory Expertise: Toward the Democratization of Science Policy." In *Advances in Policy Studies since 1950*, ed. W. Dunn and R. M. Kelley, 351–76. New Brunswick: Transaction Books.
Fleck, Ludwik. 1979. *Genesis and Development of a Scientific Fact.* Trans. F. Bradley and T. Trenn. Chicago: University of Chicago Press. First published in 1935.
Forman, Paul. 1971. "Weimar Culture, Causality, and Quantum Theory: 1918–1927." *Historical Studies in the Physical Sciences* 3:1–115.
———. 1991. "Independence, not Transcendence, for the Historian of Science." *Isis* 82:71–86.
Foucault, Michel. 1967. *Madness and Civilization.* London: Tavistock.
———. 1970. *The Order of Things.* New York: Random House.
Fox, Robin. 1997. "State of the Art/Science in Anthropology." In Gross, Levitt, and Lewis 1997, 327–45.
Fox Keller, Evelyn. 1983. *A Feeling for the Organism.* New York: Freeman.
Frank, Phillip. 1950. *Modern Science and Its Philosophy.* Cambridge: Harvard University Press.
Fraser, Mariam. 1998. "The Face-off between Will and Fate: Artistic Identity and Neurological Style in de Kooning's Late Works." *Body and Society* 4, no. 4: 1–22.
Freeland, Richard. 1992. *Academia's Golden Age: Universities in Massachusetts, 1945–1970.* Oxford: Oxford University Press.
Freeman, Christopher. 1992. *The Economics of Hope.* London: Pinter.
Frey, Scott, Thomas Dietz, and Linda Kalof. 1992. "Characteristics of Successful American Protest Groups." *American Journal of Sociology* 98:386–87.
Friedman, Michael. 1998. "On the Sociology of Scientific Knowledge and Its Philosophical Agenda." *Studies in History and Philosophy of Science* 29:239–72.
Friedrichs, Robert. 1970. *A Sociology of Sociology.* New York: Free Press.

Frisby, David. 1992. *The Alienated Mind: The Sociology of Knowledge in Germany 1918–1933*. London: Routledge.
Fruton, Joseph. 1990. *Contrasts in Scientific Style: Research Groups in the Chemical and Biochemical Sciences*. Philadelphia: American Philosophical Society.
Frye, Northrop. 1957. *Anatomy of Criticism*. Princeton: Princeton University Press.
Fuchs, Stephan. 1992. *The Professional Quest for Truth*. Albany: SUNY Press.
Fuhrman, Ellsworth. 1980. *The Sociology of Knowledge in America: 1883–1915*. Charlottesville: University of Virginia Press.
Fujimura, Joan. 1998. "Authorizing Knowledge in Science and Anthropology." *American Anthropologist* 100:347–60.
Fukuyama, Francis. 1992. *The End of History and the Last Man*. New York: Free Press.
Fuller, Steve. 1983. "A French Science (With English Subtitles)." *Philosophy and Literature* 7:1–14.
———. 1988. *Social Epistemology*. Bloomington: Indiana University Press.
———. 1990. "They Shoot Dead Horses, Don't They? Philosophical Fear and Sociological Loathing in St. Louis." *Social Studies of Science* 20:664–81.
———. 1991. "Is History and Philosophy of Science Withering on the Vine?" *Philosophy of the Social Sciences* 21:149–74.
———. 1992a. "Being There with Thomas Kuhn: A Parable for Postmodern Times." *History and Theory* 31:241–75.
———. 1992b. "Epistemology Radically Naturalized: Recovering the Normative, the Experimental, and the Social." In Giere 1992, 427–59.
———. 1992c. "Social Epistemology and the Research Agenda of Science Studies." In Pickering 1992, 390–428.
———. 1992d. "STS as Social Movement: On the Purpose of Graduate Programs." *Science, Technology, and Society*, no. 91 (September): 1–5.
———. 1993a. *Philosophy of Science and Its Discontents*. New York: Guilford Press. First published in 1989.
———. 1993b. *Philosophy, Rhetoric, and the End of Knowledge: The Coming of Science and Technology Studies*. Madison: University of Wisconsin Press.
———. 1994a. "The Constitutively Social Character of Expertise." *International Journal of Expert Systems* 7:51–64.
———. 1994b. "Making Agency Count." *American Behavioral Scientist* 37:741–53.
———. 1994c. "The Reflexive Politics of Constructivism." *History of the Human Sciences* 7:87–94.
———. 1994d. "Rethinking the University from a Social Constructivist Standpoint." *Science Studies* 7, no. 1: 4–16.
———. 1994e. "The Social Psychology of Scientific Knowledge: Another Strong Programme." In *The Social Psychology of Science*, ed. W. Shadish and S. Fuller, 162–80. New York: Guilford Press.
———. 1994f. "The Sphere of Critical Thinking in a Post-epistemic World." *Informal Logic* (winter): 39–54.
———. 1994g. "Towards a Philosophy of Science Accounting: A Critical Rendering of Instrumental Rationality." *Science in Context* 7:591–621.

———. 1994h. "Underlaborers for Science." *Science* 264:982–83.

———. 1994i. "Why Post-industrial Society Never Came: What a False Prophecy Can Teach Us about the Impact of Technology on Academia." *Academe* 80, no. 6 (November): 22–28.

———. 1995a. "The Strong Programme in the Rhetoric of Science." In *Science, Reason, and Rhetoric*, ed. H. Krips, J. McGuire, and T. Melia, 95–118. Pittsburgh: University of Pittsburgh Press.

———. 1995b. "The Voices of Rhetoric and Politics in Social Epistemology: For a Critical-Rationalist Multiculturalism." *Philosophy of the Social Sciences* 25: 512–22.

———. 1996a. "Does Science Put an End to History, or History to Science? Or, Why Being Pro-science Is Harder than You Think." In Ross 1996, 29–60.

———. 1996b. "Enlightened Hybrids or Transcendental Mongrels? The Place of Science Studies in the Human Studies." *History of the Human Sciences* 9: 122–31.

———. 1996c. "Recent Work in Social Epistemology." *American Philosophical Quarterly* 33:149–66.

———. 1996d. "Social Epistemology and Psychology." In *Philosophy of Psychology*, ed. W. O'Donohue and R. Kitchener, 33–49. London: Sage.

———. 1996e. "Social Epistemology and the Recovery of the Normative in the Post-epistemic Era." *Journal of Mind and Behavior* 17, no. 2: 83–98.

———. 1996f. "Talking Metaphysical Turkey about Epistemological Chicken, and the Poop on Pidgins." In Galison and Stump 1996, 170–88, 468–71.

———. 1997a. "Kuhn as Trojan Horse." *Radical Philosophy* 82 (March/April): 5–7.

———. 1997b. "Putting People Back into the Business of Science: Constituting a National Forum for Setting the Research Agenda." In Collier 1997, 233–66.

———. 1997c. Review essay of *Android Epistemology* and *Knowing Machines*. *Information Society* 13:289–93.

———. 1997d. *Science*. Milton Keynes, U.K.: Open University Press; Minneapolis: University of Minnesota Press.

———. 1997e. "The Secularization of Science and a New Deal for Science Policy." *Futures* 29:483–504.

———. 1998a. "Divining the Future of Social Theory: From Theology to Rhetoric via Social Epistemology." *European Journal of Social Theory* 1:107–26.

———. 1998b. "From Content to Context: A Social Epistemology of the Structure-Agency Craze." In *What Is Social Theory?: The Philosophical Debates*, ed. A. Sica, 92–117. Oxford: Blackwell.

———. 1998c. "An Intelligent Person's Guide to Intelligent Design Theory." *Rhetoric and Public Affairs* 1:603–10.

———. 1998d. "Making Science an Experimenting Society." In *The Experimenting Society: Policy Essays in Honor of Donald T. Campbell*, ed. W. Dunn, 69–102. New Brunswick, N.J.: Transaction Books.

———. 1998e. "Society's Shifting Human-Computer Interface: A Sociology of Knowledge for the Information Age." *Information, Communication, and Society* 1:182–98.

———. 1998f. "What Does the Sokal Hoax Say about the Prospects for Positivism?" In *Positivismes*, ed. A. Despy-Meyer D. Devries, 265–84. Brussels: Brepols.
———. 1999a. "Authorizing Science Studies, or Why We Have Never Had Paradigms." *American Anthropologist* 101:379–81.
———. 1999b. *The Governance of Science: Ideology and the Future of the Open Society.* Milton Keynes, U.K.: Open University Press.
———. 1999c. "Making the University Fit for Critical Intellectuals: Recovering from the Ravages of the Postmodern Condition." *British Education Research Journal* 25:583–95.
———. 1999d. Review of Castells 1996–98. *Science, Technology, and Human Values* 24:159–66.
———. 1999e. "Social Epistemology as a Critical Philosophy of Multiculturalism." In *Multicultural Curriculum*, ed. C. McCarthy and R. Mahalingam. London: Routledge.
Fuller, Steve, and David Gorman. 1987. "Burning Libraries and the Problem of Historical Consciousness." *Annals of Scholarship* 4, no. 3: 105–22.
Gadamer, Hans-Georg. 1975. *Truth and Method.* New York: Seabury Press. First published in 1960.
Gaddis, John Lewis. 1993. "Presidential Address: The Tragedy of Cold War History." *Diplomatic History* 17:1–16.
Galbraith, John Kenneth. 1952. *American Capitalism: The Concept of Countervailing Power.* Boston: Houghton Mifflin.
Gale, Richard. 1967. "Indexical Signs, Egocentric Particulars, and Token-Reflexive Words." In *The Encyclopedia of Philosophy*, ed. P. Edwards, 4:151–55. New York: Macmillan.
Galison, Peter. 1987. *How Experiments End.* Chicago: University of Chicago Press.
———. 1990. "Aufbau/Bauhaus: Logical Positivism and Architectural Modernism." *Critical Inquiry* 16:709–52.
———. 1998. "The Americanization of Unity." *Daedalus* 127, no. 1: 45–72.
Galison, Peter, and David Stump, eds. 1996. *The Disunity of Science: Boundaries, Contexts, and Power.* Palo Alto: Stanford University Press.
Garber, Daniel. 1992. *Descartes' Metaphysical Physics.* Chicago: University of Chicago Press.
Gardner, Howard. 1987. *The Mind's New Science.* 2d ed. New York: Basic Books. 1st ed. 1985.
Gellner, Ernest. 1959. *Words and Things.* London: Victor Gollancz.
———. 1968. "The New Idealism—Cause and Meaning in the Social Sciences." In *Problems in the Philosophy of Science*, ed. I. Lakatos and A. Musgrave, 377–406. Amsterdam: North-Holland.
———. 1989. *Plough, Sword, and Book.* Chicago: University of Chicago Press.
Georgescu-Roegen, Nicholas. 1971. *The Entropy Law and the Economic Process.* Cambridge: Harvard University Press.
Geyer, Michael. 1993. "Multiculturalism and the Politics of General Education." *Critical Inquiry* 19:499–533.
Gibbons, Michael, Camille Limoges, Helga Nowotny, Simon Schwartzman, Peter Scott, and Martin Trow. 1994. *The New Production of Knowledge.* London: Sage.

Giddens, Anthony. 1990. *The Consequences of Modernity.* Cambridge: Polity Press.
Giere, Ronald, ed. 1992. *Cognitive Models of Science.* Minneapolis: University of Minnesota Press.
———. 1995. "Viewing Science." In *PSA 1994*, vol 2., ed. D. Hull, M. Forbes, and R. Burian, 3–16. East Lansing: Philosophy of Science Association.
Gieryn, Thomas. 1995. "Boundaries of Science." In Jasanoff et al. 1995, 393–443.
Gilbert, G. Nigel, and Michael Mulkay. 1984. *Opening Pandora's Box.* Cambridge: Cambridge University Press.
Gildea, Robert. 1996. *France since 1945.* Oxford: Oxford University Press.
Gilliatt, Stephen. 1995. "Disliking Politics: Philosophical Foundations for a Sociology of the Apolitical." *International Sociology* 10:283–98.
Ginzburg, Carlo. 1989. *Clues, Myths, and the Historical Method.* Baltimore: Johns Hopkins University Press.
Gjertsen, Derek. 1984. *The Classics of Science: Twelve Enduring Scientific Works.* New York: Lilian Barber Press.
———. 1989. *Science and Philosophy: Past and Present.* Harmondsworth, U.K.: Penguin.
Glaser, Barney, and Anselm Strauss. 1967. *The Discovery of Grounded Theory: Strategies for Qualitative Research Practice.* Chicago: Aldine.
Godelier, Maurice. 1986. *The Mental and the Material.* London: Verso. First published in 1984.
Goldman, Alvin. 1986. *Epistemology and Cognition.* Cambridge: Harvard University Press.
Golinski, Jan. 1998. *Making Natural Knowledge: Constructivism and the History of Science.* Cambridge: Cambridge University Press.
Gombrich, Ernst. 1979. *The Sense of Order: A Study in the Psychology of Decorative Art.* London: Phaidon Press.
González Garcia, M. I., J. A. López Cerezo, and J. L. Luján. 1996. *Ciencia, tecnologia y sociedad: Una introducción al estudio social de la ciencia y la tecnologia.* Madrid: Tecnos.
Gooday, Graeme. 1990. "Precision Measurement and the Genesis of Physics Teaching Laboratories." *British Journal for the History of Science* 23:25–52.
Goodman, Nelson. 1954. *Fact, Fiction, and Forecast.* Cambridge: Harvard University Press.
Goodson, Ivor. 1988. *The Making of Curriculum.* London: Falmer Press.
Goodwin, Craufurd. 1991. "National Security and Classical Political Economy." In *Economics and National Security: A History of their Interaction*, ed. C. Goodwin, 23–35. Durham: Duke University Press.
Goody, Jack. 1995. *The Expansive Moment: Anthropology in Britain and Africa, 1918–1970.* Cambridge: Cambridge University Press.
Gorz, Andre. 1988. *Critique of Economic Reason.* London: New Left Books.
Gouldner, Alvin. 1965. *Enter Plato: Classical Greece and the Origins of Social Theory.* London: Routledge & Kegan Paul.
———. 1968. "The Sociologist as Partisan: Sociology and the Welfare State." *American Sociologist* 3:103–16.

———. 1970a. "Anti-Minotaur: The Myth of a Value-Free Sociology." In *The Relevance of Sociology*, ed. J. Douglas, 64–84. New York: Appleton-Century-Crofts.
———. 1970b. *The Coming Crisis in Western Sociology*. New York: Basic Books.
———. 1979. *The Future of Intellectuals and the Rise of the New Class*. London: Macmillan.
Graham, Gordon. 1996. *The Shape of the Past*. Oxford: Oxford University Press.
Graham, Loren. 1981. *Between Science and Values*. New York: Columbia University Press.
———. 1985. "The Socio-political Roots of Boris Hessen: Soviet Marxism and the History of Science." *Social Studies of Science* 15:705–22.
Granovetter, Mark. 1973. "The Strength of Weak Ties." *American Journal of Sociology* 78:1360–80.
Greenberg, Daniel. 1967. *The Politics of Pure Science*. New York: New American Library.
Gregory, Frederick. 1992. "Theologians, Science, and Theories of Truth in the Nineteenth Century." In *The Invention of Physical Science*, ed. M. J. Nye et al., 81–96. Dordrecht: Kluwer.
Grint, Keith, and Steve Woolgar. 1995. "On Some Failures of Nerve in Constructivist and Feminist Analyses of Technology." *Science, Technology, and Human Values* 20:286–310.
Gross, Paul. 1997. "Characterizing Scientific Knowledge." *Science* 275:142.
Gross, Paul, and Norman Levitt. 1994. *Higher Superstition: The Academic Left and Its Quarrels with Science*. Baltimore: Johns Hopkins University Press.
Gross, Paul, Norman Levitt, and Martin Lewis, eds. 1997. *The Flight from Science and Reason*. Baltimore: Johns Hopkins University Press.
Gunnell, John. 1986. *Between Philosophy and Politics: The Alienation of Political Theory*. Amherst: University of Massachusetts Press.
Gutting, Gary, ed. 1979. *Paradigms and Revolutions*. South Bend: University of Notre Dame Press.
Habermas, Jürgen. 1989. *The New Conservatism: Cultural Criticism and the Historians' Debate*. Cambridge: MIT Press.
Hacking, Ian. 1975. *The Emergence of Probability*. Cambridge: Cambridge University Press.
———. 1979a. "Lakatos's Philosophy of Science." *British Journal for the Philosophy of Science* 30:381–410.
———. 1979b. "Review of *The Essential Tension*." *History and Theory* 18:223–36.
———. 1982. "Language, Truth, and Reason." In Hollis and Lukes 1982, 48–66.
———. 1983. *Representing and Intervening*. Cambridge: Cambridge University Press.
———. 1984a. "Five Parables." In *Philosophy in History*, ed. R. Rorty, J. B. Schneewind, and Quentin Skinner, 103–24. Cambridge: Cambridge University Press.
———. 1984b. "Wittgenstein Rules." *Social Studies of Science* 14:469–76.
———. 1992. " 'Style' for Historians and Philosophers." *Studies in History and Philosophy of Science* 23:1–20.
———. 1993. "Working in a New World." In Horwich 1993, 275–310.

Hagstrom, Warren. 1965. *The Scientific Community*. New York: Basic Books.
Hakfoort, Caspar. 1992. "Science Deified: Wilhelm Ostwald's Energeticist World-View and the History of Scientism." *Annals of Science* 49:525–44.
Hall, A. Rupert. 1963. "Merton Revisited." *British Journal for the History of Science* 2:1–16.
———. 1984. "Beginnings in Cambridge." *Isis* 75:22–25.
Hamblin, C. L. 1970. *Fallacies*. London: Methuen.
Hanson, Norwood Russell. 1958. *Patterns of Discovery*. Cambridge: Cambridge University Press
———. 1962. "Scientists and Logicians: A Confrontation." *Science* 138:1311–14.
———. 1965. "A Note on Kuhn's Method." *Dialogue* 4:371–75.
———. 1971. *What I Do Not Believe and Other Essays*. Ed. S. Toulmin and H. Woolf. Dordrecht: Reidel.
Hanson, Robin. 1995. "Comparing Peer Review to Information Prizes—A Possible Economics Experiment." *Social Epistemology* 9:49–55.
Haraway, Donna. 1991. *Simians, Cyborgs, and Women*. London: Free Association Books.
———.1997.*Modest_Witness@Second_Millenium.FemaleMan_Meets_OncoMouse*. London: Routledge.
Harding, Sandra. 1986. *The Science Question in Feminism*. Ithaca: Cornell University Press.
———. 1991. *Whose Science? Whose Knowledge?* Ithaca: Cornell University Press.
———, ed. 1993. *The Racial Economy of Science*. Bloomington: Indiana University Press.
Harre, Rom. 1970. *The Principles of Scientific Thinking*. Chicago: University of Chicago Press.
———. 1986. *Varieties of Realism*. Oxford: Blackwell.
Harre, Rom, and Paul Secord. 1972. *The Explanation of Social Behaviour*. Oxford: Blackwell.
Harris, Roy. 1987. *Reading Saussure*. La Salle, Ill.: Open Court Press.
Harrisville, Roy, and Walter Sundberg. 1995. *The Bible in Modern Culture: Theology and Historical-Critical Method from Spinoza to Kaesemann*. Grand Rapids: William Eerdmans.
Hart, Joan. 1993. "Erwin Panofsky and Karl Mannheim: A Dialogue on Interpretation." *Critical Inquiry* 19:534–66.
Harvey, Lee. 1990. *Critical Social Research*. London: Unwin Hyman.
Harwood, Jonathan. 1994. "Institutional Innovation in *Fin de Siecle* Germany." *British Journal for the History of Science* 27:197–211.
Hatfield, Gary. 1990a. "Metaphysics and the New Science." In *Reappraisals of the Scientific Revolution*, ed. D. Lindberg and R. Westman, 93–166. Cambridge: Cambridge University Press.
———. 1990b. *The Natural and the Normative*. Cambridge: MIT Press.
Hayakawa, S. I. 1939. *Language, Thought, and Action*. New York: Harcourt, Brace and Jovanovich.
Hayek, Friedrich von. 1945. "The Use of Knowledge in Society." *American Economic Review* 35:519–30.

———. 1952. *The Counter-revolution in Science.* Chicago: University of Chicago Press.
———. 1960. *The Constitution of Liberty.* Chicago: University of Chicago Press.
———. 1978. *New Studies in Philosophy, Politics, and Economics.* London: Routledge & Kegan Paul.
Heelan, Patrick. 1983. *Space-Perception and the Philosophy of Science.* Berkeley and Los Angeles: University of California Press.
Heidegger, Martin. 1996. *Being and Time.* Trans. Joan Stambaugh. Albany: SUNY Press. First published in 1927.
Heilbron, John. 1986. *Dilemmas of an Upright Man: Max Planck as Spokesman for German Science.* Berkeley and Los Angeles: University of California Press.
Heims, Steve. 1991. *Constructing a Social Science for Postwar America: The Cybernetics Group, 1946–1953.* Cambridge: MIT Press.
Held, David. 1987. *Models of Democracy.* Cambridge: Polity Press.
Hempel, Carl G. 1942. "The Function of General Laws in History." *Journal of Philosophy* 39:35–48.
———. 1965. *Aspects of Scientific Explanation.* New York: Free Press.
Herf, Jeffrey. 1984. *Reactionary Modernism: Technology, Culture, and Politics in Weimar and the Third Reich.* Cambridge: Cambridge University Press.
Herman, Arthur. 1997. *The Idea of Decline in Western History.* New York: Free Press.
Hershberg, James. 1993. *James B. Conant: Harvard to Hiroshima and the Making of the Nuclear Age.* New York: Alfred Knopf.
Hess, David. 1993. *Science in the New Age: The Paranormal, Its Defenders, and Debunkers.* Madison: University of Wisconsin Press.
———. 1997. *Science Studies: An Advanced Introduction.* New York: New York University.
Hesse, Mary. 1970a. "Duhem, Quine, and a New Empiricism." In Royal Institute of Philosophy, *Knowledge and Necessity,* 191–209. London: Macmillan.
———. 1970b. "Hermeticism and Historiography: An Apology for the Internal History of Science." In *Historical and Philosophical Perspectives on Science,* ed. Stuewer 1970, 134–60.
———. 1980. *Revolutions and Reconstructions in the Philosophy of Science.* Brighton, U.K.: Harvester.
Heyl, Barbara. 1968. "The Harvard 'Pareto Circle.'" *Journal of the History of the Behavioral Sciences* 4:316–34.
Hiebert, Erwin. 1990. "The Transformation of Physics." In *Fin de Siecle and Its Legacy,* ed. M. Teich and R. Porter, 235–53. Cambridge: Cambridge University Press.
Hirsch, Fred. 1976. *The Social Limits to Growth.* Cambridge: Cambridge University Press.
Hirschmann, Albert. 1977. *The Passions and the Interests.* Princeton: Princeton University Press.
Hjelmslev, Louis. 1961. *Prolegomenon to a Theory of Language.* Madison: University of Wisconsin Press.
Hodge, M. J. S. 1991. "The History of the Earth, Life, and Man: Whewell and Palaetiological Science." In Fisch and Schaffer 1991, 253–88.

Hofstadter, Richard, and Walter Metzger. 1955. *The Development of Academic Freedom in the United States.* New York: Columbia University Press.
Hollinger, David. 1990. "Free Enterprise and Free Inquiry: The Emergence of Laissez-Faire Communitarianism in the Ideology of Science in the United States." *New Literary History* 21:897–919.
———. 1995. "Science as a Weapon in the Kulturkaempfe in the United States and After World War II." *Isis* 86:440–54.
Hollis, Martin, and Steven Lukes, eds. 1982. *Rationality and Relativism.* Cambridge: MIT Press.
Holmes, Frederic L. 1997. "Writing about Scientists of the Near Past." In Soederqvist 1997, 165–78.
Holton, Gerald. 1978. *The Scientific Imagination.* Cambridge: Cambridge University Press.
———. 1993. *Science and Anti-science.* Cambridge: Harvard University Press.
———, ed. 1967. *Science and Culture.* Boston: Beacon Press.
Holub, Robert. 1991. *Juergen Habermas: Critic in the Public Sphere.* London: Routledge.
Hooks, Gregory. 1991. *Forging the Military-Industrial Complex.* Urbana: University of Illinois Press.
Horgan, John. 1991. "Profile: Reluctant Revolutionary—Thomas S. Kuhn Unleashed 'Paradigm' on the World." *Scientific American,* May, 40, 49.
———. 1996. *The End of Science: Facing the Limits of Knowledge in the Twilight of the Scientific Age.* Reading, Mass.: Addison Wesley.
Horowitz, Irving Louis. 1968. *Radicalism and the Revolt against Reason.* Carbondale: Southern Illinois University Press. First published in 1961.
Horwich, P., ed. 1993. *World Changes: Thomas Kuhn and the Nature of Science.* Cambridge: MIT Press.
Hovland, Carl, A. Lumsdaine, and F. Sheffield. 1949. *Experiments on Mass Communication: Studies in Social Psychology in World War II.* Vol. 3. Princeton: Princeton University Press.
Hoyningen-Huene, Paul. 1993. *Reconstructing Scientific Revolutions.* Chicago: University of Chicago Press.
———. 1995. "Two Letters of Paul Feyerabend to Thomas Kuhn on a Draft of *The Structure of Scientific Revolutions.*" *Studies in History and Philosophy of Science* 26:353–88.
Huff, Toby. 1993. *The Rise of Early Modern Science: Islam, China, and the West.* Cambridge: Cambridge University Press.
Hughes, Thomas. 1987. "The Evolution of Large Technological Systems." In Bijker, Hughes, and Pinch 1987, 51–82.
Hull, David. 1988. *Science as a Process.* Chicago: University of Chicago Press.
Hull, David, Peter Tessner, and Arthur Diamond. 1978. "Planck's Principle." *Science* 202:717–23.
Hultberg, Jon. 1991. *A Tale of Two Cultures: The Image of Science in C. P. Snow.* Report 165. Gothenburg: Department of Theory of Science, Gothenburg University.

Humphrey, George. 1951. *Thinking: An Introduction to Its Experimental Psychology.* London: Methuen.
Husserl, Edmund. 1970. *The Crisis of European Sciences and Transcendental Phenomenology.* Trans. D. Carr. Evanston: Northwestern University Press. First published in 1936.
Inkster, Ian. 1991. *Science and Technology in History: An Approach to Industrial Development.* London: Macmillan.
Irzik, Guerol, and Theo Gruenberg. 1995. "Carnap and Kuhn: Arch Enemies or Close Allies?" *British Journal for the Philosophy of Science* 46:285–307.
James, William. 1948. "Pragmatism's Conception of Truth." In *Essays in Pragmatism.* New York: Hafner Publishing Co. First published in 1907.
——. 1956. *The Will to Believe and Other Essays in Popular Philosophy.* New York: Dover. First published in 1897.
Janik, Allan, and Stephen Toulmin. 1973. *Wittgenstein's Vienna.* New York: Simon and Schuster.
Janis, Irving. 1982. *Groupthink.* 2d ed. Boston: Houghton Mifflin. 1st ed. 1972.
Jarvie, Ian. 1988. "Explanation, Reduction, and the Sociological Turn in the Philosophy of Science: or Kuhn as Ideologue for Merton's Theory of Science." In *Centripetal Reason,* ed. G. Radnitzky, 299–320. New York: Paragon House.
Jasanoff, Sheila, Gerald Markle, James Petersen, and Trevor Pinch, eds. 1995. *Handbook of Science and Technology Studies.* Thousand Oaks, Calif.: Sage.
Jervis, Robert. 1976. *Perception and Misperception in International Politics.* Princeton: Princeton University Press.
Johnson, Conrad. 1985. "The Authority of the Moral Agent." *Journal of Philosophy* 82:391–413.
Johnson, Harry. 1965. *The World Economy at the Crossroads.* Oxford: Clarendon Press.
Johnson, Jeffrey. 1990. *The Kaiser's Chemists: Science and Modernization in Imperial Germany.* Chapel Hill: University of North Carolina Press.
Jones, Caroline, and Peter Galison, eds. 1998. *Picturing Science, Producing Art.* London: Routledge.
Jones, Greta. 1988. *Science, Politics, and the Cold War.* London: Routledge & Kegan Paul.
Jones, Robert Alun. 1994. "The Positive Science of Ethics in France: German Influences on *De la division du travail social.*" *Sociological Forum* 9:37–57.
Jonsen, Albert, and Stephen Toulmin. 1988. *The Abuse of Casuistry.* Berkeley and Los Angeles: University of California Press.
Journet, Debra. 1995. "Synthesizing Disciplinary Narratives: George Gaylord Simpson's *Tempo and Mode in Evolution.*" *Social Epistemology* 9:113–50.
Kaldor, Mary. 1982. *The Baroque Arsenal.* London: Deutsch.
Katznelson, Ira. 1997. "The Subtle Politics of Developing Emergency: Political Science as Liberal Guardianship." In Chomsky et al. 1997, 233–59.
Keegan, John. 1993. *A History of Warfare.* New York: Alfred Knopf.
Keith, William, and Kenneth Zagacki. 1992. "Rhetoric and Paradox in Scientific Revolutions." *Southern Communication Journal* 57:165–77.

Kelley, Donald. 1970. *Foundations of Modern Historical Scholarship: Language, Law, and History in the French Renaissance*. New York: Columbia University Press.

———. 1984. *Historians and the Law in Post-revolutionary France*. Princeton: Princeton University Press.

———. 1990. *The Human Measure: Social Thought in the Western Legal Tradition*. Cambridge: Harvard University Press.

Kevles, Daniel. 1977. "The National Science Foundation and the Debate over Postwar Research Policy: 1942–1945." *Isis* 68:5–26.

———. 1995. *Physicists: The History of a Scientific Community in Modern America*. Cambridge: Harvard University Press. First published in 1971.

Kim, Kyung-Man. 1996. "Hierarchy of Scientific Consensus and the Flow of Dissensus over Time." *Philosophy of Social Sciences* 26:3–25.

Kinneavy, James. 1986. "*Kairos*: A Neglected Concept in Classical Rhetoric." In *Rhetoric and Practice*, ed. J. Moss, 79–105. Washington: Catholic University Press.

King, M. D. 1971. "Reason, Tradition, and the Progressiveness of Science." *History and Theory* 10:3–32.

Kitcher, Philip. 1982. *Abusing Science: The Case against Creationism*. Cambridge: MIT Press.

———. 1985. *Vaulting Ambition*. Cambridge: MIT Press.

———. 1993. *The Advancement of Science*. Oxford: Oxford University Press.

Klein, Martin, Abner Shimony, and Trevor Pinch. 1979. "Paradigm Lost?" *Isis* 70:429–40.

Kleinman, Daniel. 1995. *Politics on the Endless Frontier: Postwar Research Policy in the United States*. Durham: Duke University Press.

Knight, David. 1994. *Ideas in Chemistry: A History of the Science*. London: Athlone.

Knorr-Cetina, Karin. 1981. *The Manufacture of Knowledge*. Oxford: Pergamon.

———. 1999. *Epistemic Cultures*. Cambridge: Harvard University Press.

Koehler, Wolfgang. 1971. *Selected Papers*. Ed. Mary Henle. New York: Liveright.

Koffka, Kurt. 1935. *Principles of Gestalt Psychology*. New York: Harcourt Brace & World.

Kohler, Robert. 1982. *From Medical Chemistry to Biochemistry*. Cambridge: Cambridge University Press.

———. 1991. *Partners in Science: Foundations and Natural Scientists, 1900–1945*. Chicago: University of Chicago Press.

Kolakowski, Leszek. 1972. *Positivist Philosophy: From Hume to the Vienna Circle*. Harmondsworth, U.K.: Penguin.

Koyré, Alexandre. 1945. *Discovering Plato*. New York: Columbia University Press.

———. 1957. *From the Closed World to the Infinite Universe*. Baltimore: Johns Hopkins University Press.

———. 1963. "Commentary of Henry Guerlac's 'Some Historical Assumptions of the History of Science.'" In Crombie 1963, 847–57.

———. 1965. *Newtonian Studies*. London: Chapman and Hall.

———. 1978. *Galileo Studies*. Trans. J. Mepham. Atlantic Highlands, N.J.: Humanities Press. First published in 1939.

Kragh, Helge. 1987. *An Introduction to the Historiography of Science.* Cambridge: Cambridge University Press.

Krause, Elliott. 1996. *Death of the Guilds: Professions, States, and the Advance of Capitalism, 1930 to the Present.* New Haven: Yale University Press.

Krohn, Wolfgang, and Wolf Schaefer. 1976. "The Origins and Structure of Agricultural Chemistry." In *Perspectives on the Emergence of Scientific Disciplines,* ed. G. Lemaine et al. Chicago: Aldine.

Krueger, Lorenz, Lorraine Daston, and Michael Heidelberger, eds. 1987. *The Probabilistic Revolution.* 2 vols. Cambridge: Cambridge University Press.

Kuhn, Thomas S. 1957. *The Copernican Revolution.* Cambridge: Harvard University Press.

———. 1963. "The Function of Dogma in Scientific Research." In Crombie 1963, 347–69.

———. 1968. "The History of Science." In *International Encyclopedia of the Social Sciences,* 14:74–83. New York: Collier Macmillan.

———. 1970a. "Reflections on My Critics." In Lakatos and Musgrave 1970, 231–78.

———. 1970b. *The Structure of Scientific Revolutions.* 2d ed. Chicago: University of Chicago Press. 1st ed. 1962.

———. 1971. "The Relations between History and History of Science." *Daedalus* 100:271–304.

———. 1976. Foreword to Ludwik Fleck, *Genesis and Development of a Scientific Fact.* Chicago: University of Chicago Press.

———. 1977a. *The Essential Tension.* Chicago: University of Chicago Press.

———. 1977b. "Second Thoughts on Paradigms." In Suppe 1977, 459–82.

———. 1978. *Black-Body Radiation and the Quantum Discontinuity, 1894–1912.* Oxford: Oxford University Press.

———. 1983. "Rationality and Theory Choice." *Journal of Philosophy* 80:563–70.

———. 1989. "Possible Worlds in History of Science." In *Humanities, Arts, and Sciences,* ed. S. Allen, 9–32. Berlin: Walter de Gruyter.

———. 1992. *The Trouble with the Historical Philosophy of Science: Rothschild Distinguished Lecture.* Cambridge: Department of History of Science, Harvard University.

———. 1993. "Afterwords." In Horwich 1993, 311–42.

Kuhn, Thomas; and Aristides Baltas, Kostas Gavroglu, and Vasso Kindi. 1997. "A Discussion with Thomas S. Kuhn." *Neusis* 6:143–98.

Kuhn, Thomas, John Heilbron, Paul Forman, and Lini Allen. 1967. *Sources for the History of Quantum Physics.* Philadelphia: American Philosophical Society.

Kuznick, Peter. 1987. *Beyond the Laboratory: Scientists as Political Activists in 1930s America.* Chicago: University of Chicago Press.

Labinger, Jay. 1995. "Out of the Petri Dish Endlessly Rocking." *Social Studies of Science* 25:341–48.

Lacan, Jacques. 1972. "The Insistence of the Letter in the Unconscious." In *The Structuralists from Marx to Levi-Strauss,* ed. R. and F. De George, 287–324. Garden City, N.Y.: Doubleday. First published in 1966.

LaFollette, Marcel. 1990. *Making Science Our Own: Public Images of Science 1910–1955.* Chicago: University of Chicago Press.

Lagemann, Ellen. 1989. *The Politics of Knowledge: The Carnegie Corporation, Philanthropy, and Public Policy.* Chicago: University of Chicago Press.

Lakatos, Imre. 1970. "Falsification and the Methodology of Scientific Research Programmes." In Lakatos and Musgrave 1970, 91–196.

———. 1981. "History of Science and Its Rational Reconstructions." In *Scientific Revolutions,* ed. I. Hacking, 107–27. Oxford: Oxford University Press.

Lakatos, Imre, and Alan Musgrave, eds. 1970. *Criticism and the Growth of Knowledge.* Cambridge: Cambridge University Press.

La Lumia, Joseph. 1966. *The Ways of Reason: A Critical Study of the Work of Emile Meyerson.* New York: Humanities Press.

Lambropoulos, Vassilis. 1993. *The Rise of Eurocentrism.* Princeton: Princeton University Press.

Larvor, Brendan. 1998. *Lakatos.* London: Routledge.

Lasch, Christopher. 1991. *The True and Only Heaven: Progress and Its Critics.* New York: Norton.

Latour, Bruno. 1987. *Science in Action.* Milton Keynes, U.K.: Open University Press.

———. 1988a. *The Pasteurization of France.* Cambridge: Harvard University Press.

———. 1988b. "The Politics of Explanation." In Woolgar 1988, 155–76.

———. 1993. *We Have Never Been Modern.* Cambridge: Harvard University Press.

———. 1994. "Pragmatogonies: A Mythical Account of How Humans and Nonhumans Share Properties." *American Behavioral Science* 37:791–808.

———. 1996. *Aramis, or the Love of Technology.* Cambridge: Harvard University Press.

———. 1997. "A Few Steps toward an Anthropology of the Iconoclastic Gesture." *Science in Context* 10:63–83.

Latour, Bruno, and Michel Callon. "Don't Throw the Baby Out with the Bath School! Reply to Collins and Yearley." In Pickering 1992, 343–68.

Latour, Bruno, and Steve Woolgar. 1979. *Laboratory Life: The Social Construction of Scientific Facts.* London: Sage.

Laudan, Larry. 1977. *Progress and Its Problems.* Berkeley and Los Angeles: University of California Press.

———. 1981. *Science and Hypothesis.* Dordrecht: Reidel.

———. 1982. "Science at the Bar: Causes for Concern." *Science, Technology, and Human Values* 7:16–19.

Laudan, Rachel. 1993. "Histories of Science and Their Uses: A Review to 1913." *History of Science* 31:1–34.

Lave, Jean, and Edward Wenger. 1991. *Situated Learning.* Cambridge: Cambridge University Press.

Leavis, F. R. 1963. *Two Cultures? The Significance of C. P. Snow.* London: Chatto and Windus.

Lecourt, Dominique. 1975. *Marxism and Epistemology: Bachelard, Canguilhem, Foucault.* London: Verso.

———. 1992. "The Scientist and the Citizen: A Critique of Technoscience." *Philosophical Forum* 23:174–78.

Leplin, Jarrett, ed. 1984. *Scientific Realism.* Berkeley and Los Angeles: University of California Press.

———. 1997. *A Novel Defense of Scientific Realism.* Oxford: Oxford University Press.
Leslie, Stuart. 1993. *The Cold War and American Science: The Military-Industrial-Academic Complex at MIT and Stanford.* New York: Columbia University Press.
Levin, Harry. 1969. "Two *Romanisten* in America." In *The Intellectual Migration: Europe and America, 1930–1960,* ed. D. Fleming and B. Bailyn, 467–83. Cambridge: Harvard University Press.
Levine, John. 1989. "Reaction to Opinion Deviance in Small Groups." In *The Psychology of Group Influence,* ed. P. Paulus, 187–233. 2d ed. Hillsdale, N.J.: Lawrence Erlbaum Associates. 1st ed. 1980.
Levins, Richard, and Richard Lewontin. 1985. *The Dialectical Biologist.* Cambridge: Harvard University Press.
Lévi-Strauss, Claude. 1966. *The Savage Mind.* Chicago: University of Chicago Press. First published in 1962.
Lewis, C. I. 1929. *Mind and World-Order.* New York: Scribners.
———. 1970. "The Pragmatic Element in Knowledge" (1926). In *The Collected Papers of C. I. Lewis,* ed. J. D. Goheen and J. L. Mothershead, 240–57. Palo Alto: Stanford University Press.
Lewontin, Richard. 1997. "The Cold War and the Transformation of the Academy." In Chomsky et al. 1997, 1–34.
List, Friedrich. 1904. *The National System of Political Economy.* Trans. S. S. Lloyd. London: Longmans. First published in 1845.
Locke, John. 1959. "Epistle to the Reader." In *An Essay Concerning Human Understanding,* 1:3–16. New York: Dover. First published in 1690.
Longino, Helen. 1990. *Science as Social Knowledge.* Princeton: Princeton University Press.
Luhmann, Niklas. 1983. *The Differentiation of Society.* New York: Columbia University Press.
Lynch, Michael. 1993. *Scientific Practice and Ordinary Action: Ethnomethodology and Social Studies of Science.* Cambridge: Cambridge University Press.
Lyotard, Jean-François. 1983. *The Postmodern Condition.* Trans. G. Bennington. Minneapolis: University of Minnesota Press. First published in 1979.
MacCannell, Dean. 1984. "Baltimore in the Morning . . . After: On the Forms of Post-nuclear Leadership." *Diacritics,* summer, 33–45.
Mach, Ernst. 1960. *The Science of Mechanics: A Critical and Historical Account of Its Development.* 6th U.S. ed., based on 9th German ed. (1933). La Salle, Ill.: Open Court Press. 1st ed. 1883.
Machamer, Peter. 1975. "Understanding Scientific Change." *Studies in History and Philosophy of Science* 5:373–81.
MacIntyre, Alasdair. 1970. "Is Understanding Religion Compatible with Believing?" In Wilson 1970, 62–77.
———. 1984. *After Virtue.* 2d ed. South Bend: University of Notre Dame Press. 1st ed. 1981.
———. 1990. *Three Rival Versions of Moral Inquiry.* London: Duckworth.
MacKenzie, Donald. 1995. *Knowing Machines.* Cambridge: MIT Press.
MacLeod, Roy. 1971a. "Of Medals and Men: A Reward System in Victorian Science 1826–1914." *Notes and Records of the Royal Society of London* 26:81–105.

———. 1971b. "The Royal Society and the Government Grant: Notes on the Administration of Scientific Research, 1849–1914." *Historical Journal* 14:323–58.
Mandelbaum, Maurice. 1971. *History, Man, and Reason: A Study in the Nineteenth Century.* Baltimore: Johns Hopkins University Press.
Manicas, Peter. 1966. "On Toulmin's Contribution to Logic and Argumentation." *Journal of the American Forensic Association* 3:83–94.
———. 1986. *A History and Philosophy of the Social Sciences.* Oxford: Blackwell.
Mannheim, Karl. 1936. *Ideology and Utopia.* Trans. L. Wirth and E. Shils. New York: Harcourt, Brace & World. First published in 1929.
———. 1940. *Man and Society in an Age of Reconstruction.* London: Routledge & Kegan Paul.
Marcus, George, and Michael Fischer. 1986. *Anthropology as Cultural Critique.* Chicago: University of Chicago Press.
Margolis, Howard. 1987. *Patterns, Thinking, and Cognition.* Chicago: University of Chicago Press.
Marx, Karl. 1970. *The German Ideology.* New York: International Publishers. First published in 1845.
Massey, Marilyn Chapin. 1983. *Christ Unmasked: The Meaning of "The Life of Jesus" for German Politics.* Chapel Hill: University of North Carolina Press.
Matthews, J. Rosser. 1995. *Quantification and the Quest for Medical Certainty.* Princeton: Princeton University Press.
Matthews, Michael. 1994. *Science Education: The Role of the History and Philosophy of Science.* London: Routledge.
Mayr, Ernst. 1994. "The Advance of Science and Scientific Revolutions." *Journal of the History of the Behavioral Sciences* 30:328–34.
McGuire, W., and D. Papageorgis. 1961. "The Relative Efficacy of Various Prior Belief-Defense in Producing Immunity against Persuasion." *Journal of Abnormal and Social Psychology* 62:327–37.
McMullin, Ernan. 1997. "Galileo on Science and Scripture." In *The Cambridge Companion to Galileo,* ed. P. Machamer, 271–347. Cambridge: Cambridge University Press.
Medawar, Peter. 1969. *The Art of the Soluble.* Harmondsworth, U.K.: Penguin.
Medhurst, Martin. 1994. "Reconceptualizing Rhetorical History: Eisenhower's Farewell Address." *Quarterly Journal of Speech* 80:195–218.
Meister, Robert. 1991. *Political Identity: Thinking through Marx.* Oxford: Blackwell.
Melucci, Alberto. 1996. *Challenging Codes: Collective Action in the Information Age.* Cambridge: Cambridge University Press.
Mendelsohn, Everett. 1989. "Robert K. Merton: The Celebration and Defense of Science." *Science in Context* 3:269–99.
Merton, Robert. 1942. "Science and Technology in a Democratic Social Order." *Journal of Legal and Political Sociology* 1:115–26.
———. 1970. *Science, Technology, and Society in Seventeenth Century England.* 2d ed. New York: Harper & Row. 1st ed. 1938.
———. 1973. *The Sociology of Science.* Chicago: University of Chicago Press.
———. 1977. *The Sociology of Science: An Episodic Memoir.* Carbondale: Southern Illinois University Press.

Merz, John T. 1965. *A History of European Thought in the 19th Century.* 4 vols. New York: Dover. First published in 1896–1914.
Midgley, Mary. 1992. *Science as Salvation.* London: Routledge & Kegan Paul.
Miller, David. 1972. "Back to Aristotle?" *British Journal for the Philosophy of Science* 23:69–78.
Miller, J. Hillis. 1990. "Narrative." In *Critical Terms for Literary Study,* ed. F. Lentricchia and T. McLaughlin, 66–79. Chicago: University of Chicago Press.
Miller, Jonathan. 1984. *States of Mind.* New York: Pantheon.
Mills, Clarence. 1948. "Distribution of American Research Funds." *Science* 107: 127–30.
Mills, C. Wright. 1956. *The Power Elite.* Oxford: Oxford University Press.
———. 1958. *The Causes of World War Three.* New York: Ballantine.
———. 1959. *The Sociological Imagination.* Oxford: Oxford University Press.
Minutes of the Committee on General Education. 1946–58. Harvard University Archives.
Mirowski, Philip. 1989. *More Heat than Light.* Cambridge: Cambridge University Press.
———. 1996. "A Visible Hand in the Marketplace of Ideas: Precision Measurement as Arbitrage." In *Accounting and Science,* ed. M. Power, 219–46. Cambridge: Cambridge University Press.
Montgomery, Scott. 1994. *Minds for the Making: The Role of Science in American Education, 1750–1990.* New York: Guilford Press.
———. 1995. *The Scientific Voice.* New York: Guilford Press.
Morris, Aldon, and Carol Mueller, eds. 1992. *Frontiers in Social Movement Theory.* New Haven: Yale University Press.
Mueller, Detlef. 1987. "The Process of Systematizaton: The Character of German Secondary Education." In Mueller, Ringer, and Simon 1987, 15–52.
Mueller, Detlef, Fritz Ringer, and Brian Simon, eds. 1987. *The Rise of the Modern Educational System: Structural Change and Social Reproduction, 1870–1920.* Cambridge: Cambridge University Press.
Mulkay, Michael. 1979a. "Knowledge and Utility: Implications for the Sociology of Knowledge." *Social Studies of Science* 9:69–74.
———. 1979b. *Science and the Sociology of Knowledge.* London: George Allen & Unwin.
Munevar, Gonzalo, ed. 1991. *Beyond Reason: Essays on the Philosophy of Paul K. Feyerabend.* Dordrecht: Reidel.
Naess, Arne. 1965. *Four Modern Philosophers.* Chicago: University of Chicago Press.
———. 1991. "Paul Feyerabend—A Green Hero?" In Munevar 1991, 403–16.
Nash, Leonard. 1952. "The Use of Historical Cases in Science Teaching." In Cohen and Watson 1952, 110–21.
Nelson, John. 1974. "Once More on Kuhn." *Political Methodology* 1 (spring): 73–104.
Nelson, Rodney. 1995. "Pragmatic Validity: Mannheim and Dewey." *History of the Human Sciences* 8:25–46.
Nicholson, Linda, ed. 1994. *Feminist Contentions.* London: Routledge.

Nickles, Thomas. 1980a. "Can Scientific Constraints Be Rationally Violated?" In Nickles 1980b, 1:285–315.
———, ed. 1980b. *Scientific Discovery*. 2 vols. Dordrecht: D. Reidel.
Noble, Douglas. 1991. *The Classroom Arsenal*. London: Falmer Press.
Notturno, Mark. 1999. *Science and the Open Society*. Budapest: Central European University Press.
Novick, Peter. 1988. *That Noble Dream: The "Objectivity Question" and the American Historical Profession*. Cambridge: Cambridge University Press.
Oakley, Francis. 1992. *Community of Learning*. Oxford: Oxford University Press.
Ober, Josiah. 1989. *Mass and Elite in Democratic Athens*. Princeton: Princeton University Press.
O'Connor, James R. 1973. *The Fiscal Crisis of the State*. New York: St. Martin's.
O'Donohue, William. 1993. "The Spell of Kuhn on Psychology: An Exegetical Elixir." *Philosophical Psychology* 6:267–87.
Office of Technology Assessment. 1991. *Federally Funded Research: Decisions for a Decade*. Washington: U.S. Government Printing Office.
Olby, R. C., G. N. Cantor, J. Christie, and M. J. S. Hodge, eds. 1990. *Companion to the History of Modern Science*. London: Routledge.
Olesko, Kathryn. 1993. "Tacit Knowledge and School Formation." *Osiris*, n.s., 8:16–29.
Olson, Keith. 1974. *The G.I. Bill, the Veterans, and the Colleges*. Lexington: University of Kentucky Press.
Olson, Mancur. 1965. *The Logic of Collective Action: Public Goods and the Theory of Groups*. Cambridge: Harvard University Press.
Ophir, Adi, and Steven Shapin. 1991. "The Place of Knowledge: A Methodological Survey." *Science in Context* 4:3–21.
Ostwald, Wilhelm. 1910. *Elements of Natural Philosophy*. New York: Henry Holt.
Owens, Larry. 1990. "Vannevar Bush." In *Dictionary of Scientific Biography*, ed. F. L. Holmes, 17:134–39. New York: Charles Scribner's Sons.
Parayil, Govindan. 1992. "Review of Yearley." *Social Epistemology* 6:57–64.
Parens, Joshua. 1995. *Metaphysics as Rhetoric: Alfarabi's "Summary of Plato's 'Laws.'"* Albany: SUNY Press.
Parsons, Talcott. 1937. *The Structure of Social Action*. New York: Harper & Row.
———. 1951. *The Social System*. New York: Free Press.
Passmore, John. 1966. *A Hundred Years of Philosophy*. 2d ed. Harmondsworth, U.K.: Penguin.
Paul, Harry. 1985. *From Knowledge to Power: The Rise of the Science Empire in France, 1860–1939*. Cambridge: Cambridge University Press.
Pavitt, Keith. 1991. "What Makes Basic Research Economically Useful?" *Research Policy* 20:109–19.
Pettit, Philip. 1997. *Republicanism*. Oxford: Oxford University Press.
Pfetsch, Frank. 1979. "The 'Finalization' Debate in Germany: Some Comments and Explanations." *Social Studies of Science* 9:115–24.
Piaget, Jean. 1952. *The Origins of Intelligence in Children*. New York: International Universities Press.
———. 1970. *Structuralism*. New York: Basic Books.

Pickering, Andrew, ed. 1992. *Science as Practice and Culture*. Chicago: University of Chicago Press.
———. 1995. *The Mangle of Practice*. Chicago: University of Chicago Press.
Pigou, Arthur. 1921. *The Political Economy of the War*. London: Macmillan.
Pinch, Trevor. 1988. "Reservations about Reflexivity and New Literary Forms, or Why Let the Devil Have All the Good Tunes?" In Woolgar 1988, 178–97.
Pine, Martin. 1973. "Double Truth." In *Dictionary of the History of Ideas*, ed. P. Wiener, 2:31–37. New York: Charles Scribners.
Pocock, John. 1973. *Politics, Language, and Time*. New York: Atheneum.
———. 1985. *Virtue, Commerce, and History*. Cambridge: Cambridge University Press.
Polanyi, Karl. 1944. *The Great Transformation*. Boston: Beacon Press.
Polanyi, Michael. 1958. *Personal Knowledge*. Chicago: University of Chicago Press.
———. 1962. "The Republic of Science: Its Political and Economic Theory." *Minerva* 1:54–73.
———. 1963. "Comment on Kuhn." In Crombie 1963, 377–82.
———. 1967. "The Growth of Science in Society." *Minerva* 5:533–45.
———. 1969. "The Social Determinants of Action." In *Roads to Freedom: Essays in Honor of F. A. von Hayek*, ed. E. Streissler, 165–79. New York: Augustus Kelley.
———. 1974. "On the Modern Mind." In *Scientific Thought and Social Reality*, 131–49. New York: International Univerrsity Press.
Pollner, Melvin. 1987. *Mundane Reason*. Cambridge: Cambridge University Press.
Popper, Karl. 1945. *The Open Society and Its Enemies*. 2 vols. New York: Harper & Row.
———. 1957. *The Poverty of Historicism*. London: Routledge & Kegan Paul.
———. 1959. *The Logic of Scientific Discovery*. New York: Harper & Row.
———. 1963. *Conjectures and Refutations*. New York: Harper & Row.
———. 1970. "Normal Science and Its Dangers." In Lakatos and Musgrave 1970, 51–58.
———. 1972. *Objective Knowledge: An Evolutionary Approach*. Oxford: Oxford University Press.
———. 1974. "Kuhn on the Normality of Normal Science." In *The Philosophy of Karl Popper: The Library of Living Philosophers*, ed. P. A. Schilpp, 1144–48. La Salle, Ill.: Open Court Press.
———. 1975. "The Rationality of Scientific Revolutions." In *Problems of Scientific Revolution*, ed. R. Harre, 72–101. Oxford: Oxford University Press.
———. 1994. *The Myth of the Framework*. London: Routledge. First published in 1965.
Porter, Theodore. 1986. *The Rise of Statistical Thinking, 1820–1900*. Princeton: Princeton University Press.
———. 1995. *Trust in Numbers*. Princeton: Princeton University Press.
Post, Heinz. 1971. "Correspondence, Invariance, and Heuristics." *Studies in History and Philosophy of Science* 2:213–55.
Postman, Neil, and Charles Weingartner. 1969. *Teaching as a Subversive Activity*. New York: Dell.
Prendergast, Christopher. 1986. "Alfred Schutz and the Austrian School of Economics." *American Journal of Sociology* 92:1–26.

Price, Colin. 1993. *Time, Discounting, and Value*. Oxford: Blackwell.
Price, Derek de Solla. 1978. "Toward a Model for Science Indicators." In *Toward a Metric of Science*, ed. Y. Elkana et al., 69–96. New York: Wiley-Interscience.
———. 1986. *Big Science, Little Science, and Beyond*. New York: Columbia University Press. First published in 1963.
Price, Don K. 1965. *The Scientific Estate*. Cambridge: Harvard University Press.
Prigogine, Ilya, and Isabelle Stengers. 1984. *Order out of Chaos*. New York: Bantam Books.
Proctor, Robert. 1991. *Value-Free Science? Purity and Power in Modern Knowledge*. Cambridge: Harvard University Press.
———. 1999. *The Nazi War on Cancer*. Princeton: Princeton University Press.
Purcell, Edward. 1973. *The Crisis of Democratic Theory: Scientific Naturalism and the Problem of Value*. Lexington, Ky.: University of Kentucky Press.
Putnam, Hilary. 1984. "What Is Realism?" In Leplin 1984, 140–53.
Pyenson, Lewis. 1990. "Science and Imperialism." In Olby et al. 1990, 920–33.
———. 1993. "Prerogatives of European Intellect: History of Science and the Promotion of Western Civilization." *History of Science* 31:289–315.
Quine, W. V. 1953. *From a Logical Point of View*. New York: Harper & Row.
———. 1960. *Word and Object*. Cambridge: MIT Press.
———. 1985. *The Time of My Life*. Cambridge: MIT Press.
Quine, W. V., and Joseph Ullian. 1978. *The Web of Belief*. 2d ed. New York: Random House. 1st ed. 1970.
Rabinbach, Anson. 1990. *The Human Motor: Energy, Fatigue, and the Origins of the Modernity*. New York: Basic Books.
Radder, Hans. 1992. "Normative Reflections on Constructivist Approaches to Science and Technology." *Social Studies of Science* 22:141–73.
———. 1996. *In and about the World*. Albany: SUNY Press.
———. 1998. "The Politics of STS." *Social Studies of Science* 28:325–32.
Raj, Kapil. 1988. "Images of Knowledge, Social Organization, and Attitudes in an Indian Physics Department." *Science in Context* 2:317–89.
Ralston, David. 1990. *Importing the European Army: The Introduction of the European Military Techniques and Institutions into the Extra-European World, 1600–1914*. Chicago: University of Chicago Press.
Rapaport, Anatol. 1989. *The Origins of Violence*. New York: Paragon House.
Ravetz, Jerome. 1971. *Scientific Knowledge and Its Social Problems*. Oxford: Oxford University Press.
———. 1987. "Usable Knowledge, Usable Ignorance: Incomplete Science with Policy Implications." *Knowledge* 9:87–116.
———. 1990a. *The Merger of Knowledge with Power: Essays in Critical Science*. London: Mansell.
———. 1990b. "Orthodoxies, Critiques, and Alternatives." In Olby et al. 1990, 898–908.
———. 1991. "Ideological Commitments in the Philosophy of Science." In Munevar 1991, 355–78.
Rawls, John. 1955. "Two Concepts of Rules." *Philosophical Review* 64:3–32.
Redner, Harry. 1987. *The Ends of Science*. Boulder: Westview Press.

Rehg, William. 1999. "Argumentation and the Philosophy of Science since Kuhn." *Inquiry* 42:229–58.
Reich, Robert. 1990. *The Work of Nations*. New York: Alfred Knopf.
Reichenbach, Hans. 1938. *Experience and Prediction*. Chicago: University of Chicago Press.
Reingold, Nathan. 1991. *Science American Style*. New Brunswick: Rutgers University Press.
———. 1994. "Science and Government in the United States since 1945." *History of Science* 32:361–86.
Reisch, George. 1991. "Did Kuhn Kill Logical Empiricism?" *Philosophy of Science* 58:264–77.
Rescher, Nicholas. 1984. *The Limits of Science*. Berkeley and Los Angeles: University of California Press.
Restivo, Sal. 1983. "The Myth of the Kuhnian Revolution." In *Sociological Theory*, ed. R. Collins, 293–305. San Francisco: W. H. Freeman.
———. 1985. *The Social Relations of Physics, Mysticism, and Mathematics*. Dordrecht: Kluwer.
Restivo, Sal, and Julia Loughlin. 1987. "Critical Sociology of Science and Scientific Validity." *Knowledge* 8:486–508.
Richards, I. A. 1936. *The Philosophy of Rhetoric*. Oxford: Oxford University Press.
Ringer, Fritz. 1979. *Education and Society in Modern Europe*. Bloomington: Indiana University Press.
Ritzer, George. 1975. *Sociology: A Multiple Paradigm Science*. Boston: Allyn & Bacon.
Rocke, Alan. 1993. "Group Research in German Chemistry." *Osiris* 8:53–79.
Rorty, Richard. 1972. "The World Well Lost." *Journal of Philosophy* 69:649–65.
———. 1979. *Philosophy and the Mirror of Nature*. Princeton: Princeton University Press.
———. 1995. "The End of Leninism and History as Comic Frame." In *History and the Idea of Progress*, ed. A. Meltzer, J. Weinberger, and M. Zinman, 211–26. Ithaca: Cornell University Press.
Rose, Hilary, and Steven Rose. 1970. *Science and Society*. Harmondsworth, U.K.: Penguin.
Rosen, Edward. 1970. "Was Copernicus a Hermetist?" In Stuewer 1970, 163–71.
Rosenbaum, E., and A. J. Sherman. 1979. *M. M. Warburg & Co. 1798–1938: Merchant Bankers of Hamburg*. London: C. Hurst & Co.
Rosenthal, Joel. 1991. *Righteous Realists: Political Realism, Responsible Power, and American Culture in the Nuclear Age*. Baton Rouge: Louisiana State University Press.
Ross, Andrew, ed. 1996. *Science Wars*. Durham: Duke University Press.
Ross, Dorothy. 1991. *The Origins of American Social Science*. Cambridge: Cambridge University Press.
Ross, Stephen David. 1989. *Metaphysical Aporia and Philosophical Heresy*. Albany, N.Y.: SUNY Press.
Rossiter, Margaret. 1984. "The History and Philosophy of Science Program at the National Science Foundation." *Isis* 75:95–104.

Rostow, Walt. 1960. *The Stages of Economic Growth: A Non-communist Manifesto.* Cambridge: Cambridge University Press.
Roszak, Theodore, ed. 1967. *The Dissenting Academy.* New York: Random House.
———. 1969. *The Making of a Counter Culture: Reflections on the Technocratic Society and Its Youthful Opposition.* Garden City, N.Y.: Doubleday.
———. 1972. *Where the Wasteland Ends: Politics and Transformation in Postindustrial Society.* Garden City, N.Y.: Doubleday.
Roth, Paul. 1987. *Meaning and Method in the Social Sciences.* Ithaca: Cornell University Press.
Rouse, Joseph. 1987. *Knowledge and Power.* Ithaca: Cornell University Press.
———. 1991. "The Politics of Postmodern Philosophy of Science." *Philosophy of Science* 58:607–27.
———. 1996. *Engaging Science.* Ithaca: Cornell University Press.
Rubin, Joan Shelley. 1992. *The Making of Middle Brow Culture.* Chapel Hill: University of North Carolina Press.
Rueschemeyer, Dietrich, and Theda Skocpol, eds. 1996. *States, Social Knowledge, and the Origins of Modern Social Policies.* Princeton: Princeton University Press.
Ruse, Michael. 1979. *The Darwinian Revolution: Science Red in Tooth and Claw.* Chicago: University of Chicago Press.
Said, Edward. 1978. *Orientalism.* New York: Random House.
Santayana, George. 1905. *The Life of Reason.* Vol. 1. *Reason and Commonsense.* New York: Scribners.
———. 1936. *Obiter Scripta.* New York: Scribners.
Sardar, Ziauddin. 1997. "The Return of the Repressed." *Nature* 389:451–52.
Sarton, George. 1948. *The Life of Science: Essays in the History of Civilization.* New York: Henry Schuman.
Sassower, Raphael. 1985. *Philosophy of Economics: A Critique of Demarcation.* Lanham, Md.: University Press of America.
Schaefer, Wolf, ed. 1984. *Finalization in Science.* Dordrecht: Reidel.
Schaffer, Simon. 1991. "The History and Geography of the Intellectual World: Whewell's Politics of Language." In Fisch and Schaffer 1991, 201–32.
———. 1996. "Contextualizing the Canon." In Galison and Stump 1996, 207–30.
Schaffner, Kenneth. 1967. "Approaches to Reduction." *Philosophy of Science* 34:137–47.
Schapiro, Meyer. 1953. "Style." In *Anthropology Today,* ed. Alfred Kroeber, 278–303. Chicago: University of Chicago Press.
Scheffler, Israel. 1963. *The Anatomy of Inquiry.* Indianapolis: Hackett.
———. 1967. *Science and Subjectivity.* Indianapolis: Bobbs-Merrill.
Schilpp, Paul, ed. 1968. *The Philosophy of C. I. Lewis.* Vol. 13 of *The Library of Living Philosophers.* La Salle, Ill.: Open Court Press.
Schlick, Moritz. 1974. *The General Theory of Knowledge.* Berlin: Springer-Verlag. First published in 1925.
Schmitt, Carl. 1996. *The Concept of the Political.* Trans. G. Schwab. Chicago: University of Chicago Press. First published in 1932.
Schmitt, Frederick, ed. 1994. *Socializing Epistemology.* Lanham, Md.: Rowman & Littlefield.

Schnaedelbach, Herbert. 1984. *Philosophy in Germany, 1831–1933*. Cambridge: Cambridge University Press.
Schroeder-Gudehus, Brigitte. 1990. "Nationalism and Internationalism." In Olby et al. 1990, 909–19.
Schumpeter, Joseph. 1950. *Capitalism, Socialism, and Democracy*. 2d ed. New York: Harper & Row. 1st ed. 1945.
Schutz, Alfred. 1964. *Collected Papers*. Vol. 2. The Hague: Martinus Nijhoff.
Scott, Robert, and Arnold Shore. 1979. *Why Sociology Does Not Apply: A Study of the Use of Sociology in Public Policy*. New York: Elsevier.
Searle, John. 1983. "The World Turned Upside Down." *New York Review of Books* 30, no. 16: 74–79.
Senghaas, Dieter. 1991. "Friedrich List and the Basic Problems of Modern Development." *Review* 14:451–67.
Serres, Michel. 1980. *Le parasite*. Paris: Grasset.
Serres, Michel, and Bruno Latour. 1995. *Conversations on Science, Culture, and Time*. Ann Arbor: University of Michigan Press.
Servos, John. 1993. "Research Schools and Their Histories." *Osiris*, n.s., 8:3–15.
Shapere, Dudley. 1960. "Mathematical Ideals and Metaphysical Concepts." *Philosophical Review* 69:376–85.
———. 1963. "Space, Time, and Language." In *Philosophy of Science: The Delaware Seminar*, ed. B. Baumrin, 139–70. New York: Interscience Publishers.
———. 1964. "The Structure of Scientific Revolutions." *Philosophical Review* 73: 383–94.
———. 1966. "Meaning and Scientific Change." In *Mind and Cosmos*, ed. R. Colodny, 41–85. Pittsburgh: University of Pittsburgh Press.
Shapin, Steven. 1992a. "Discipline and Bounding: The History and Sociology of Science as Seen through the Externalism-Internalism Debate." *History of Science* 30:333–69.
———. 1992b. "A Magician's Cloak Cast Off for Clarity." *The Times Higher Education Supplement* (London, 14 February), 15.
———. 1992c. "Why the Public Ought to Understand Science-in-the-Making." *Public Understanding of Science* 1:27–30.
———. 1994. *A Social History of Truth*. Chicago: University of Chicago Press.
———. 1996. *The Scientific Revolution*. Chicago: University of Chicago Press.
Shapin, Steven, and Simon Schaffer. 1985. *Leviathan and the Air-Pump*. Princeton: Princeton University Press.
Shapley, Deborah. 1993. *Promise and Power: The Life and Times of Robert MacNamara*. Boston: Little, Brown.
Shils, Edward, ed. 1968. *Criteria for Scientific Development*. Cambridge: MIT Press.
———. 1997. *Portraits: A Gallery of Intellectuals*. Chicago: University of Chicago Press.
Siebers, Tobin. 1993. *Cold War Criticism and the Politics of Skepticism*. Oxford: Oxford University Press.
Siegel, Harvey. 1991. *Educating Reason*. London: Routledge.
Sigurdsson, Skuli. 1990. "The Nature of Scientific Knowledge: An Interview with Thomas Kuhn." *Harvard Science Review*, winter, 18–25.

———. 1992. "Einsteinian Fixations." *Annals of Science* 49:577–83.
Simon, Brian. 1987. "Systematization and Segmentation in Education: The Case of England." In Mueller, Ringer, and Simon 1987, 88–110.
Simon, Herbert. 1991. *Models of My Life*. New York: Basic Books.
Singer, Peter. 1975. *Animal Liberation*. New York: Random House.
Skinner, Quentin. 1998. *Liberty before Liberalism*. Cambridge: Cambridge University Press.
Smith, Michael Joseph. 1986. *Realist Thought from Weber to Kissinger*. Baton Rouge: Louisiana State University Press.
Smith, Wilfred Cantrell. 1977. *Belief and History*. Charlottesville: University of Virginia Press.
Smith, Woodruff. 1991. *Politics and the Sciences of Culture in Germany, 1840–1920*. Oxford: Oxford University Press.
Smocovitis, V. Betty. 1995. "Contextualizing Science: From Science Studies to Cultural Studies." In *PSA 1994*, ed. D. Hull, M. Forbes, and R. Burian, 2:402–12. East Lansing: Philosophy of Science Association.
Snow, C. P. 1959. *The Two Cultures and the Scientific Revolution*. Cambridge: Cambridge University Press.
Soederqvist, Thomas, ed. 1997. *The Historiography of Contemporary Science and Technology*. Amsterdam: Harwood Academic Publishers.
Sokal, Alan. 1996. "Transgressing the Boundaries: Towards a Transformative Hermeneutics of Quantum Gravity." *Social Text* 46/47:217–52.
Sokal, Alan, and Jean Bricmont. 1998. *Intellectual Impostures*. London: Phaidon.
Sorell, Tom. 1992. *Scientism*. London: Routledge.
Soros, George. 1998. *The Crisis of Global Capitalism: Open Society Endangered*. London: Little, Brown.
Sowell, Thomas. 1980. *Knowledge and Decisions*. New York: Harper & Row.
Sperber, Dan. 1982. "Apparently Irrational Beliefs." In Hollis and Lukes 1982, 149–80.
———. 1996. *Explaining Culture*. Oxford: Blackwell.
Spiegel-Roesing, Ina, and Derek de Solla Price, eds. 1977. *Science, Technology, and Society: A Cross-Disciplinary Perspective*. London: Sage.
Stalker, Douglas, ed. 1994. *Grue!: The New Riddle of Induction*. La Salle, Ill.: Open Court Press.
Star, Susan Leigh, ed. 1995. *Ecologies of Knowledge: Work and Politics in Science and Technology*. Albany: SUNY Press.
Stehr, Nico. 1994. *Knowledge Societies*. London: Sage.
Stinchcombe, Arthur. 1990. *Information and Organizations*. Berkeley and Los Angeles: University of California Press.
Stocking, George. 1968. *Race, Culture, and Evolution*. Chicago: University of Chicago Press.
Stove, David. 1982. *Popper and After*. Oxford: Pergamon Press.
Strauss, Leo. 1952. *Persecution and the Art of Writing*. New York: Free Press.
Stuewer, R., ed. 1970. *Historical and Philosophical Perspectives on Science*. Minneapolis: University of Minnesota Press.
Sulloway, Frank. 1996. *Born to Rebel*. New York: Pantheon Books.

Suppe, Fred, ed. 1977. *The Structure of Scientific Theories*. 2d ed. Urbana: University of Illinois Press. 1st ed. 1973.
Taylor, Charles A. 1982. "Rationality." In Hollis and Lukes 1982, 87–105.
———. 1996. *Defining Science*. Madison: University of Wisconsin Press.
Thackray, Arnold, and Robert Merton. 1975. "George Sarton." In *Dictionary of Scientific Biography*, ed. C. Gillispie, 12:107–14. New York: Charles Scribners.
Thayer, H. S. 1968. *Meaning and Action: A Critical History of Pragmatism*. Indianapolis: Bobbs-Merrill.
Thompson, Michael, Richard Ellis, and Aaron Wildavsky. 1990. *Cultural Theory*. Boulder, Colo.: Westview Press.
Tibbetts, Paul. 1975. "On a Proposed Paradigm Shift in the Social Sciences." *Philosophy of the Social Sciences* 5:289–96.
Time. 1973. "Reaching beyond the Rational." 23 April, 83–86.
Tompkins, Jane. 1980. "The Reader in History: The Changing Shape of Literary Response." In *Reader-Response Criticism*, ed. J. Tompkins, 201–32. Baltimore: Johns Hopkins University Press.
Toulmin, Stephen. 1951. *Philosophy of Science: An Introduction*. London: Hutchinson.
———. 1958. *The Uses of Argument*. Cambridge: Cambridge University Press.
———. 1959. "Concerning the Philosophy Which Holds That the Conclusions of Science Are Never Final." *Scientific American*, May, 189–96.
———. 1961. *Foresight and Understanding*. Bloomington: Indiana University Press.
———. 1964. "The Complexity of Scientific Choice: A Stocktaking." *Minerva* 2:343–59.
———. 1966. "The Complexity of Scientific Choice: Culture, Overheads, or Tertiary Industry?" *Minerva* 4:155–69.
———. 1972. *Human Understanding*. Princeton: Princeton University Press.
———. 1977. "From Form to Function: Philosophy and History of Science in the 1950s and Now." *Daedalus* 106, no. 3:143–62.
———. 1986. *The Place of Reason in Ethics*. 2d ed. Chicago: University of Chicago Press. 1st ed. 1950.
———. 1990. *Cosmopolis: The Hidden Agenda of Modernity*. New York: Free Press.
Tully, James, ed. 1988. *Meaning and Context: Quentin Skinner and His Critics*. Princeton: Princeton University Press.
Turner, Stephen. 1994. *The Social Theory of Practices*. Chicago: University of Chicago Press.
Turner, Stephen, and Jonathan Turner. 1990. *The Impossible Science: An Institutional Analysis of American Sociology*. Newbury Park, Calif.: Sage.
Turney, Jon, ed. 1984. *Sci-Tech Report: Everything You Need to Know about Science and Technology in the 80s*. New York: Pantheon.
Uebel, Thomas. 1996. "Anti-foundationalism and the Vienna Circle's Revolution in Philosophy." *British Journal for the Philosophy of Science* 47:415–40.
Urry, John. 1995. *Consuming Places*. London: Routledge.
van Eemeren, F., R. Grootendorst, and T. Kruiger, eds. 1987. *Handbook of Argumentation Theory*. Amsterdam: Floris.

Van Vleck, John. 1962. *The So-Called Age of Science*. Cherwell Simon Memorial Lecture. London: Oliver & Boyd.
Wagner, Peter, Bjorn Wittrock, and Richard Whitley, eds. 1991. *Discourses on Society: The Shaping of the Social Science Disciplines*. Dordrecht: Kluwer.
Wainwright, Hilary, and Dave Elliott. 1982. *The Lucas Plan: A New Trade Unionism in the Making?* London: Allison and Busby.
Wallace, William. 1983. "Aristotelian Influences on Galileo's Thought." In *Aristotelismo veneto e scienza moderna*, ed. Luigi Oliveri, 349–403. Padua, Italy: Atenore.
Wallerstein, Immanuel. 1996. *Open the Social Sciences*. Palo Alto: Stanford University Press.
Waxman, Chaim, ed. 1968. *The End of Ideology Debate*. New York: Funk and Wagnalls.
Weatherford, Jack. 1993. "Early Andean Experimental Agriculture." In *The Racial Economy of Science*, ed. S. Harding, 64–83. Bloomington: Indiana University Press.
Weber, Max. 1958. "Science as a Vocation." In *From Max Weber*, ed. H. Gerth and C. W. Mills, 129–58. Oxford: Oxford University Press. First published 1918.
———. 1964. *The Sociology of Religion*. Boston: Beacon Press.
Weinberg, Steven. 1992. *Dreams of a Final Theory*. New York: Pantheon.
———. 1996. "Sokal's Hoax." *New York Review of Books*, 8 August, 11–15.
Weingart, Peter. 1986. "T. S. Kuhn: Revolutionary or Agent Provocateur?" In Deutsch, Markovits, and Platt 1986, 265–77.
Werskey, Gary. 1988. *The Visible College*. 2d ed. London: Free Association Books. 1st ed. 1978.
Westman, Robert. 1994. "Two Cultures or One? A Second Look at Kuhn's *The Copernican Revolution*." *Isis* 85:79–115.
White, Hayden. 1973. *Metahistory: The Historical Imagination in Nineteenth Century Europe*. Baltimore: Johns Hopkins University Press.
White, Morton. 1956. *Toward Reunion in Philosophy*. Cambridge: Harvard University Press.
———. 1957. *Social Thought in America: The Revolt against Formalism*. Boston: Beacon Press.
White, Theodore. 1992. *Theodore H. White at Large: The Best of His Magazine Writing, 1939–1986*. Ed. E. T. Thompson. New York: Pantheon.
Whitehead, Alfred North. 1926. *Science and the Modern World*. Cambridge: Cambridge University Press.
———. 1949. *The Aims of Education*. New York: Mentor Books. First published in 1929.
Whitley, Richard. 1984. *The Intellectual and Social Organization of the Sciences*. Oxford: Clarendon Press.
Will, Frederick. 1988. *Beyond Deduction: Ampliative Aspects of Philosophical Reflection*. London: Routledge.
Willard, Charles Arthur. 1983. *Argumentation and the Social Grounds of Knowledge*. Tuscaloosa: University of Alabama Press.
Williams, Bernard. 1981. *Moral Luck*. Cambridge: Cambridge University Press.

Williams, Howard, David Sullivan, and Gwynn Matthews. 1997. *Francis Fukuyama and the End of History*. Cardiff: University of Wales Press.
Williams, Raymond. 1961. *The Long Revolution*. London: Chatto and Windus.
Wilson, Bryan, ed. 1970. *Rationality*. Oxford: Blackwell.
Winch, Peter. 1958. *The Idea of a Social Science*. London: Routledge & Kegan Paul.
Wise, Norton. 1983. "On the Relation of Physical Science to History in Late 19th Century Germany." In *Functions and Uses of Disciplinary Histories*, ed. L. Graham, W. Lepenies, and P. Weingart, 7:3–34. Dordrecht: D. Reidel.
Wittgenstein, Ludwig. 1922. *Tractatus Logico-Philosophicus*. Trans. D. Pears and B. McGuinness. London: Routledge & Kegan Paul.
——. 1953. *Philosophical Investigations*. Trans. E. Anscombe. London: Macmillan.
Woellflin, Heinrich. 1932. *Principles of Art History*. New York: Dover. First published in 1915.
Woldring, Henk. 1986. *Karl Mannheim: The Development of His Thought*. New York: St. Martin's Press.
Wolin, Sheldon. 1968. "Paradigms and Political Theories." In *Politics and Experience*, ed. P. King and B. Parekh, 125–52. Cambridge: Cambridge University Press.
——. 1986. "History and Theory: Methodism Redivivus." In *Tradition, Interpretation, and Science: Political Theory in the American Academy*, ed. J. Nelson, 43–67. Albany: SUNY Press.
Woodmansee, Martha. 1984. "The Genius and the Copyright." *Eighteenth Century Studies* 17:425–48.
Woolgar, Steve, ed. 1988. *Knowledge and Reflexivity*. London: Sage.
Wrong, Dennis. 1961. "The Oversocialized Conception of Man." *American Sociological Review* 26:184–93.
Wuthnow, Robert. 1989. *Communities of Discourse*. Cambridge: Harvard University Press.
Yates, Frances. 1964. *Giordano Bruno and the Hermetic Tradition*. London: Routledge & Kegan Paul.
——. 1966. *The Art of Memory*. London: Routledge & Kegan Paul.
——. 1975. *Shakespeare's Last Plays*. London: Routledge & Kegan Paul.
Yearley, Steven. 1988. *Science, Technology, and Social Change*. London: Unwin Hyman.
Yeo, Richard. 1993. *Defining Science: William Whewell, Natural Knowledge, and Public Debate in Early Victorian Britain*. Cambridge: Cambridge University Press.
Young, Michael, ed. 1971. *Knowledge and Control*. London: Collier Macmillan.
Young, Robert Maxwell. 1975. "The Historiographic and Ideological Contexts of the Nineteenth-Century Debate on Man's Place in Nature." In *Changing Perspectives in the History of Science: Essays in Honour of Joseph Needham*, ed. M. Teich and R. Young, 344–438. London: Heinemann.
——. 1985. *Darwin's Metaphor*. Cambridge: Cambridge University Press.
Zagacki, Kenneth, and William Keith. 1992. "Rhetoric, *Topoi*, and Scientific Revolution." *Philosophy and Rhetoric* 25:59–78.

Index

academic freedom, 47, 120–21, 129, 160, 253, 413
acritical (noncritical) perspective on science, xv, 315, 317, 347n
activism, 115, 255, 356, 359–60, 405, 409n, 414. *See also* movements, social
actor-network theory (Callon, Latour), 323, 343, 365, 367–68, 372–78
adjudicative function of philosophy, 82, 148, 225n, 297–99. *See also* constitutive function of philosophy; evaluative function of philosophy
agency, 347n, 365, 374–77, 385, 395
alienation, 24, 75, 242–43, 247n, 354
Althusser, Louis, 14n, 344n, 370
analytic-synthetic distinction, 269–70
anarchistic epistemology (Feyerabend), 314n, 410
anthropology, xv, 19n, 20n, 35, 131n, 146–49, 177, 195, 202–4, 236n, 270n, 291n, 296, 329–30, 335n, 342, 351n
Aristotelianism, 64, 100, 143–44, 150n, 153n, 191, 202–5, 302, 346, 390n, 396, 413
Aristotle, 7n, 17, 40, 45n, 51n, 67, 83, 115, 132n, 187, 190n, 202–4, 207, 266, 282, 300, 390n
Aron, Raymond, 62, 105, 223, 423
art history, 33, 45, 52n, 53n, 55–61
Aryan science, 19n, 22, 84. *See also* Jewish science
Athens, 27n, 33, 39–40, 44n, 44–45, 49–52, 69n, 313n
atomic bomb, 5, 43, 66n, 75, 77, 108n, 128n, 141n, 151, 154, 161, 171, 174, 216, 233, 379
autonomy (of science), 4n, 7, 9n, 11, 13n, 16, 33–35, 47, 64–65, 71, 75–78, 93–96, 106n, 111, 129, 139, 142, 151–55, 166, 182, 189, 211, 214, 245, 253n, 256–59, 280, 297, 310, 316n, 318–19, 334, 351–52, 377n, 399, 410, 413, 418n, 422
Ayer, A. J., 42n, 311, 335n, 390n, 391n

Bachelard, Gaston, 12, 14n, 316n, 344–46, 392–94
Barnes, Barry, 178n, 224n, 246n, 319n, 323n, 328
Baudrillard, Jean, 105n, 235n
Beard, Charles, 163, 239n, 241n, 272n
Becker, Howard, 232n, 363–65
Bell, Daniel, 7, 35, 106n, 137, 162, 165n, 174n, 175n, 199n, 233n, 239–48, 368
Bergson, Henri, 68n, 91n, 276n, 356n, 393–94
Bernal, John Desmond, 12, 98n, 151n, 158–64, 252, 324–28, 330n
Big Science (Price), 10, 22, 33, 73, 122n, 125, 138, 157, 201, 217, 222, 225n, 235, 237n, 294, 304–5, 346, 352n
biology, xi, 14, 19, 51n, 55n, 56n, 67, 76, 80–81, 84n, 90, 98, 112, 128, 150n, 151n, 156n, 177–78, 188–93, 253–54, 261–62, 264n, 289n, 299n, 319n, 324n, 346n, 355, 410n, 416n. *See also* evolutionary theory; genetics
Bismarck, Otto von, 106n, 167, 195, 294n
Bloor, David, 246n, 282–83, 319n, 328–35, 343, 391n
Bohr, Niels, 4, 9–10, 66n, 154, 185–86
Bourdieu, Pierre, 236n, 337n, 350, 370, 377n, 383n
Boyle, Robert, 65, 125, 154, 215, 260, 332n, 393n
Bradley, Francis Herbert, 122n, 383n
Brinton, Crane, 164, 168–69, 238
Bruner, Jerome, 17n, 135n, 219–21

463

Brush, Stephen, 24n, 28n, 201n, 221n, 303
Burke, Kenneth, 52n, 313
Bush, Vannevar, 107n, 150–56, 157n, 167
Butterfield, Herbert, 23, 25n, 325n

Callon, Michel, 351, 365–71, 374n, 377n
Cambridge University, 15n, 23, 78, 80, 108n, 157–58, 194, 255, 311, 313n, 324, 325n, 390n
Campbell, Donald, 239n, 358n
Carnap, Rudolf, 13n, 97n, 110, 113n, 119, 211n, 257, 278, 286–88, 292n, 296n, 305n, 391n
Carnegie Foundation, 9n, 153, 217, 220n
Cassirer, Ernst, 12, 68, 52n, 78n, 264n
Catholicism, 8n, 39, 123, 189, 203, 208, 213, 278, 282n, 330n, 404
Cavell, Stanley, 8n, 54n, 68n
chemical revolution (eighteenth century), 101, 181n, 312n
chemistry, 43, 55, 60, 73–76, 99–103, 111, 116, 124, 139–43, 197, 211, 214, 216, 220, 253, 270, 417
Christianity, 23, 25n, 38, 41–42, 48, 79, 122–23, 168, 239n, 282, 409–13, 416n. See also Catholicism; Protestantism
cognitivism, 14n, 16–17, 28–30, 42n, 70n, 89, 113n, 119, 124, 135–41, 148, 175, 177, 184–88, 208, 211, 227, 230n, 264, 271, 280–84, 304, 307n, 313n, 335n, 342n, 351n, 352, 398, 403–4, 412. See also internal history of science
Cohen, I. B., 164n, 206, 219n, 286n, 392
Cold War, 5–8, 10n, 21, 33, 58, 63, 74n, 77, 105–7, 108n, 150, 156, 163n, 169–80, 219–21, 239, 252–54, 258, 319, 324, 332, 391, 410–21
Collingwood, Robin, 69–70, 122n, 312n, 325n
Collins, Harry, 4n, 20n, 86n, 141n, 145n, 322n, 326n, 343, 350, 419–20
Columbia University, 1, 233n, 239–41, 384n
Committee on General Education (Harvard), xiv, 5, 8n, 10–11, 17n, 31–32, 42, 163n, 165, 167n, 173n, 179, 213–21, 224n, 239, 324, 379–84, 391
Communism, 5–6, 63, 107, 168, 171n, 171–74, 179–80, 255, 258, 326, 331, 376, 382
computers, 77, 79, 113–14, 148n, 159n, 174n, 226n, 240–47, 342, 352n, 353, 366–67, 372, 377
Comte, Auguste, 11, 38, 121, 194, 281–82, 291n, 333, 368n, 378
Conant, James Bryant, xiv, xvii, 5–6, 9–11, 17, 31–36, 42–43, 60, 67n, 70, 75–78, 84n, 90–91, 106, 133n, 138, 150–54, 160–65, 168–74, 179–83, 187, 194–204, 212–24, 227–30, 236–37, 239, 256, 286n, 318, 321, 379–84, 386, 391–92, 398
connoisseurship, 54, 215
Constitution (U.S.), 157, 157n, 239n, 297
constitutive function of philosophy, 296–99, 324n. See also adjudicative function of philosophy; evaluative function of philosophy
constructivism, xvi, 99–101, 105n, 207, 232n, 290, 312, 320n, 338, 341–42, 345, 348, 372–73, 393–94, 415
contextualism, 21n, 22, 24, 60n, 69, 93, 97, 247n, 273n, 336–42
contingency, 37, 65, 232n, 268, 279, 338–42, 365
Copernican Revolution, The (Kuhn), xiv n, 5, 173, 183, 219n, 382–83
Copernicus, Nicolaus, 9n, 12n, 81, 187–92, 203n, 207–10, 248, 271–72, 300, 421
Creationism, 121, 262, 360
Criticism and the Growth of Knowledge (Lakatos and Musgrave), xv, 200n, 288, 292
criticism, xv, 46n, 54n, 58, 228, 352n, 357–58, 383n
 of Kuhn, 287n, 289
 lack of, in science, 8n, 177, 172n, 175n, 226n, 305n, 421
 philosophical, 16n, 36, 114n, 141, 263n, 292, 295, 304, 401
 science, 67n, 123, 245, 262, 291, 294, 314n, 323n, 347n, 360n
 social, 51n, 171–72, 316
 of STS, 322n, 323n, 358
 See also Enlightenment (eighteenth century); Popper, Karl
cultural studies, 320n, 355–56
culturopathy, Kuhn's, 398–99
cybernetics, 77, 264

Darwin, Charles, xi, 54n, 57, 76n, 79, 154, 178n, 189n, 191, 243, 246, 256, 289n, 406, 416
Darwinism, 14n, 55n, 76n, 122, 148, 151n, 166n, 177, 178n, 185n, 228n, 256n, 314n, 416–17
Davidson, Donald, 52n, 260n, 273–74
De Gaulle, Charles, 62, 367–69, 372
deconstruction, xv, 3, 26, 57n, 357n, 373n, 415
democracy, xv, 8–11, 40, 43–50, 63–65, 70, 72, 92, 103, 109–11, 130, 137, 155–56, 168, 173–74, 180, 217, 223–26, 239n, 240n, 243, 262, 313n, 325, 327n, 338–39, 383n, 389n, 407, 418, 422
Depression, Great (1929), 106, 155–58, 162–67, 331

INDEX / 465

Derrida, Jacques, 17, 29, 62n, 291n, 356n, 357, 414n
Descartes, René, 12n, 25–26, 64, 81n, 263n, 284, 293n, 399
Deutsch, Karl, 85n, 227n, 418n
Dewey, John, 12, 66, 83n, 212–13, 239n, 263, 266, 272–74, 275n
dirty hands (in ethics), 40, 67n, 171, 218, 236, 237n
discovery, context of, 34, 43, 45n, 81–85, 86n, 89–92, 100, 154, 180, 208, 231, 277n, 288n, 307n, 395, 412, 415–19
disunification (of science), 264–65, 286n
double truth doctrine, 27, 32–33, 41n, 53, 92–93, 104n, 171, 380, 406
Douglas, Mary, 146n, 328–30
Drucker, Peter, 105n, 156n, 165n
Duhem, Pierre, 12, 24, 82n, 84n, 95, 100–2, 125, 187n, 197, 258, 278, 312n, 333, 344n, 352, 359n, 413n
Durkheim, Emile, 104n, 148–49, 165, 211, 282, 333

economics of science, xiv, 9n, 12n, 42, 51n, 66n, 80n, 88, 99n, 104–6, 111–23, 136, 145–48, 153–59, 161n, 163, 222–23, 235–37, 241, 247n, 252–58, 264, 266n, 293–94, 298n, 300n, 304n, 305n, 310, 322–24, 332, 349n, 366–69, 370n, 374, 389n, 402–5, 409n, 421n
Edge, David, 320n, 322n, 328–30
Edinburgh School, 319–20, 329–30, 350, 373. See also Barnes, Barry; Bloor, David; Edge, David; MacKenzie, Donald; Shapin, Steven; sociology of scientific knowledge
Einstein, Albert, xi, 10, 13n, 66n, 84, 100, 110, 123, 140n, 151, 154, 199n, 320n, 332n, 345, 393n, 406
elitism, xi, xv, 27, 32–33, 41–42, 53, 79, 82, 92, 103, 114, 130–31, 139, 144, 151, 154, 156, 159–69, 187, 205, 208, 220, 247, 251, 255, 315, 326, 329, 368–74, 382, 386–87, 414, 422
embushelment (of reason), 33, 38–39, 256, 415
empiricism, 42n, 81n, 213, 223, 278, 322n, 342
ends of science, 8n, 14, 98, 103, 110, 112, 116, 130, 138, 213, 238, 259, 305n
energy, 7n, 22, 93n, 100, 115, 121n, 197n, 223, 358, 366, 395, 397, 401
engineering, 8n, 51n, 66n, 98n, 200, 234n, 252, 304, 318, 351, 365, 370, 372, 377
Enlightenment (eighteenth century), 9n, 18n, 38, 40, 42, 47, 50, 53, 67n, 79, 80n, 82, 122, 130, 160n, 194, 212n, 241, 261, 265, 267, 282, 289–93, 297–98, 303, 356n, 401–2, 407–14
Epistemological Chicken (controversy in STS), 350, 419–20
epistemology, 14, 51n, 53n, 62, 134n, 263, 267, 275, 280, 283n, 293n, 295, 298, 314n, 317, 334, 337n, 361, 389n, 390n, 396n, 399, 405, 409, 415. See also anarchistic epistemology; naturalized epistemology; social epistemology
essentialism, 178n, 276n, 291n
"essential tension" (in science generally), 167, 308–9, 399, 403–6, 409
Essential Tension, The (Kuhn) 167n, 176, 267, 332n
ethics (of scientific research), 27n, 62, 136, 265–67, 295, 298, 311, 361. See also dirty hands (in ethics)
ethnography, 3, 54n, 147, 183n, 221–22, 226, 315–16, 346n, 347n, 351, 411
ethnomethodology, 18n, 20n, 225n, 226n
evaluative function of philosophy, 18, 28–30, 78, 83n, 88, 197, 215, 218, 238, 264, 279, 296, 299–300, 307, 314, 336, 350, 381. See also adjudicative function of philosophy; constitutive function of philosophy
Evans-Pritchard, Edward, 146–49, 177, 330n, 363
evolutionary theory, 14n, 57, 90, 112, 120–21, 136, 177–78, 188n, 191, 235n, 244, 251, 256, 259, 262, 269n, 270n, 289n, 306, 320, 328n, 346n, 375, 406, 416. See also Creationism; Darwin, Charles; Darwinism
expertise, 32, 37, 66n, 69n, 76, 80, 98n, 114, 124, 129, 138–42, 172, 182, 218–25, 233, 240, 245, 272, 315–17, 328, 350, 354, 361–62
external history of science, 16n, 20n, 34, 56, 59, 61, 65, 73, 76n, 86–87, 107, 142, 154, 164n, 166, 178, 181, 184–85, 201, 211, 268, 294n, 296, 307n, 331–33, 395, 397, 414. See also internal history of science

Faraday, Michael, 10, 79, 81n, 134, 154
Fascism, 165, 168, 171n, 326, 376–77. See also Nazism
feminism, 143, 324n, 335, 347, 356n, 373n, 405, 414. See also women in science
Feyerabend, Paul, 5, 10n, 11, 14n, 15n, 20n, 36, 64n, 71–74, 83, 87n, 121, 261–62, 285n, 287–88, 291n, 294, 304–6, 314n, 341n, 390, 395, 410
finalization (German school of Kuhnians), 35, 251–54, 258, 310

Fleck, Ludwik, 19n, 59–60, 211n, 295n, 392
Forman, Paul, 16n, 94–95, 103n, 185n
Foucault, Michel, 1n, 14n, 56n, 337, 344n, 357, 373n, 378, 414n
Franco-Prussian War, 26, 104–7, 120, 137, 230
Frankfurt School, 114n, 218n, 288n, 290n
French Revolution (1789), 22, 50, 52, 62, 194, 298n, 308n
Freud, Sigmund, 40, 228, 240, 246
Freudianism, 148, 182, 248n, 353, 397. *See also* Lacan, Jacques; psychoanalysis
Fukuyama, Francis, 12, 31n, 35, 63n, 256–59, 310

Galilei, Galileo, 7n, 12n, 26n, 63–64, 68–70, 80–81, 103, 111, 123, 187–92, 203–10, 214–15, 231, 243, 246, 282, 303, 305n, 314n, 356n, 381n, 392, 396, 413n
Gellner, Ernest, 20n, 147n, 291n, 356n, 390n
gemeinschaft/gesellschaft (Toennies), 211, 213, 316n
genetic fallacy, 83–84
genetics, 19, 51n, 76n, 84n, 89, 98n, 143n, 151n, 234n, 327–28, 331, 416
genius, xi–xii, 9, 54n, 82, 108, 203, 209, 213, 246–47, 321, 333, 346, 400
Gestalt psychology, 134–35, 273n
Gestalt switch, 8n, 17n, 175, 191–92, 204–5, 220, 227, 273n, 295n
Giere, Ronald, 17n, 82n, 290n
God, 25n, 30, 38n, 40–41, 52n, 63–65, 67n, 78, 80, 112, 126, 178n, 208, 263–64, 272, 274, 278, 310, 316n, 335n, 383n, 411. *See also* religion; theology
Goethe, Johann Wolfgang von, 49, 67n, 124, 127, 307n
Goodman, Nelson, 8n, 87–89
Gouldner, Alvin, xvii, 35, 44n, 106n, 159n, 164n, 229–34, 250n, 363–65, 403, 412

Habermas, Jürgen, 7, 23n, 49, 252, 262, 290n, 291n, 311n, 314n, 372, 414n
habitus (Bourdieu), 381–87, 391n, 399
Hacking, Ian, xiii n, 56n, 94n, 177n, 264n, 287n, 345n
Hanson, Norwood Russell, 5, 15n, 83, 192n, 234n, 276, 285n, 287–88, 306–7, 391n
Harre, Rom, 276n, 390n
Harvard University
  history of science at, xiv n, 9n, 164n, 219n, 386n
  Kuhn at, xiv n, 381–87, 391n
  philosophy at, 8, 36, 57, 87, 266–67, 284–85, 302n
  scientists at, 194, 197, 218–20, 261, 324

Society of Fellows, 6, 165, 381
strategy for resisting the New Deal, 162–69
students at, 179, 182, 217–20, 239, 242
See also Conant, James Bryant; Committee on General Education; Pareto Circle
Hayek, Friedrich von, 19n, 225n, 288n, 298n, 305n, 349n, 356n, 368n
Hegel, Georg Wilhelm Friedrich, 11, 50, 61–63, 83n, 122n, 144n, 206–9, 251, 264n, 296n, 334, 375n, 408n, 413
Hegelianism, 11n, 34, 40n, 53, 58–59, 62–63, 69n, 71n, 122, 206–10, 218n, 231, 256, 275, 307n, 362, 411
Heidegger, Martin, xii, 27n, 62n, 114n, 265n, 305n
Heisenberg, Werner, 4, 66n, 110, 128, 154, 185–86, 254
Helmholtz, Hermann von, 134n, 194, 209–10, 396n
Hempel, Carl, 13n, 14n, 262n, 275–76, 287n
Henderson, Lawrence J., 164–67
hermeticism, 59, 308, 396
hermeneutics, 11, 20n, 27, 30, 46n, 204–6, 409, 411n
Hertz, Heinrich, 177n, 197n
High-Low Church (of STS), 409–12, 415
historicism, 13, 15n, 46n, 68, 83–86, 97, 101n, 234n, 285–88, 296n, 301, 306–7, 415–16, 418n
historicity, awareness of one's own, 11, 19n, 122, 206, 209. *See also* reflexivity
historiography, 4, 9, 12, 18, 23–27, 32–37, 59n, 69n, 93n, 104n, 108n, 115n, 124, 127, 148n, 166, 170, 186n, 193–95, 197n, 201–3, 238, 249, 258n, 296n, 309, 314n, 331–36, 359n, 389, 396. *See also* external history of science; internal history of science; Orwellian history; Prig history; Tory history; Whig history
Hitler, Adolf, 12n, 140, 170, 226, 325, 377. *See also* Nazism
Hobbes, Thomas, 12n, 64, 125, 249–51, 263n, 313n, 322n, 330n, 393n
Holton, Gerald, xiv n, 253, 386n
Humboldt, Wilhelm von, 47n, 130–33, 414
Hume, David, 14, 42, 264n, 270, 293n, 298n, 312, 341n, 423n
Husserl, Edmund, 64n, 114n, 225n, 320n, 356n

iconography, 33, 43, 52–62, 70, 93, 195, 200, 215, 251, 383n
idealism, 41, 50, 69, 122, 144n, 264n, 345, 413

ideology, 19n, 22, 28, 29n, 49, 54n, 61–62, 64, 66, 73n, 106, 122, 151, 165, 244, 250, 263, 284, 298, 310, 327, 362, 372, 376–77, 404–5, 412
  conflict of, 2, 9n, 114n, 160, 174, 245, 345, 349n, 401, 418
  end of, 7, 9n, 114n, 174–75, 245, 345, 349n, 401, 418 (*see also* Bell, Daniel)
  natural science as, 31, 71n, 81n, 120–21, 138, 143–44, 174, 181, 255–56, 259, 323n, 332–34, 348, 388, 390, 395, 417–18
  See also Communism; Fascism; Mannheim, Karl; Marx, Karl
imperialism, 13n, 65, 98n, 105, 147, 176, 259, 316n, 339–40, 377n, 389n
incommensurability (Kuhn), 2, 4, 7–8, 52n, 105, 170, 176–77, 184, 188, 201–2, 205, 207, 264n, 268–71, 286n, 296n, 307n, 309, 312, 389, 397
information technology. *See* computers
instrumentalism, 35, 68, 96–99, 110–11, 117, 122–25, 129, 176, 189, 230n, 259, 300, 389, 400
intellectual property. *See* patents
internal history of science, 59, 64, 78, 86, 164n, 332–33, 340, 343. *See also* cognitivism; external history of science
Islam, 41, 265n, 410, 418n

James, William, 8n, 164n, 192n, 260, 271, 277, 392–93, 400
Japan, 43n, 63, 65, 105n, 107, 113n, 174, 386n, 418n
Jewish science, 19n, 40–41, 58n, 69n, 84, 375, 395. *See also* Aryan science
Johnson, Lyndon, 239n, 258
Judaism. *See* Jewish science
justification, context of. *See* discovery, context of

kairos, 45, 86, 93–94, 207, 313n
Kaiser-Wilhelm Institutes, 107–8, 111, 122n, 140
Kant, Immanuel, 14, 29, 56, 81n, 94n, 99, 144n, 192n, 268–70, 293n, 296–99, 389n, 393, 396n, 413. *See also* neo-Kantianism
Keynes, John Maynard, 106, 145–46, 157, 225, 254
Kitcher, Philip, 84n, 262, 315n
Knorr-Cetina, Karin, 226, 346n
"knowledge society," 103, 106, 145n, 156n, 239
Koyré, Alexandre, 17n, 23, 26n, 33, 40n, 42–43, 60–70, 78, 81n, 203, 205, 254, 325n, 333, 344n, 356n, 359n, 381, 392, 395–98
Kuhnification (of science), 36, 300, 314n, 317–18, 322, 324, 331, 336–37, 342, 354, 363

Lacan, Jacques, 62, 64n, 353, 356–57, 381n
Lakatos, Imre, xv, 15n, 36, 83–84, 249n, 287–88, 295–96, 300, 302–3, 307, 341n, 356n, 390
Lamarckianism, 19, 55n, 57, 109n, 151n, 328n
Latour, Bruno, xv, 226, 236n, 316n, 343–51, 356–57, 365–77, 409n, 419–20
Laudan, Larry, 15n, 16n, 83–84, 127n, 199n, 262n, 285n, 287n, 300n, 346n
Lazarsfeld, Paul, 226, 233n
legislative function of philosophy, 36, 41, 156, 159, 224, 298–99. *See also* prescriptive function of philosophy; regulative function of philosophy
Leibniz, Gottfried von, 25, 64, 81n, 113n, 126, 263n, 264n, 393
Lenin, Vladimir, 114n, 175n, 255, 309
Lewis, Clarence Irving, 36, 266–80, 297
liberal arts, 79–81, 108n, 150n, 183n, 194, 225n, 318, 383, 390, 414, 422
liberalism, 18–20, 49, 82, 109–12, 122n, 129–30, 167, 174, 211n, 223, 228n, 240, 248, 250–51, 256, 259, 288n, 304n, 316n, 324n, 326, 356n, 363, 370, 374, 377n, 405
linguistics, 20n, 56–57, 97, 100n, 195n, 204n, 207n, 232n, 270n, 285n, 286n, 316, 339, 353, 399
Locke, John, 260–61, 349n
logical positivism, 13n, 14n, 20n, 29–30, 42, 82–83, 91n, 97–98, 119, 122, 203n, 211n, 218, 260–62, 264n, 265n, 266–69, 273n, 275–76, 278, 280, 282n, 284–89, 293–96, 298n, 301–4, 305n, 307, 311–14, 335n, 390–94. *See also* Carnap, Rudolf; Hempel, Carl; Reichenbach, Hans; Neurath, Otto; positivism; Schlick, Moritz
London School of Economics (LSE), 51n, 288, 305n, 391n
Longino, Helen, 264n, 324n
Low Church (of STS). *See* High-Low Church (of STS)
Luhmann, Niklas, 109, 137
Lukács, György, 231, 295n, 340n
Lynch, Michael, 131n, 226, 324n
Lyotard, Jean-François, 31, 291n, 335n, 371–72
Lysenko, Tremfil, 151, 328n

Mach, Ernst, 11, 15n, 24, 34, 38, 66n, 67n, 84, 94n, 96–139, 142, 176, 193, 199n, 209, 212–13, 259, 266n, 272n, 300, 312n, 345, 350, 352, 359n, 408n, 419, 422n
Machiavelli, Niccolò, 44, 51n, 165–67, 368–69, 373n, 409
MacIntyre, Alasdair, 27n, 132n, 147n, 335n, 413
MacKenzie, Donald, 372–73, 377n
Mannheim, Karl, 19n, 35, 55, 58n, 60n, 229–33, 246n, 282–84, 334, 340n
Marx, Karl, xiii, 20n, 33, 40, 122, 165, 168, 175n, 228, 240, 249–52, 293–94, 321, 325, 344, 349n, 373, 404, 411, 413
Marxism, 85n, 114n, 120n, 130n, 151n, 153n, 158–59, 163–64, 225n, 231, 239, 249–50, 254–55, 288n, 305n, 319, 327, 329, 332–33, 335, 345, 347, 349n, 370, 372–73, 414
mathematics, 19n, 42, 68–69, 78, 81n, 108n, 111, 113n, 116, 118, 124, 153–54, 189–90, 194–95, 213, 218n, 230, 235n, 264–65, 278, 320n, 355–56, 390n
Massachusetts Institute of Technology (MIT), 4n, 7n, 150, 163, 258, 399
Matthew effect (Merton), 38n, 384
McNamara, Robert, 183n, 242
medicine, 10, 60, 76n, 106n, 140n, 153–54, 171, 204, 211, 245, 328n, 355, 413
Mendel, Gregor, 19n, 89, 151n
Merton, Robert, xiv n, 17n, 38n, 84n, 161–64, 173n, 224n, 235–36, 332, 384, 392
metascience, 124, 188, 191, 226n, 265n, 315–16, 319n
Meyerson, Emile, 12, 356n, 392–97
military-industrial complex, 26, 35, 73–75, 105–6, 157–59, 229, 241, 244, 247, 304, 387
Mill, John Stuart, 79, 121, 154, 158n, 247, 265, 356n
Mills, C. Wright, 35, 229–34, 240n
MIT. *See* Massachusetts Institute of Technology
Moltke, Helmut von, 105, 107, 109, 230
Moore, G. E., 84n, 390n
movements, social, 29, 35, 37, 46, 53, 61, 73–74, 91n, 101n, 158n, 162, 217–18, 225, 240n, 247–48, 251, 264n, 296n, 305n, 319, 336, 347n, 351, 357–58, 364, 373–75, 399–409, 412, 414–15, 417–20
multiculturalism, 20n, 25n, 114–15, 414
Muslim science. *See* Islam
mythos. *See kairos*

Nagel, Ernest, 83n, 285n, 287n
Napoleon Bonaparte, 50–51, 62–63, 195, 378

National Science Foundation (NSF), 4n, 44n, 150, 153, 155–56, 160–63, 167n, 173n, 202, 220, 218n, 285n
naturalized epistemology, 84n, 167n, 173n, 263
nature (as explanatory category in science), 100n, 144n, 204, 270, 352–54, 357n, 395
Naturphilosophie, 99, 124, 128
Nazism, 19n, 23n, 52, 58, 60n, 62, 66n, 82, 84, 151, 173n, 214, 233n, 356n, 360, 376n, 382, 390n, 396
Needham, Joseph, 324n, 325n, 327, 359n
neo-Kantianism, 19n, 62, 62n, 264n, 265n, 304n
Neurath, Otto, 113n, 119–20, 279
New Deal (U.S.), 29n, 33, 106, 150n, 155–57, 162–63, 165n, 202, 230n
Newton, Isaac, xi–xii, 7n, 9n, 17, 25, 67–70, 76n, 78, 103, 108n, 111, 116, 126, 154, 191–92, 195, 208, 210, 246, 251n, 260–61, 264n, 271, 282, 298, 300, 308–10, 332n
Newtonianism, xi, 66–67, 78–84, 88n, 114, 119–24, 127–28, 187, 194, 203n, 209, 307n, 312n, 332, 346, 396, 416–17
Nietzsche, Friedrich, 29, 40, 63n, 130n, 240, 247, 277, 357n
nonhumans (as objects of STS study), 144, 356n, 374–76
normal science (Kuhn), xvi, 2, 5–7, 23n, 71n, 73, 76n, 81, 89, 91, 113n, 126, 172n, 175, 182, 187–201, 205, 210, 216, 221–22, 228, 233n, 243–49, 264, 277, 280, 283n, 286n, 292–93, 299, 301–3, 315, 321–24, 329n, 333n, 356, 359, 372, 380–81, 387, 407, 421
normative function of philosophy, 7–8, 13–14, 18–19, 33, 36, 38n, 44, 57, 68n, 72–74, 84n, 92, 146n, 183n, 185–86, 193, 206–7, 224n, 226, 237, 248, 250, 257, 260, 265n, 267, 272, 274–75, 282n, 284, 286n, 289–91, 295–302, 307, 310, 312, 320n, 328, 335n, 346, 348–50, 358–60, 380, 383–84, 418n. *See also* adjudicative function of philosophy; constitutive function of philosophy; evaluative function of philosophy; legislative function of philosophy; prescriptive function of philosophy; regulative function of philosophy

Orwellian history, 31, 86, 96, 126, 172n, 174–75, 193, 206, 235n, 302–3, 319, 380, 401, 423

INDEX / 469

Ostwald, Wilhelm, 98, 100–2, 109n, 115, 129, 142–44, 196–97, 199n, 352
Oxford University, 42n, 69, 80, 122n, 146, 172n, 263n, 324, 390n, 413n

Panofsky, Erwin, 43, 55, 60–61
"paradigmitis," 318, 342–43, 350, 361, 414n
parasite, STS researcher as, 361–62, 371
Pareto Circle (Harvard, 1932–42), 34, 164
Pareto, Vilfredo, 165–68, 247, 346n, 368, 373, 377n
Paris School of STS. *See* actor-network theory; Callon, Michel; Latour, Bruno
Parsons, Talcott, 164–65, 167n, 250n
patents, xii, 25, 82–83, 86n, 150n, 154–57, 188, 203, 275, 416
peer review, 160–61, 218, 321
Peirce, Charles Sanders, 11, 126, 180, 267, 307n
phenomenology, xii, 20n, 64n, 111, 114n, 119, 124, 128, 175n, 225–26, 246n, 265n, 391n. *See also* Heidegger, Martin; Husserl, Edmund; Schutz, Alfred
philosophical history, 11–13, 15, 31, 33, 65n, 206
physics, 51n, 66n, 67n, 76n, 81, 84, 91–118, 130, 133, 141n, 146n, 188–91, 197n, 212n, 253, 261, 264n, 268, 278, 282, 285n, 286n, 287n, 314–15, 320n, 324n, 332, 381n, 390n
  Kuhn's historiography of, xiv n, 4, 10n, 17, 60, 126, 178n, 185, 187, 194–97, 203, 226, 235n, 302n, 321, 380
  post–World War II, 75–77, 151, 154, 201, 379–80, 388, 411
  as scientific standard, 12–14, 108n, 133–35, 143, 188n, 196n, 266, 284–85, 352
  See also Aristotle; atomic bomb; Bohr, Niels; Einstein, Albert; Heisenberg, Werner; Newton, Isaac; Planck, Max; quantum mechanics; Weinberg, Steven
Piaget, Jean, 17, 230n, 280, 356n, 392, 396–97, 419n
Pigou, Arthur, 157–59
Planck effect (Kuhn), 19n, 169, 289, 301, 306, 309, 313, 330n
Planck, Max, 16n, 34–35, 38, 60, 66n, 84, 90, 96, 98–114, 116–39, 142–43, 193, 199n, 209–10, 212, 231, 235n, 252, 259, 300, 350, 419
Plato, 16–17, 30, 32–33, 38–50, 132n, 136, 182, 190, 203, 249, 300, 310, 312–14, 399
Platonism, 14n, 26–28, 35, 39–43, 51–72, 92–93, 102–3, 133, 144, 171n, 191, 205, 218n, 220n, 256, 259, 342n, 396–97, 412
pluralism, xv, 3, 188, 224–25, 237n, 264
Polanyi, Michael, 5, 35, 71n, 73n, 86n, 139–53, 172n, 199, 210–11, 218, 225n, 231, 252, 310, 316, 326, 328–30, 390n, 400
Popper, Karl, 11, 13n, 14n, 15n, 36, 51n, 82n, 88, 91, 110, 120n, 140n, 209–10, 231, 252–53, 257–58, 261, 265n, 290n, 291n, 295, 304n, 341n, 356n, 395, 407
  critical approach, 38, 49, 135, 172, 213, 218n, 262, 282n, 290, 401
  criticisms of Kuhn, 8n, 74n, 89, 92, 113n, 194, 200, 266, 273n, 281, 288, 290, 292, 303–5, 307n, 314n, 316, 333–34, 390–93
  criticisms of logical positivism, 8, 269, 275–77, 289, 292–93
  defense of the "open society," 91–92, 224, 286n, 288n, 305n, 371n, 393
positivism, xiii n, xv, 2–3, 8n, 38, 60, 68, 94n, 121, 142, 212, 217, 225n, 250, 254, 258, 281–83, 290–93, 323n, 340, 344, 349n, 378, 393, 395. *See also* Comte, Auguste; logical positivism; Mach, Ernst
postindustrial society (Bell), 106, 137, 236n, 240–47, 367–68
postmodernism, xii, xv, 13n, 18n, 31, 51n 71n, 105n, 128, 143, 224n, 235n, 268, 272, 274, 291n, 296n, 320n, 323n, 348, 357, 368, 371–72, 380, 418
pragmatism, 8n, 36, 66, 100n, 119, 180, 212–13, 217, 233n, 260, 267–68, 272–78, 285n, 400. *See also* Dewey, John; James, William; Lewis, Clarence Irving; Peirce, Charles Sanders; Rorty, Richard
praxis (science as), 77n, 130–33
prescriptive function of philosophy, 36, 45, 71, 154n, 238, 262, 292, 296, 299–300, 307, 313n. *See also* legislative function of philosophy; regulative function of philosophy
Price, Derek de Solla, 10n, 235–57, 258–59, 294, 331
Prig history (Brush), 22, 24–26, 33, 37, 130n, 249, 389
progress, 2, 51–52, 59, 64, 78, 84, 97, 100n, 127n, 138–39, 148, 152n, 158n, 182–83, 193, 209, 213, 223, 263, 269, 283n, 292, 355, 417, 420n
  Kuhn's conception of, 13–14, 16, 22, 28, 55n, 96, 132n, 177, 254, 259, 302n, 320, 389, 397, 401, 423

progress (continued)
  obstacles to/skepticism of, 29, 33–34, 123–25, 168–69, 203n, 238, 259n, 269
  social, 19n, 175n, 249–50, 257, 310, 372, 404
  technology as mark of, 66, 199n, 200n, 217, 366
Protestantism, 228n, 304n, 399–401, 407, 409–10
psychoanalysis, 26, 64n, 141n, 248, 353, 356n, 381n, 397
Ptolemy, 187–92, 203n, 207, 271, 300
Putnam, Hilary, 15n, 88n, 203

quantum mechanics, 4, 11n, 12n, 13n, 94, 122, 127n, 185–86, 203n, 268, 273n
Quine, Willard van Orman, 8n, 29, 57–58, 85–89, 95, 100n, 172n, 180, 278–80, 270n, 285n, 391n

racism, 19, 22, 54n, 82, 84, 175n
Ravetz, Jerome, 73–74, 253, 255
realism
  legal, 298–99
  political, 21, 33, 67n, 90, 168–78
  scientific, 35, 90, 96–139, 176, 189, 230n, 232n, 257–59, 263n, 265n, 275, 284, 299, 312, 340–45, 355, 389, 400
reflexivity, 11n, 18n, 58n, 93, 126–27, 138n, 156n, 191, 207, 227n, 232, 283, 322n, 350, 354, 381, 392, 398. *See also* historicity, awareness of one's own
regulative function of philosophy, 296–99. *See also* legislative function of philosophy; prescriptive function of philosophy
Reichenbach, Hans, 13n, 60n, 82n, 126, 392
relativism, xvi, 4, 13n, 17–24, 31, 40, 45, 52n, 126, 130n, 139, 147–49, 170, 177, 221, 249, 255n, 265n, 282–85, 312, 323n, 330n, 337n, 341–42, 348, 363, 377n
religion, 41–42, 65, 91n, 122–23, 130, 210, 263n, 264n, 335n, 393, 407–9
  opposed to science, 9n, 34, 80, 80n, 180, 208, 277, 360n
  See also God; theology
Renaissance (sixteenth-century Europe), 12n, 21, 52n, 81n, 104n, 302n
republicanism, 49, 91, 415n
research and development, 105, 158, 237
revolutionary science (Kuhn), 2, 17–18, 28, 61, 70, 72, 75, 90, 92, 94, 107, 125, 127n, 140–41, 181, 185, 203n, 231, 234n, 235n, 243n, 274, 285–86, 289, 292–95, 299, 301–3, 307n, 308n, 312–13, 327, 333–34, 338, 344n, 372, 395–97, 400, 403, 413n

founders of, xi–xii, 54n, 91n, 172n, 178n, 188–92, 204–5, 243, 246, 248, 273n, 289n, 308, 321, 390n
Hegelian account of, 34, 206–9
"permanent revolution," 50, 172, 177, 280–81, 304–6, 407 (*see also* Popper, Karl)
restorationist account of, 34, 164, 167–69
in social science, 235–36, 243–46, 248–51, 255
See also Scientific Revolution
rhetoric, 7, 29n, 42n, 73–74, 81n, 130, 145n, 163, 181, 188n, 191, 199, 203–4, 227–28, 244–45, 247, 253, 276, 294n, 312–14, 319, 320n, 327n, 330n, 353, 366, 369, 375n
  art as, 52, 56
  history as, 29, 70, 92–93
  Kuhn as, xii n, 308n, 321–22, 389
  philosophy as, 2, 41, 45, 69n, 88n, 91n, 309–12
  science as, 36, 43n, 55, 81, 102, 122, 140, 142, 209, 328, 352, 411
  STS as, 322n, 334, 343, 347, 350, 360n
Richards, I. A., 29n, 219, 313n
Rockefeller Foundation, 153, 167, 399
romanticism, 82, 105, 218n, 233n, 245, 246–47, 308–9
Roosevelt, Franklin Delano (FDR), 66n, 106n, 155–56, 163, 165n, 179, 230n
Rorty, Richard, 31n, 36, 88n, 191, 205–6, 268, 272, 285n, 291n, 414n
Roszak, Theodore, 36, 241–48
Rousseau, Jean-Jacques, 50, 72n
Russell, Bertrand, 12, 126, 356n, 390n, 391n
Russia. *See* Soviet Union

Sarton, George, 9n, 164n, 325n, 333
Scheffler, Israel, 18n, 87n, 307n
Schlick, Moritz, 98, 136–38, 395
Scholastics, medieval, 26, 42, 47, 119, 180n, 207n, 241n, 396
Schumpeter, Joseph, 159–60, 164, 166, 294, 373
Schutz, Alfred, 175n, 225–26
science and technology studies (STS), 3–4, 18n, 25n, 35–37, 144–45, 173, 175n, 216, 230–31, 247n, 253n, 290, 301, 316n, 319–24, 331, 336–74, 377–78, 409–11, 415, 420
science policy, 5n, 31, 34, 65n, 73n, 96, 99, 104, 107–8, 116, 136–37, 140, 146, 150n, 155–56, 159–60, 173–74, 202–5, 233, 236–38, 252–55, 259, 310, 320, 323n, 324n, 331, 367, 374, 376, 398, 420
science studies. *See* science and technology studies (STS)

"Science Wars," 145n, 261n, 319, 354–55, 386n, 411, 422
Scientific Revolution (seventeenth-century Europe), 23, 59, 64, 66, 68, 108n, 190, 203, 223, 265n, 281, 324, 327n, 334, 343, 359n, 396–97, 400, 413n
scientometrics, 236–38, 320n, 367n
secularization, 37, 67n, 228n, 265, 335n, 365n, 399–400, 409–11
Sellers, Peter (Kuhnian icon), xii, 233–34
Shapere, Dudley, 15n, 90n, 285–88, 306, 316
Shapin, Steven, 145n, 147, 265n, 322n, 331, 359n, 393n
Simon, Herbert, 80n, 135n, 186n, 242n, 245
Skinner, Quentin, 49n, 70n
Smith, Adam, 51n, 104, 158n, 264n, 298n
Snow, C. P., 37, 173, 324–27, 382, 386n
social epistemology (Fuller), 6–7, 38n, 49n, 284n, 315n, 317, 323n, 324n, 341n, 350, 365, 372, 393n, 401
social movements. *See* movements, social
sociology of knowledge (Mannheim), 8n, 19n, 22, 38n, 55, 60n, 82n, 148, 230–33, 255n, 282–83, 305n, 320n, 340n, 343, 366, 372
sociology of scientific knowledge (SSK), 3, 20n, 36, 46n, 54n, 125, 141n, 147n, 221, 282, 285n, 286n, 287n, 290, 292n, 319, 324, 328, 391n
Soviet Union, 31n, 62, 151n, 163n, 167n, 170, 172, 173n, 214, 233, 325
Spencer, Herbert, 11, 106n. *See also* evolutionary theory
Sputnik (Soviet satellite, 1957), 74n, 107n, 167n, 219–20, 391
Stalin, Josef, 63, 151n, 170, 326, 328n, 332, 368, 377
Stoicism, 46–48, 51n, 226n, 387
Strauss, David Friedrich, 42n, 122n, 411
Strauss, Leo, 27, 30, 40–41, 48, 62, 64, 251
structuralism (French intellectual movement, 1960s–1970s), xv, 14n, 344, 351n, 357, 396

technocracy, 37, 240–47, 252–53, 326, 329n, 368, 378
technology, xii, 73n, 111, 114, 191, 213, 265n, 318–19, 327, 335n, 337, 340–41, 352, 354, 366, 369–77, 402, 405, 409, 412
  excluded from science, 10, 34, 66n, 208–10, 218n
  information technology. *See* computers
  resulting from science, 105, 114n, 256–59, 418n
  science as "off the shelf" technology, 114n, 138, 189, 254
theology, 41–42, 48, 78, 80–81, 102, 109n,
122–23, 171, 188–91, 194, 208, 228, 263, 265, 266n, 278, 280, 304n, 316n, 335n, 352, 365n, 390, 394, 404, 409–13, 416n. *See also* God; religion
Theory Lite (character of STS), 301, 342, 377n
Thucydides, 25, 44, 166
Tory history (Butterfield), 24–26, 29n
totalitarianism, 48, 58, 62, 303, 327, 374, 376–77
Toulmin, Stephen, 5, 10n, 12n, 15n, 36, 70, 71n, 73n, 83, 178n, 234n, 276, 285, 287–88, 296n, 306–14, 390
two cultures problem (arts vs. sciences), 29n, 37, 173, 202, 324–25, 326n, 382, 386n. *See also* Snow, C. P.

underdetermination, 85–87, 95, 100n, 141n, 172n, 176–77, 180, 208, 270n, 350n, 389. *See also* Duhem, Pierre; Quine, Willard van Orman
underlaborers (philosophers as), 14n, 260–65, 280, 300, 315–16
UNESCO (United Nations Education and Scientific Organization), 163n, 324n
unionization (of scientists), 111, 163, 369
university, changing character of, 2, 13n, 16n, 22–23, 37, 47, 50, 78–82, 98, 103, 108n, 120, 122n, 130–33, 140, 144n, 150–54, 156n, 162, 173, 179, 183n, 194, 197, 199n, 208, 211n, 228n, 235n, 240–41, 244–45, 253n, 321, 336, 356n, 362, 369–72, 370n, 375, 383, 412–22
University of Berlin, 100n, 111, 130, 134, 194, 209
University of Chicago, 140, 13n, 150n, 286n
USSR. *See* Soviet Union
utilitarianism, 81n, 90n, 164n, 201, 265, 296n, 297, 299, 326n, 383n

Vienna Circle, 28, 42n, 98, 119, 212n, 218n, 279n, 390, 391, 395. *See also* logical positivism
Vietnam War, 240, 242, 258

Waddington, Conrad, 327, 328n, 331
Warburg, Aby, 52–54, 59, 104n, 215
Weber, Max, 12, 39n, 85n, 121, 130n, 132–33, 142, 144, 148, 164n, 165–66, 175n, 195, 199n, 211, 232, 233n, 246, 251, 304n, 310, 365n, 371, 407
Weimar Republic (Germany, 1918–33), 19n, 60n, 94, 103, 122, 142, 376n
Weinberg, Steven, 98, 99n, 128, 185n, 191, 193n

welfare state, 159n, 250n, 253n, 254, 294n, 310, 363, 406
Whewell, William, 11, 34, 60, 78–82, 85–92, 122n, 139, 150, 154, 161, 176, 193, 208–12, 281, 296n, 312n, 345, 394n, 415
Whig history (Butterfield), 11–12, 22–33, 49–50, 97, 126, 320n, 333, 383, 389
White, Hayden, 13n, 93n, 398
White, Morton, 8n, 272n, 285n, 298n
Whitehead, Alfred North, 12, 67n, 267, 302n, 423
Wilson, Harold, 254n, 318, 319n, 327, 366
Wittgenstein, Ludwig, 8n, 20n, 25, 57n, 72, 100n, 110, 122, 192n, 273n, 274n, 282, 285n, 296n, 311, 330n, 334–35, 390n
Wolin, Sheldon, 249–51
women in science, 109, 143n, 243, 381n, 405. *See also* feminism

Woolgar, Steve, 18n, 236n, 373n
World War I, 9n, 10, 19n, 26, 34–35, 43, 51, 55, 60, 66, 75, 92–93, 94n, 102–6, 110, 129, 133n, 139–42, 151, 155n, 157–58, 159n, 168n, 187, 197, 211, 217–18, 239, 241n, 281, 324n, 381n
World War II, xiv n, 5, 19n, 23, 26, 58, 60n, 61, 75, 77, 106–7, 108n, 128n, 150n, 151, 153, 155, 162, 165n, 170–73, 179, 182, 183n, 185n, 199–202, 218n, 231, 235n, 237n, 239–40, 275–76, 286n, 305n, 325n, 326, 332, 343, 356n, 359n, 379, 388
Wuthnow, Robert, 401–3, 407, 408n

Zen (attitude toward science), 25, 37, 361–62